Navjeet Kaur

Lawesson's Reagent
in Heterocycle Synthesis

 Springer

Navjeet Kaur
Department of Chemistry
Banasthali Vidyapith (University)
Jaipur, Rajasthan, India

ISBN 978-981-16-4657-7   ISBN 978-981-16-4655-3 (eBook)
https://doi.org/10.1007/978-981-16-4655-3

© The Editor(s) (if applicable) and The Author(s), under exclusive license to Springer Nature Singapore Pte Ltd. 2022
This work is subject to copyright. All rights are solely and exclusively licensed by the Publisher, whether the whole or part of the material is concerned, specifically the rights of translation, reprinting, reuse of illustrations, recitation, broadcasting, reproduction on microfilms or in any other physical way, and transmission or information storage and retrieval, electronic adaptation, computer software, or by similar or dissimilar methodology now known or hereafter developed.
The use of general descriptive names, registered names, trademarks, service marks, etc. in this publication does not imply, even in the absence of a specific statement, that such names are exempt from the relevant protective laws and regulations and therefore free for general use.
The publisher, the authors and the editors are safe to assume that the advice and information in this book are believed to be true and accurate at the date of publication. Neither the publisher nor the authors or the editors give a warranty, expressed or implied, with respect to the material contained herein or for any errors or omissions that may have been made. The publisher remains neutral with regard to jurisdictional claims in published maps and institutional affiliations.

This Springer imprint is published by the registered company Springer Nature Singapore Pte Ltd.
The registered company address is: 152 Beach Road, #21-01/04 Gateway East, Singapore 189721, Singapore

Lawesson's Reagent in Heterocycle Synthesis

Preface

Heterocycles have always been the center of attraction because of their applications in medicinal chemistry. The heterocyclic chemistry research comprises a significant part of the organic chemistry research in the world. The huge quantity of bioactive organic compounds that possess heterocyclic frameworks plays an important role in the medicinal field.

Today, analogues of heterocyclic compounds and their derivatives have become strong interest in pharmaceutical research field due to their valuable biological and pharmacological activities. Because of the extensive importance of heterocycles, the formation of these compounds has always been the most important research field in synthetic chemistry. The chemistry of heterocycles is as logical as the chemistry of aromatic or aliphatic compounds. The study of heterocyclic structures is of great attention both from the theoretical and practical point of view. The versatile synthetic use and biological action of these heterocyclic compounds have motivated the pharmacologist to plan, design, and execute new methodologies for the synthesis of novel drugs.

Due to the properties that are similar to drugs, libraries of different heterocyclic compounds are usually used in high-throughput screening at initial stages of drug design system. The competition in the area of drug design has helped to identify the speed of synthesis as a top preference in drug design. Subsequently, techniques that could enhance and promote both synthesis and screening of compounds are highly required.

Lawesson's reagent is a commercially accessible reagent and has been extensively used in organic synthesis to achieve the transformation of carbonyl compounds to thiocarbonyl compounds, which are important functional groups to achieve different organic reactions or to utilize them as end products in medicinal, material chemistry, etc.

The LR has remained the most important reagent in thionation chemistry and is followed by P_4S_{10}. The P_4S_{10} has been an indispensable commercially accessible reagent, specifically for transforming almost all types of oxo groups to thio groups. This has been employed more commonly for replacing the oxygen atom of a furan ring with sulfur atom. Almost all types of heterocycles containing sulfur atom(s) are synthesized using P_4S_{10}.

A number of publications are being published every year on both reagents. Both reagents have their own advantages and disadvantages over specific reactions, and both of them deserve to be utilized.

It could be a benefit to the synthetic researchers to use Lawesson's reagent in synthetic pathways for heterocycles to provide the best and surprising results.

Jaipur, India   Navjeet Kaur

Contents

1 **Five-Membered *N*-Heterocycle Synthesis** 1
   1.1   Introduction 1
   1.2   Synthesis of Five-Membered *N*-Heterocycles 2
   1.3   Synthesis of Five-Membered *N,N*-Heterocycles 15
   1.4   Synthesis of Five-Membered *N,N,N*-Heterocycles 19
   References 27

2 **Thiazole Synthesis** 35
   2.1   Introduction 35
   2.2   Synthesis of Thiazoles 36
   2.3   Synthesis of Benzothiazoles 40
   2.4   Synthesis of Thiazoles from Dicarbonyl Compounds 42
   References 58

3 **Thiazole Synthesis by Thionation of C=O to C=S** 63
   3.1   Introduction 63
   3.2   Synthesis of Thiazoles 64
   3.3   Synthesis of Benzothiazoles 78
   3.4   Synthesis of Fused Thiazoles 100
   3.5   Synthesis of Fused Benzothiazoles 102
   References 105

4 **Thiadiazole Synthesis** 115
   4.1   Introduction 115
   4.2   Synthesis of 1,2,3-Thiadiazoles 116
   4.3   Synthesis of 1,2,4-Thiadiazoles 118
   4.4   Synthesis of 1,3,4-Thiadiazoles 121
   References 141

5 **Five-Membered *S*-Heterocycle Synthesis** 149
   5.1   Introduction 149
   5.2   Synthesis of Thiophenes 150
   5.3   Synthesis of Benzothiophenes 158
   5.4   Synthesis of Fused Thiophenes 162

5.5	Synthesis of Dithioles	165
5.6	Synthesis of Trithioles	169
	References	171

6 S-Heterocycle Synthesis ... 175
- 6.1 Introduction ... 175
- 6.2 Synthesis of Five-Membered S,N-Heterocycles ... 176
- 6.3 Synthesis of Five-Membered S,N-Polyheterocycles ... 185
- 6.4 Synthesis of Five-Membered Fused S,N-Heterocycles ... 187
- 6.5 Synthesis of Six-Membered S-Heterocycles ... 191
- 6.6 Synthesis of Six-Membered Fused S-Heterocycles ... 193
- 6.7 Synthesis of Six-Membered S,S- and S,S,S-Heterocycles ... 198
- 6.8 Synthesis of Six-Membered S,N-Heterocycles ... 204
- 6.9 Synthesis of Six-Membered S,N,N-Heterocycles ... 205
- 6.10 Synthesis of Seven-Membered S-Heterocycles ... 206
- References ... 208

7 O- and N-Heterocycles Synthesis ... 215
- 7.1 Introduction ... 215
- 7.2 Synthesis of Five-Membered O-Heterocycles ... 215
- 7.3 Synthesis of Five-Membered O,N-Heterocycles ... 216
- 7.4 Synthesis of Five-Membered O,N,N-Heterocycles ... 220
- 7.5 Synthesis of Six-Membered N-Heterocycles ... 221
- 7.6 Synthesis of Six-Membered N,N-Heterocycles ... 230
- 7.7 Synthesis of Seven-Membered N-Heterocycles ... 235
- 7.8 Synthesis of Seven-Membered O-Heterocycles ... 236
- References ... 239

8 Phosphorus Pentasulfide in Heterocycle Synthesis ... 245
- 8.1 Introduction ... 245
- 8.2 P_2S_5 in Heterocycle Synthesis ... 246
- 8.3 P_4S_{10} in Heterocycle Synthesis ... 252
  - 8.3.1 Synthesis of Five-Membered Heterocycles ... 252
  - 8.3.2 Synthesis of Six-Membered Heterocycles ... 287
  - 8.3.3 Synthesis of Seven-Membered Heterocycles ... 296
- References ... 298

Conclusion ... 307

About the Author

Dr. Navjeet Kaur was born in Punjab, India. She received her B.Sc. from Panjab University Chandigarh (Punjab, India) in 2008. In 2010, she completed her M.Sc. in chemistry from Banasthali Vidyapith. She was awarded with Ph.D. in 2014 by the same university, under the supervision of **Prof. D. Kishore**. Presently, she is working as an Assistant Professor in Department of Chemistry, Banasthali Vidyapith and has entered into a specialized research career focused on the synthesis of 1,4-benzodiazepine-based heterocyclic compounds (organic synthetic and medicinal chemistry). With 10 years of teaching experience, she has published over 160 scientific research papers, review articles, book chapters, and monographs in the field of organic synthesis in national and international reputed journals. She has published three books *Palladium Assisted Synthesis of Heterocycles*, *Metals and Nonmetals: Five-Membered N-Heterocycle Synthesis* with CRC Press, Taylor & Francis group and *Metal- and Nonmetal-Assisted Synthesis of Six-Membered Heterocycles* with Elsevier. Her name featured in the **WORLD RANKING OF TOP 2% SCIENTISTS** released on 16 October, 2020 in a subject-wise analysis conducted by a team of scientists at **Stanford University**, USA. She was among top 2% scientists of world in both full career-wise **(Rank: 424 in World and 04 in India)** and single year-wise **(2019, Rank: 03 in World and 01 in India)** published lists. This story of achieving top ranking in such short span (7–8 years) of her career was covered by numerous newspapers. She was presented the Prof. G. L. Telesara Award in 2011 by Indian Council of Chemists (Agra, Uttar Pradesh) at Osmania

University (Hyderabad), and the Best Paper Presentation Award in National Conference on "Emerging Trends in Chemical and Pharmaceutical Sciences" (Banasthali Vidyapith, Rajasthan). She has attended about 40 conferences, workshops, and seminars. She has delivered many invited lectures and radio talks. Apart from all this, she has been working as NSS Program Officer since 2016 and member of UBA (Unnat Bharat Abhiyan) since 2018. Dr. Navjeet finds interest in Sikh literature and has completed a two-year Sikh Missionary course from Sikh Missionary College (Ludhiana, Punjab).

Dr. Navjeet Kaur is currently guiding five research scholars—Meenu Devi, Yamini Verma, Pooja Grewal, Pranshu Bhardwaj, and Neha Ahlawat—as their Ph.D. supervisor.

Abbreviations

(-)-pHTX	Philanthotoxin
4-CR	Four-component reaction
AIBN	Azobisisobutyronitrile
BD	Benzene diacetate
BEDT-TTF	Bis(ethylenedithio)tetrathiafulvalene
Boc	*t*-butoxycarbonyl
BOM	Benzyloxymethyl
BTDT	2,1,3-benzenethiadiazole-5,6-dithiolate
CAN	Ceric ammonium nitrate
Cbz	Carboxybenzyl
CDI	Carbonyldiimidazole
CSA	Camphorsulfonic acid
CSI	Chlorosulfonyl isocyanate
DABCO	1,4-diazabicyclo[2.2.2]octane
DAST	Diethylaminosulfur trifluoride
Dba	Dibenzylideneacetone
DBN	1,5-diazabicyclo[4.3.0]non-5-ene
DBU	1,8-diazabicyclo[5.4.0]undec-7-ene
DCB	Dichlorobenzene
DCC	N,N'-dicyclohexylcarbodiimide
DCE	Dichloroethane
DCM	Dichloromethane
DCPB	1,4-bis(dicyclohexylphosphino)butane
DCR	Dipolar cycloaddition reaction
DDC	N,N'-dicyclohexylcarbodiimide
DDQ	2,3-dichloro-5,6-dicyanobenzoquinone
DDT	Dichlorodiphenyltrichloroethane
DEACM-MN	Diethylaminocoumarylidenemalononitrilemethyl
DHB	2-(3,5-dihydroxyphenyl)hydroxybenzothiazole
DIBAL	Diisobutylaluminum hydride
DIBAL-H	Diisobutylaluminum hydride
DIC	Diisopropylcarbodiimide

DIPEA	*N,N*-diisopropylethylamine
DMAP	4-dimethylaminopyridine
DME	Dimethoxyethane
DMF	Dimethylformamide
DMP	Dess–Martin periodinane/dimethoxypyridine/2,9-dimethyl-1,10-phenanthroline
DMSO	Dimethylsulfoxide
Dppf	1,1'-bis(diphenylphosphino)ferrocene
DSC	Disuccinimide carbonate
DTT	Dithienothiophene
DTTs	Dithienothiophenes
EBP	Ethyl bromopyruvate
EDC	1-ethyl-3-(3-dimethylaminopropyl)carbodiimide
EDCI	1-ethyl-3-(3-dimethylaminopropyl)carbodiimide
EDOT	Ethylenedioxythiophene
ERG	Electron releasing group
EWG	Electron withdrawing group
FAAH	Fatty acid amide hydrolase
Fmoc	9-fluorenylmethoxycarbonyl
F-SPE	Fluorous solid-phase extraction
GCE	Glassy carbon electrode
HATU	Hexafluorophosphate azabenzotriazole tetramethyluranium
HDNIB	[hydroxyl-(2,4-dinitrobenzene)-sulfonyloxy)iodo]benzene
HFA(Asp)	Hexafluoroacetone aspartate
HFA(Ida)	Hexafluoroacetone iminodiacetic acid
HMDO	Hexamethyldisiloxane
HMPA	Hexamethylphosphoramide
HOAt	1-hydroxy-7-azabenzotriazole
HOBt	1-hydroxybenzotriazole
HPLC	High-performance liquid chromatography
HWE	Horner–Wadsworth–Emmons
IBD	Iodobenzene diacetate
LAH	Lithium aluminum hydride
LDA	Lithium diisopropylamide
LHMDS	Lithium bis(trimethylsilyl)amide
LiHMDS	Lithium hexamethyldisilamide
LR	Lawesson's reagent
MBT	Mercaptobenzothiazole
MMPs	Matrix metalloproteinase
MOM	Methoxymethyl
MPLC	Medium-pressure liquid chromatography
MW	Microwave
MWI	Microwave irradiation
NBS	*N*-bromosuccinimide
NFSI	*N*-fluorobenzenesulfonimide

NIS	*N*-iodosuccinimide
NMM	*N*-methylmaleimide
NMP	*N*-methylpyrrolidinone
NMR	Nuclear magnetic resonance
NOESY	Nuclear Overhauser effect spectroscopy
OTf	Trifluoromethanesulfonate
PBD	Pyrrolobenzodiazepine
PBDT	Triazolopyrrolo[2,1-*c*][1,4]benzodiazepin-8-one
PCC	Pyridinium chlorochromate
PEG	Poly(ethylene glycol)
PET	Positron emission tomography
Phth	Phthalimide
PIB	Polyisobutylene
PITN	Poly(isothianaphthene)
PMP	Polymethylpentene
PPA	Polyphosphoric acid
PPTS	Pyridinium *p*-tolylsulfonate
PS-DIEA	Diisopropylaminomethyl polystyrene
p-TSA	*p*-toluenesulfonic acid
RCY	Radiochemical yield
S_N2	Bimolecular nucleophilic substitution
S_NAr	Aromatic nucleophilic substitution
SOD	Superoxide dismutase
TBAF	Tetrabutylammonium fluoride
TBAI	Tetrabutylammonium iodide
TBAT	Tetrabutylammonium triphenyldifluorosilicate
TBDMS	*t*-butyldimethylsilyl
TBDPS	*t*-butyldiphenylsilyl
TBS	*t*-butyldimethylsilyl
TBTU	tetramethyluronium tetrafluoroborate
TDA	4,4'-thiodianiline
TEA	Triethylamine
Teoc	2-(trimethylsilyl)ethoxycarbonyl
TES	Triethylsilyl
TFA	Trifluoroacetic acid
TFAA	Trifluoroacetic anhydride
TFE	Tetrafluoroethylene
TFP	Tri(2-furyl)phosphine
THF	Tetrahydrofuran
TMEDA	Tetramethylethylenediamine
TMS	Trimethylsilyl/tetramethylsilane
TMSCl	Trimethylsilyl chloride
TMSD	Trimethylsilyldiazomethane
TMSI	Trimethylsilyl iodide
TMSOTf	Trimethylsilyl trifluoromethanesulfonate

TPMA	Tris(2-pyridylmethyl)amine
TPy	Thiazolo[4,5-*b*]pyrazine
Ts	Tosyl/toluene-*p*-sulfonyl
UV	Ultraviolet

Chapter 1
Five-Membered *N*-Heterocycle Synthesis

1.1 Introduction

The heterocyclic compound is one of the most encountered frameworks in medicinal and pharmaceutically related materials. Due to the properties that are similar to drugs, libraries of different heterocyclic compounds are usually used in high-throughput screening at initial stages of drug design system. The competition in the area of drug design has helped to identify the speed of synthesis as a top preference in drug design. Subsequently, techniques that could enhance and promote both synthesis and screening of compounds are highly required [1–3].

The heteroaromatic structures are important and present in many natural and synthetic alkaloids that are employed in the field of agrochemicals, medicine, or cosmetics. Among the molecules related to this class of compounds, condensed heteroaromatic compounds bearing at least one nitrogen atom such as indoles and quinolones are undoubtedly the most appropriate because they generally affect the health of humans. The rich structural diversity encountered in these compounds, along with their biological and pharmaceutical importance, has encouraged more than 100 years of research aiming at developing efficient, economical, and selective synthetic approaches for such compounds [4–7].

Lawesson's reagent is a commercially accessible reagent and has been extensively used in organic synthesis in order to achieve the transformation of carbonyl compounds to thiocarbonyl compounds under different reaction conditions like utilization of polar solvents or base catalysts [8].

It is broadly recognized that organometallic reagents play an important role in organic synthesis, but the importance of main group-based reagents is not well appreciated. The LR has been extensively utilized in organic chemistry as a reagent for the transformation of ketones, esters, and amides into their corresponding thio analogues [9, 10]. This perception is concerned with the synthesis and new organic chemistry of this and related thionation reagents and their metal complexes [11–13].

1.2 Synthesis of Five-Membered *N*-Heterocycles

A conventional and well-known pathway for the synthesis of imidates/thioimidates is through the alkylation of amides and thioamides, respectively. Scheme 1.1 shows a procedure starting from commercially accessible Cbz-protected 1-aminocyclopropane-1-carboxylic acid [14]. Firstly, the Cbz-protected 1-aminocyclopropane-1-carboxylic acid underwent transformation to amide utilizing 1-ethyl-3-(3-dimethylaminopropyl)carbodiimide chloride and ammonium chloride in dimethylformamide and subsequent reaction with LR afforded thioamide. The reaction of thioamide with MeI in CH_3COCH_3 at 60 °C afforded thioimidate intermediate, via a rapid rearrangement process, which ultimately provided cyclic thioimidate.

Many strategies have been employed for the synthesis of enaminones [15–23], out of which the most adaptable is the Eschenmoser sulfide contraction of thiolactams, originally explained by Eschenmoser et al. [24–26]. The thiolactams were synthesized either through thionating lactams prepared from primary amines and bifunctional reagents or through a suitable conjugate addition of secondary thiolactams to acrylate esters, acrylonitrile, and analogous acceptors (Scheme 1.2) [27–44].

There are sixteen new compounds composed of unique five-membered heterocyclic compounds (thiophene, pyrrole, or furan) related to thiazole, imidazole, or quinoline rings. The Stetter reaction using thiazolium salt as a catalyst provided 1,4-diketones intermediate. The cyclization of 1,4-dicarbonyl compounds with H_3PO_4, CH_3COONH_4, or LR through Paal–Knorr synthesis afforded 2,5-disubstituted furan,

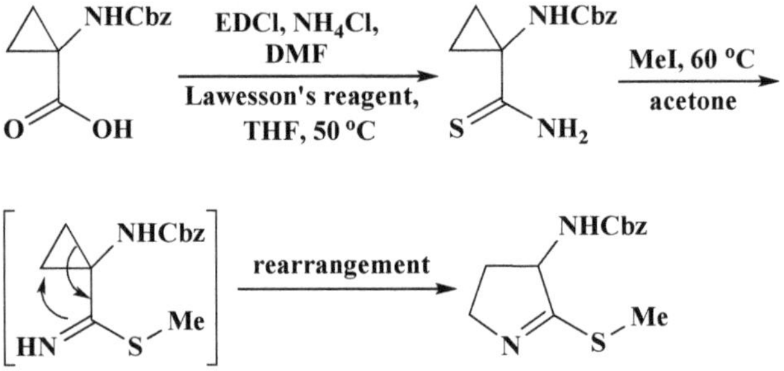

Scheme 1.1 Synthesis of cyclic thioimidate

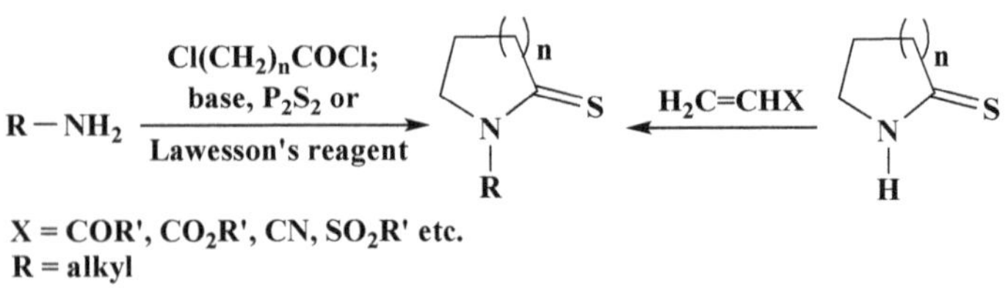

Scheme 1.2 Synthesis of thiolactams

1.2 Synthesis of Five-Membered N-Heterocycles

pyrrole, or thiophene compounds in moderate yields. The reaction of heteroaromatic aldehydes and 3-buten-2-one using thiazolium ylide as a catalyst and triethylamine as a base afforded Stetter-type products in moderate yields. The dicarbonyl compounds were transformed into pyrroles, thiophenes, and furans in moderate yield utilizing standard Paal–Knorr processes (Schemes 1.3, 1.4, 1.5, and 1.6) [45].

A direct amidation of 4-oxo-(2-thienyl)butanoic acid [46] with arylamines using DCC-BtOH afforded aryl-4-(2'-thienyl)-4-oxobutanamides [47]. No secondary products were observed, and the yields ranged from fair to good (30–64%) based on the nucleophilicity of the arylamine. Different effects of substituents in the utilized anilines were remarkable. The aryl-4-(2'-thienyl)-4-oxobutanamides were observed to be good starting compound for the formation of pyrroles by reaction with LR at reflux in toluene (Scheme 1.7). Attempts to transform the aryl-4-(2'-thienyl)-4-oxobutanamides into 5-arylamino-2,2'-bithiophenes afforded only thienylpyrroles (49%) or a mixture of thienylpyrroles (32–58%) and bithiophene derivatives in low yields (7–19%), pyrroles being the major compounds [48]. The 1-aryl-2-(2'-thienyl)pyrroles were prepared from aryl-4-(2'-thienyl)-4-oxobutanamides as major compounds via combination of Friedel–Crafts and Lawesson reactions. The bithiophene derivatives were also afforded as by-products, generally in low yields.

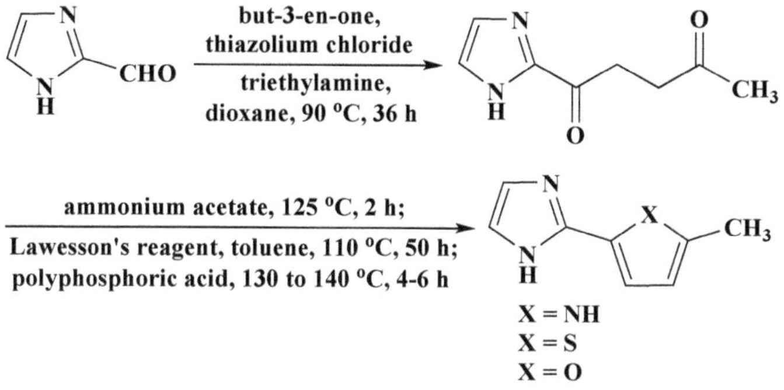

Scheme 1.3 Synthesis of pyrrole, thiophene, and furan

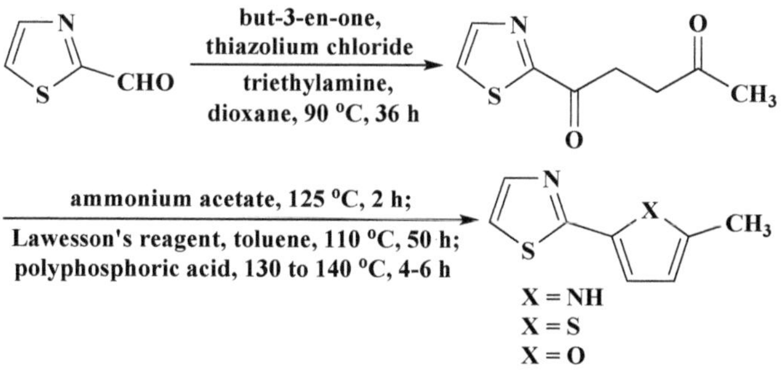

Scheme 1.4 Synthesis of pyrrole, thiophene, and furan

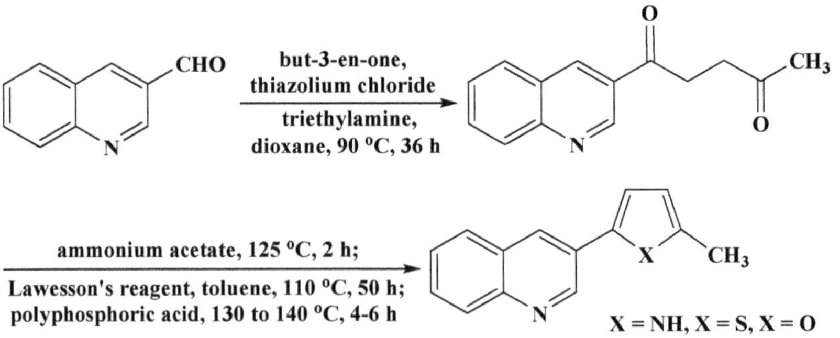

Scheme 1.5 Synthesis of pyrrole, thiophene, and furan

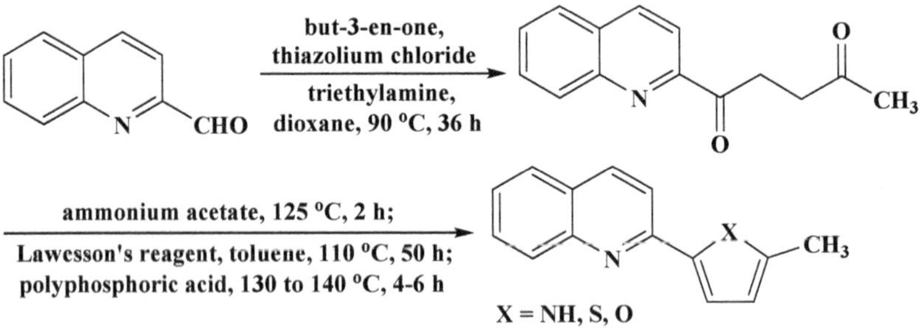

Scheme 1.6 Synthesis of pyrrole, thiophene, and furan

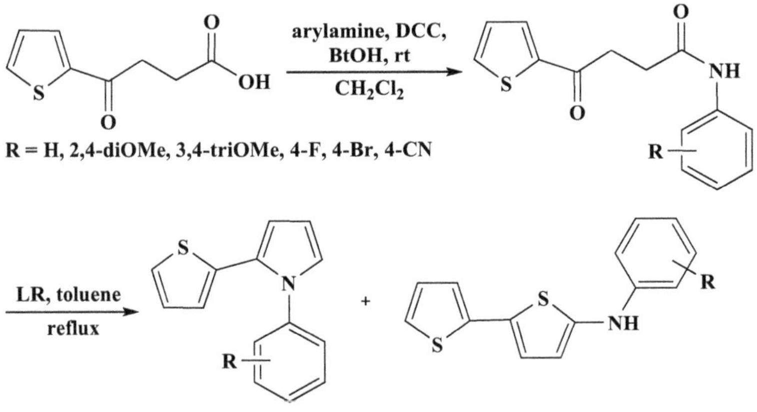

Scheme 1.7 Synthesis of 1-aryl-2-(2′-thienyl)pyrroles and bithiophene derivatives

1.2 Synthesis of Five-Membered N-Heterocycles

The reaction of 1,2-dioxines with LR, *n*-butylamine, or $(NH_4)_2CO_3$ was further examined (Scheme 1.8). The dimethylformamide was utilized as a solvent for the formation of thiophenes as it was expected that the Kornblum-de la Mare rearrangement of 1,2-dioxines to isomeric 1,4-diketones would be assisted in this solvent. The reaction of 1,2-dioxines with LR in dimethylformamide at 100 °C afforded thiophenes in good yields, respectively (entries 1, 2, and 3). The reaction of 1,2-dioxines with LR afforded thiophenes in lower yield under analogous conditions (entry 4). This may be because of instability of thiophene at enhanced temperatures. The 1,2,5-trisubstituted pyrroles were synthesized under different reaction conditions, which were employed for the preparation of thiophenes. The *n*-butylamine reagent could be employed, and dimethylformamide was not needed for the rearrangement of 1,2-dioxine into its 1,4-diketone. Therefore, refluxing CH_3OH was utilized instead of dimethylformamide. The 3,5-dihydro-1,2-dioxines were reacted with excess *n*-butylamine under these conditions in refluxing CH_3OH for 16 h to afford the 1,2,5-trisubstituted pyrroles, respectively, in good yields (entries 5 and 6) [49]. The reaction of 1,2-dioxines with $(NH_4)_2CO_3$ in refluxing CH_3OH afforded only isomeric 1,4-diketones, both in near quantitative yields (entries 7 and 9). This was rectified when the lower boiling CH_3OH was substituted by dimethylformamide

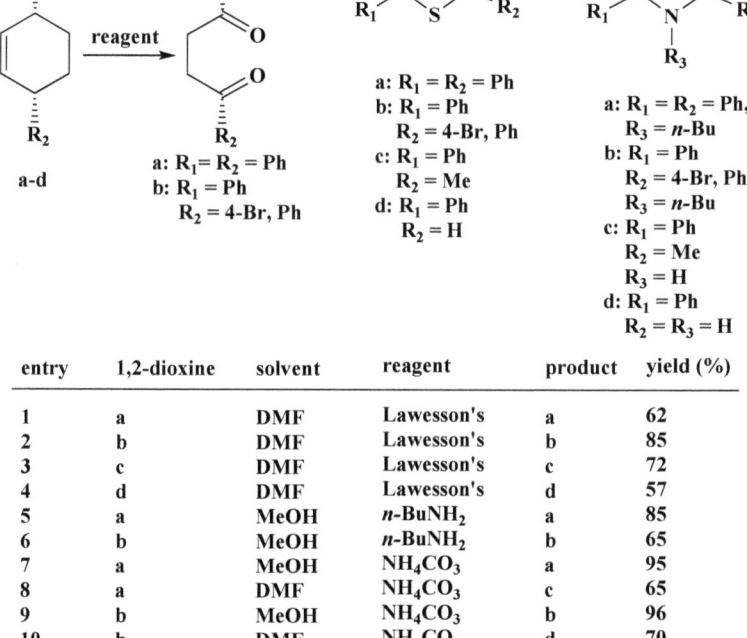

entry	1,2-dioxine	solvent	reagent	product	yield (%)
1	a	DMF	Lawesson's	a	62
2	b	DMF	Lawesson's	b	85
3	c	DMF	Lawesson's	c	72
4	d	DMF	Lawesson's	d	57
5	a	MeOH	*n*-BuNH$_2$	a	85
6	b	MeOH	*n*-BuNH$_2$	b	65
7	a	MeOH	NH$_4$CO$_3$	a	95
8	a	DMF	NH$_4$CO$_3$	c	65
9	b	MeOH	NH$_4$CO$_3$	b	96
10	b	DMF	NH$_4$CO$_3$	d	70

Scheme 1.8 Synthesis of 2,5-disubstituted pyrroles

as the solvent, affording 2,5-disubstituted pyrroles in moderate yields of 65 and 70%, respectively (entries 8 and 10).

The γ-ketoamides afforded pyrroles and thiophenes (Scheme 1.9) [11, 48].

Since studies had reported that malononitrile functionalization further redshifts the DEACM absorption profile [50], the diethylaminocoumarylidene-malononitrilemethyl (DEACM-MN) bearing linker was also formed in eight steps (Scheme 1.10). The aldehyde was reacted with nitromethane using N,N,N',N'-tetramethylethylenediamine to provide the nitroalcohol. The TBDMS-protected alcohol was transformed into thiocoumarin with LR, and the dicyanocoumarin was obtained by silver(I)-assisted malononitrile condensation. The NO_2 group was reduced with Zn and CH_3COOH, and the formed amine was reacted with 6-(2-chloroacetamido)-hexanoic acid to afford the intermediate. The alcohol deprotection and DSC coupling further afforded final DEACM-MN linker [51].

The ozonolysis afforded ketone, which was further reduced and the formed alcohol was protected to afford the diacetate (Scheme 1.11). This lactam was exposed to flash vacuum thermolysis at 600 °C followed by ethanolysis to afford the alcohol in 45% yield. The Swern oxidation of alcohol followed by Wittig olefination afforded alkene, a key intermediate in the formation of peduncularine. The last step in the synthesis involved the formation of thiolactam and alkylation with MeI to produce the iminium salt. The Grignard addition followed by reduction with $NaBH_3CN$ provided a modest yield of 35% (42% of C7-isomer). The Fischer indole formation utilizing phenylhydrazine afforded (−)-peduncularine in 19 steps and 0.7% overall yield [52].

It was described that the reaction of 2-acylbenzamides with LR afforded different products based on the groups attached to the starting compounds (Schemes 1.12, 1.13, and 1.14) [12, 48].

The steps involved in the overall formation of (−)-indolizidine (−)-209B [44] are described in Scheme 1.15. The absolute stereocontrol resulted from Davies's protocol [53, 54] where homochiral amine, formed from t-butyl (E)-oct-2-enoate and (R)-N-benzyl-1-phenylethylamine, was transformed into primary amine and further into thiolactam in multiple steps. The Eschenmoser sulfide contraction [55, 56] with ethyl bromoacetate afforded key enaminone intermediate and the chemoselective reduction of saturated ester synthesized alcohol. The bicyclic nucleus of alkaloid was further constructed by a cycloalkylation that took the benefit of nucleophilic reactivity

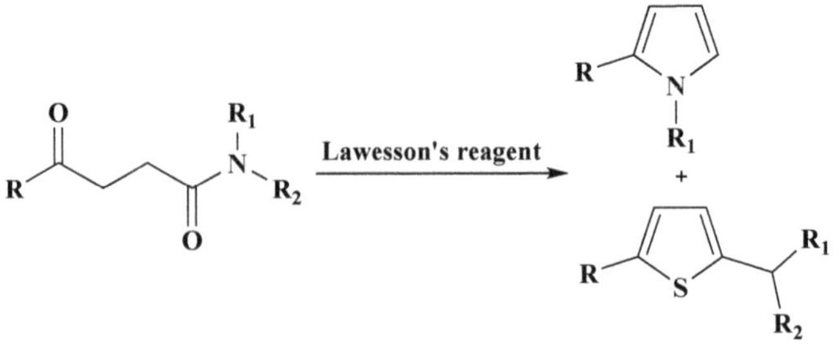

Scheme 1.9 Synthesis of pyrroles and thiophenes

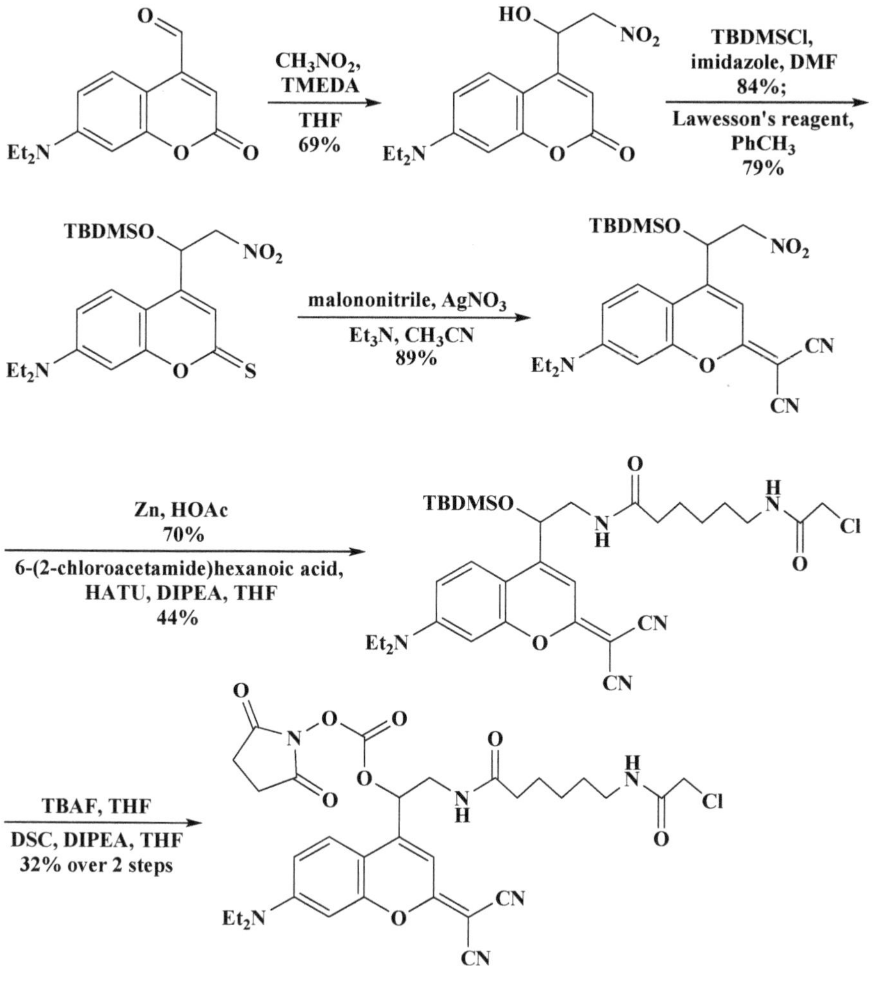

1.2 Synthesis of Five-Membered N-Heterocycles

Scheme 1.10 Synthesis of DEACM-MN linker

of enaminone, after that a chemoselective and reasonably diastereoselective (88:12) reduction of the alkene bond of bicyclic enaminone set up the desired stereochemistry at C-8 and C-8a. The epimerization of ester in the reduced compound synthesized a compound, which was reduced to afford the alcohol. Holmes et al. [57] demonstrated that the reduction of methanesulfonate with $Li(C_2H_5)_3BH$ completed the overall formation of (−)-indolizidine 209B [58].

The phosphonate was synthesized starting from *t*-butyl bromoacetate and triethyl phosphite. The phosphonate underwent a Horner–Wadsworth–Emmons reaction to afford the Michael acceptor in 63% yield over two steps. The alkenoate underwent a stereoselective aza-Michael reaction with dibenzylated chiral amine. The enantiomerically pure amine was isolated in 52% yield over two steps after debenzylation

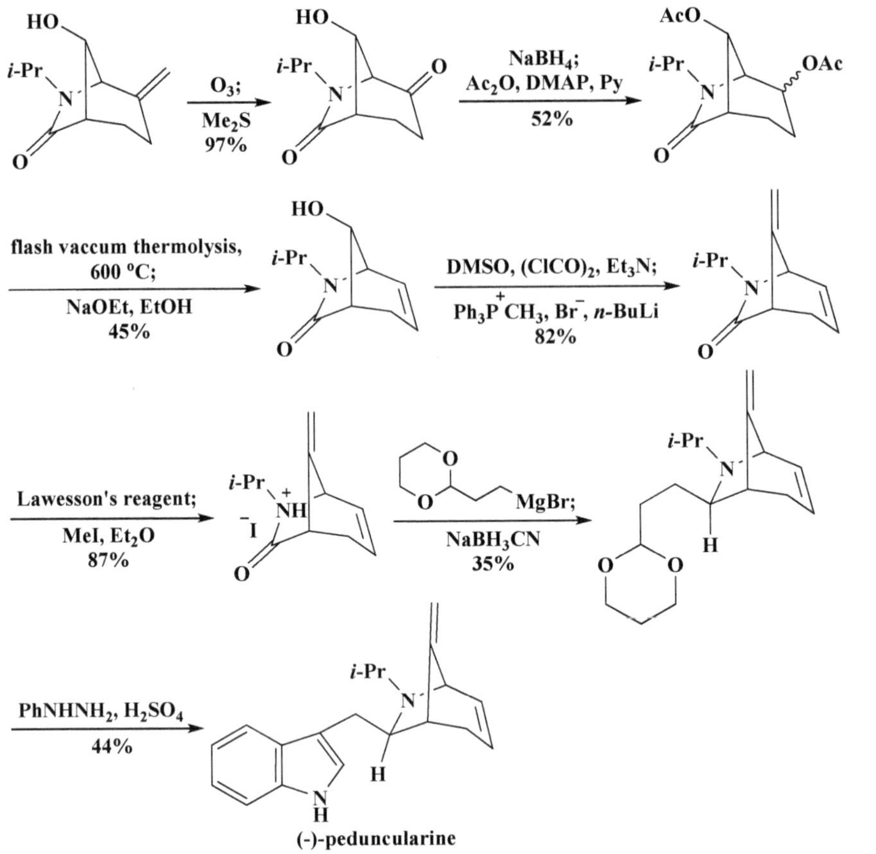

Scheme 1.11 Synthesis of (-)-peduncularine

Scheme 1.12 Synthesis of isoindolines

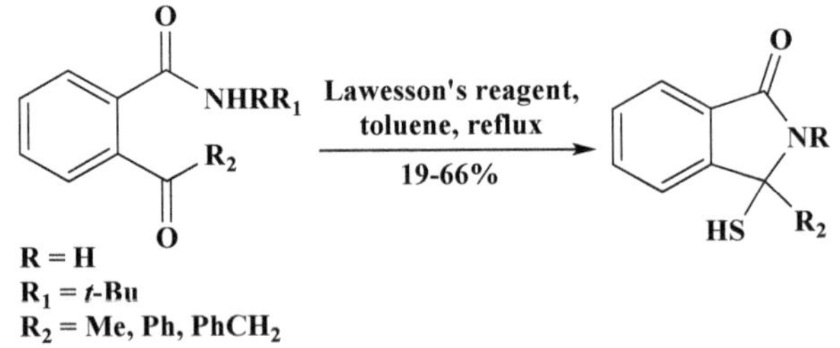

with 10% Pd/C and H$_2$ gas in CH$_3$COOH. This amine was treated with chlorobutyryl chloride and cyclized to obtain the lactam in 56% yield. The lactam was thionated with LR to afford the thiolactam in 85% yield. The enaminone was formed in 79% yield when thiolactam underwent an Eschenmoser sulfide contraction reaction. The enaminone was transformed to mixed anhydride through the carboxylic acid, and this

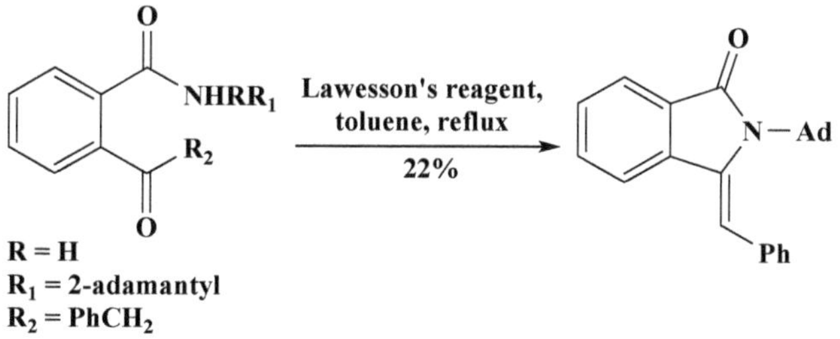

1.2 Synthesis of Five-Membered *N*-Heterocycles

Scheme 1.13 Synthesis of isoindolines

Scheme 1.14 Synthesis of isoindolines

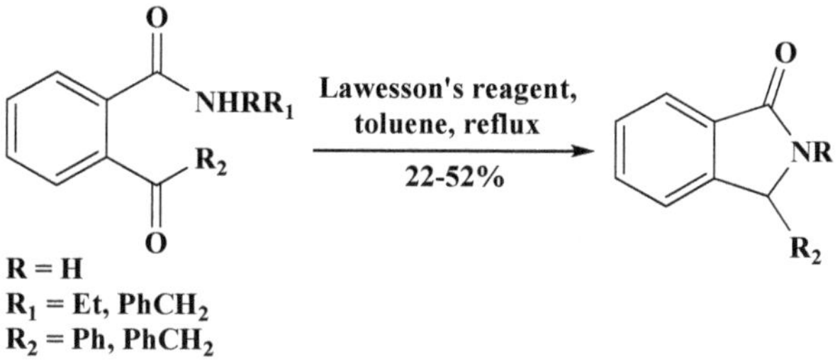

assisted the acylative ring-closure of enaminone to provide the hexahydroindolizidinone in 55% yield. Standard reactions were utilized for the defunctionalization of ester, enaminone, and ketone to afford the (−)-indolizidine 167B in fifteen steps and 1.5% yield (Scheme 1.16) [41].

The formation of racemic indolizidine 167B by Michael and Gravestock [59] has been modified to afford the alkaloid's (−)-enantiomer (Scheme 1.17) [42]. The desired absolute configuration at the site adjacent to nitrogen was assured with Davies protocol, which included the highly diastereoselective conjugate addition of anion of (*R*)-*N*-benzyl-1-phenylethylamine to a conjugated ester, in this case (−)-menthyl (*E*)-hex-2-enoate. The amino ester product was transformed into thiolactam in four steps. This intermediate was reacted via reaction sequence (Eschenmoser sulfide contraction, acylative ring-closure, hydrolysis, and decarboxylation) to obtain the bicyclic vinylogous amide, which was carefully reduced with LiAlH$_4$ to afford the volatile indolizidinone as a single diastereoisomer. This completed the formal formation of target compound, i.e., (−)-alkaloid [60, 61].

An aldol reaction of starting compound with (*R*)-3-(triethylsilyloxy)butanal synthesized hydroxyketone. A four-step reaction of hydroxyketone afforded α,β-unsaturated ester which then underwent double bond reduction, cleavage of Cbz group, lactamization, and reduction of lactam carbonyl group to provide the first formation of grandisine A (Scheme 1.18) [62].

The 2-methylresorcinol was utilized for the synthesis of 2-benzyloxy-6-bromo-4-methoxy-3-methylaniline in six steps. According to a process proposed by Raphael and Ravenscroft [63], the 2-methylresorcinol was reacted with NaNO$_2$ in acidic

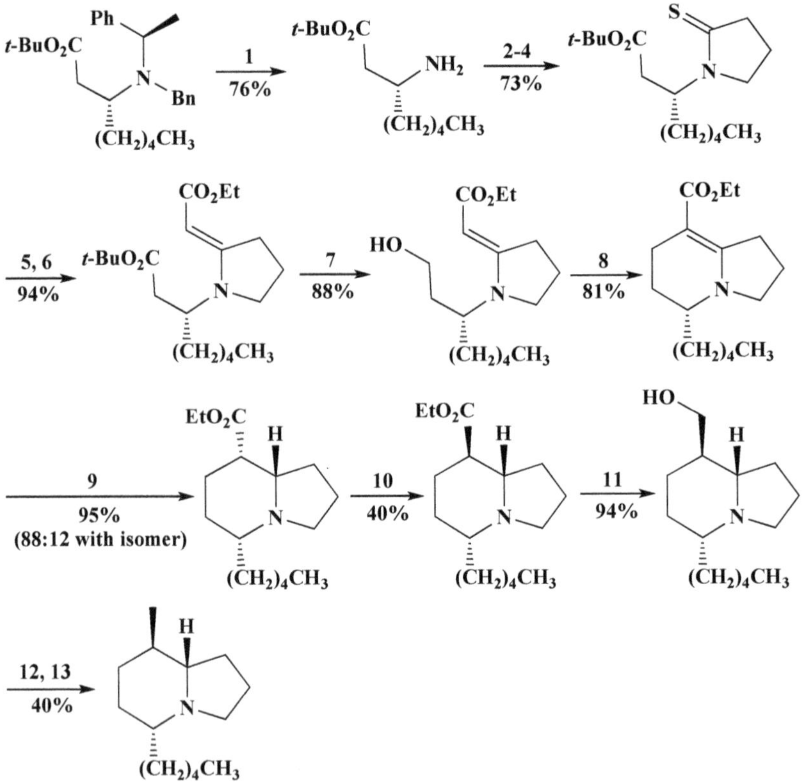

Reagents and conditions: (1) 7 atm H_2, 10% Pd/C, AcOH, rt, (2) Cl$(CH_2)_3$COCl, NaHCO$_3$, CHCl$_3$, reflux, (3) t-BuOK, t-BuOH, rt, (4) Lawesson's reagent, PhMe, reflux, (5) BrCH$_2$CO$_2$Et, MeCN, rt, (6) Ph$_3$P, Et$_3$N, MeCN, rt, (7) LiAlH$_4$, THF, rt, (8) I$_2$, imidazole, Ph$_3$P, PhMe, 110 °C, (9) 1 atm H$_2$, PtO$_2$, AcOH, rt, (10) NaOEt (catalyst), EtOH, reflux, (11) LiAlH$_4$, THF, 0 °C to rt, (12) MeSO$_2$Cl, NEt$_3$, CH$_2$Cl$_2$, 0 °C to rt, (13) LiEt$_3$BH, THF, 0 °C.

Scheme 1.15 Synthesis of (-)-indolizidine 209B

medium, and the nitroso intermediate was then oxidized to nitro product with 70% concentrated HNO$_3$ (Scheme 1.19). The crystalline product 3-methoxy-2-methyl-6-nitrophenol was prepared by regioselective methylation of less hindered OH group with dimethyl sulfate at ambient rt. The doubly protected nitroresorcinol was synthesized by benzylation of second OH group under more dynamic conditions, which was then reduced to sensitive aniline with hydrazine hydrate over Raney Ni in boiling CH$_3$OH. The crude C$_6$H$_5$NH$_2$ was instantly brominated with Br$_2$ in a mixture of CH$_2$Cl$_2$ and CH$_3$COOH to complete the formation of bromoaniline in 77% yield over two steps. The complete reaction sequence from 2-methylresorcinol to bromoaniline could also be carried out on a scale of ca. 25 g without purification of the

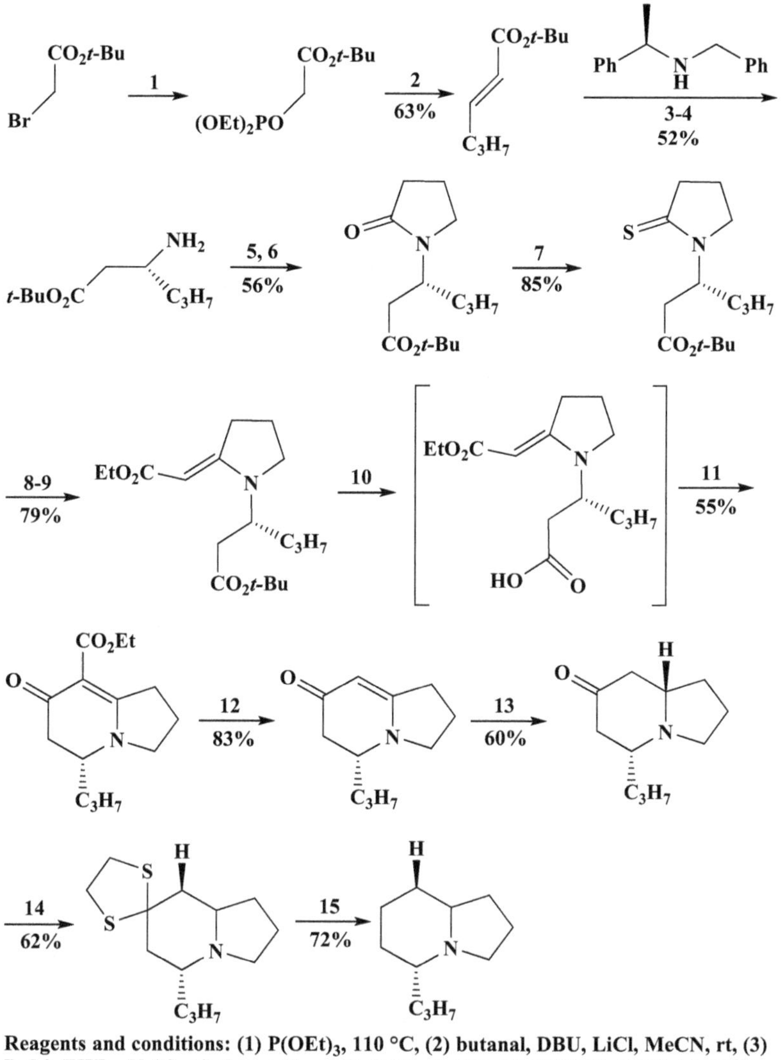

1.2 Synthesis of Five-Membered N-Heterocycles

Reagents and conditions: (1) P(OEt)$_3$, 110 °C, (2) butanal, DBU, LiCl, MeCN, rt, (3) BuLi, THF, -78 °C, (4) 7 atm H$_2$, 10% Pd/C, AcOH, (5) Cl(CH$_2$)$_3$COCl, NaHCO$_3$, CHCl$_3$, rt, (6) t-BuOK, t-BuOH, (7) Lawesson's reagent, toluene, reflux, (8) BrCH$_2$CO$_2$Et, MeCN, rt, (9) Ph$_3$P, Et$_3$N, MeCN, rt, (10) Me$_3$SiI, CCl$_4$, rt, (11) Ac$_2$O, MeCN, 50 °C, (12) KOH, H$_2$O, reflux, then HCl, reflux, (13) LiAlH$_4$, THF, rt, (14) HS(CH$_2$)$_3$SH, BF$_3$·Et$_2$O, CF$_3$CO$_2$H, rt, (15) Raney-Ni, EtOH, reflux.

Scheme 1.16 Synthesis of (-)-indolizidine 167B

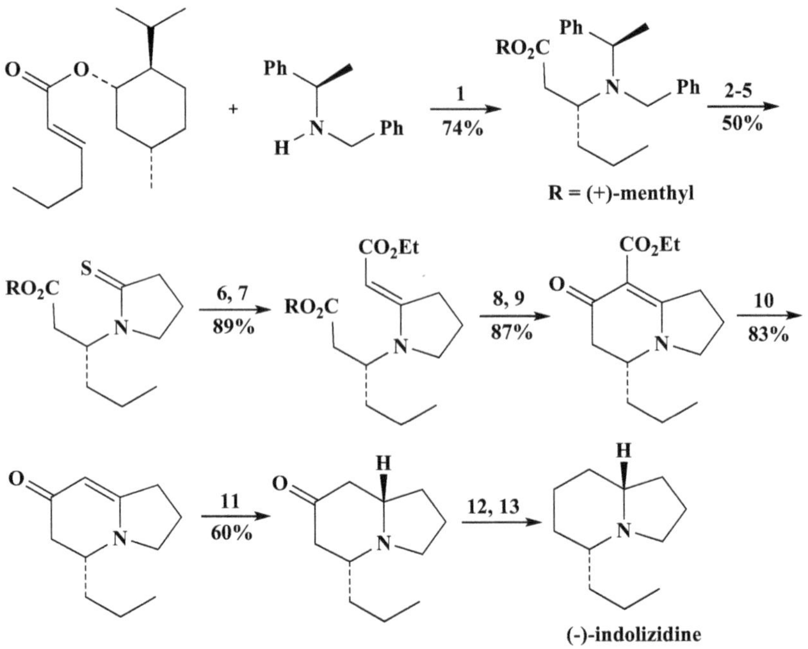

Reagents and conditions: (1) BuLi, THF, -78 °C, (2) 7 atm H_2, 10% Pd/C, HOAc, rt, (3) Cl(CH$_2$)$_3$COCl, NaHCO$_3$, CHCl$_3$, rt, (4) t-BuOK, t-BuOH, rt, (5) Lawesson's reagent, PhMe, reflux, (6) BrCH$_2$CO$_2$Et, MeCN, rt, (7) Ph$_3$P, Et$_3$N, MeCN, rt, (8) KOH, EtOH, reflux, (9) Ac$_2$O, MeCN, 50 °C, (10) KOH, H$_2$O, reflux, then HCl, reflux, (11) LiAlH$_4$, THF, rt, (12) HS(CH$_2$)$_3$SH, BF$_3$.Et$_2$O, TFA, rt, (13) Raney-Ni W-2, EtOH, reflux.

Scheme 1.17 Synthesis of indolizidine

intermediates in approximately 45% yield. The absolute configuration of aziridinomitosene was derived from the second reaction partner, (−)-2,3-O-i-propylidene-D-erythronolactone, which was formed in 77% yield by an approach that required oxidative cleavage of D-isoascorbic acid with H_2O_2 followed by ketal exchange with 2,2-dimethoxypropane [64]. Since it was known that this lactone was reacted poorly with 2-bromoaniline unless the latter was deprotonated first, a solution of bromoaniline in THF was reacted with C$_2$H$_5$MgCl at −50 °C before adding (−)-2,3-O-i-propylidene-D-erythronolactone and allowing the mixture to warm to rt. The alcohol intermediate was purified for characterization purposes and was transformed directly into lactam. The cyclization was completed by mesylation followed by reaction with NaH in a mixture of THF and N,N-dimethylformamide at rt. The transformation of bromoaniline into lactam took place on a 20 g scale without purification of intermediates in 90% yield. The product was observed to exist as a 1:1 mixture of two separable rotamers that could be independently characterized, although this was not required in view of the later convergence of these intermediates to a single product. With both rings A and C in place, the next aim was to complete the formation of

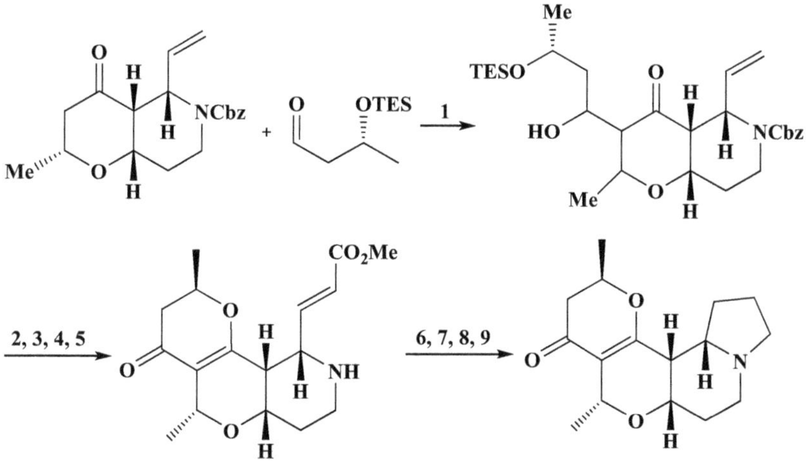

1.2 Synthesis of Five-Membered N-Heterocycles

Reagents and conditions: (1) LiHMDS, ZnCl$_2$, THF, -78 °C, -78 to -50 °C, 3.5 h, (2) Dess-Martin periodinane, CH$_2$Cl$_2$, (3) TFA, CH$_2$Cl$_2$, 73% over 3 steps, (4) O$_3$, MeOH, Sudan III (indicator), -78 °C, then Me$_2$S, -78 to 25 °C, (5) methyl (triphenylphosphoranylidene)acetate, benzene, 60 to 40 °C, 9.5 h, 80% over 2 steps, (6) 10% Pd/C, 1 atm H$_2$, MeOH, (7) PhMe, reflux, 24 h, 98% over 2 steps, (8) Lawesson's reagent, PhMe, 65 °C, 98%, (9) Raney-Ni, THF, 25 °C, 94%.

Scheme 1.18 Synthesis of grandisine A

pyrrolo[1,2-*a*]indole skeleton through the tandem Reformatsky–Heck sequence. At last, the rotameric mixture of lactams was thionated with LR in boiling toluene to afford the thiolactam as a mixture of rotamers in 73% yield. Both rotameric mixtures of thiolactam were obtained when lactam rotamers were separated and were independently exposed to thionation. The transformation of thiolactam into vinylogous urethane was attained through extended reaction in boiling THF with an excess of organozinc reagent synthesized by sonicating activated Zn powder with bromoacetate using I$_2$ as a catalyst. The product, yet again a mixture of two rotamers, was obtained in reproducibly good yields of above 85% yield on scales as large as 5 g. The crucial intramolecular Heck cyclization occurred by adapting conditions developed by Tietze and Petersen [65] in which a carefully balanced mixture of solvent, base, ligand, and additives was used. In this case, the reactant was heated at reflux with 0.3 eq. Pd(OAc)$_2$, tri-*o*-tolylphosphine, and TEA in the mixed solvent system of DMF, CH$_3$CN, and H$_2$O (5:5:1) to afford the desired compound (R = Bn) in 82% yield [66]. In one case, the benzyl protecting group was also removed using an excess of 1.6 eq. Pd(OAc)$_2$ to afford the free phenol (R = H) in 90% yield.

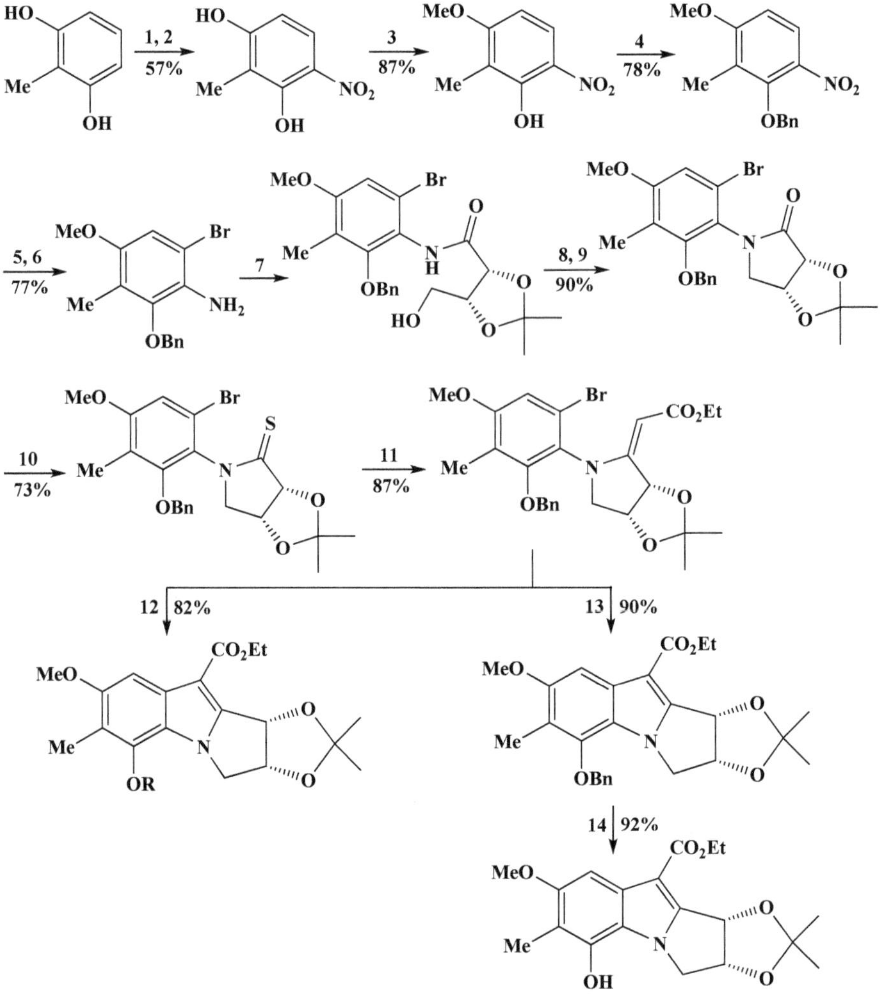

Reagents and conditions: (1) NaNO$_2$, H$_2$SO$_4$, H$_2$O, rt, 45 min, (2) HNO$_3$ (70%), H$_2$O, rt, 70 h, (3) Me$_2$SO$_4$, K$_2$CO$_3$, acetone, rt, 20 h, (4) BnBr, K$_2$CO$_3$, acetone, reflux, 41 h, (5) NH$_2$NH$_2$·H$_2$O, Raney-Ni, MeOH, reflux, 1.5 h, (6) Br$_2$, CH$_2$Cl$_2$-AcOH, rt, 6 h, (7) EtMgCl, THF, -50 °C, 50 min, then 14, -50 °C to rt, 22 h, (8) MsCl, Et$_3$N, CH$_2$Cl$_2$, 0 °C to rt, 7 d, (9) NaH, THF-DMF, rt, 20 h, (10) Lawesson's reagent, toluene, reflux, 2.5 h, (11) BrZnCH$_2$CO$_2$Et (from BrCH$_2$CO$_2$Et, Zn, I$_2$, THF, ultrasound), THF, reflux, 120 h, (12) 0.3 eq. Pd(OAc)$_2$ P(o-Tol)$_3$, Et$_3$N, DMF, MeCN, H$_2$O, reflux, 4 h, (13) 1.6 eq. Pd(OAc)$_2$, P(o-Tol)$_3$, Et$_3$N, DMF, MeCN, H$_2$O, reflux, 4 h, (14) 10% Pd/C, H$_2$, EtOH, HCl (1 drop), rt, 2 h.

Scheme 1.19 Synthesis of dioxolopyrroloindole

1.3 Synthesis of Five-Membered *N,N*-Heterocycles

Dodd et al. [67] have described a solid-supported formation of 5-*N*-alkylamino- and 5-*N*-arylaminopyrazoles. The β-ketoesters were heated with resin-bound amines in resin-compatible solvents such as NMP or toluene using 4-dimethylaminopyridine to afford the resin-immobilized β-ketoamides. The β-ketoamides, aryl or alkyl hydrazines, and LR were suspended in a mixture of tetrahydrofuran/pyridine and heated at 50–55 °C in order to provide the resin-bound 5-aminopyrazoles. The free 5-aminopyrazoles were liberated from solid support through reaction with trifluoroacetic acid (Scheme 1.20) [68].

The LR was employed as a substitute to P_2S_5. Chebil and Jouil [69] in 2012 described the formation of thiopyrazolone from β-phosphoryl-β'-carbethoxyhydrazones using LR (Scheme 1.21).

The substituted 5-alkylamino- and 5-(arylamino)pyrazoles could be synthesized by a one-pot protocol from ketoamide, aryl or alkyl hydrazine, and LR (Scheme 1.22) [70]. The hydrazones were probable intermediates. This approach was also employed for the solid-supported formation of 5-(*N*-monosubstituted amino)pyrazoles [67].

This example [71] is a somewhat different pathway for the synthesis of 3-amino-1*H*-indazoles (Scheme 1.23). Three other examples of formation of 3-amino-1*H*-indazoles can be observed in Refs. [72–78] and Ref. [79] which provided additional examples of synthesis of different substituted 1*H*-indazoles.

A 1,3,4-thiadiazole ring was incorporated at 17 position of the androstane frame (Scheme 1.24). In this case, a D-ring condensed pyrazolidine-3-thione, synthesized through intramolecular 1,4-addition to C=C bond, was identified as a major product

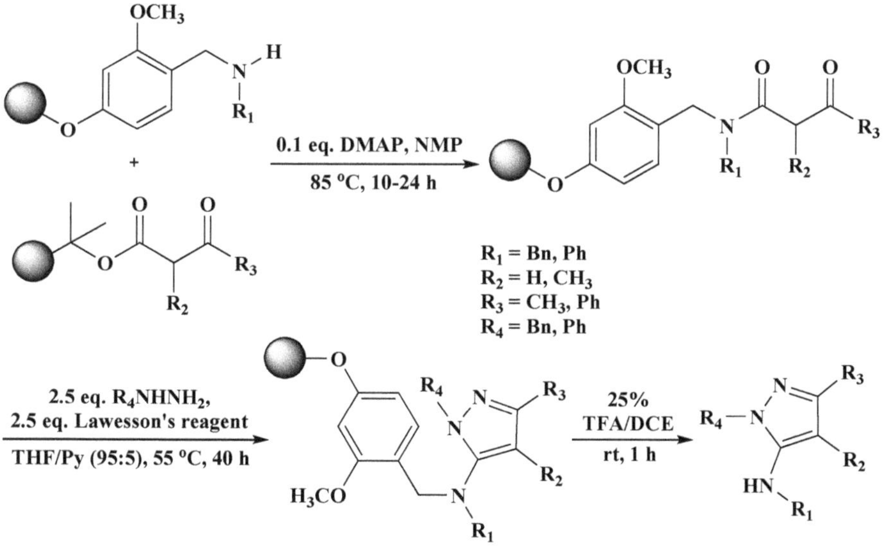

Scheme 1.20 Synthesis of aminopyrazoles

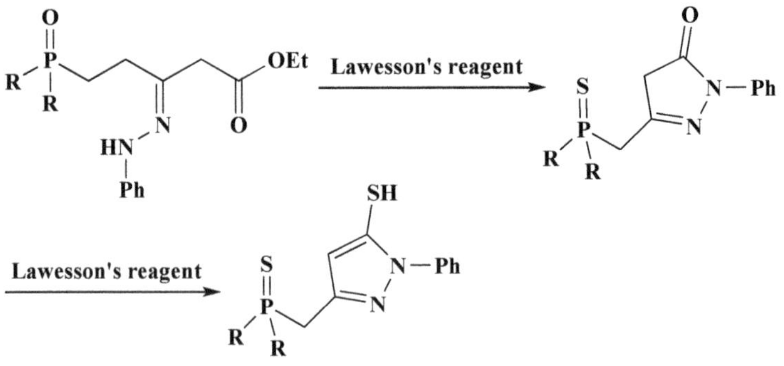

Scheme 1.21 Synthesis of thiopyrazolone

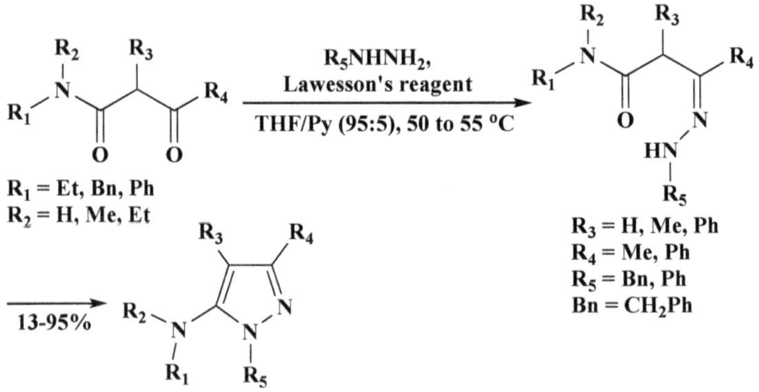

Scheme 1.22 Synthesis of aminopyrazoles

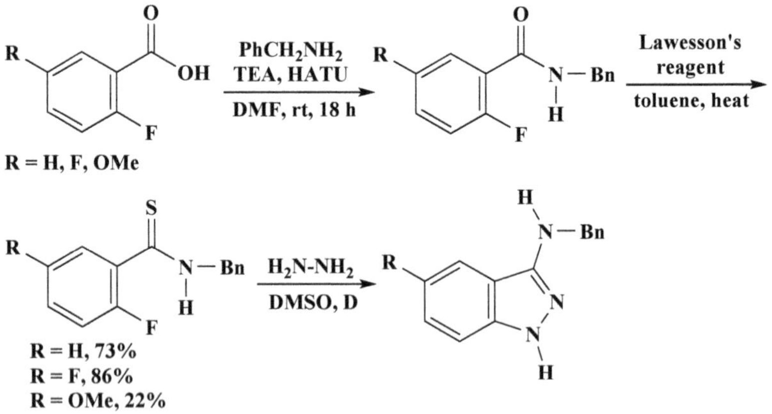

Scheme 1.23 Synthesis of 3-amino-1H-indazoles

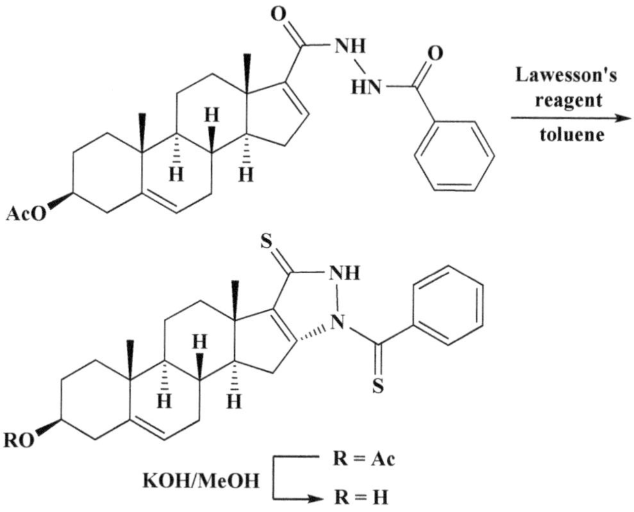

1.3 Synthesis of Five-Membered N,N-Heterocycles

Scheme 1.24 Synthesis of pyrazolidinethione

and was exposed to deacetylation in basic medium. The 16α,17α-*cis* junction of heteroring was established via NOESY spectrum, NMR, and MS measurements on final compound (R = Ac), and its deacetylated analogue (R = H) confirmed that oxygen → sulfur exchange took place on both the oxygen atoms of CO groups directly attached to nitrogen atoms, whereas the ester group at C-3 (R = Ac) remained unaffected [80, 81].

The 2-bromobenzoyl chloride and 2-fluorobenzoylchloride were treated with methanethiolate sodium salt to provide the 2-halo-thiobenzoic acid *S*-methyl esters in almost quantitative yield (Scheme 1.25). These compounds were subsequently reacted with LR to afford the dithioesters in excellent yield [82]. The imidazoline ring was constructed by condensation of dithioesters with a chiral diamine. The driving force of reaction was the precipitation of mercury sulfide through the utilization of a desulfurizing agent like HgO [83]. Thus, the transformation of dithioesters (R = Br) with (1*R*,2*R*)-diphenylethylene diamine afforded imidazoles (R = Br) in 66% yield. The condensation of dithioesters (R = F) with chiral diamine operated significantly better, and the ring-closure occurred in only half an hour to afford the imidazoles (R = F) in 92% yield. In this case, the total yield was 86%.

The compounds bearing diimidazoline and dipyrimidine moieties were prepared (Scheme 1.26) [84]. The alkanedinitriles were reacted with propylenediamine and ethylenediamine in toluene (dry) at 90 °C for 10 h to afford the dipyrimidines and diimidazolines, respectively, using a small amount of P_4S_{10}. Analogous results were obtained when LR, S_8, or $Na_2S \cdot 9H_2O$ was employed instead of P_4S_{10} [85].

In the formation of a potent adrenergic agent, Lawesson's reagent was employed to synthesize its imidazole ring [86]. The transformation of aminolactam with

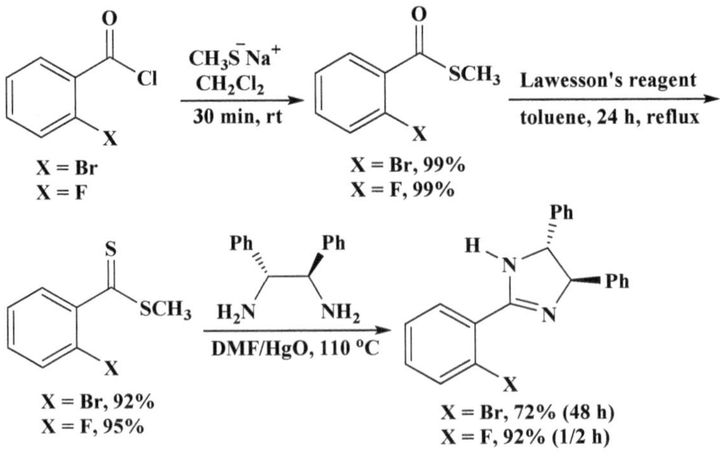

Scheme 1.25 Synthesis of imidazoles

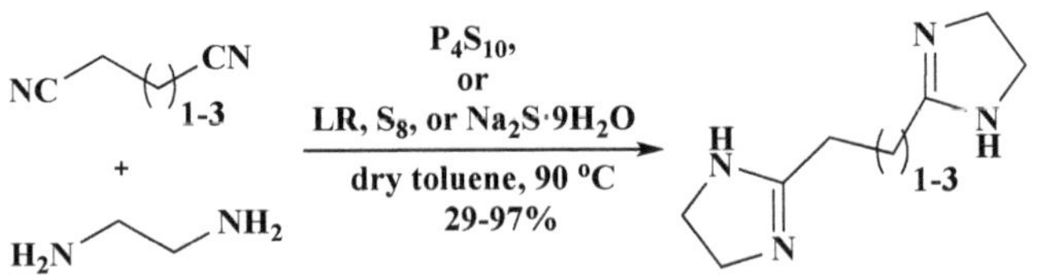

Scheme 1.26 Synthesis of diimidazolines

Lawesson's reagent using D-10 camphorsulfonic acid in refluxing xylene for 72 h under nitrogen atmosphere provided imidazole ring in 38% yield (Scheme 1.27) [12].

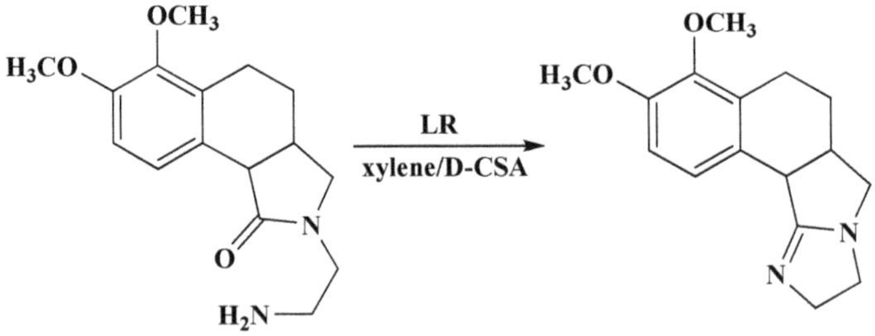

Scheme 1.27 Synthesis of imidazoisoindole

1.4 Synthesis of Five-Membered *N,N,N*-Heterocycles

The esterification of acid was completed employing concentrated H_2SO_4 in C_2H_5OH to provide the ethyl ester in 92% yield [87]. Also, ethyl ester was directly condensed with propargyl alcohol in concentrated H_2SO_4 to afford the sulfide terminal alkyne in 85% yield, which was instantly reacted with (3-azidopropyl)cyclohexane, sodium ascorbate, and $CuSO_4 \cdot 5H_2O$ in tetrahydrofuran/water (v/v, 1:1) or (4-azidobutyl)benzene, sodium ascorbate, and $CuSO_4 \cdot 5H_2O$ in tetrahydrofuran/water (v/v, 1:1) to synthesize the triazoles in 95% and 97% yield, respectively [88]. Similarly, oxidation of ethyl ester occurred with 30% hydrogen peroxide in acidic media in alcohol solvent which was hydrolyzed in situ with sodium hydroxide in water and ethanol to afford the acid in 70% yield. The reaction of acid with propargyl alcohol and ethyl(dimethylaminopropyl)carbodiimide (EDC) using 4-dimethylaminopyridine (DMAP) in *N,N*-dimethylformamide afforded sulfonyl terminal alkyne in 88% yield, which was further reacted with (3-azidopropyl)cyclohexane, sodium ascorbate, and $CuSO_4 \cdot 5H_2O$ in tetrahydrofuran/water (v/v, 1:1) or (4-azidobutyl) benzene, sodium ascorbate, and $CuSO_4 \cdot 5H_2O$ in tetrahydrofuran/water (v/v, 1:1) for 10 min to obtain the triazoles in 96% and 95% yield, respectively. These triazole syntheses through copper(I)-catalyzed 1,3-dipolar cycloaddition reaction (1,3-DCR) of azido moieties and terminal alkyne proceeded easily under very mild conditions [89]. Disappointingly, condensation of acid with propargyl alcohol in concentrated H_2SO_4 did not result in the synthesis of terminal alkyne, and this reaction condition resulted in mainly detected staring compounds and/or decomposed product. The synthesis of sulfonyl terminal alkyne via oxidation of sulfide terminal alkyne with LR in toluene was successfully used for the preparation of desired product in 70% yield (Scheme 1.28) [90].

Lactam was transformed into thiolactam, which was further reacted with acetyl hydrazide and $HgCl_2$ to afford the intermediate. The yields were quite moderate, 65% for the first step and 47% for the second step; however, the procedure was straightforward to achieve the smooth purifications. It was expected that the following intramolecular condensation of hydrazide carbonyl group with bridged nitrogen atom would be little difficult. A variety of known conditions was screened under both conventional and MW heating, but all were unsuccessful. In most of the cases, unreacted starting compound was recovered even under forcing conditions (toluene, 200 °C, MW). Upon examination of other solvents than those usually used in the literature to carry out this cyclization, the breakthrough came with pyridine. Heating a solution of intermediate in pyridine for 30 min at 130 °C under MWI afforded desired triazole in 30% yield along with unreacted starting compound. More forcing conditions (180 °C, MW) improved the conversion and also resulted in partial cleavage of the Boc group. The reaction at 240 °C under MW [91] heating led to complete consumption of intermediate with the desired deprotected product spiropiperidine being formed in 80% yield (Scheme 1.29).

The 3-methyl-2,6-diphenylpiperidin-4-one was utilized as starting compound for the formation of new triazolodiazepines. The piperidin-4-one on reaction with NaN_3

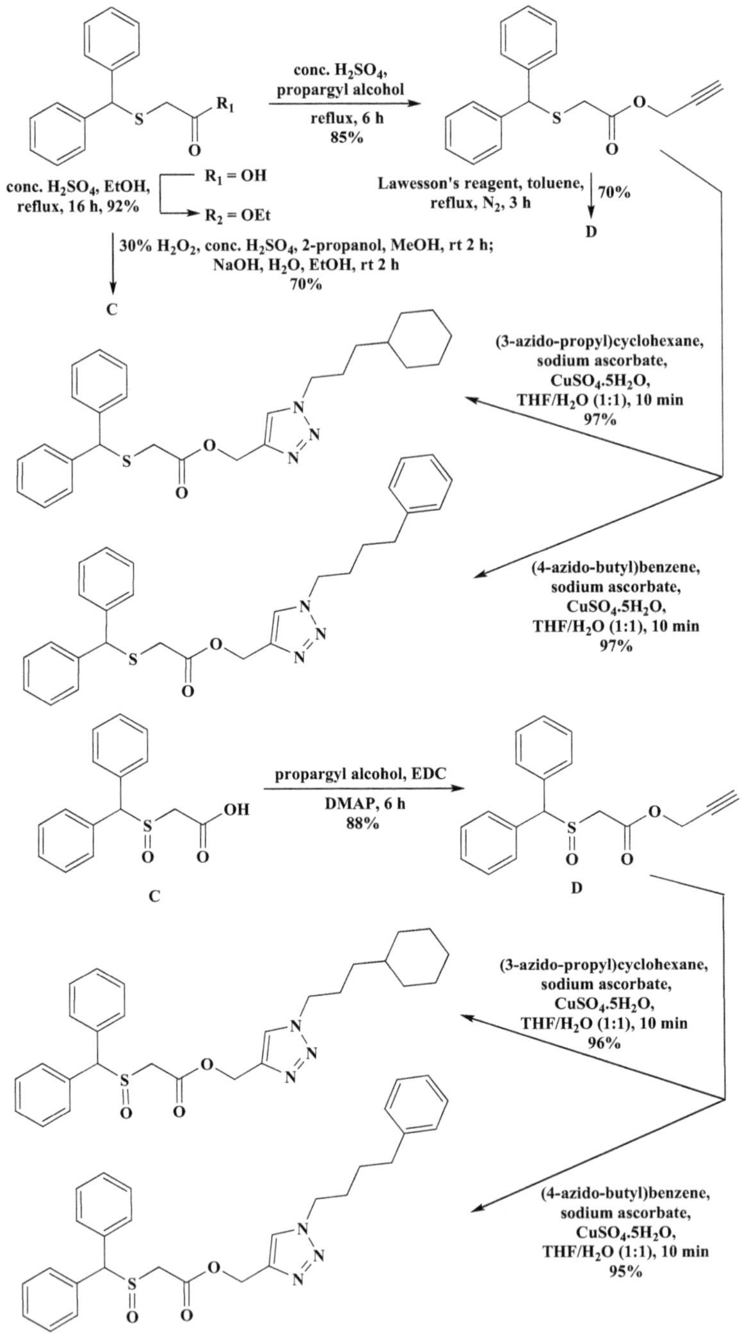

Scheme 1.28 Synthesis of triazoles

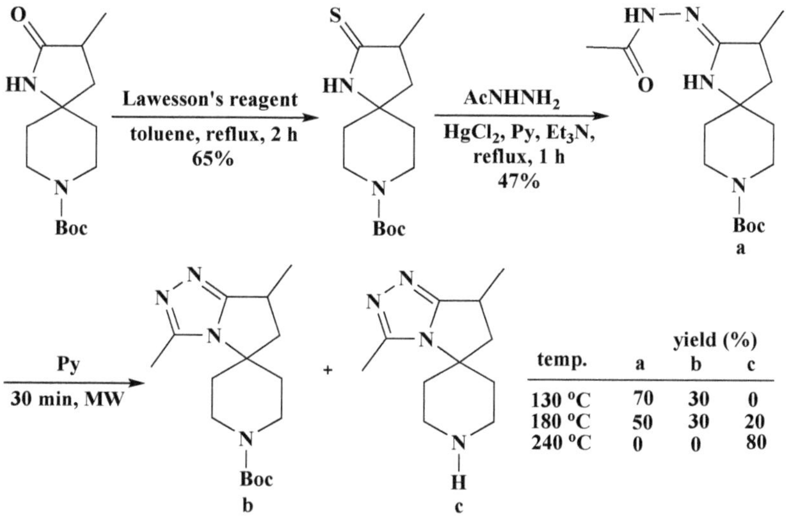

1.4 Synthesis of Five-Membered N,N,N-Heterocycles

Scheme 1.29 Synthesis of dihydrospiropiperidine pyrrolotriazole

and concentrated sulfuric acid afforded diazepam-5-one. The diazepams were reacted with LR to afford the thiones. The target triazolodiazepines were obtained when thiones were reacted with acetyl hydrazide (Scheme 1.30) [92].

The 6-ethyltriazolobenzodiazepines were synthesized from benzodiazepinone (Scheme 1.31) [93]. The transformation of benzodiazepinone to thiolactam could not be completed under standard conditions [94] (phosphorus pentasulfide in Py or other high boiling solvents), but proceeded employing LR in anhydrous tetrahydrofuran under nitrogen atmosphere. The target compounds were prepared by condensation of thiolactam with carboxylic acid hydrazides. The residual hydrazides were observed to be poorly separable from the desired compounds through column chromatography. This challenge was solved by transforming the hydrazides into H_2O-soluble condensation products on stirring with glucose solution prior to extraction with an organic solvent [95, 96].

Lattmann et al. [97] have prepared a series of 4H-triazolo-1,4-benzodiazepines from p-chloroaminobenzophenone (Scheme 1.32). This commercially accessible ketone was transformed into 1,4-benzodiazepine, which on reaction with LR afforded thioamides. These reactive intermediates were reacted with acetyl hydrazide to afford the 4H-triazolo-1,4-benzodiazepine.

Imine was intramolecularly cyclized under weak acidic conditions following the HBTU condensation to afford the triazole benzodiazepines. Further, the triazole target compound was prepared by cyclization utilizing triethylorthoformate, LR,

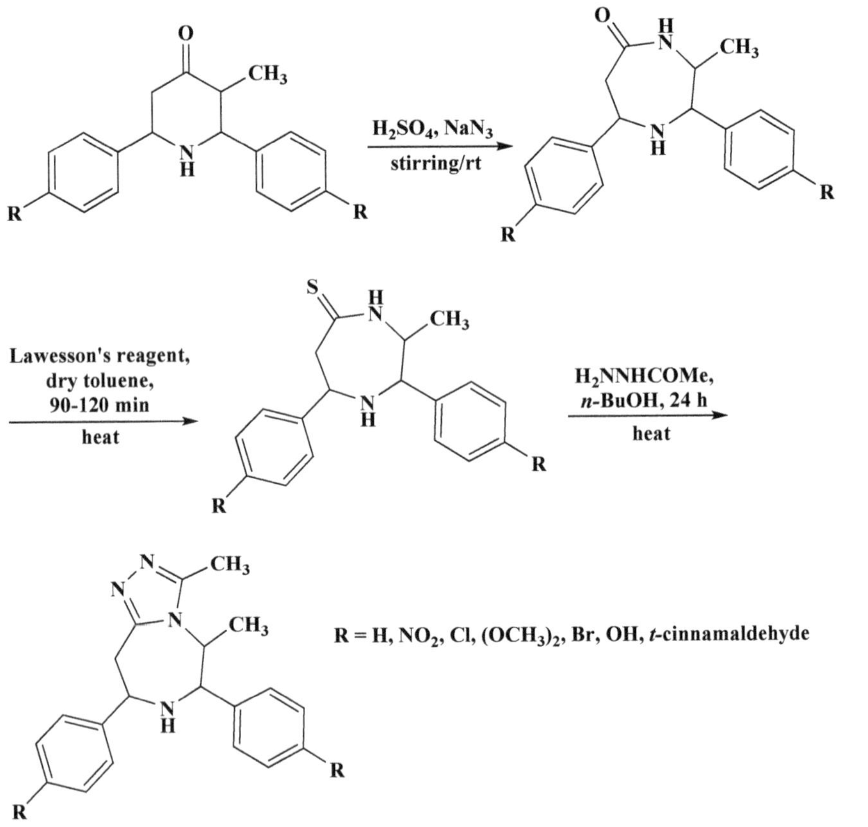

Scheme 1.30 Synthesis of triazolodiazepines

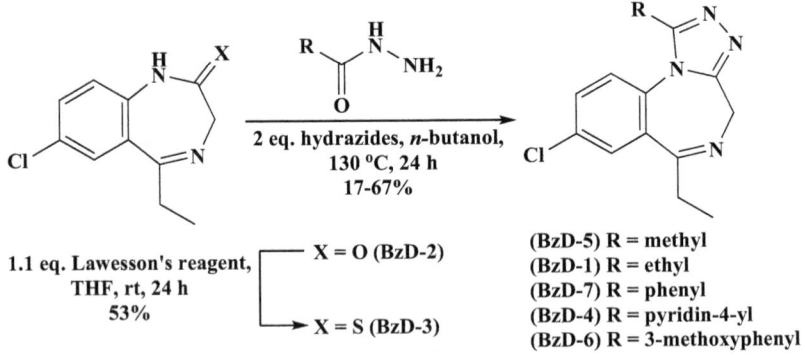

Scheme 1.31 Synthesis of triazolobenzodiazepines

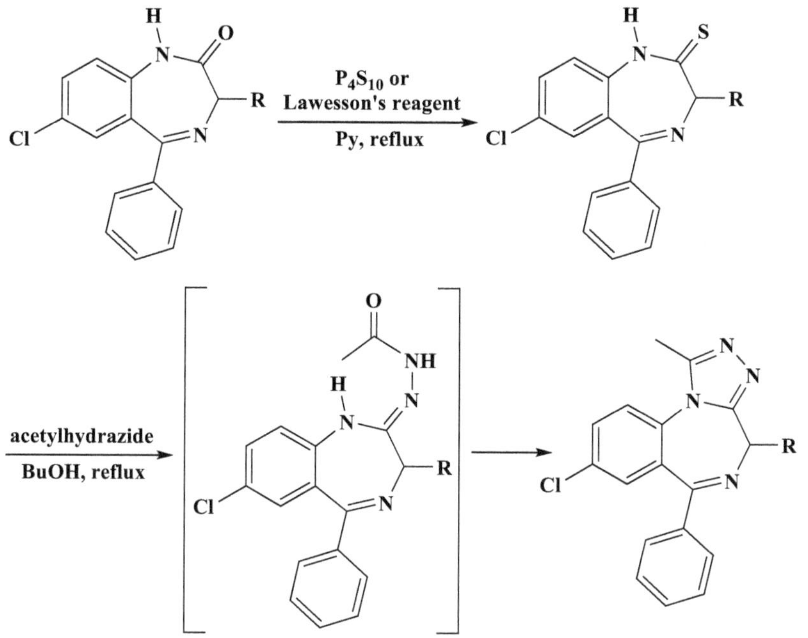

1.4 Synthesis of Five-Membered N,N,N-Heterocycles

Scheme 1.32 Synthesis of triazolo-1,4-benzodiazepines

and hydrazine hydrate. The triazole compound was further reduced with sodium cyanoborohydride to give the desired product (Scheme 1.33) [98].

A one-pot approach [99] for the transformation of amide to JQ1 was described. The one-pot method (Scheme 1.34) began with the reaction of amide with LR in tetrahydrofuran at 80 °C for 2 h (observed by thin layered chromatography) and after that addition of excess 10 eq. hydrazine hydrate at 0 °C. The reaction mixture was stirred for 30 min (monitored by thin layered chromatography) to afford the amidrazone, which was utilized for the next step directly after aqueous work-up. The amidrazone was heated to 110 °C for 2 h in a mixture of trimethyl orthoacetate and toluene (2:3) to afford the desired compound (±)-JQ1 in 60% yield over three steps. It was observed that a one-pot method (thionation and amidrazone synthesis) greatly minimized sulfur-related concerns (the strong, unpleasant odor of sulfur side-products). The purification procedure was also facile, and the reaction proceeded with a little improved yield (60%). The reaction was performed in four batches to decrease sulfur-related odors but can be achieved on a larger scale. A related method for the formation of bromodomain inhibitors including LR was reported in a patent [100].

The tetracyclic 1-amido-substituted triazolopyrrolo[2,1-c][1,4]benzodiazepin-8-ones were synthesized as depicted in Scheme 1.35. The key intermediate was synthe-

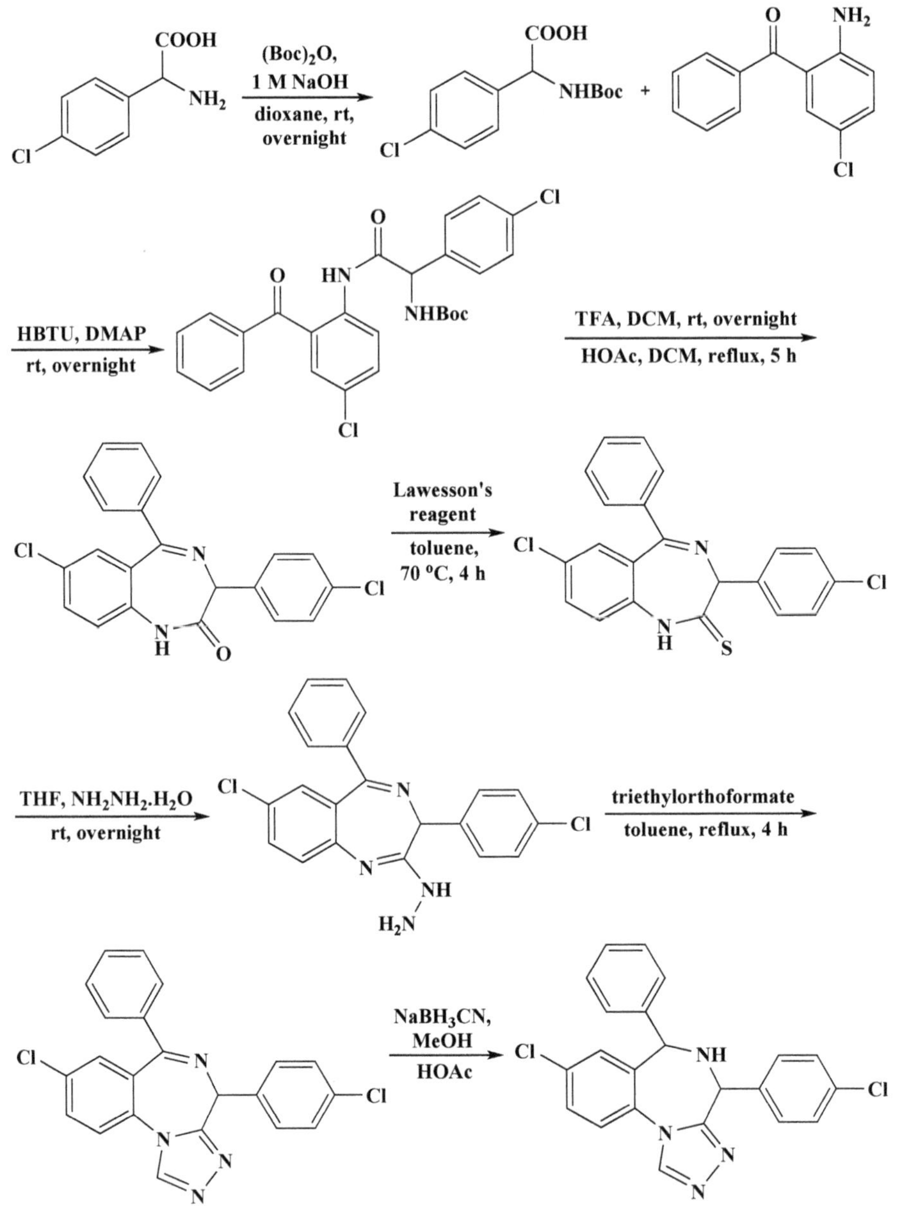

Scheme 1.33 Synthesis of triazolobenzodiazepine

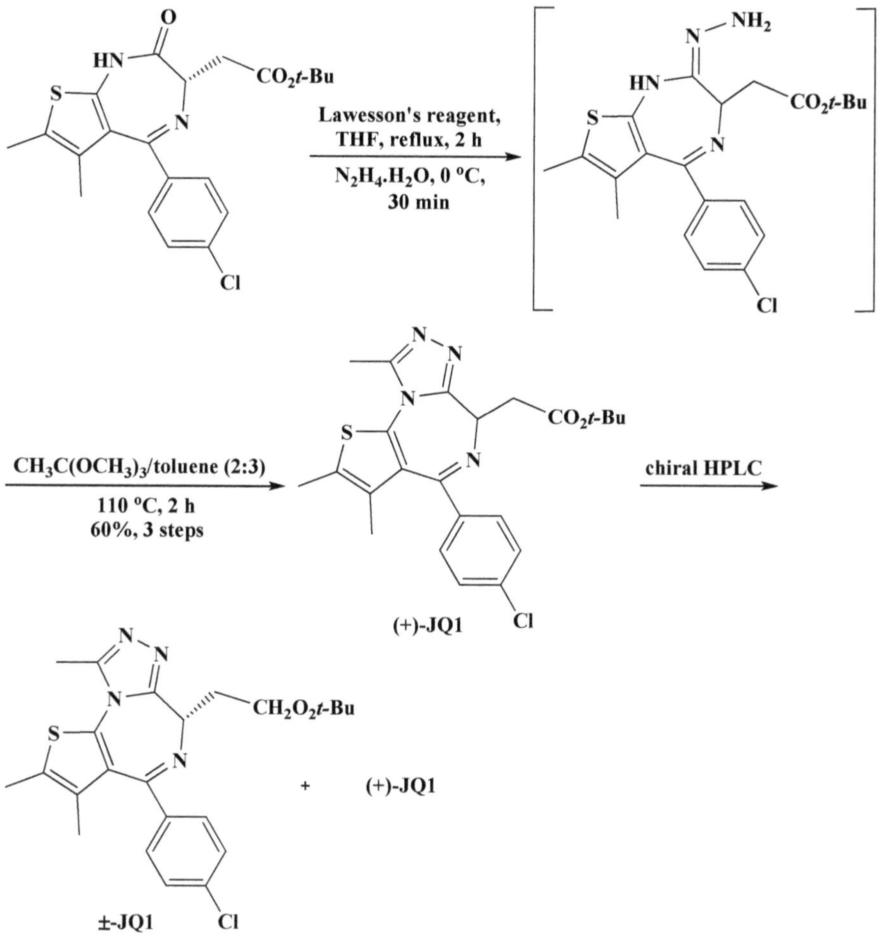

1.4 Synthesis of Five-Membered N,N,N-Heterocycles

Scheme 1.34 Synthesis of triazolobenzodiazepine

sized in four steps. First, pyrrolobenzodiazepinedione was synthesized by cyclocondensation of L-proline with isatoic anhydride. The monothiolactam was synthesized by thiation using LR in toluene at 70 °C [101]. Subsequently, the thiolactam was transformed into 11-hydrazinopyrrolo[2,1-c][1,4]benzodiazepine employing hydrazine hydrate in EtOH at room temperature, and cyclization with CNBr provided triazolopyrrolobenzodiazepinedione [102, 103]. Further, amido-substituted triazolo-fused pyrrolobenzodiazepinediones were prepared from tetracyclic intermediate by reacting with carboxylic acids employing HATU as a coupling agent and Hünig's base [104].

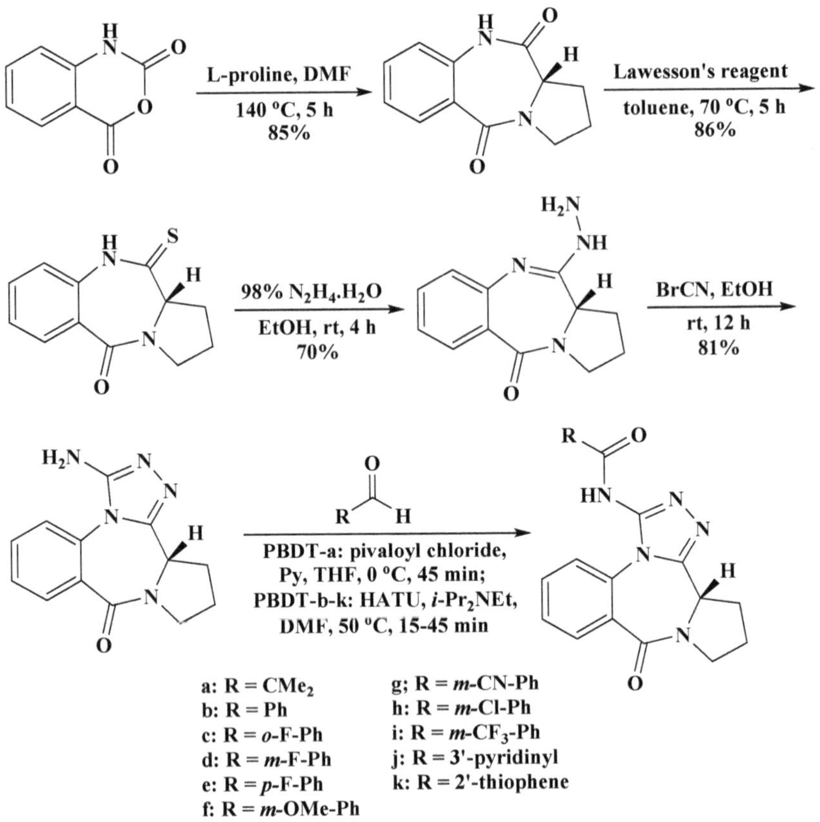

Scheme 1.35 Synthesis of tetracyclic 1-amido-substituted triazolopyrrolo[2,1-c][1,4]benzodiazepin-8-ones

a: R = CMe₃
b: R = Ph
c: R = o-F-Ph
d: R = m-F-Ph
e: R = p-F-Ph
f: R = m-OMe-Ph
g: R = m-CN-Ph
h: R = m-Cl-Ph
i: R = m-CF₃-Ph
j: R = 3'-pyridinyl
k: R = 2'-thiophene

A fused heterocyclic compound was introduced at N3,C4-position via stepwise processes analogous to those demonstrated for 1,4-benzodiazepines. In the first step, 3,5-dihydro-4H-2,3-benzodiazepin-5-ones were activated by thiolation with LR [105]. The 2,3-benzodiazepinthione was reacted with ethyl carbazate to provide the 11H-[1,2,4]triazolo[4,5-c][2,3]benzodiazepin-3(2H)-ones. The condensation of 2,3-benzodiazepinthione with hydrazine provided hydrazinyl derivatives, which were transformed into 2,12-dihydro[1,2,4]triazino[4,3-c][2,3]benzodiazepine-3,4-diones by reacting with (COCl)₂ (Scheme 1.36) [106].

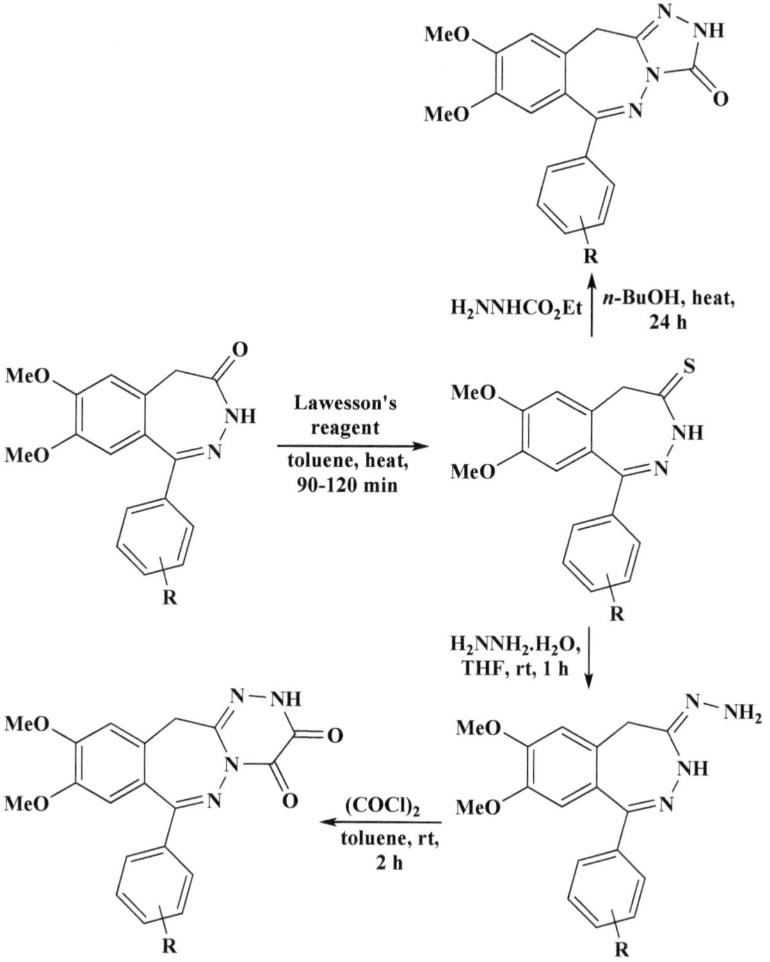

1.4 Synthesis of Five-Membered *N,N,N*-Heterocycles

Scheme 1.36 Synthesis of 11*H*-[1,2,4]triazolo[4,5-*c*][2,3]benzodiazepin-3(2*H*)-ones and 2,12-dihydro[1,2,4]triazino[4,3-*c*][2,3]benzodiazepine-3,4-diones

References

1. (a) A. Aimi, M. Nishimura, A. Miwa, H. Hoshino, S. Sakai and J. Higiniwa. 1989. Pumiloside and deoxypumiloside; plausible intermediates of camptothecin biosynthesis. Tetrahedron Lett. 30: 4991–994. (b) N. Kaur and D. Kishore. 2012. Montmorillonite: An efficient, heterogeneous, and green catalyst for organic synthesis. J. Chem. Pharm. Res. 4: 991–1015. (c) N. Kaur. 2018. Synthesis of six- and seven-membered heterocycles under ultrasound irradiation. Synth. Commun. 48: 1235–1258.
2. (a) J. Tois, R. Franzén and A. Koskinen. 2003. Synthetic approaches towards indoles on solid phase - recent advances and future directions. Tetrahedron 59: 5395–5405. (b) J. Dwivedi, N. Kaur, D. Kishore, S. Kumari and S. Sharma. 2016. Synthetic and biological aspects of

thiadiazoles and their condensed derivatives: An overview. Curr. Top. Med. Chem. 16: 2884–2920. (c) N. Kaur and D. Kishore. 2013. An insight into hexamethylenetetramine: A versatile reagent in organic synthesis. J. Iran. Chem. Soc. 10: 1193–1228. (d) N. Kaur, J. Dwivedi and D. Kishore. 2014. Solid-phase synthesis of nitrogen containing five-membered heterocycles. Synth. Commun. 44: 1671–1729.
3. (a) L. Joucla and L. Djakovitch. 2009. Transition metal-catalysed, direct and site-selective N1-, C2- or C3-arylation of the indole nucleus: 20 Years of improvements. Adv. Synth. Catal. 351: 673–714. (b) N. Kaur. 2013. An insight into medicinal and biological significance of privileged scaffold: 1,4-Benzodiazepine. Int. J. Pharm. Biol. Sci. 4: 318–337. (c) N. Kaur. 2013. Solid-phase synthetic approach to the synthesis of azepine heterocycles of medicinal interest. Int. J. Pharm. Biol. Sci. 4: 357–372. (d) N. Kaur. 2018. Ruthenium catalysis in six-membered O-heterocycles synthesis. Synth. Commun. 48: 1551–1587.
4. D.L. Hughes. 1993. Progress in the Fischer indole reaction - a review. Org. Prep. Proced. Int. 25: 609–632.
5. R.D. Clark and D.B. Repke. 1984. The Leimgruber-Batcho indole synthesis. Heterocycles 22: 195–221.
6. S.W. Wright, L.D. McClure and D.L. Hageman. 1996. A convenient modification of the Gassman oxindole synthesis. Tetrahedron Lett. 37: 4631–4634.
7. D.A. Wacker and P. Kasireddy. 2002. Efficient solid-phase synthesis of 2,3-substituted indoles. Tetrahedron Lett. 43: 5189–5191.
8. A. Sudalai, S. Kanagasabapathy and B.C. Benicewicz. 2000. Phosphorus pentasulfide: A mild and versatile catalyst/reagent for the preparation of dithiocarboxylic esters. Org. Lett. 2: 3213–3216.
9. M.P. Cava and M.I. Levinson. 1985. Thionation reactions of Lawesson's reagents. Tetrahedron 41: 5061–5087.
10. R.A. Cherkasov, G.A. Kutyrev and A.N. Pudovik. 1985. Tetrahedron report number 186: Organothiophosphorus reagents in organic synthesis. Tetrahedron 41: 2567–2624.
11. M.S.J. Foreman and J.D. Woollins. 2000. Organo-P-S and P-Se heterocycles, organo-P-S and P-Se heterocycles. J. Chem. Soc. Dalton Trans. 10: 1533–1543.
12. T. Ozturk, E. Ertas and O. Mert. 2007. Use of Lawesson's reagent in organic syntheses. Chem. Rev. 107: 5210–5278.
13. H.Z. Lecher, R.A. Greenwood, K.C. Whitehouse and T.H. Chao. 1956. The phosphonation of aromatic compounds with phosphorus pentasulfide. J. Am. Chem. Soc. 78: 518–522.
14. S.D. Kuduk, C. Ng, R.K. Chang and M.G. Bock. 2003. Synthesis of 2,3-diaminodihydropyrroles via thioimidate cyclopropane rearrangement. Tetrahedron Lett. 44: 1437-1440.
15. M. Natsume, M. Takahashi, K. Kiuchi and H. Sugaya. 1971. A Wittig reaction of N-sulfonyl lactam. Chem. Pharm. Bull. 19: 2648–2651.
16. J.P. Célérier, E. Deloisy, G. Lhommet and P. Maitte. 1979. Lactam ether chemistry. Cyclic beta-enamino ester synthesis. J. Org. Chem. 44: 3089–3089.
17. J.P. Célérier, M.G. Richard, G. Lhommet and P. Maitte. 1983. Imidoylation reactions: A simple direct synthesis of 3-amino-2-alkenoic esters (β-aminoesters). Synthesis 3: 195–197.
18. M.M. Gugelchuk, D.J. Hart and Y. Tsai. 1981. Methods for converting N-alkyl lactams to vinylogous urethanes and vinylogous amides via (methylthio)alkylideniminium salts. J. Org. Chem. 46: 3671–3675.
19. M. Yamaguchi and I. Hirao. 1985. A direct synthesis of [(tert-butoxycarbonyl)methylidene]azacycloalkanes. J. Org. Chem. 50: 1975–1977.
20. K. Kobayashi and H. Suginome. 1986. Synthesis of 2-pyrrolidinylideneacetates by means of a reaction of magnesium ester enolate with γ-cyanoalkyl tosylates. Bull. Chem. Soc. Jpn. 59: 2635–2636.
21. A. Brandi, S. Carli and A. Goti. 1988. High regio- and stereoselective cycloaddition of a nitrone to alkylidenecyclopropanes. Heterocycles 27: 17–20.
22. E.G. Occhiato, A. Guarna, A. Brandi, A. Goti and F. de Sarlo. 1992. N-Bridgehead polycyclic compounds by sequential rearrangement-annulation of isoxazoline-5-spirocyclopropanes.

6. A general synthetic method for 5,6-dihydro-7(8H)- and 2,3,5,6-tetrahydro-7(1H)-indolizinones. J. Org. Chem. 57: 4206–4211.
23. H.M.C. Ferraz, E.O. de Oliveira, M.E. Payret-Arrua and C.A. Brandt. 1995. A new and efficient approach to cyclic beta-enamino esters and beta-enamino ketones by iodine-promoted cyclization. J. Org. Chem. 60: 7357–7359.
24. A. Eschenmoser. 1970. Centenary lecture. (Delivered November 1969). Roads to corrins. Quarterly Review. Chem. Soc. 24: 366–415.
25. P. Dubs, E. Götschi, M. Roth and A. Eschenmoser. 1970. Sulfide contraction via alkylative coupling - a method for synthesizing beta dicarbonyl systems. Chimia 24: 34.
26. M. Roth, P. Dubs, E. Götschi and A. Eschenmoser. 1971. Sulfidkontraktion via alkylative kupplung: Eine methode zur darstellung von β-dicarbonylderivaten. Über synthetische methoden, 1. Mitteilung. Helv. Chim. Acta 54: 710–734.
27. G.C. Gerrans, A.S. Howard and B.S. Orlek. 1975. General methods of alkaloid synthesis. Ambident nucleophilicity of vinylogous urethanes. Synthesis of (±)-lupinine and a functionalised hydrojulolidine derivative. Tetrahedron Lett. 40: 4171–4172.
28. A.S. Howard, G.C. Gerrans and C.A. Meerholz. 1980. Vinylogous urethanes in alkaloid synthesis: Formal syntheses of *Elaeocarpus* alkaloids. Tetrahedron Lett. 21: 1373–1374.
29. A.S. Howard, G.C. Gerrans and J.P. Michael. 1980. Use of vinylogous urethanes in alkaloid synthesis: Formal synthesis of ipalbidine. J. Org. Chem. 45: 1713–1715.
30. A.S. Howard, R.B. Katz and J.P. Michael. 1983. Thiolactams in alkaloid synthesis: A particularly short synthesis of Δ^7-mesembrenone. Tetrahedron Lett. 24: 829–830.
31. R. Ghirlando, A.S. Howard, R.B. Katz and J.P. Michael. 1984. The application of the sulphide contraction to the synthesis of some simple pyrrolidine alkaloids. Tetrahedron 40: 2879–2884.
32. J.P. Michael, G.D. Hosken and A.S. Howard. 1988. Syntheses of alkyl (E)-(1-aryl-2-pyrrolidinylidene)acetates. Tetrahedron 44: 3025–3036.
33. J.P. Michael, A.S. Parsons and R. Hunter. 1989. Synthesis of two pyrrolidine alkaloids, peripentadenine and dinorperipentadenine. Tetrahedron Lett. 30: 4879–4880.
34. J.P. Michael, A.S. Howard, R.B. Katz and M.I. Zwane. 1992. Synthesis of hexahydroindol-6-ones by cycloacylation of vinylogous urethanes. Tetrahedron Lett. 33: 4751–4754.
35. J.P. Michael and M.I. Zwane. 1992. Synthesis of hexahydroindol-6-ones by reaction of 2-methylthiopyrrolinium salts with Nazarov reagents. Tetrahedron Lett. 33: 4755–4758.
36. J.P. Michael, A.S. Howard, R.B. Katz and M.I. Zwane. 1992. Formal syntheses of (±)-mesembrine and (±)-dihydromaritidine. Tetrahedron Lett. 33: 6023–6024.
37. C.M. Jungmann and J.P. Michael. 1992. New syntheses of (±)-lamprolobine and (±)-epilamprolobine. Tetrahedron 48: 10211–10220.
38. J.P. Michael, S.S.-F. Chang and C. Wilson. 1993. Synthesis of pyrrolo[1,2-a]indoles by intramolecular Heck reaction of N-(2-bromoaryl) enaminones. Tetrahedron Lett. 34: 8365–8368.
39. J.P. Michael and A.S. Parson. 1993. Formal synthesis of *Elaeocarpus* alkaloids - elaeocarpine and isoelaeocarpine. S. Afr. J. Chem. 46: 65–69.
40. J.P. Michael and A.S. Parsons. 1996. Chemoselective reactions of vinylogous amides, and the synthesis of two *Peripentadenia* alkaloids. Tetrahedron 52: 2199–2216.
41. J.P. Michael and D. Gravestock. 1996. Synthesis of (±)-indolizidine 209B and a new 209B diastereoisomer. Synlett 10: 981–982.
42. P. Michael and D. Gravestock. 1998. An enantioselective synthesis of (-)-indolizidine 167B, a skin alkaloid from a *Neotropical dendrobatid* frog. S. Afr. J. Chem. 51: 146–157.
43. J.P. Michael, C.B. de Koning, D. Gravestock, G.D. Hosken, A.S. Howard, C.M. Jungmann, R.W.M. Krause, A.S. Parsons, S.C. Pelly and T.V. Stanbury. 1999. Enaminones: Versatile intermediates for natural product synthesis. Pure Appl. Chem. 71: 979–988.
44. J.P. Michael and D. Gravestock. 2000. Vinylogous urethanes in alkaloid synthesis. Applications to the synthesis of racemic indolizidine 209B and its (5R*,8S*,8aS*)-(±) diastereomer, and to (−)-indolizidine 209B. J. Chem. Soc. Perkin Trans. 1 12: 1919–1928.
45. N. Kaniskan, D. Elmali and P.U. Civcir. 2008. Synthesis and characterization of novel heterosubstituted pyrroles, thiophenes, and furans. ARKIVOC (xii): 17–29.

46. L. Fieser and R.G. Kennelly. 1935. A comparison of heterocyclic systems with benzene. IIV. Thionaphthenequinones. J. Am. Chem. Soc. 57: 1611–1616.
47. M.M.M. Raposo and G. Kirsch. 2001. A combination of Friedel-Crafts and Lawesson reactions to 5-substituted 2,2'-bithiophenes. Heterocycles 55: 1487–1498.
48. T. Nishio. 1998. Sulfur-containing heterocycles derived by reaction of ω-keto amides with Lawesson's reagent. Helv. Chim. Acta 81: 1207–1214.
49. C.E. Hewton, M.C. Kimber and D.K. Taylor. 2002. A one-pot synthesis of thiophene and pyrrole derivatives from readily accessible 3,5-dihydro-1,2-dioxines. Tetrahedron Lett. 43: 3199–3201.
50. L. Fournier, I. Aujard, T. LeSaux, S. Maurin, S. Beaupierre, J.B. Baudin and L. Jullien. 2013. Coumarinylmethyl caging groups with red shifted absorption. Chemistry 19: 17494–17507.
51. S. Yamazoe, Q. Liu, L.E. McQuade, A. Deiters and J.K. Chen. 2014. Sequential gene silencing using wavelength-selective caged morpholines. Angew. Chem. Int. Ed. 53: 10114–10118.
52. M. Betou. 2013. Semipinacol rearrangement of cis-fused β-lactam diols into bicyclic lactams. PhD Thesis, The University of Birmingham.
53. S.G. Davies and O. Ichihara. 1991. Asymmetric synthesis of R-β-amino butanoic acid and S-β-tyrosine: Homochiral lithium amide equivalents for Michael additions to α,β-unsaturated esters. Tetrahedron: Asymmetry 2: 183–186.
54. J.F. Costello, S.G. Davies and O. Ichihara. 1994. Origins of the high stereoselectivity in the conjugate addition of lithium(α-methylbenzyl)benzylamide to t-butyl cinnamate. Tetrahedron: Asymmetry 5: 1999–2008.
55. K. Agapiou, D.F. Cauble and M.J. Krische. 2004. Copper-catalyzed tandem conjugate addition-electrophilic trapping: Ketones, esters, and nitriles as terminal electrophiles. J. Am. Chem. Soc. 126: 4528–4529.
56. K. Shiosaki. 1991. The Eschenmoser coupling reaction. B.M. Trost (Ed.). Comprehensive organic synthesis. Oxford: Pergamon Press, 2: 865–892.
57. A.B. Holmes, A.L. Smith, S.F. Williams, L.R. Hughes, Z. Lidert and C. Swithenbank. 1991. Stereoselective synthesis of (+-)-indolizidines 167B, 205A, and 207A. Enantioselective synthesis of (-)-indolizidine 209B. J. Org. Chem. 56: 1393–1405.
58. J.P. Michael, C. Accone, C.B. de Koning and C.W. van der Westhuyzen. 2008. Analogues of amphibian alkaloids: Total synthesis of (5R,8S,8aS)-(-)-8-methyl-5-pentyloctahydroindolizine (8-epi-indolizidine 209B) and [(1S,4R,9aS)-(-)-4-pentyloctahydro-2H-quinolizin-1-yl]methanol. Beilstein J. Org. Chem. 4: 1–7.
59. J.P. Michael and D. Gravestock. 1998. An expeditious synthesis of the dendrobatid indolizidine alkaloid 167B. Eur. J. Org. Chem. 5: 865–870.
60. M. Weymann, W. Pfrengle, D. Schollmeyer and H. Kunz. 1997. Enantioselective syntheses of 2-alkyl-, 2,6-dialkylpiperidines and indolizidine alkaloids through diastereoselective Mannich-Michael reactions. Synthesis 10: 1151–1160.
61. J.P. Michael. 2000. Indolizidine and quinolizidine alkaloids. Nat. Prod. Rep. 17: 579–602.
62. D.J. Maloney and S.J. Danishefsky. 2007. Conformational locking through allylic strain as a device for stereocontrol - total synthesis of grandisine A. Angew. Chem. Int. Ed. 46: 7789–7792.
63. R.A. Raphael and P. Ravenscroft. 1988. Synthesis of indolin-2-ones (oxindoles) related to mitomycin A. J. Chem. Soc. Perkin Trans. 1 7: 1823–1828.
64. N. Cohen, B.L. Banner, A.J. Laurenzano and L. Carozza. 1985. 2,3-O-Isopropylidene-D-erythronolactone. Org. Synth. 63: 127–132.
65. L.F. Tietze and S. Petersen. 2000. Stereoselective total synthesis of a novel D-homosteroid by a two fold Heck reaction. Eur. J. Org. Chem. 9: 1827–1830.
66. J.P. Michael, C.B. de Koning, T.T. Mudzunga and R.L. Petersen. 2006. Formal asymmetric synthesis of a 7-methoxyaziridinomitosene. Synlett 19: 3284–3288.
67. D.S. Dodd, R.L. Martinez, M. Kamau, Z. Ruan, K. van Kirk, C.B. Cooper, M.A. Hermsmeier, S.C. Traeger and M.A. Poss. 2005. Solid-phase synthesis of 5-substituted amino pyrazoles. J. Comb. Chem. 7: 584–588.

References

68. R. Aggarwal, V. Kumar, R. Kumar and S.P. Singh. 2011. Approaches towards the synthesis of 5-aminopyrazoles. Beilstein J. Org. Chem. 7: 179–197.
69. E. Chebil and S. Jouil. 2012. Unusual course of the reaction of Lawesson's reagent with β-phosphoryl-β'-carbethoxyhydrazones: First synthesis of 5-mercapto-3-(methylthiophosphoryl) pyrazoles. Lett. Org. Chem. 9: 320–324.
70. D.S. Dodd and R.L. Martinez. 2004. One-pot synthesis of 5-(substituted-amino)pyrazoles. Tetrahedron Lett. 45: 4265–4267.
71. M.J. Burke and B.M. Trantow. 2008. An efficient route to 3-aminoindazoles and 3-amino-7-azaindazoles. Tetrahedron Lett. 49: 4579–4581.
72. D. Bauer, D.A. Whittington, A. Coxon, J. Bready, S.P. Harriman, V.F. Patel, A. Polverino and J.-C. Harmange. 2008. Evaluation of indazole-based compounds as a new class of potent KDR/VEGFR-2 inhibitors. Bioorg. Med. Chem. Lett. 18: 4844–4848.
73. T. Yakaiah, B.P.V. Lingaiah, B. Narsaiah, B. Shireesha, B.A. Kumar, S. Gururaj, T. Parthasarathy and B. Sridhar. 2007. Synthesis and structure-activity relationships of novel pyrimido[1,2-b]indazoles as potential anticancer agents against A-549 cell lines. Bioorg. Med. Chem. Lett. 17: 3445–3453.
74. A.P. Piccionello, A. Pace, I. Pibiri, S. Buscemi and N. Vivona. 2006. Synthesis of fluorinated indazoles through ANRORC-like rearrangement of 1,2,4-oxadiazoles with hydrazine. Tetrahedron 62: 8792–8797.
75. K.W. Woods, J.P. Fisher, A. Clairborne, T. Li, S.A. Thomas, G.-D. Zhu, R.B. Diebold, X. Liu, Y. Shi, V. Klinghofer, E.K. Han, R. Guan, S.R. Magnone, E.F. Johnson, J.J. Bouska, A.M. Olson, R. de Jong, T. Oltersdorf, Y. Luo, S.H. Rosenberg, V.L. Giranda and Q. Li. 2006. Synthesis and SAR of indazole-pyridine based protein kinase B/Akt inhibitors. Bioorg. Med. Chem. 14: 6832–6846.
76. S. Caron and E. Vasquez. 2001. The synthesis of a selective PDE4/TNFα inhibitor. Org. Process Res. Dev. 5: 587–592.
77. R.J. Steffan, E. Matelan, M.A. Ashwell, W.J. Moore, W.R. Solvibile, E. Trybulski, C.C. Chadwick, S. Chippari, T. Kenney, A. Eckert, L. Borges-Marcucci, J.C. Keith, Z. Xu, L. Mosyaz and D.C. Harnish. 2004. Synthesis and activity of substituted 4-(indazol-3-yl)phenols as pathway-selective estrogen receptor ligands useful in the treatment of rheumatoid arthritis. J. Med. Chem. 47: 6435–6438.
78. X. Li, S. Chu, V.A. Feher, M. Khalili, Z. Nie, S. Marosiak, V. Nikulin, J. Levin, K.G. Sprankle, M.E. Tedder, R. Almassy, K. Appelt and K.M. Yager. 2003. Structure-based design, synthesis, and antimicrobial activity of indazole-derived SAH/MTA nucleosidase inhibitors. J. Med. Chem. 46: 5663–5673.
79. Y.-K. Lee, D.J. Parks, T. Lu, T.V. Thieu, T. Markotan, W. Pan, D.F. McComsey, K.L. Milkiewicz, C.S. Crysler, N. Ninan, M.C. Abad, E.C. Giardino, B.E. Maryanoff, B.P. Damiano and M.R. Player. 2008. 7-Fluoroindazoles as potent and selective inhibitors of factor Xa. J. Med. Chem. 51: 282–297.
80. D. Kovács, G. Mótyán, J. Wölfling, I. Kovács, I. Zupkó and E. Frank. 2014. A facile access to novel steroidal 17-2'-(1',3',4')-oxadiazoles, and an evaluation of their cytotoxic activities in vitro. Bioorg. Med. Chem. Lett. 24: 1265–1268.
81. D. Kovács, J. Wölfling, N. Szabó, M. Szécsi, R. Minorics, I. Zupkó and E. Frank. 2015. Efficient access to novel androsteno-17-(1',3',4')-oxadiazoles and 17β-(1',3',4')-thiadiazoles via N-substituted hydrazone and N,N'-disubstituted hydrazine intermediates, and their pharmacological evaluation in vitro. Eur. J. Med. Chem. 98: 13–29.
82. E. Montenegro, R. Echarri, C. Claver, S. Castillón, A. Moyano, M.A. Pericàs and A. Riera. 1996. New camphor-derived sulfur chiral controllers: Synthesis of (2R-exo)-10-methylthio-2-bornanethiol and (2R-exo)-2,10-bis(methylthio)bornane. Tetrahedron: Asymmetry 7: 3553–3558.
83. F. Merchán, J. Garín, V. Martínez and E. Meléndez. 1982. Synthesis of 2-aryliminoimidazolidines and 2-arylaminobenzimidazoles from methyl N-aryldithiocarbamates. Synthesis 6: 482–484.

84. M. Machaj, M. Pach, A. Wolek, A. Zabrzenska, K. Ostrowska, J. Kalinowska-Tluscik and B. Oleksyn. 2007. Succinonitrile activated by thiating agents as precursor of bis-cyclic amidines, tectons for molecular engineering. Monatsh. Chem. 138: 1273–1277.
85. T. Ozturk, E. Ertas and O. Mert. 2010. A Berzelius reagent, phosphorus decasulfide (P_4S_{10}), in organic syntheses. Chem. Rev. 110: 3419–3478.
86. F.Z. Basha and J.F. DeBernardis. 1987. Synthesis of hexahydro-5H-benz[g]imidazo[2,1-a]isoindole via an intramolecular Diels-Alder reaction and a novel Lawesson's reagent mediated cyclization. J. Heterocycl. Chem. 24: 789–791.
87. T. Prisinzano, J. Podobinski, K. Tidgewell, M. Luo and D. Swenson. 2004. Synthesis and determination of the absolute configuration of the enantiomers modafinil. Tetrahedron: Asymmetry 15: 1053–1058.
88. V.V. Rostovtsev, L.G. Green, V.V. Fokin and K.B. Sharpless. 2002. A stepwise Huisgen cycloaddition process: Copper(I)-catalyzed regioselective "ligation" of azides and terminal alkynes. Angew. Chem. Int. Ed. 41: 2596–2599.
89. F. Amblard, J.H. Cho and R.F. Schinazi. 2009. Cu(I)-Catalyzed Huisgen azide-alkyne 1,3-dipolar cycloaddition reaction in nucleoside, nucleotide, and oligonucleotide chemistry. Chem. Rev. 109: 4207–4220.
90. J.-C. Jung, Y. Lee, J.-Y. Son, E. Lim, M. Jung and S. Oh. 2011. Convenient synthesis and biological evaluation of modafinil derivatives: Benzhydrylsulfanyl and benzhydrylsulfinyl[1,2,3]triazol-4-yl-methyl esters. Molecules 16: 10409–10419.
91. P. Mullen, H. Miel and M.A. McKervey. 2010. N-Boc 4-nitropiperidine: Preparation and conversion into a spiropiperidine analogue of the eastern part of maraviroc. Tetrahedron Lett. 51: 3216–3217.
92. S. Sathishkumar and H.P. Kavitha. 2015. Synthesis, characterization and anti-inflammatory activity of novel triazolodiazepine derivatives. J. Appl. Chem. 8: 47–52.
93. S.C. Bell and S.J. Childress. 1973. 1,4-Benzodiazepine-2-ones and intermediates. United States Patent 3714145.
94. J.B. Hester, A.D. Rudzik and B.V. Kamdar. 1971. 6-Phenyl-4H-s-triazolo[4,3-a][1,4]benzodiazepines which have central nervous system depressant activity. J. Med. Chem. 14: 1078–1081.
95. P.H. Richter and U. Scheefeldt. 1991. Synthesis and biological activity of 5-phenyl-1,3,4-benzotriazepines. 25. Synthesis of (1,2,4)triazolo(4,3-a)(1,3,4)benzotriazepines and related tricyclics. Pharmazie 46: 701–705.
96. P. Filippakopoulos, S. Picaud, O. Fedorov, M. Keller, M. Wrobel, O. Morgenstern, F. Bracher and S. Knapp. 2012. Benzodiazepines and benzotriazepines as protein interaction inhibitors targeting bromodomains of the BET family. Bioorg. Med. Chem. 20: 1878–1886.
97. E. Lattmann, J. Sattayasai, D.C. Billington, D.R. Poyner, P. Puapairoj, S. Tiamkao, W. Airarat, H. Singh and M. Offel. 2002. Synthesis and evaluation of N1-substituted-3-propyl-1,4-benzodiazepine-2-ones as cholecystokinin (CCK2) receptor ligands. J. Pharm. Pharmacol. 54: 827–834.
98. Z. Yu, C. Zhuang, Y. Wu, Z. Guo, J. Li, G. Dong, J. Yao, C. Sheng, Z. Miao and W. Zhang. 2014. Design, synthesis and biological evaluation of sulfamide and triazole benzodiazepines as novel p53-MDM2 inhibitors. Int. J. Mol. Sci. 15: 15741–15753.
99. B.K. Albrecht, J.E. Audia, A. Cote, V.S. Gehling, J.-C. Harmange, M.C. Hewitt, C.G. Naveschuk, A.M. Taylor and R.G. Vaswani. Bromodomain inhibitors and uses there of. WO 2012075456A.
100. S.S. Syeda, S. Jakkaraj and G.I. Georg. 2015. Scalable syntheses of the BET bromodomain inhibitor JQ1. Tetrahedron Lett. 56: 3454–3457.
101. S. Kumaraswamy, K. Mukkanti and P. Srinivas. 2012. Palladium catalyzed synthesis of quinazolino[1,4]benzodiazepine alkaloids and analogous. Tetrahedron 68: 2001–2006.
102. M.-P. Foloppe, I. Rault, S. Rault and M. Robba. 1993. Pyrrolo[2,1-c][1,4]benzodiazepines: Synthesis of N-substituted amidines. Heterocycles 36: 63–69.
103. J.B. Hester, C.G. Chidester and J. Szmuszkovicz. 1974. Synthesis and chemistry of N-methyl-6-phenyl-4H-s-triazolo[4,3-a][1,4]benzodiazepinium derivatives. J. Org. Chem. 44: 2688–2693.

104. K. Sorra, C.-F. Chang, S. Pusuluri, K. Mukkanti, M.-C. Laiu, B.-Y. Bao, C.-H. Su and T.-H. Chuang. 2012. Synthesis and cytotoxicity testing of new amido-substituted triazolopyrrolo[2,1-c][1,4]benzodiazepine (PBDT) derivatives. Molecules 17: 8762–8772.
105. M. Jesberger, T.P. Davis and L. Barner. 2003. Applications of Lawesson's reagent in organic and organometallic syntheses. Synthesis 13: 1929–1958.
106. R. Gitto, V. Orlando, S. Quartarone, G. de Sarro, A. de Sarro, E. Russo, G. Ferreri and G. Chimirri. 2003. Synthesis and evaluation of pharmacological properties of novel annelated 2,3-benzodiazepine derivatives. J. Med. Chem. 46: 3758–3761.

Chapter 2
Thiazole Synthesis

2.1 Introduction

The heterocyclic chemistry research comprises a significant part of the organic chemistry research in the world. The huge quantity of bioactive organic compounds that possess heterocyclic frameworks plays an important role in the medicinal field. It is generally described that heterocyclic compounds bearing nitrogen or sulfur atoms or both of them are the general features present in the structures of most of the pharmaceutical and natural compounds [1, 2]. They also serve as multidentate ligands for various metals because of the presence of sulfur and nitrogen atoms and are therefore utilized widely in coordination chemistry to construct new scaffolds with efficient bioactivity [3, 4].

The heterocyclic compounds which contain sulfur and nitrogen atoms have a massive effect in medicinal and pharmaceutical chemistry fields. These have been reported to have different biological properties like anti-inflammatory, antifungal, antihypertensive, and antibacterial [5–10].

The LR has become now an indispensable reagent for sulfur chemistry especially in order to transform the oxo groups to thio groups, which are important functional groups to achieve different organic reactions or to utilize them as end products in medicinal, material chemistry, etc. The LR's rapid and slow reactions toward the functional groups like ketones, alcohols, esters, and amides offer the synthetic researchers with a tool of designing their synthetic strategies. The LR is employed for the formation of almost all heterocyclic compounds having sulfur atom(s). Its range varies from thiophene to thiadiazine, thiazole, thiazine, pyrazoles, thiadiazole, and dithiin. It finds wide uses in thionation reaction of purines, peptides, pyrimidines, and nucleosides. Another valuable reaction of Lawesson's reagent is reduction of sulfoxides to sulfides. The Lawesson's reagent is a reagent which surprisingly gives unexpected reactions, consequences of which lead the chemists to novel strategies and reactions [11–13].

2.2 Synthesis of Thiazoles

The 4-[(methylthio)-(het)arylmethylene]-2-phenyl/(2-thienyl)-5-oxazolone pioneers and 4-bis(methylthio)methylene derivatives were prepared in good yields [14, 15]. The 2-phenyl-5-(het)arylthiazole-4-carboxylates and 5-(methylthio)thiazole-4-carboxylates were synthesized in high yields through the thionation–cyclization of enamino esters by nucleophilic ring-opening of oxazolones with different sodium alkoxides (Scheme 2.1). The thionation–cyclization of enamide ester to thiazole-4-carboxylate was attempted, which acted as a model substrate for the optimization of reaction conditions. The refluxing enamide ester with 1 eq. LR in tetrahydrofuran for a long time afforded only unreacted starting compound without any trace of thiazole (or thioamide). However, it was observed that a higher temperature reflux in toluene for 12 h resulted in thionation and intramolecular cyclization of enamide ester to provide the ethyl 2-phenyl-4-(methoxyphenyl)thiazole-4-carboxylate in 68% yield. The enamide ester was treated with 2 eq. LR in refluxing toluene for 2 h to provide the thiazole-4-carboxylate in 70% yield. This optimized protocol (with 2 eq. LR) for the transformation of enamide ester to thiazole-4-carboxylate was utilized throughout for the formation of other 5-(het)arylthiazole-4-carboxylates. The reaction was equally facile for the formation of other 5-arylthiazole-4-carboxylates that have both electron-withdrawing and electron-donating substituents on the 5-aryl group. The thionation–cyclization of enamide *t*-butyl carboxylate proceeded easily without any side reactions to afford the *t*-butyl thiazole-4-carboxylate in 75% yield. Likewise, enaminone carboxylic esters having het(aryl) groups were also transformed into 2-phenyl-5-(2-furyl)/(2-*N*-methylpyrrolyl)/(3-*N*-methylindolyl)thiazoles in good yields under same conditions that needed prolonged refluxing (12 h). Further diversity at the 2- and 5-positions of thiazoles could be obtained through the formation of *n*-butyl 2,5-bis(2-thienyl)thiazole-4-carboxylate in 80% yield by thionation–cyclization of enamino ester formed by ring-opening of 2-thienyl-4-[methylthio(2-thienyl)methylene]-5-oxazolone with sodium *n*-butoxide [16]. The extension of protocol to bis(methylthio)enamide carboxylate (obtained by ring-opening of 4-bis(methylthio)methylene derivative with NaOEt) also provided ethyl 2-phenyl-5-(methylthio)thiazole-4-carboxylate in 70% yield [17, 18].

This protocol afforded 2,5-(het)-arylthiazole-4-(*N*-aryl/alkyl)carboxamides through a one-step thionation–cyclization of enamides having a secondary amide functionality that was easily formed by ring-opening of oxazolones with primary aliphatic amines, aromatic amines, or amino acid esters [14–17]. The enamide amide pioneers have two secondary amide functionalities (however, they are electronically diverse), and their conversion to thiazole-4-(*N*-substituted) carboxamides using LR was more challenging. The chemoselective thionation of enamide benzoylamino group generated enamide monothioamide intermediates, which underwent intramolecular cyclization to provide the thiazoles. The enamide anilide was designated as a model substrate for the examination of the optimal conditions for

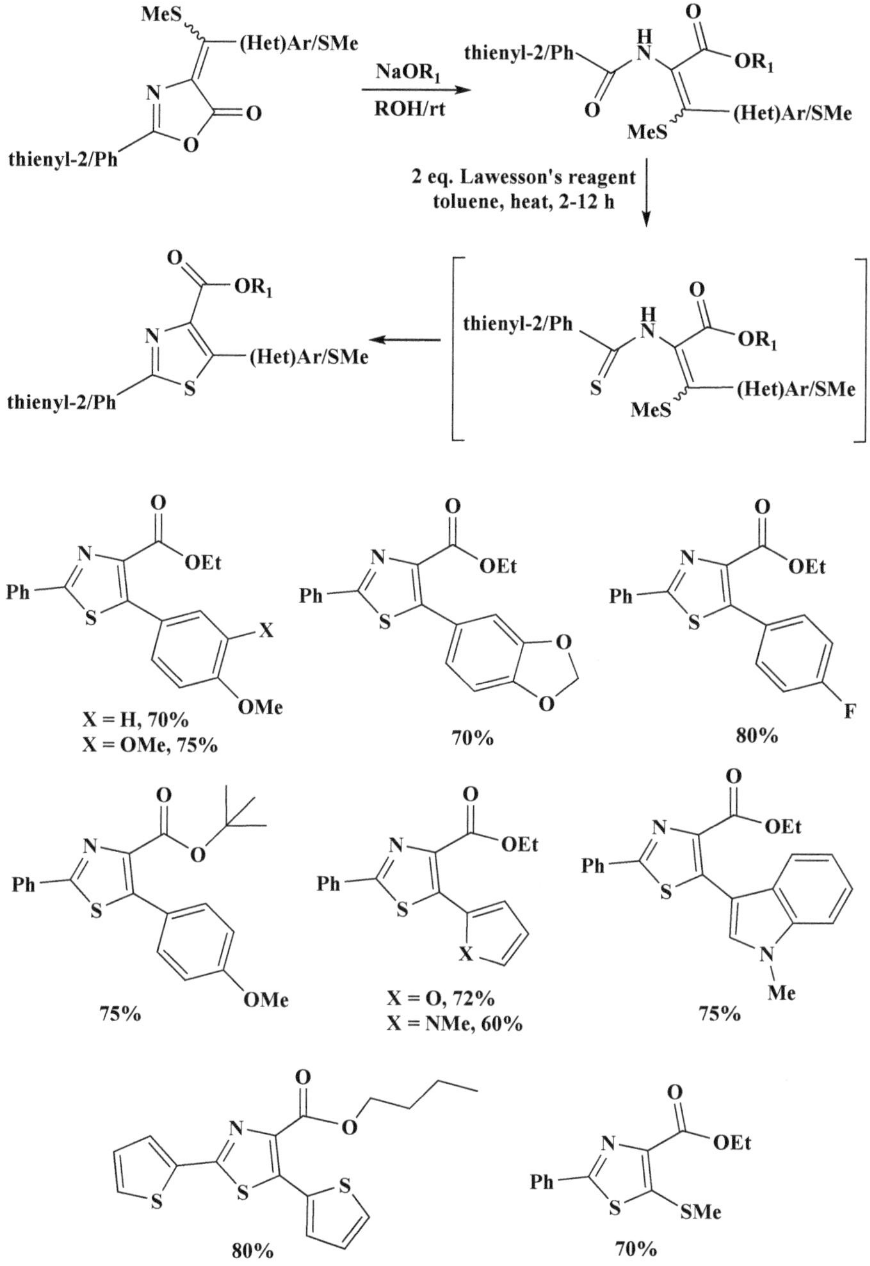

2.2 Synthesis of Thiazoles

Scheme 2.1 Synthesis of thiazoles

its chemoselective thionation–cyclization to provide the thiazole. The thionation–cyclization of enamide anilide took place under reflux in toluene using 2 eq. LR. However, enamide amide intermediates were insoluble in toluene, and the cyclization of enamide anilide to thiazole was attempted under toluene reflux for extended time (20 h) which provided only unreacted starting material. The reaction of enamide anilide with 2 eq. LR under reflux in tetrahydrofuran for 12 h afforded a mixture containing single product exclusively in reasonably good yield (65%), which was observed to be 2-pheny-5-(4-methoxyphenyl)-thiazole-4-(N-phenyl)carboxamide [18]. Other enamide anilide substrates, derived from ring-opening of oxazolones with 4-fluoroaniline, also underwent facile chemoselective monothionation–cyclization with LR to provide the 2-phenyl-5-(het)arylthiazole-4-carboxyanilides in good yield. The 2,5-bis(2-thienyl)thiazole-4-carboxyanilide could also be prepared in 75% yield by thionation–cyclization of enamide (formed by ring-opening of 2-(2-thienyl)-4-[(methylthio)(2-thienyl)-methylene]-5-oxazolone with 3,4,5-trimethoxyaniline). The scope and versatility of this chemoselective monothionation–cyclization protocol was then demonstrated by the efficient synthesis of 2-phenyl/(2-thienyl)-5-(het)arylthiazole-4-(N-alkyl) carboxamides in good yields from enamide-N-(alkyl)amides under same conditions. This methodology was efficiently used for the formation of 2,5-(het)arylthiazole-4-(N-aryl/alkyl)carboxamides, then this reaction was extended for the formation of thiazole-based peptidomimetics [19, 20]. The open-chain peptidoenamide substrates were easily converted into thiazole-based peptidomimetics with a variety of 5-(het)aryl groups in good yields under these optimized reaction conditions. The bis[(methylthio)methylene]enamide anilide (formed by ring-opening of 4-bis(methylthio)methyleneoxazolone with 4-fluoroaniline) also afforded 2-phenyl-5-(methylthio)thiazole-4-(N-4-fluorophenyl)carboxyanilide in 70% yield (Scheme 2.2a,b).

The formation of 4-phenylthiazole derivative started with the serine-derived amide, which was then transformed to thioamide in the presence of LR (Scheme 2.3). The α-iminothioketone, synthesized by base-induced S-alkylation of thioamide, was

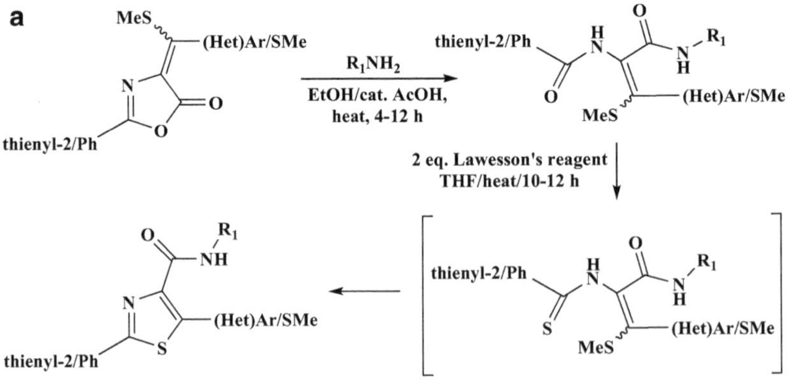

Scheme 2.2 Synthesis of 2,5-(het)-arylthiazole-4-(N-aryl/alkyl)carboxamides

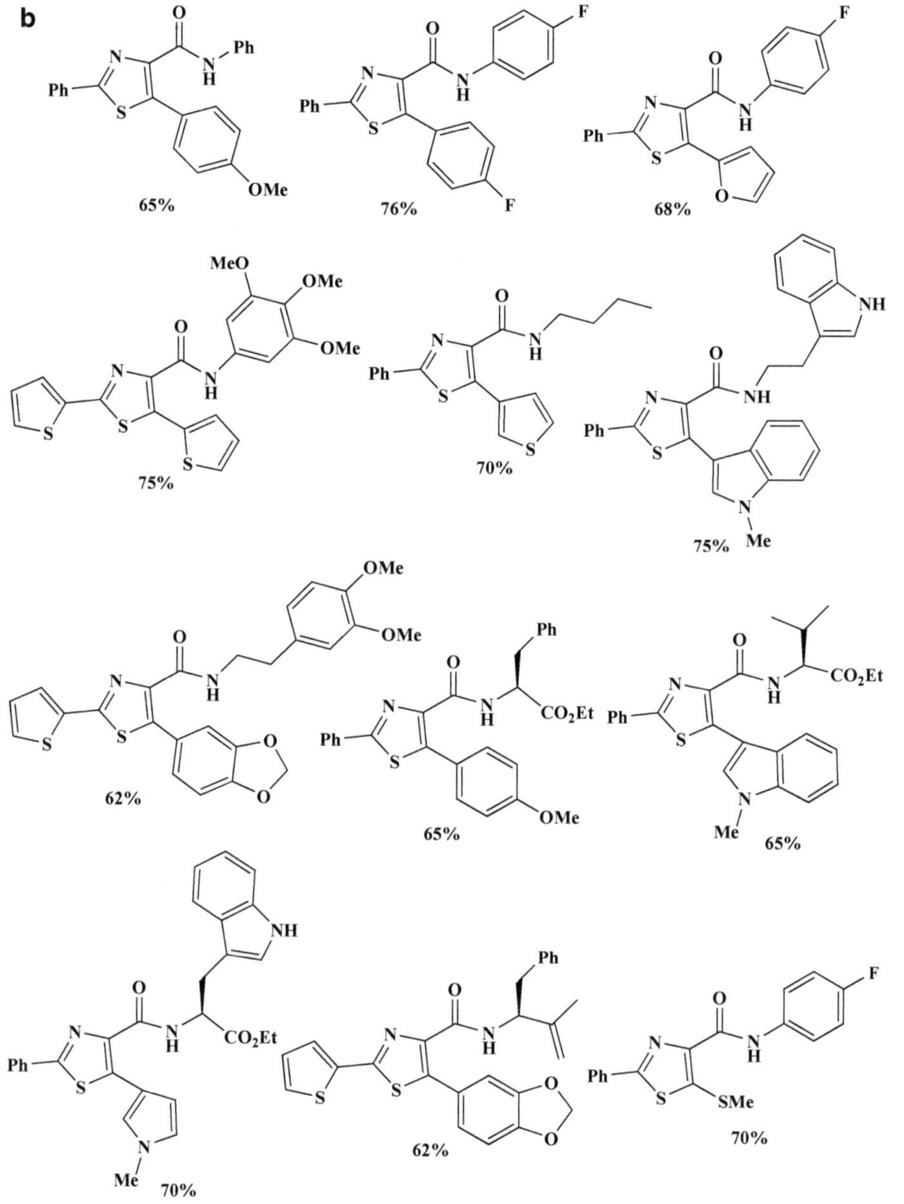

2.2 Synthesis of Thiazoles

b

Scheme 2.2 (continued)

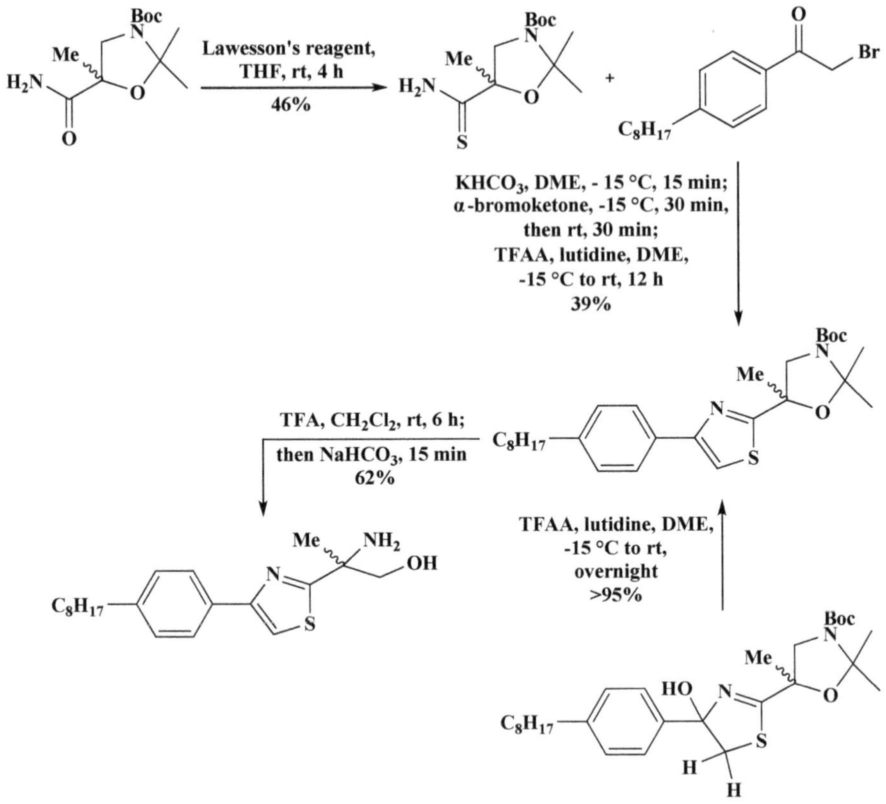

Scheme 2.3 Synthesis of 2-amino-2-(4-(4-octylphenyl)thiazol-2-yl)propan-1-ol

dehydrated in situ to afford an isolable mixture of thiazole and incomplete dihydrothiazole [21, 22]. This one-pot reaction was not optimized, but dehydration occurred in excellent yields on retreatment of dihydrothiazole intermediate with dry lutidine and TFAA. The thiazole was deprotected with trifluoroacetic acid and neutralized to afford the aminoalcohol VPC45214 [23].

2.3 Synthesis of Benzothiazoles

The compound DHB is a drug like analogue of resveratrol, which is well-known dietary polyphenol offering significant cancer chemopreventive properties [24]. The 2-(3,5-dihydroxyphenyl)hydroxybenzothiazole is most effective antiproliferative agent and also displayed the highest levels of vasorelaxing potency and efficiency in rat aortic rings precontracted with potassium chloride 60 mM since it is widely known that high levels of membrane depolarization because of the high

2.3 Synthesis of Benzothiazoles

extracellular concentrations of potassium ions can nonspecifically and dramatically decrease the vasorelaxing effects of vasorelaxing agents acting via the activation of every type of membrane potassium channels [25]. The above controversial observations on the cytotoxicity activity and vasorelaxing potency of DHB led to the synthesis of it following the method described by Bertini et al. [26] (Scheme 2.4). This compound was further investigated and studied in details including the dose and time-dependent antiproliferation and morphological study and clonogenic activity of DHB on PC-3 prostate cancer cell lines, which were originally derived from advanced androgen-independent bone metastasized prostate cancer [27].

The importance of green chemistry in organic synthesis has motivated researchers to look for the applications of MWI in organic reactions. Since the past few years, MWI has raised as a great source of energy for a wide range of organic reactions with less reaction time and high yield of products with high purity [28–41]. Therefore, an analysis has been undertaken under MWI for the condensation of o-aminothiophenol with ethyl cyanoformate utilizing LR under solvent- and catalyst-free conditions (Scheme 2.5).

Seizas et al. [42] reported that LR acted as an efficient promoter in MW-assisted formation of 2-substituted benzothiazoles from carboxylic acids and 2-aminothiophenol without using solvent (Scheme 2.6).

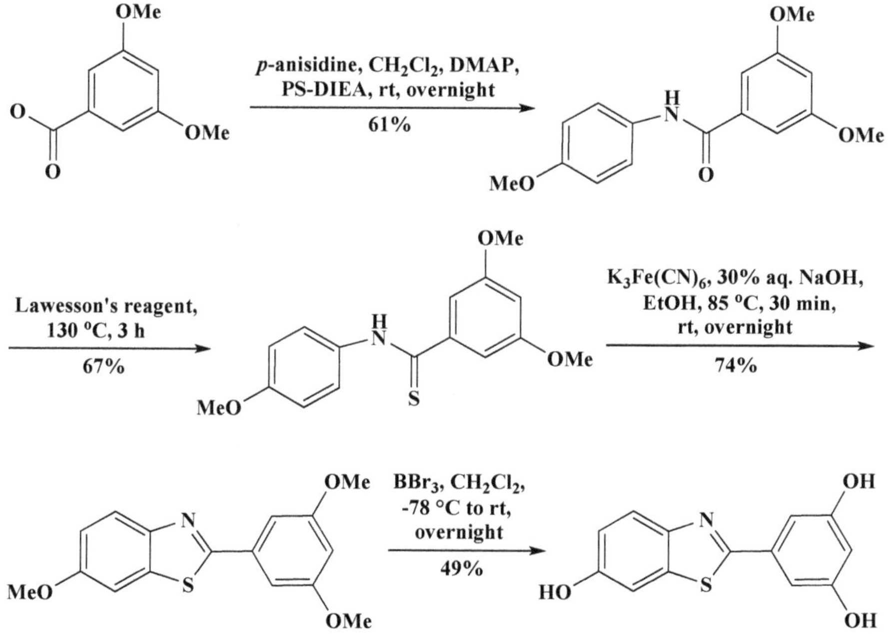

Scheme 2.4 Synthesis of 2-(3,5-dihydroxyphenyl)hydroxybenzothiazole

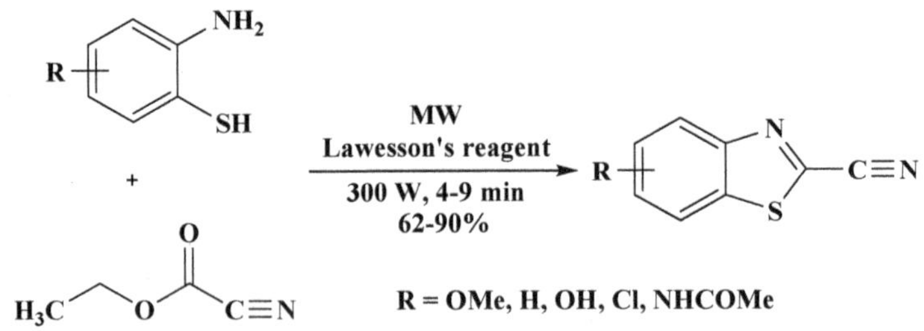

Scheme 2.5 Synthesis of benzothiazole-2-carbonitriles

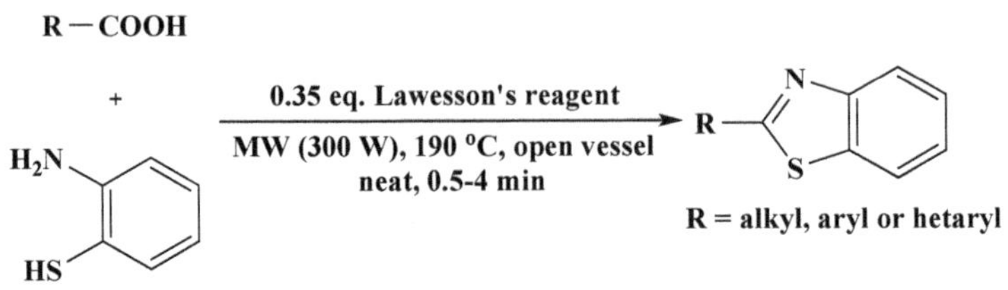

Scheme 2.6 Synthesis of 2-substituted benzothiazoles

2.4 Synthesis of Thiazoles from Dicarbonyl Compounds

As described in Scheme 2.7, acylamino esters were prepared via esterification of glycine with alcohol [43] and then *N*-acylation with 4-bromobenzoyl chloride. The work-up process was little modified to avoid the gross adulteration from the starting alcohol in the isolated product; a small amount of alcohol that remained was removed smoothly via recrystallization. The 5-alkoxy-1,3-thiazoles could be synthesized in high yield by reacting an acylamino ester with LR under MWI [44]. First effort for this reaction was in THF at rt since, in earlier unpublished work, good success was found in synthesizing 2,5-diaryl-1,3-thiazoles using LR under such conditions. However, thioamide was the only product in place of 1,3-thiazole when amides were treated with LR in tetrahydrofuran at rt. Analogous conditions have earlier been applied for the synthesis of thioamides [45, 46]. In contrast, amides and thioamides could be transformed to 1,3-thiazole employing LR in refluxing toluene with good yield (92% and 83% over two steps, respectively). The LR was mandatory for the reaction of thioamide to afford the 1,3-thiazole. Only starting thioamide was recovered after heating thioamide in both refluxing tetrahydrofuran and toluene for 5 h without using Lawesson's reagent. While LR is helpful for synthesizing *S*-heterocycles, a common problem with the utilization of LR is the removal of LR-based by-products. This difficulty may be eliminated using fluorous derivative of LR and subsequent fluorous solid-phase extraction [47, 48]. However, simple washing of crude reaction mixture with aqueous potassium hydroxide followed by recrystallization from ethanol allowed the isolation of 5-alkoxy-1,3-thiazole compounds in pure form with

2.4 Synthesis of Thiazoles from Dicarbonyl Compounds

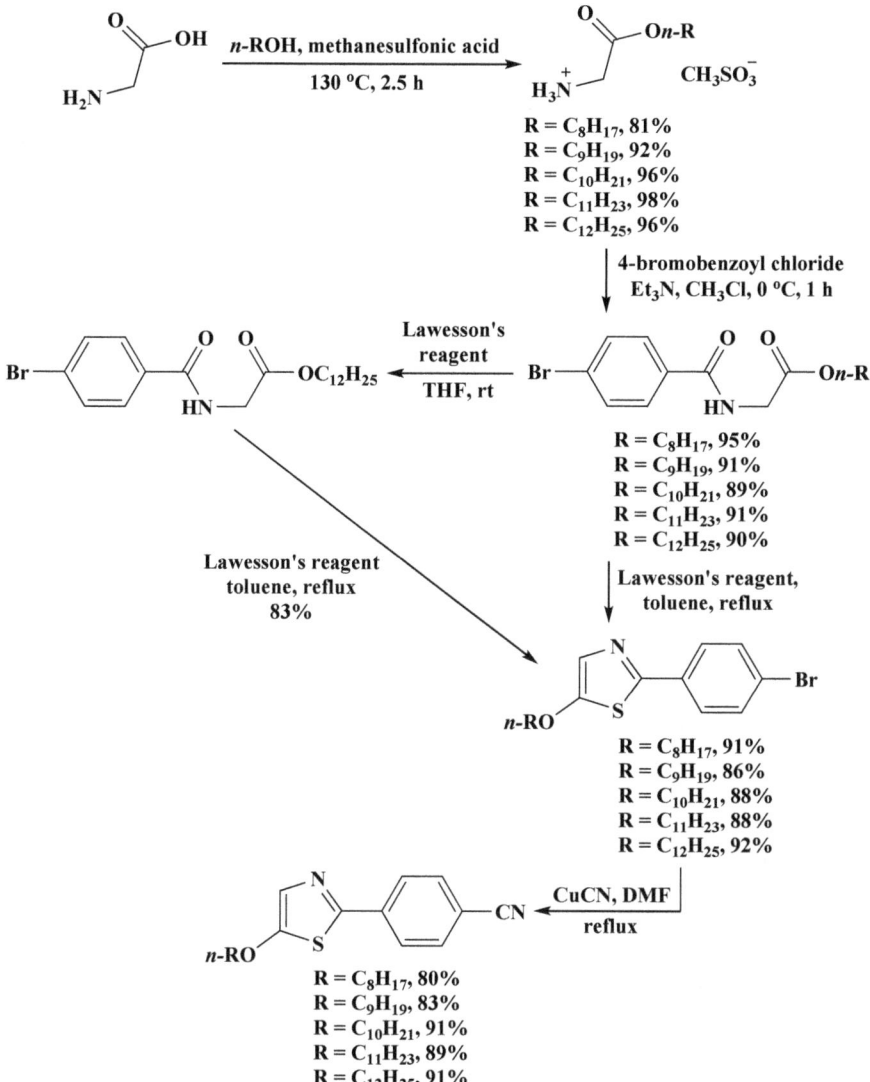

Scheme 2.7 Synthesis of 1,3-thiazoles

less or no loss of product or generation of side-products. The aryl bromides so formed were efficiently transformed to final nitrile targets utilizing a modification of a process designed by Friedman and Shechter [49] where aryl bromides were treated with CuCN in refluxing dimethylformamide [50–54].

The 5-alkoxy-1,3-thiazoles are generally produced via ring-closure of a suitably substituted acylamino carbonyl compound with either phosphorus pentasulfide or LR (Scheme 2.8) [55].

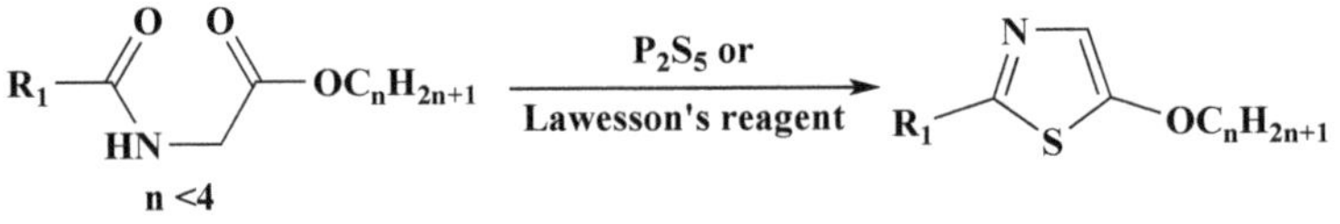

Scheme 2.8 Synthesis of 5-alkoxy-1,3-thiazoles

Since the 2- and 5-positions of 1,3-thiazole are known to be acidic and will react with *n*-BuLi, a simple 2,5-disubstituted 1,3-thiazole was desired which could be used as a test substrate for lithiation but would not create any potential complications from competing sites of lithiation. The 2,5-dibromo-1,3-thiazole served as an excellent building block for the synthesis of desired substrate 4-bromo-2,5-bis(4-methoxyphenyl)-1,3-thiazole. Therefore, a Lawesson's reagent-mediated ring-closing strategy was employed for its synthesis. Starting from commercially inexpensive 4-methoxyacetophenone, the methyl ketone was brominated with elemental bromine to generate the α-bromo-4-methoxyacetophenone in good yield [56]. The α-bromo-4-methoxyacetophenone was converted into α-azido-4-methoxyacetophenone [57] and was subsequently reduced to hydrochloride salt in excellent yield. The reaction of α-amino-4-methoxyacetophenone hydrochloride with 4-methoxybenzoyl chloride, which was prepared from 4-methoxybenzoic acid and $SOCl_2$, in the presence of pyridine and solid $NaHCO_3$ afforded acylamino carbonyl compound in excellent yield [58]. The 2,5-disubstituted 1,3-thiazole was obtained in quantitative yield when 4-methoxy-*N*-2-(4-methoxyphenyl)-2-oxoethylbenzamide was reacted with LR. The conditions for the synthesis of 4-bromo-2,5-bis(4-methoxyphenyl)-1,3-thiazole were chosen carefully to avoid the bromination of highly activated 4-methoxyphenyl rings. Fortunately, the literature contained an example of a 1,3-thiazole being selectively brominated at the 4-position of 1,3-thiazole with NBS at low temperature even in the presence of a 4-methoxyphenyl ring [59]. Application of this procedure to 2,5-bis(4-methoxyphenyl)-1,3-thiazole generated desired 4-bromo-2,5-bis(4-methoxyphenyl)-1,3-thiazole in good yield. Lithiation of 4-bromo-2,5-bis(4-methoxyphenyl)-1,3-thiazole followed by quenching with NFSI afforded desired 4-fluoro-2,5-bis(4-methoxyphenyl)-1,3-thiazole in moderate yield (50%) (Scheme 2.9), which was the first instance for the synthesis of 4-fluoro-1,3-thiazole outside the patent literature.

The reaction of *N*-acylaminoalcohols afforded thiazoline derivatives (Scheme 2.10) [60].

The carboxylic acid [61, 62] was reacted with 2-amino-1-(furan-2-yl)ethanone hydrochloride, 1-ethyl-3-(3-dimethylaminopropyl)carbodiimide, hydroxybenzotriazole, and *N*-methylmorpholine to afford the amide in 82% yield. The amide underwent thionation easily followed by subsequent cyclization with LR in refluxing tetrahydrofuran to afford the thiazole. The benzyl protection was completely removed with trimethylsilyl iodide at mild heating for 15 h to afford the thiazole in 73% yield (Scheme 2.11) [63].

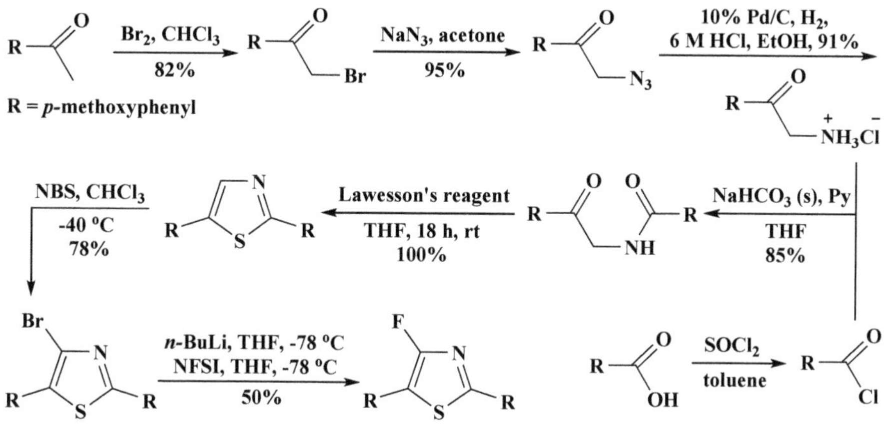

2.4 Synthesis of Thiazoles from Dicarbonyl Compounds

Scheme 2.9 Synthesis of 4-fluoro-1,3-thiazoles

Scheme 2.10 Synthesis of thiazolines

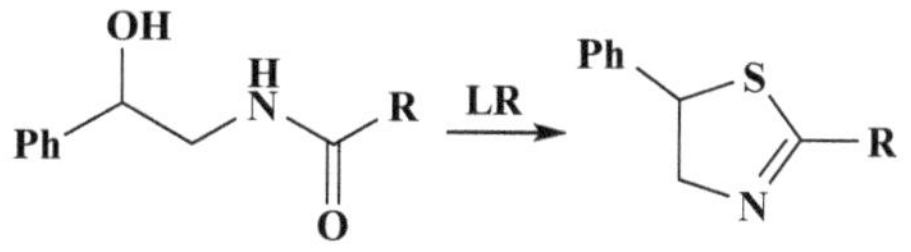

R = H, 4-Me, 4-Cl, 4-OMe, 4-Br

The MW-assisted cyclization of different 1,4-dicarbonyl compounds with LR afforded 2-alkoxythiazoles in 90% yield (Scheme 2.12) [44, 64].

Both P_2S_5 and LR can be utilized as sulfurating agent for the formation of 5-aminothiazoles from amides (Scheme 2.13) [65].

The ketoamido intermediates were reacted with LR and PPh_3/I_2 to afford the thiazoles and oxazoles, respectively (Scheme 2.14) [66].

Sanz-Cervera et al. [67] prepared a small library of compounds with thiazole frameworks and structural diversity in both 2- and 5-positions in the presence of Lawesson's reagent. A double acylation of a protected glycine provided intermediate α-amido-β-ketoesters, which were reacted with LR to provide the benzyl-2,5-substituted thiazole-4-carboxylate (Scheme 2.15).

The 2,4,4-trisubstituted 1,3-thiazole-5(4H)-thiones were synthesized by thionation of N,N-disubstituted carboxamides. The single product 1,3-thiazole-5(4H)-thione was afforded by the reaction of N-acylated α,α-disubstituted α-amino acid amide in toluene/pyridine at 100 °C [68]. An analogous reaction with LR afforded 1,3-thiazol-5(4H)-one exclusively (Scheme 2.16) [69–71].

Thompson and Chen [72, 73] synthesized 2,4-disubstituted 5-aminothiazoles by a sequential Ugi/deprotection/thionation/cyclization methodology in which both R_1 and R_2 positions could be varied easily (Scheme 2.17). The linear

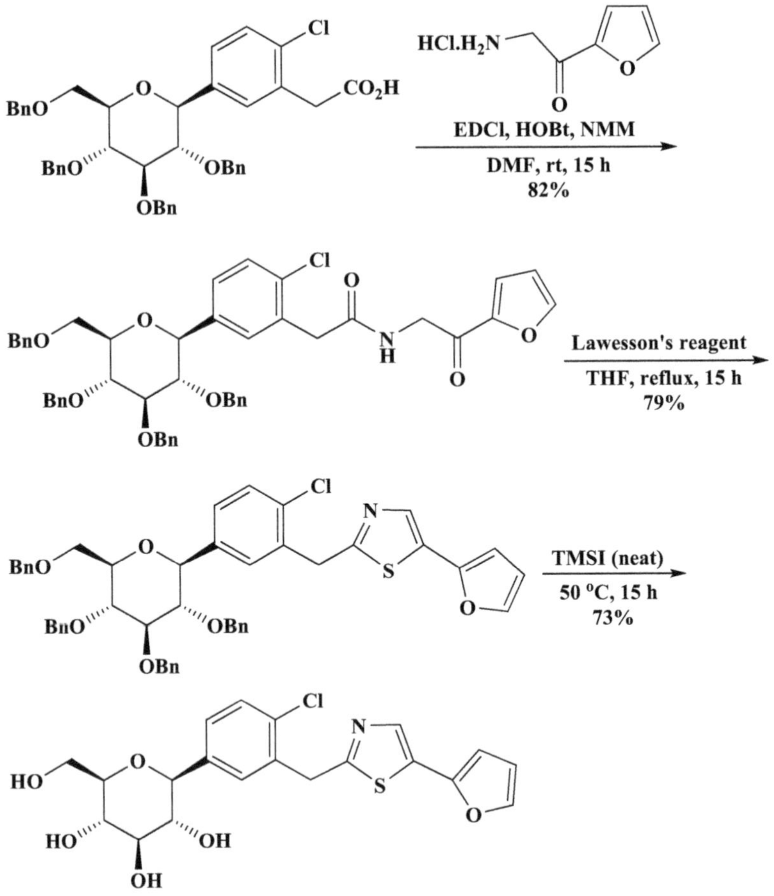

Scheme 2.11 Synthesis of thiazole

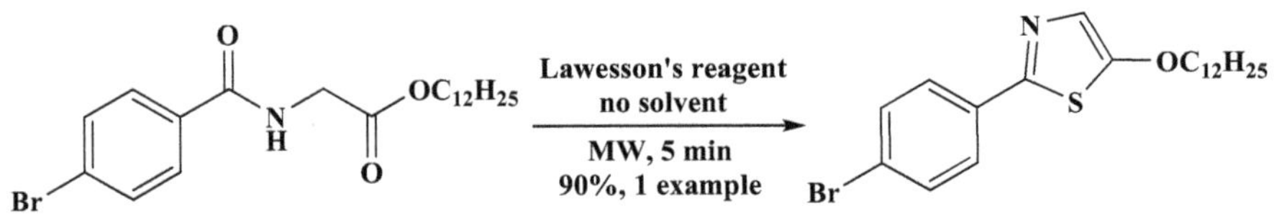

Scheme 2.12 Synthesis of 2-alkoxythiazole

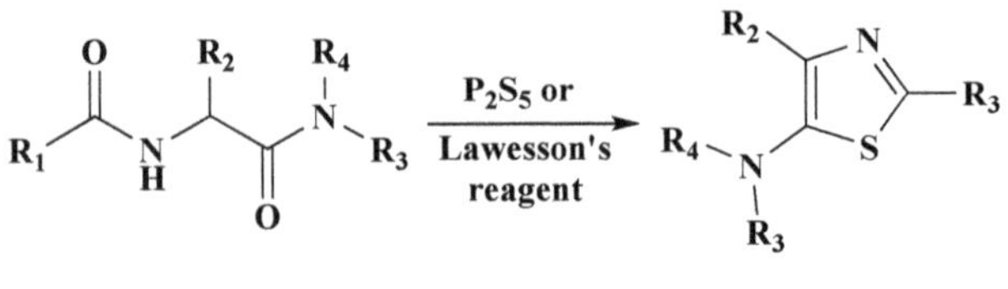

Scheme 2.13 Synthesis of 5-aminothiazoles

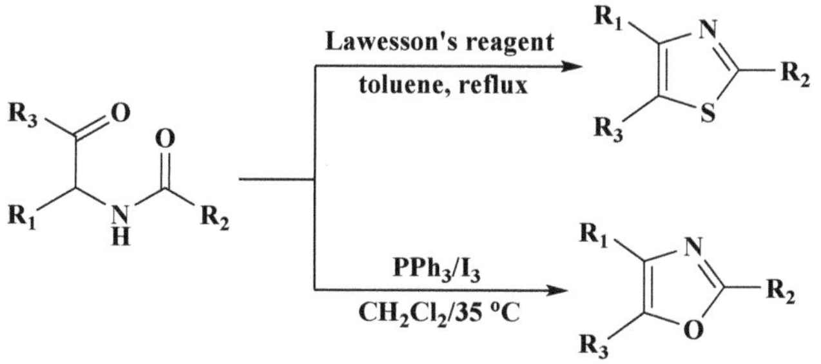

2.4 Synthesis of Thiazoles from Dicarbonyl Compounds   47

Scheme 2.14 Synthesis of thiazoles and oxazoles

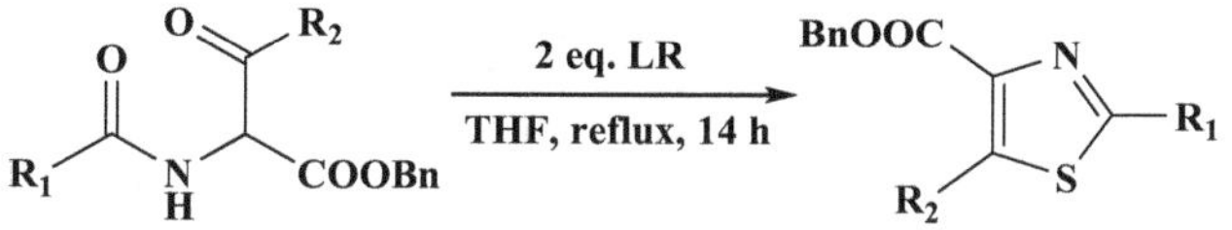

Scheme 2.15 Synthesis of benzyl-2,5-substituted thiazole-4-carboxylates

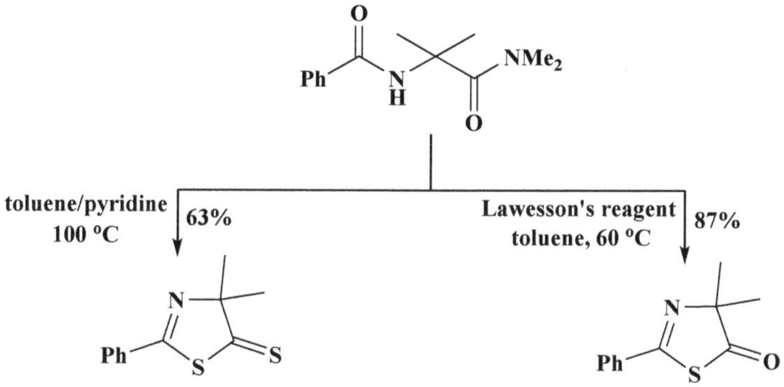

$R_1 = NH_2, CH_3, C_6H_5$
$R_2 = CH_3, C_6H_5, p\text{-}ClC_6H_4, p\text{-}CH_3C_6H_4, p\text{-}NO_2C_6H_4$

Scheme 2.16 Synthesis of 1,3-thiazol-5(4H)-thione/one

dipeptide was constructed by Ugi 4-CR using Walborsky reagent (1,1,3,3-tetramethylbutyl isocyanide) as a cleavable isocyanide input, DMB-NH$_2$ (2,4-dimethoxybenzylamine), diverse carboxylic acids, and aldehydes. Then, reaction with TFA furnished substrate, which was reacted with LR and an intramolecular cyclization synthesized thiazole derivative. A second trifluoroacetic acid cleavage of N-(1,1,3,3-tetramethylbutyl) group afforded 5-aminothiazole peptidomimetics in sufficient yields (5–13%) [74].

This N–H insertion reaction has been widely used to construct the oxazole and thiazole building blocks for cyclic peptides utilizing single-enantiomer amides

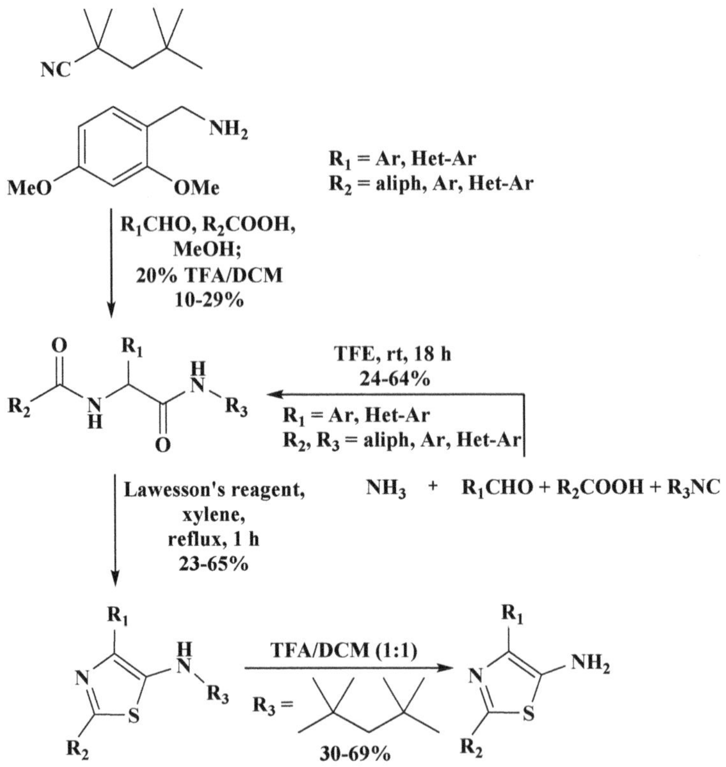

Scheme 2.17 Synthesis of 2,4-disubstituted 5-aminothiazoles

derived from amino acids [75, 76]. An illustrative case including a reaction of simple alkanamide and diazoketoester for the synthesis of five-membered heterocyclic compounds under Rh(II) acetate catalysis is described in Scheme 2.18. The dicarbonyl adduct was easily dehydrated to afford an oxazole; alternatively, reaction with LR afforded thiazole, and reaction with ammonium acetate provided imidazole [77].

The adducts of carboxamides and phenylglyoxal and its analogues were utilized to prepare a series of functional derivatives of azoles and azines [78–80]. The adducts of carboxamides and phenylglyoxal could be transformed into new derivatives of 1,3-thiazole-4-thiol. The key step was the reaction of S-amidophenacylation products and similar compounds in the presence of P_2S_5 or LR (Scheme 2.19) [12, 81]. The transition included the sulfurization of S-amidophenacylation products and their cyclization [82–84].

The thiazole was prepared from commercially accessible N-Boc-L-aspartic acid 4-benzyl ester (Scheme 2.20). The amide underwent N–H insertion reaction in the presence of Rh carbene derived from methyl 2-diazo-3-oxobutanoate to afford the 1,4-dicarbonyl compound in good yield, a reaction earlier utilized as a key step in a

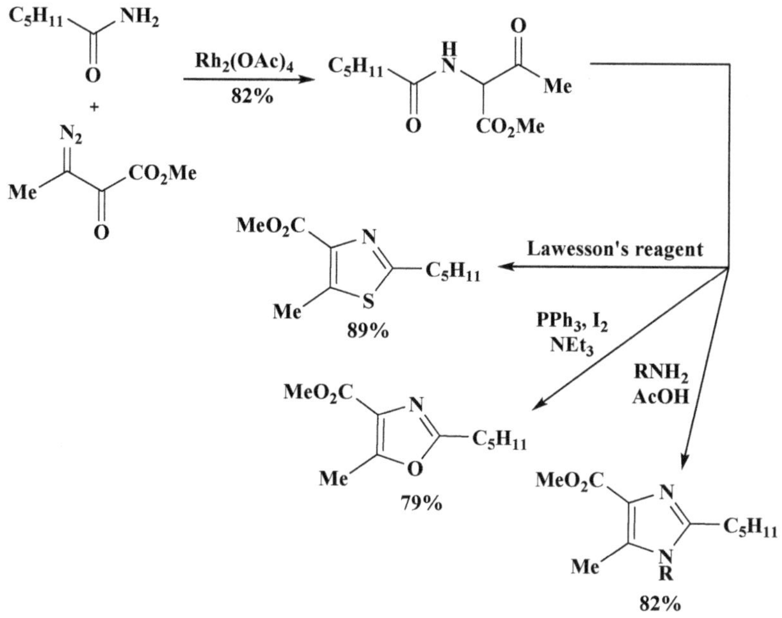

2.4 Synthesis of Thiazoles from Dicarbonyl Compounds

Scheme 2.18 Synthesis of thiazole, oxazole, and imidazoles

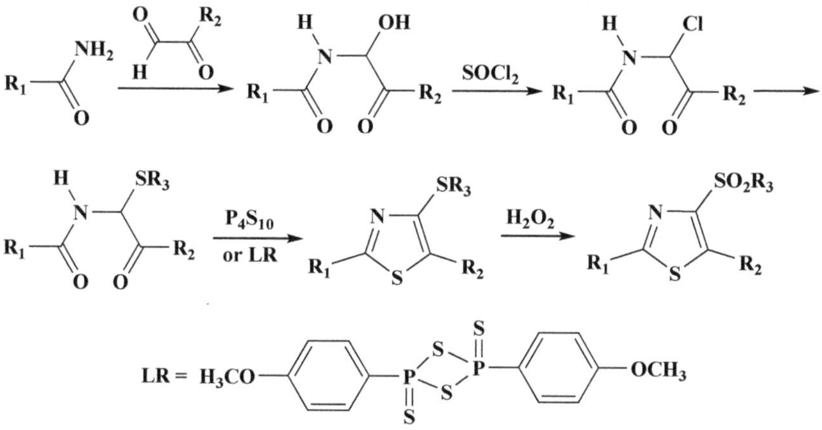

R_1 = H, R_2 = 4-CH$_3$C$_6$H$_4$; R_1 = CH$_3$, R_2 = C$_6$H$_5$; R_1 = CH$_3$, R_2 = 4-CH$_3$C$_6$H$_4$
R_1 = R_2 = C$_6$H$_5$; R_1 = C$_6$H$_5$, R_2 = 4-CH$_3$C$_6$H$_4$; R_1 = C$_6$H$_5$, R_2 = 2-thienyl
R_1 = 4-ClC$_6$H$_4$, R_2 = 4-FC$_6$H$_4$; R_1 = 4-CH$_3$OC$_6$H$_4$, R_2 = C$_6$H$_5$
R_1 = 2-furyl, R_2 = C$_6$H$_5$; R_1 = 2-furyl, R_2 = 2-thienyl; R_1 = 2-thienyl, R_2 = C$_6$H$_5$
R_3 = C$_2$H$_5$, C$_6$H$_5$CH$_2$, C$_6$H$_5$, 4-ClC$_6$H$_4$, 4-CH$_3$C$_6$H$_4$

Scheme 2.19 Synthesis of 1,3-thiazole-4-thiols

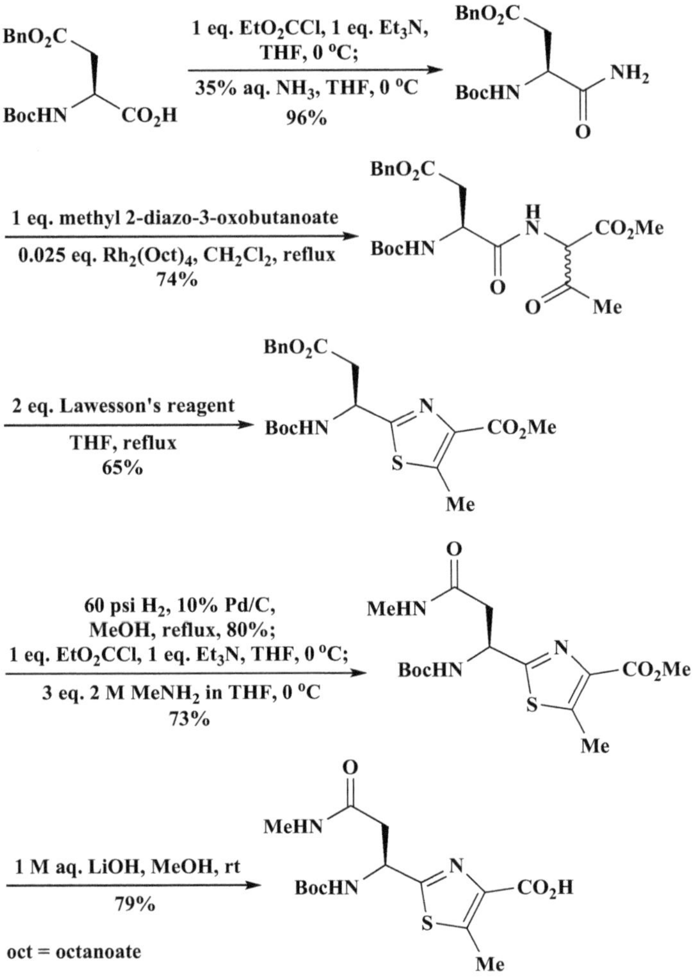

Scheme 2.20 Synthesis of thiazole

path to oxazole building blocks of nostocyclamide and promothiocin A. However, in this case instead of dehydrating the ketoamide to afford an oxazole, it was reacted with LR to afford the thiazole [19]. The correct side-chain was installed by hydrogenolysis of benzyl ester and amide synthesis to afford the thiazole; and after that alkaline hydrolysis exposes the free thiazole-4-carboxylic acid for subsequent coupling reaction [85].

The 1,4-diamides afforded thiazolethiones on reacting with LR in refluxing toluene (Scheme 2.21) [11, 68, 71, 86].

2.4 Synthesis of Thiazoles from Dicarbonyl Compounds

Scheme 2.21 Synthesis of thiazolethiones

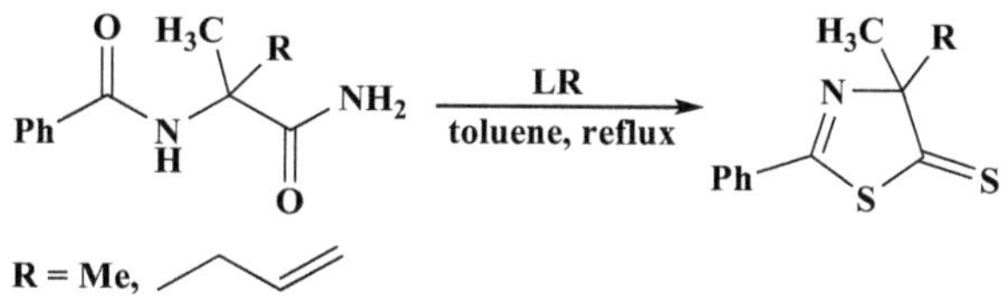

The methyl ester of N-acrylthreonine in the presence of LR in refluxing toluene afforded thiazolone and 4-methoxycarbonyl thiazoline (Scheme 2.22) [11].

An attempt to prepare the corresponding thiol amide with 0.6 eq. Lawesson's reagent afforded ring-closed product as the only product in 80% yield (entry 2). The amido alcohols were subsequently directly converted into desired 5-substituted Δ^2-thiazolines in 71–80% yield under these conditions (Scheme 2.23) [87].

The Steglich esterification and the $SnCl_2 \cdot 2H_2O$-promoted azide reduction/O → N acyl migration proceeded well for azido alcohols, while the following ring-closure to 5-substituted Δ^2-thiazolines with Lawesson's reagent was troublesome for benzyl-substituted compound that was obtained in only 24% yield. However, the reaction proceeded smoothly with other aryl/heteroaryl-substituted amido alcohols to afford

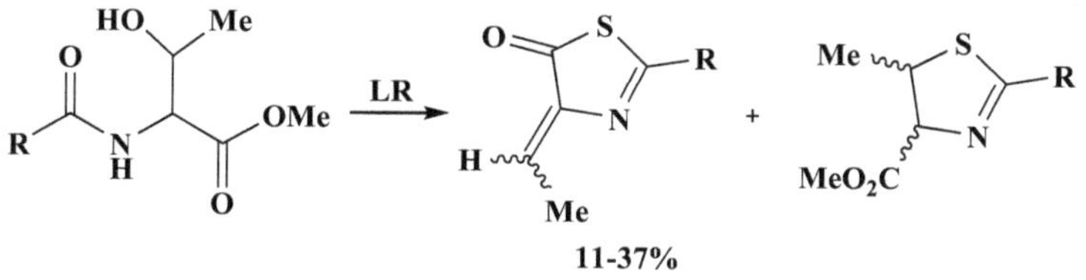

Scheme 2.22 Synthesis of thiazolones and 4-methoxycarbonyl thiazolines

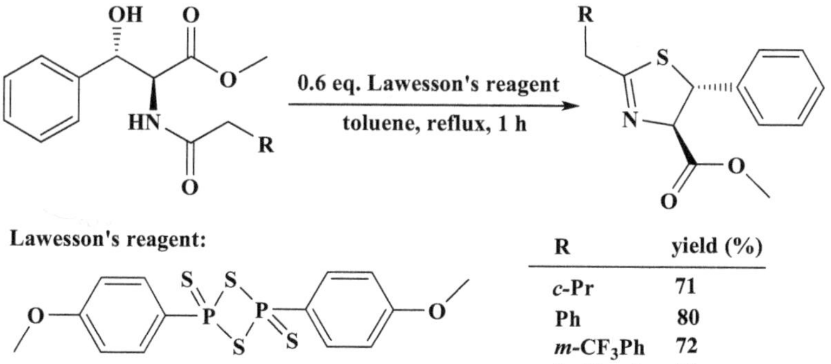

R	yield (%)
c-Pr	71
Ph	80
m-CF$_3$Ph	72
2-thienyl	74

Scheme 2.23 Synthesis of 5-substituted Δ^2-thiazolines

the 5-substituted Δ²-thiazolines in 71%, 80%, and 65% isolated yield, respectively (Scheme 2.24) [87].

The epimerization took place in the ring-closing reaction with Lawesson's reagent. Subsequently, a set of reactions in varying amounts of Lawesson's reagent was performed; the amido alcohol was selected as a model substance. Lower amounts of Lawesson's reagent gave lower yield and *ee* (entry 1). However, both the yield and the *ee* improved significantly using 1 eq. Lawesson's reagent (entry 4) (Scheme 2.25) [87].

The thiazoles were synthesized from α-amido-β-ketoesters via intermediates in the presence of LR. The thiazoles were formed in 60–97% yield except in the case of thiazoles, which were formed in 30% yield only. This drop in yield might be caused by steric hindrance between the Lawesson's reagent and the trifluoromethyl group in the *ortho*-position of the R_2 group. The ester deprotection was carried out by palladium-catalyzed hydrogenolysis and hydrolysis with lithium hydroxide in tetrahydrofuran/water (Scheme 2.26) [67].

The fluorous 1,3-thiazoles were synthesized in high yields (54–82%) by cyclization of fluorous amido-β-ketoesters in the presence of LR, which are only slightly lower than those obtained for the similar nonfluorous 1,3-thiazoles (67–97%) [67]. Again, the last stage was basic hydrolysis with lithium hydroxide in tetrahydrofuran/water (4:1) to deprotect the carboxylic moiety, which afforded 2,5-disubstituted 1,3-thiazoles in high yield (95–99%) (Scheme 2.27).

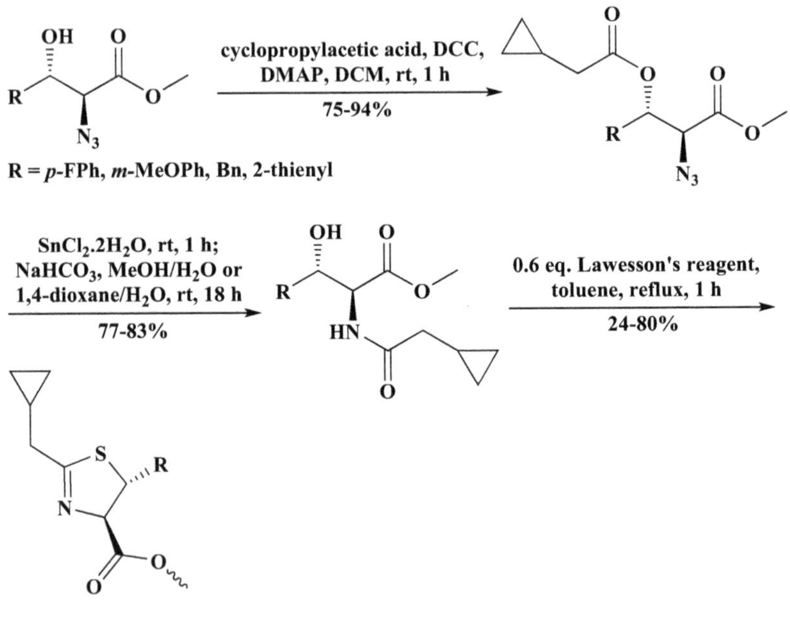

Scheme 2.24 Synthesis of 5-substituted Δ²-thiazolines

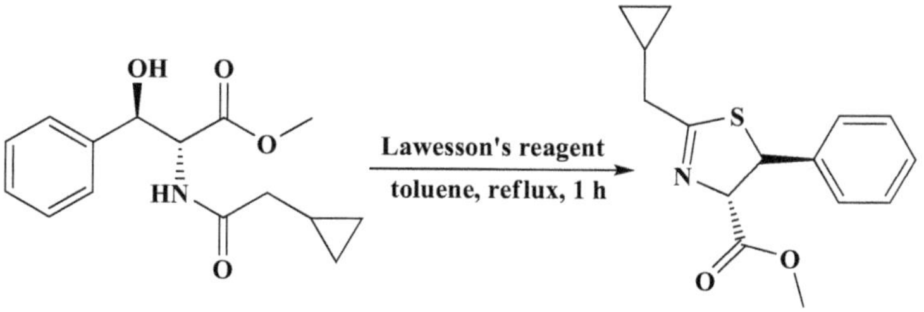

2.4 Synthesis of Thiazoles from Dicarbonyl Compounds

entry	LR (eq.)	temp.	yield (%)	ee (%)
1	0.5	reflux	50	65
2	0.6	rt-reflux	75	73
3	0.6	reflux	78	75
4	1.0	reflux	90	97

Scheme 2.25 Synthesis of methyl (4S,5R)-2-(cyclopropylmethyl)-5-phenyl-4,5-dihydrothiazole-4-carboxylate

During the total synthesis of amythiamycin D [85, 88], Davies et al. [89] used Rh carbene N–H insertion strategy to provide the aspartate-derived thiazole (Scheme 2.28). Amide underwent chemoselective N–H insertion with Rh carbenoid, synthesized by the reaction of dirhodium tetraoctanoate and 2-diazo-3-oxobutanoate. The formed ketoamide was heated with LR to afford the thiazole.

Davies et al. [89] reported a common method to synthesize the trisubstituted thiazoles and oxazoles that depends on the carbenoids chemistry. The sequence began with rhodium(II)-assisted formal insertion of N–H bond of amide into the carbene formed via deazoniation of methyl diazoacetoacetate. The formed compound cyclized to thiazole with LR [90, 91] without loss of chirality at the α-position (Scheme 2.29). The procedure was not employed for the formation of 5H-thiazoles because of the problems associated with the formation of diazo compound.

The amide was synthesized by the reaction of pyridyl carboxylic acid with $SOCl_2$ in C_6H_6 and after that treatment with glycine methyl ester in HCl and Et_3N in chloroform. Further reaction with LR resulted in the synthesis of methylpropyltryptamine via a ring-closure procedure [92]. The derivation of MPT framework with a halogen atom, either bromine or iodine, at the 4-position of thiazole, rendered the MPT active and able to conjugate other groups. An incorporation of halogen atom on the thiazole was completed via reaction with either N-iodosuccinimide or N-bromosuccinimide to synthesize the 4-bromo-5-methoxy-2-(2-pyridyl)-thiazole or 4-iodo-5-methoxy-2-(2-pyridyl)-thiazole. Then, depriving the halogen atom on the thiazole with a strong base provided a nucleophilic MPT anion intermediate, which was treated with perfluorocyclopentene moiety to afford the symmetric or asymmetric photoactive diarylethene compounds [93, 94]. The fluorescence behavior is different in ring-open and ring-closed isomers due to different energy and charge transfer phenomena in both ground and excited states resulting from the diverse extensions of π-nature frontier orbitals. The formed symmetric diarylethene (1,2-bis[5-methoxy-2-(2-pyridyl)thiazolyl]perfluoropentene is fluorescent in ring-open form but

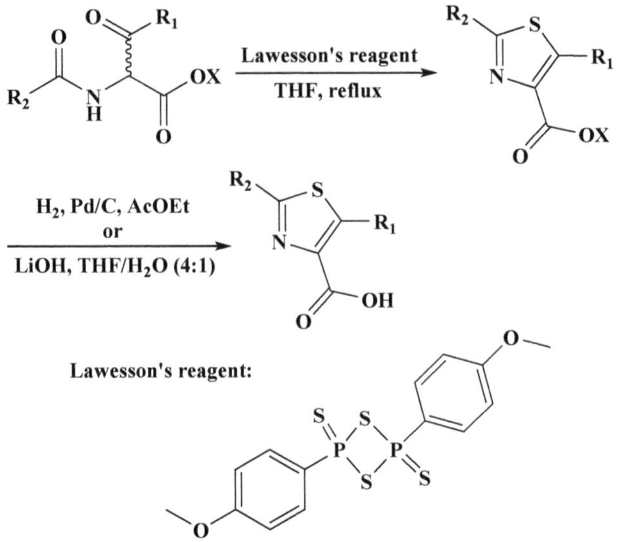

entry	R$_1$	R$_2$	yield (%)
1	C$_6$H$_5$	C$_6$H$_5$	99
2	C$_6$H$_5$	o-FC$_6$H$_4$	99
3	C$_6$H$_5$	o-MeOC$_6$H$_4$	96
4	C$_6$H$_5$	o-CF$_3$OC$_6$H$_4$	97
5	C$_6$H$_5$	m-MeOC$_6$H$_4$	97
6	C$_6$H$_5$	m-CF$_3$OC$_6$H$_4$	96
7	C$_6$H$_5$	p-MeOC$_6$H$_4$	99
8	C$_6$H$_5$	p-CF$_3$OC$_6$H$_4$	99
9	C$_6$H$_5$	3,4,5-(OMe)$_3$C$_6$H$_2$	96
10	C$_6$H$_5$	piperonyl	96
11	C$_6$H$_5$	p-NO$_2$C$_6$H$_4$	97
12	C$_6$H$_5$	p-CF$_3$OC$_6$H$_4$	97
13	C$_6$H$_5$	Me	99
14	C$_6$H$_5$	t-Bu	97
15	o-MeOC$_6$H$_4$	C$_6$H$_5$	99
16	o-MeOC$_6$H$_4$	o-MeOC$_6$H$_4$	99
17	p-MeOC$_6$H$_4$	C$_6$H$_5$	97
18	p-MeOC$_6$H$_4$	o-MeOC$_6$H$_4$	96
19	t-Bu	C$_6$H$_5$	97
20	i-Pr	C$_6$H$_5$	97
21	Me	C$_6$H$_5$	98
22	p-MeC$_6$H$_4$	C$_6$H$_5$	98
23	p-ClC$_6$H$_4$	p-ClC$_6$H$_4$	99
24	p-NO$_2$C$_6$H$_4$	C$_6$H$_5$	97

Scheme 2.26 Synthesis of thiazoles

2.4 Synthesis of Thiazoles from Dicarbonyl Compounds

Scheme 2.27 Synthesis of 1,3-thiazoles

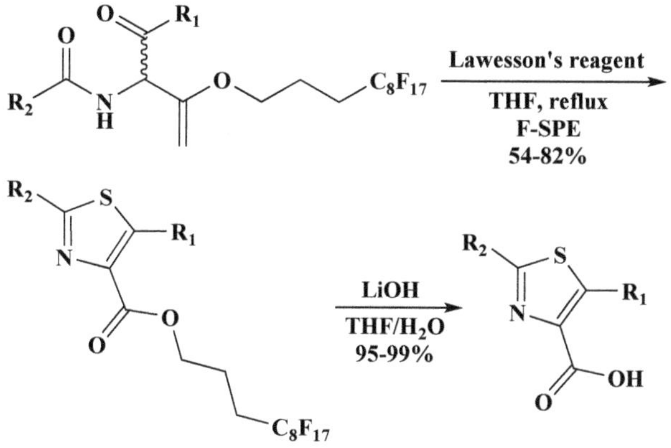

entry	R_1	R_2	yield (%)
1	C_6H_5	C_6H_5	99
2	C_6H_5	$o\text{-FC}_6H_4$	95
3	C_6H_5	$o\text{-MeOC}_6H_4$	97
4	C_6H_5	$o\text{-CF}_3OC_6H_4$	99
5	C_6H_5	$m\text{-MeOC}_6H_4$	99
6	C_6H_5	$m\text{-CF}_3OC_6H_4$	99
7	C_6H_5	$p\text{-MeOC}_6H_4$	99
8	C_6H_5	$p\text{-CF}_3OC_6H_4$	99
9	C_6H_5	$3,4,5\text{-(OMe)}_3C_6H_2$	98

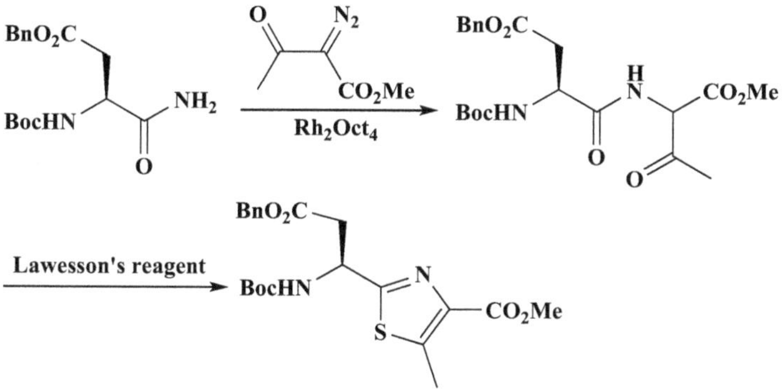

Scheme 2.28 Synthesis of thiazole

nonfluorescent in ring-closed form, while for the asymmetric 1-{4-(5-methoxy-2-(2-pyridyl)thiazolyl)}-2-{3-(2-methylbenzo[b]thiophenyl)}hexafluorocyclopentene, both the ring-open form and the ring-closed form are fluorescent, which was attributed to the presence of an electron-releasing 1-{4-(5-methoxy-2-(2-pyridyl)thiazolyl)} and energy transfer process (Scheme 2.30) [95].

Ring forming chemistry has been extended to the formation of 5-aminothiazoles from diamides (An = $p\text{-MeOC}_6H_4$) (Scheme 2.31) [65, 96].

The synthesis of interesting compounds by the reactions of unsaturated ketones (all bearing a 4-oxothiazolidine ring) with LR was demonstrated [97, 98]. The synthesis

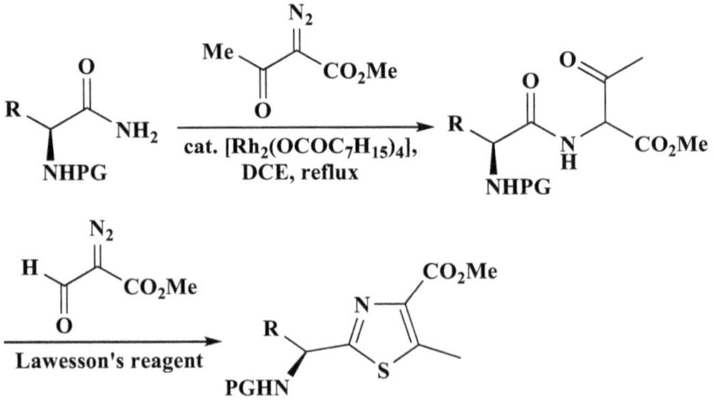

Scheme 2.29 Synthesis of thiazoles

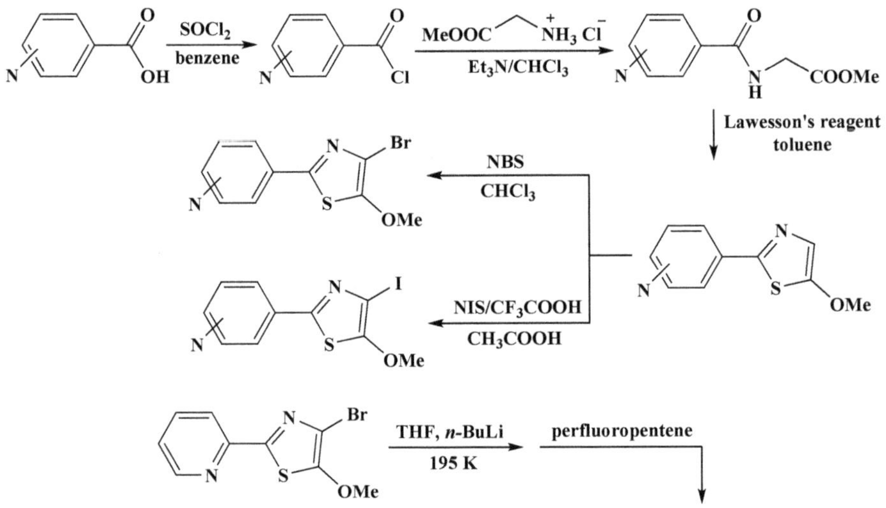

Scheme 2.30 Synthesis of thiazoles

of new dithiazole (Scheme 2.32) and thiazole (Scheme 2.33) compounds could be attributed to the presence of diverse functional groups adjacent to the CO group. The ester or amide groups resulted in the synthesis of thiazole or dithiazole heterocyclic compounds, respectively [11].

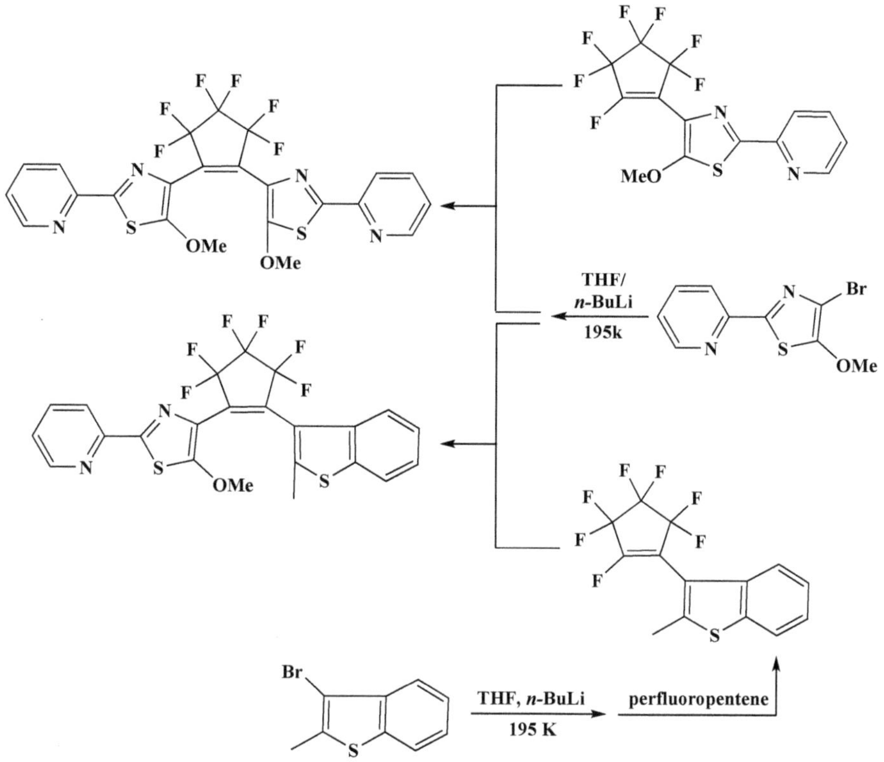

2.4 Synthesis of Thiazoles from Dicarbonyl Compounds

Scheme 2.30 (continued)

Scheme 2.31 Synthesis of benzothiazoloazepine

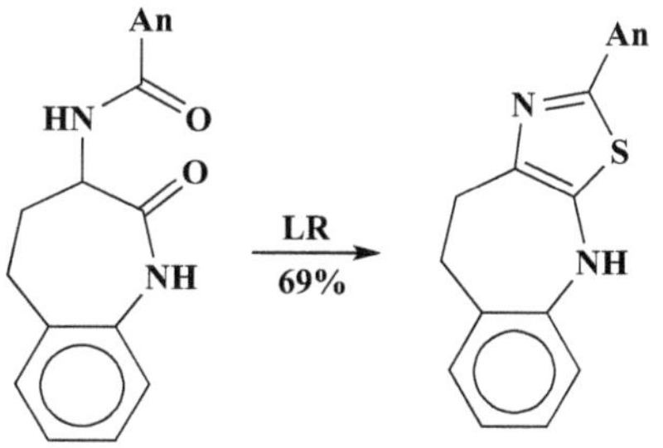

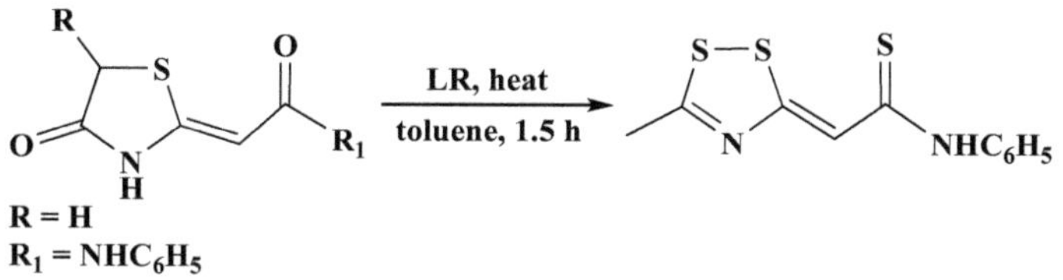

Scheme 2.32 Synthesis of dithiazole

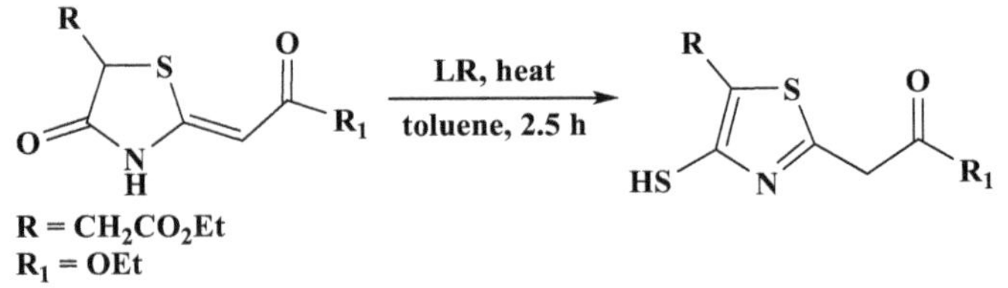

Scheme 2.33 Synthesis of thiazoles

R = CH$_2$CO$_2$Et
R$_1$ = OEt

References

1. (a) G. Deepika, P. Gopinath, G. Kranthi, C. Nagamani, Y.V. Jayasree, N.V. Naidu and S. Enaganti. 2012. Synthesis and antibacterial activity of some new thiazine derivatives. J. Pharm. Res. 5: 1105–1107. (b) N. Kaur. 2015. Review of microwave-assisted synthesis of benzo-fused six-membered N,N-heterocycles. Synth. Commun. 45: 300–330. (c) N. Kaur. 2017. Applications of gold catalysts for the synthesis of five-membered O-heterocycles. Inorg. Nano Met. Chem. 47: 163–187. (d) N. Kaur, Aditi and D. Kishore. 2016. A facile synthesis of face 'D' quinolino annulated benzazepinone analogues with its quinoline framework appended to oxadiazole, triazole and pyrazole heterocycles. J. Heterocycl. Chem. 53: 457–460.
2. (a) A.T. Chaviara, P.J. Cox, K.H. Repana, R.M. Papi, K.T. Papazisis, D. Zambouli, A.H. Kortsaris, D.A. Kyriakidis and C.A. Bolos. 2004. Copper(II) Schiff base coordination compounds of diene with heterocyclic aldehydes and 2-amino-5-methyl-thiazole: Synthesis, characterization, antiproliferative and antibacterial studies. Crystal structure of CudienOOCl2. J. Inorg. Biochem. 98: 1271–1283. (b) P. Sharma, N. Kaur, R. Sirohi and D. Kishore. 2013. Microwave assisted facile one pot synthesis of novel 5-carboxamido substituted analogues of 1,4-benzodiazepin-2-one of medicinal interest. Bull. Chem. Soc. Ethiop. 27: 301–307. (c) N. Kaur, Y. Verma, N. Ahlawat, P. Grewal, P. Bhardwaj and N.K. Jangid. 2020. Copper-assisted synthesis of five-membered O-heterocycles. Inorg. Nano Met. Chem. 50: 705–740. (d) S. Caron and E. Vazquez. 2003. Efficient synthesis of [6-chloro-2-(4-chlorobenzoyl)-1H-indol-3-yl]acetic acid, a novel COX-2 inhibitor. J. Org. Chem. 68: 4104–4107.
3. H. Joshi, P. Upadhyay and A.J. Baxi. 1990. Studies on 4-thiazolidinones. Synthesis and antimicrobial activity of 1,4-bis(2'-aryl-5'(H)-4'-thiazolidinone-3'-ylamino)phthalazine. J. Indian Chem. Soc. 67: 779–780.
4. A. Al-Mulla. 2017. A review: Biological importance of heterocyclic compounds. Der Pharm. Chem. 9: 141–147.
5. J. Matysiak. 2006. Synthesis, antiproliferative and antifungal activities of some 2-(2,4-dihydroxyphenyl)-4H-3,1-benzothiazines. Bioorg. Med. Chem. 14: 2613–2619.
6. A. Macchiarulo, G. Costantino, D. Fringuelli, A. Vecchiarelli, F. Schiaffella and R. Fringuelli. 2002. 1,4-Benzothiazine and 1,4-benzoxazine imidazole derivatives with antifungal activity: A docking study. Bioorg. Med. Chem. 10: 3415–3423.
7. R. Fringuelli, F. Schiaffella and A. Vecchiarelli. 2001. Antifungal and immunomodulating activities of 1,4-benzothiazine azole derivatives: Review. J. Chemother. 13: 9–14.
8. B.S. Rathore and M. Kumar. 2006. Synthesis of 7-chloro-5-trifluoromethyl/7-fluoro/7-trifluoromethyl-4H-1,4-benzothiazines as antimicrobial agents. Bioorg. Med. Chem. 14: 5678–5682.
9. Y. Hirokawa, H. Kinoshita, T. Tanaka, T. Nakamura, K. Fujimoto, S. Kashimoto, T. Kojima and S. Kato. 2009. Pleuromutilin derivatives having a purine ring. Part 2: Influence of the central spacer on the antibacterial activity against Gram-positive pathogens. Bioorg. Med. Chem. Lett. 19: 170–174.
10. P.K. Sharma and G. Kaur. 2017. A review on antimicrobial activities of important thiazines based heterocycles. Drug Invent. Today 9: 23–25.

References

11. T. Ozturk, E. Ertas and O. Mert. 2007. Use of Lawesson's reagent in organic syntheses. Chem. Rev. 107: 5210–5278.
12. M.P. Cava and M.I. Levinson. 1985. Thionation reactions of Lawesson's reagents. Tetrahedron 41: 5061–5087.
13. H.Z. Lecher, R.A. Greenwood, K.C. Whitehouse and T.H. Chao. 1956. The phosphonation of aromatic compounds with phosphorus pentasulfide. J. Am. Chem. Soc. 78: 518–522.
14. N.C. Misra and H. Ila. 2010. 4-Bis(methylthio)methylene-2-phenyloxazol-5-one: Versatile template for synthesis of 2-phenyl-4,5-functionalized oxazoles. J. Org. Chem. 75: 5195–5202.
15. V. Amareshwar, N.C. Misra and H. Ila. 2011. 2-Phenyl-4-bis(methylthio)methyleneoxazol-5-one: Versatile template for diversity oriented synthesis of heterocycles. Org. Biomol. Chem. 9: 5793–5801.
16. S.V. Kumar, B. Saraiah, N.C. Misra and H. Ila. 2012. Synthesis of 2-phenyl-4,5-substituted oxazoles by copper-catalyzed intramolecular cyclization of functionalized enamides. J. Org. Chem. 77: 10752–10763.
17. S. Yugandar, A. Acharya and H. Ila. 2013. Synthesis of 2,5-bis(hetero)aryl 4'-substituted 4,5'-bisoxazoles via copper(I)-catalyzed domino reactions of activated methylene isocyanides with 2-phenyl- and 2-(2-thienyl)-4-[(aryl/heteroaryl)(methylthio)methylene]oxazol-5(4H)-ones. J. Org. Chem. 78: 3948–3960.
18. S.V. Kumar, G. Parameshwarappa and H. Ila. 2013. Synthesis of 2,4,5-trisubstituted thiazoles via Lawesson's reagent-mediated chemoselective thionation-cyclization of functionalized enamides. J. Org. Chem. 78: 7362–7369.
19. T.D. Gordon, J. Singh, P.E. Hansen and B.A. Morgan. 1993. Synthetic approaches to the 'azole' peptide mimetics. Tetrahedron Lett. 34: 1901–1904.
20. N. Desroy, F. Moreau, S. Briet, G. LeFralliec, S. Floquet, L. Durant, V. Vongsouthi, V. Gerusz, A. Denis and S. Escaich. 2009. Towards Gram-negative antivirulence drugs: New inhibitors of HldE kinase. Bioorg. Med. Chem. 17: 1276–1289.
21. E. Aguliar and A.I. Meyers. 1994. Reinvestigation of a modified Hantzsch thiazole synthesis. Tetrahedron Lett. 35: 2473–2476.
22. M.W. Bredenkamp, C.W. Holzapfel and W.J. van Zyl. 1990. The chiral synthesis of thiazole amino acid enantiomers. Synth. Commun. 20: 2235–2249.
23. F.W. Foss, T.P. Mathews, Y. Kharel, P.C. Kennedy, A.H. Snyder, M.D. Davis, K.R. Lynch and T.L. MacDonald. 2009. Synthesis and biological evaluation of sphingosine kinase substrates as sphingosine-1-phosphate receptor prodrugs. Bioorg. Med. Chem. 17: 6123–6136.
24. F. Wolter and J. Stein. 2002. Biological activities of resveratrol and its analogs. Drugs Future 27: 949–959.
25. G.J. Karabatsos and N. Hsi. 1967. Structural studies by nuclear magnetic resonance - XI. Tetrahedron 23: 1079–1095.
26. S. Bertini, V. Calderone, I. Carboni, R. Maffei, A. Martelli, A. Martinelli, F. Minutolo, M. Rajabi, L. Testai, T. Tuccinardi, R. Ghidoni and M. Macchia. 2010. Synthesis of heterocycle-based analogs of resveratrol and their antitumor and vasorelaxing properties. Bioorg. Med. Chem. 18: 6715–6724.
27. J. Mehrzad, M. Rajabi and M.A. Khalilzadeh. 2010. Design and antiproliferative activity of 2-(3,5-dihydroxyphenyl)-6-hydroxybenzothiazole (DHB) on PC-3 prostate cancer cell line. Iran. J. Org. Chem. 2: 487–489.
28. R.A. Irgashev, A.A. Karmatsky, P.A. Slepukhin, G.L. Rusinov and V.N. Charushin. 2013. A convenient approach to the design and synthesis of indolo[3,2-c]coumarins via the microwave-assisted Cadogan reaction. Tetrahedron Lett. 54: 5734–5738.
29. T.M. Potewar, K.T. Petrova and M.T. Barros. 2013. Efficient microwave assisted synthesis of novel 1,2,3-triazole-sucrose derivatives by cycloaddition reaction of sucrose azides and terminal alkynes. Carbohydr. Res. 379: 60–67.
30. R.C. Lian, M.H. Lin, M.H. Liao, J.J. Fu, Y.C. Wu, F.R. Chang, C.C. Wu, M.J. Wu and P.S. Pan. 2014. Direct synthesis of the arylboronic acid analogues of phenylglycine via microwave-assisted four-component Ugi reaction. Tetrahedron 70: 1800–1804.

31. M.Y. Mentese, H. Bayrak, Y. Uygun, A. Mermer, S. Ulker, S.A. Karaoglu and N. Demirbas. 2013. Microwave assisted synthesis of some hybrid molecules derived from norfloxacin and investigation of their biological activities. Eur. J. Med. Chem. 67: 230–242.
32. F. Messina and O. Rosati. 2013. Superheated water as solvent in microwave assisted organic synthesis of compounds of valuable pharmaceutical interest. Curr. Org. Chem. 17: 1158–1178.
33. P. Appukkuttan, V.P. Mehta and E.V. van der Eycken. 2010. Microwave-assisted cycloaddition reactions. Chem. Soc. Rev. 39: 1467–1477.
34. X. Zhang, H. Jiang, D. Ye, H. Sun and H. Liu. 2009. Microwave-assisted synthesis of quinazolinone derivatives by efficient and rapid iron-catalyzed cyclization in water. Green Chem. 11: 1881–1888.
35. B. Maiti, K. Chanda, M. Selvaraju, C.C. Tseng and C.M. Sun. 2013. Multicomponent solvent-free synthesis of benzimidazolyl imidazo[1,2-*a*]pyridine under microwave irradiation. ACS Comb. Sci. 15: 291–297.
36. A. Walia, S. Kang and R.B. Silverman. 2013. Microwave-assisted protection of primary amines as 2,5-dimethylpyrroles and their orthogonal deprotection. J. Org. Chem. 78: 10931–10937.
37. E.F. Dimauro and J.M. Kennedy. 2007. Rapid synthesis of 3-amino-imidazopyridines by a microwave-assisted four-component coupling in one pot. J. Org. Chem. 72: 1013–1016.
38. R.B. Sparks and A.P. Combs. 2004. Microwave-assisted synthesis of 2,4,5-triaryl-imidazole; a novel thermally induced *N*-hydroxyimidazole N-O bond cleavage. Org. Lett. 6: 2473–2475.
39. H.D. Patel, S.M. Divatia and E. de Clercq. 2013. Synthesis of some novel thiosemicarbazone derivatives having anti-cancer, anti-HIV as well as anti-bacterial activity. Indian J. Chem. Sect. B 52: 535–545.
40. S.M. Prajapati, R.H. Vekariya, K.D. Patel, S.N. Panchal, H.D. Patel, D.P. Rajani and S. Rajani. 2014. Synthesis and in vitro antibacterial and antifungal evaluation of quinoline analogue azetidine and thiazolidine derivatives. Int. Lett. Chem. Phys. Astron. 20: 195–210.
41. N.P. Prajapati, R.H. Vekariya and H.D. Patel. 2015. Microwave induced facile one-pot access to diverse 2-cyanobenzothiazole - a key intermediate for the synthesis of firefly luciferin. Int. Lett. Chem. Phys. Astron. 44: 81–89.
42. J.A. Seijas, M.P. Vázquez-Tato, M.R. Carballido-Reboredo, J. Crecente-Campo and L. Romar-López. 2007. Lawesson's reagent and microwaves: A new efficient access to benzoxazoles and benzothiazoles from carboxylic acids under solvent-free conditions. Synlett 2: 313–317.
43. C.L. Penney, P. Shah and S. Landi. 1985. A simple method for the synthesis of long-chain alkyl esters of amino acids. J. Org. Chem. 50: 1457–1459.
44. A.A. Kiryanov, P. Sampson and A.J. Seed. 2001. Synthesis of 2-alkoxy-substituted thiophenes, 1,3-thiazoles, and related *S*-heterocycles via Lawesson's reagent-mediated cyclization under microwave irradiation: Applications for liquid crystal synthesis. J. Org. Chem. 66: 7925–7929.
45. M. Yokoyama, Y. Menjo, M. Watanabe and H. Togo. 1994. Synthesis of oxazoles and thiazoles using thioimidates. Synthesis 12: 1467–1470.
46. T.P. Andersen, A.-B.A.G. Ghattas and S.-O. Lawesson. 1983. Studies on amino acids and peptides - IV. Tetrahedron 39: 3419–3427.
47. Z. Kaleta, G. Tarkanyi, A. Gomory, F. Kalman, T. Nagy and T. Soos. 2006. Synthesis and application of a fluorous Lawesson's reagent: Convenient chromatography-free product purification. Org. Lett. 8: 1093–1095.
48. Z. Kaleta, B.T. Makowski, T. Soos and R. Dembinski. 2006. Thionation using fluorous Lawesson's reagent. Org. Lett. 8: 1625–1628.
49. L. Friedman and H. Shechter. 1961. Dimethylformamide as a useful solvent in preparing nitriles from aryl halides and cuprous cyanide; improved isolation techniques. J. Org. Chem. 26: 2522–2524.
50. M.M. Murza, T.R. Prosochina, M.G. Safarov and E.A. Kantor. 2001. Synthesis and quantum-chemical study of liquid-crystal derivatives of thiazole. Chem. Heterocycl. Compd. (Engl. Transl.) 37: 1258–1265.
51. Z.K. Kuvatov, M.G. Safarov and M.M. Murza. 2004. New derivatives of thiazole with mesomorphous properties. Chem. Heterocycl. Compd. (Engl. Transl.) 40: 500–502.

52. A.S. Golovanov, M.M. Murza and M.G. Safarov. 1997. Novel mesomorphic Schiff bases. Chem. Heterocycl. Compd. 33: 1350–1351.
53. M.M. Murza, A.S. Golovanov and M.G. Safarov. 1996. New liquid crystal derivatives of thiazole. Chem. Heterocycl. Compd. (Engl. Transl.) 32: 477–478.
54. A.M. Grubb, S. Hasan, A.A. Kiryanov, P. Sampson and A.J. Seed. 2009. The synthesis and physical evaluation of 5-alkoxy-1,3-thiazoles prepared via Lawesson's reagent-mediated cyclisation of α-benzamido esters. Liq. Cryst. 36: 443–453.
55. M. Jesberger, T.P. Davies and L. Barner. 2003. Applications of Lawesson's reagent in organic and organometallic syntheses. Synthesis 13: 1929–1958.
56. D.L.J. Clive, S. Hisaindee and D.M. Coltart. 2003. Derivatized amino acids relevant to native peptide synthesis by chemical ligation and acyl transfer. J. Org. Chem. 68: 9247–9254.
57. T. Patonay, E. Juhasz-Toth and A. Benyei. 2002. Base-induced coupling of α-azido ketones with aldehydes - an easy and efficient route to trifunctionalized synthons 2-azido-3-hydroxyketones, 2-acylaziridines, and 2-acylspiroaziridines. Eur. J. Org. Chem. 2: 285–295.
58. N.J. Gilmore, S. Jones and M.P. Muldowney. 2004. Synthetic applicability and in situ recycling of a B-methoxy oxazaborolidine catalyst derived from cis-1-amino-indan-2-ol. Org. Lett. 6: 2805–2808.
59. S. Takami and M. Irie. 2004. Synthesis and photochromic properties of novel yellow developing photochromic compounds. Tetrahedron 60: 6155–6161.
60. H.Z. Lecher, R.A. Greenwood, K.C. Whitehouse and T.H. Cho. 1956. The phosphonation of aromatic compounds with phosphorus pentasulfide. J. Am. Chem. Soc. 78: 5018–5022.
61. J. Lee, S.-H. Lee, H.J. Seo, E.-J. Son, S.H. Lee, M.E. Jung, M. Lee, H.-K. Han, J. Kim, J. Kang and J. Lee. 2010. Novel C-aryl glucoside SGLT2 inhibitors as potential antidiabetic agents: 1,3,4-Thiadiazolylmethylphenyl glucoside congeners. Bioorg. Med. Chem. 18: 2178–2194.
62. M.J. Kim, J. Lee, S.Y. Kang, S.-H. Lee, E.-J. Son, M.E. Jung, S.H. Lee, K.-S. Song, M. Lee, H.-K. Han, J. Kim and J. Lee. 2010. Novel C-aryl glucoside SGLT2 inhibitors as potential antidiabetic agents: Pyridazinylmethylphenyl glucoside congeners. Bioorg. Med. Chem. Lett. 20: 3420–3425.
63. K.-S. Song, S.H. Lee, M.J. Kim, H.J. Seo, J. Lee, S.-H. Lee, M.E. Jung, E.-J. Son, M.W. Lee, J. Kim and J. Lee. 2011. Synthesis and SAR of thiazolylmethylphenyl glucoside as novel C-aryl glucoside SGLT2 inhibitors. ACS Med. Chem. Lett. 2: 182–187.
64. T. Besson and V. Thiery. 2006. Microwave-assisted synthesis of sulfur and nitrogen-containing heterocycles. Top. Heterocycl. Chem. 1: 59–78.
65. O. Uchikawa, K. Fukatsu and T. Aono. 1994. Aminothiazole derivatives. I. A convenient synthesis of monocyclic and condensed 5-aminothiazole derivatives. J. Heterocycl. Chem. 31: 877–887.
66. T. Vojkovsky. 1995. Detection of secondary amines on solid phase. Pept. Res. 8: 236–237.
67. J.F. Sanz-Cervera, R. Blasco, J. Piera, M. Cynamon, I. Ibanez, M. Murguia and S. Fustero. 2009. Solution versus fluorous versus solid-phase synthesis of 2,5-disubstituted 1,3-azoles. Preliminary antibacterial activity studies. J. Org. Chem. 74: 8988–8996.
68. P. Wipf, C. Jenny and H. Heimgartner. 1987. 2,4-Bis(4-methylpheylthio)-1,3,2λ^5,4λ^5-dithiadiphosphetan-2,4-dithion: Ein neues reagens zur Schwefelung von N,N-disubstituierten amiden. Helv. Chim. Acta 70: 1001–1011.
69. D. Obrecht and H. Heimgartner. 1982. A convenient synthesis of 2-oxazolin-5-ones and related compounds via amide cyclization. Chimia 36: 78–81.
70. D. Obrecht, R. Prewo, J.H. Bieri and H. Heimgartner. 1982. 1,3-Dipolare cycloadditionen von 2-(benzonitrilio)-2-propanid mit 4,4-dimethyl-2-phenyl-2-thiazolin-5-thion und schwefelkohlenstoff. Helv. Chim. Acta 65: 1825–1836.
71. C. Jenny and H. Heimgartner. 1986. Synthese von 4,4-disubstituierten 1,3-thiazol-5(4H)-thionen. Helv. Chim. Acta 69: 374–388.
72. M.J. Thompson and B. Chen. 2008. Versatile assembly of 5-aminothiazoles based on the Ugi four-component coupling. Tetrahedron Lett. 49: 5324–5327.
73. M.J. Thompson and B. Chen. 2009. Ugi reactions with ammonia offer rapid access to a wide range of 5-aminothiazole and oxazole derivatives. J. Org. Chem. 74: 7084–7092.

74. G. Koopmanschap, E. Ruijter and V.A.O. Romano. 2014. Isocyanide-based multicomponent reactions towards cyclic constrained peptidomimetics. Beilstein J. Org. Chem. 10: 544–598.
75. M.C. Bagley, K.E. Bashford, C.L. Hesketh and C.J. Moody. 2000. Total synthesis of the thiopeptide promothiocin A. J. Am. Chem. Soc. 122: 3301–3313.
76. J.R. Davies, P.D. Kane and C.J. Moody. 2005. The diazo route to diazonamide A. Studies on the indole bis-oxazole fragment. J. Org. Chem. 70: 7305–7316.
77. A. Ford, H. Miel, A. Ring, C.N. Slattery, A.R. Maguire and M.A. McKervey. 2015. Modern organic synthesis with α-diazocarbonyl compounds. Chem. Rev. 115: 9981–10080.
78. B.S. Drach, I.Y. Dolgushina and A.V. Kirsanov. 1973. Reaction of omega-chlorine-omega-acylamino-acetophenones with thioacetamide. Zh. Org. Khim. 9: 414–419.
79. B.S. Drach, I.Y. Dolgushina and A.D. Sinitsa. 1974. Some cyclization reactions of ω-chloro-ω-acylamidoacetophenones. Khim. Geterotsikl. Soedin. 7: 928–931.
80. B.S. Drach, I.Y. Dolgushina and A.D. Sinitsa. 1975. Verwendung von omega-chlor-omega-acylaminoacetophenon zur synthese phosphorylierter oxazole. Zh. Obshch. Khim. 45: 1251–1255.
81. R.A. Cherkasov, G.A. Kutyrev and A.N. Pudovik. 1985. Tetrahedron report number 186. Organothiophosphorus reagents in organic synthesis. Tetrahedron 41: 2567–2624.
82. S.I. Zav'yalov, T.K. Budkova and M.N. Larionova. 1976. New method for conversion of N-acyl-α-aminoketones to substituted thiazoles. Bull. Acad. Sci. USSR Div. Chem. Sci. 25: 1353–1356.
83. T. Nishio and M. Ori. 2001. Thionation of ω-acylamino ketones with Lawesson's reagent: Convenient synthesis of 1,3-thiazoles and 4H-1,3-thiazines. Helv. Chim. Acta 84: 2347–2354.
84. A.G. Belyuga, V.S. Brovarets and B.S. Drach. 2004. Phosphorus pentasulfide and Lawesson reagent in synthesis of 1,3-thiazole-4-thiol derivatives. Russ. J. Gen. Chem. 74: 1418–1422.
85. R.A. Hughes, S.P. Thompson, L. Alcaraz and C.J. Moody. 2004. Total synthesis of the thiopeptide amythiamicin D. Chem. Commun. 8: 946–948.
86. C. Jenny and H. Heimgartner. 1989. Bildung von 5,6-dihydro-1,3(4H)-thiazin-4-carbonsäure-estern aus 4-allyl-1,3-thiazol-5(4H)-onen. Helv. Chim. Acta 72: 1639–1646.
87. C. Bengtsson. 2013. Synthesis of substituted ring-fused 2-pyridones and applications in chemical biology. PhD Thesis, Umea University.
88. R.A. Hughes, S.P. Thompson, L. Alcaraz and C.J. Moody. 2005. Total synthesis of the thiopeptide antibiotic amythiamicin D. J. Am. Chem. Soc. 127: 15644–15651.
89. J.R. Davies, P.D. Kane and C.J. Moody. 2004. N-H Insertion reactions of rhodium carbenoids. Part 5: A convenient route to 1,3-azoles. Tetrahedron 60: 3967–3977.
90. I. Thomsen, K. Clausen, S. Scheibye and S.O. Lawesson. 1984. Thiation with 2,4-bis(4-methoxyphenyl)-1,3,2,4-dithiaphosphetane 2,4-disulfide: N-Methylthiopyrrolidone. Org. Synth. 62: 158–164.
91. J.P. Freeman. 1990. John Wiley & Sons: New York, NY, 372.
92. M.H. Zheng, J.Y. Jin, W. Sun and C.H. Yan. 2006. A new series of fluorescent 5-methoxy-2-pyridylthiazoles with a pH-sensitive dual-emission. New J. Chem. 30: 1192–1196.
93. L.G. Lee, C.H. Chen and L.A. Chiu. 1986. Thiazole orange: A new dye for reiculocyte analysis. Cytometry 7: 508–517.
94. Z.X. Li, L.Y. Liao, W. Sun, C.H. Xu, C. Zhang, C.J. Fang and C.H. Yan. 2008. Re-configurable cascade circuit in a photo- and chemical-switchable fluorescent diarylethene derivative. J. Phys. Chem. C 112: 5190–5196.
95. X.C. Hu, S. Wei, Z. Chao, B.Y. Chun, F.C. Jie, L.W. Tao, H.Y. Yi and Y.C. Hua. 2009. Chemical approaches for mimicking logic functions within fluorescent MPT dyes. Sci. Chin. Ser. B: Chem. 52: 700–714.
96. M.S.J. Foreman and J.D. Woollins. 2000. Organo-P-S and P-Se heterocycles. J. Chem. Soc. Dalton Trans. 10: 1533–1543.
97. R. Markovic, A. Rasovic, M. Baranac, M. Stojanovic, P.J. Steel and S. Jovetic. 2004. Thionation of N-methyl- and N-unsubstituted thiazolidine enaminones. J. Serb. Chem. Soc. 69: 909–918.
98. R. Markovic, M. Baranac and S. Jovetic. 2003. A novel and efficient 4-oxothiazolidine-1,2-dithiole rearrangement induced by Lawesson's reagent. Tetrahedron Lett. 44: 7087–7090.

Chapter 3
Thiazole Synthesis by Thionation of C=O to C=S

3.1 Introduction

The chemistry of heterocycles is as logical as the chemistry of aromatic or aliphatic compounds. The study of heterocyclic structures is of great attention both from the theoretical and practical point of view. The heterocyclic compounds also play an important role in the construction and exploration of novel physiological/biologically active compounds. The five-membered aromatic compounds bearing three heteroatoms at the symmetrical positions have been studied due their important physiological activities [1–3].

Disappointingly, there is no straightforward mechanism for understanding the difference between two well-known thionating reagents, phosphorus pentasulfide, and Lawesson's reagent. On the other hand, although both reagents are broadly utilized in organic synthesis, considering the number of papers publishing every year, it looks though Lawesson's reagent is popular among the researchers. So conclusion is that Lawesson's reagent is better than phosphorus pentasulfide, specifically in terms of better yields. Whereas this view may change with the current developments, which indicate that the employment of HMDO (hexamethyldisiloxane) together with phosphorus pentasulfide affords better or analogous yields to those reported with Lawesson's reagent. This mixture is now known as "Curphey reagent" [4–6]. It has been claimed that this approach has the benefit of smoothly removing the reagent-derived side-products. The experimental and nuclear magnetic resonance analysis showed that during the reaction in starting phosphorus pentasulfide transforms the carbonyl groups into thiocarbonyls and, then, before the obtained reactive electrophilic polythiophosphates cause any side reactions, hexamethyldisiloxane serves as a scavenger for them, which provides higher yields because of the lesser side reactions [7].

The transformation of a carbonyl functional group into thiocarbonyl has a significant interest to synthetic organic researchers for several years. Two reagents, Lawesson's reagent and P_4S_{10}, are the most broadly utilized agents for these types of conversions as well as for the formation of broad spectrum of heterocycles bearing sulfur atom. On the other hand, Lawesson's reagent has been the most broadly utilized reagent since the starting of the last quarter of the twentieth century, and due to its significant uses in synthetic organic chemistry, it has frequently been studied [8–14].

3.2 Synthesis of Thiazoles

It was suspected that the protection of alcohol functionality in starting compound would enable the synthesis of thiazole ring. Therefore, the OH group in amide starting material was protected as a *t*-butyldimethylsilyl ether in 96% yield (Scheme 3.1); a *t*-butyldimethylsilyl-protecting group was chosen as this would allow the global deprotection of ultimate target molecule in a single step. As for the introduction of protecting group before the thiation step, it was expected that would also afford improved yields of thioamide over the unprotected thioamide. The synthesis of 2-*t*-butyldimethylsilyloxy thioacetamide was most effective in refluxing dioxane (67%); toluene at 65 °C afforded thioamide in 49% yield and only 14% yield under reflux conditions. Subsequently, addition of 1,3-dichloroacetone to thioamide afforded deprotected thiazole derivate exclusively in 50% yield. The 1.2 eq. pyridine was introduced in reaction mixture to neutralize the HCl produced during the reaction, which allowed the isolation of *t*-butyldimethylsilyl-protected thiazole derivate (10%) along with deprotected thiazole derivate (20%). An addition of an excess of 10 eq. pyridine resulted in complete decomposition of both the products [15].

The first overall formation of virenamides A and D was described using *N*-(*t*-butoxycarbonyl)-L-phenylalanyl-*N*-[(*S*)-(-)-1-(thiazole-2-yl)-2-methylpropyl]-L-valinamide, which was synthesized from (*S*)-(-)-*N*-(*t*-butoxycarbonyl)-1-(2-thiazolyl)-2-methylpropylamine, as a key intermediate. As described in Scheme 3.2, the cyclization of thiamide [16–20] with bromoacetaldehyde to synthesize the thiazole is the key step for the formation of virenamides A and D and is important as it is prone to epimerization at the α-stereogenic center. When bromoacetaldehyde affected the reaction, the deprotection of Boc group of (*S*)-(-)-*N*-(*t*-butoxycarbonyl)-1-(2-thiazolyl)-2-methylpropylamine was found because of the acidic conditions where the simultaneous release of hydrogen bromide occurred during the cyclization. The cyclization reaction mixture was reacted with

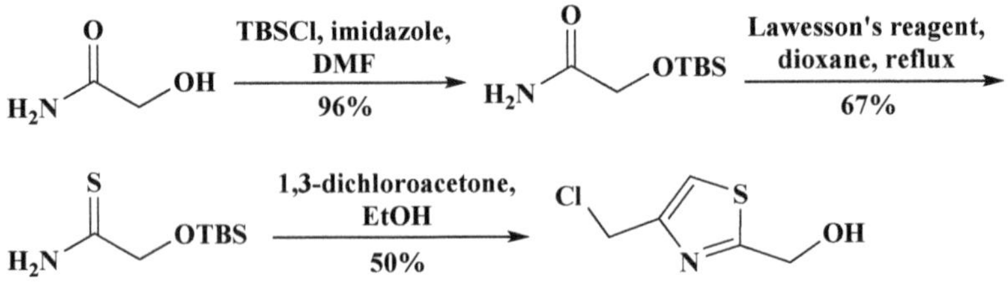

Scheme 3.1 Synthesis of thiazole

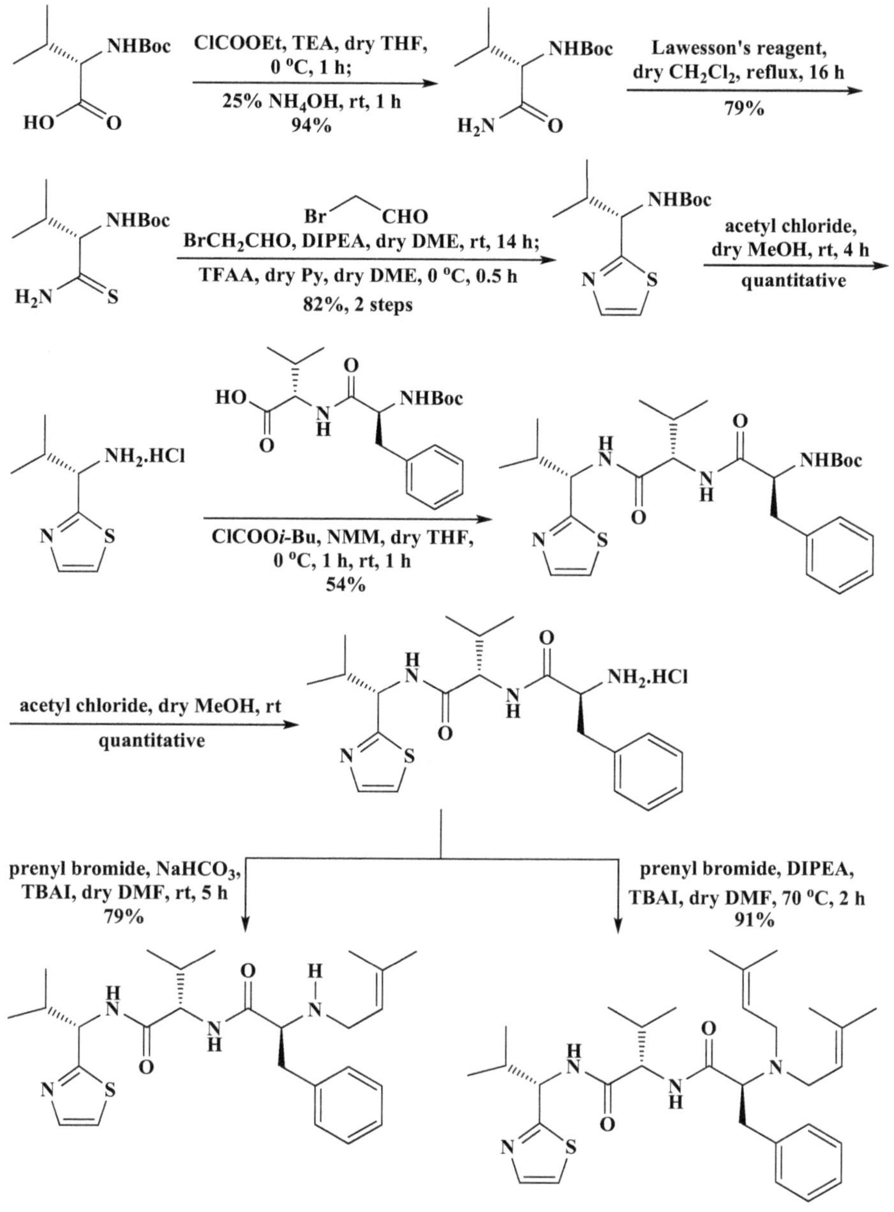

3.2 Synthesis of Thiazoles

Scheme 3.2 Synthesis of virenamide A and virenamide D

Boc$_2$O/triethylamine to provide the (S)-(-)-N-(t-butoxycarbonyl)-1-(2-thiazolyl)-2-methylpropylamine in moderate yield, but hydrogen bromide in the reaction mixture resulted in complete racemization of the product. Depending on what the literature showed [21–25], various inorganic and organic bases were tried to synthesize the thiazoline intermediate, which was utilized for next step without purification. The dehydration of thiazoline intermediate provided (S)-(-)-N-(t-butoxycarbonyl)-1-(2-thiazolyl)-2-methylpropylamine in different yields and enantiomeric excesses. Although inorganic base could afford high yield, it was observed that organic base afforded much better enantiomeric excess. The N,N-di-i-propylethylamine was verified to be the best acid trapper, which afforded 82% yield and 94.5% enantiomeric excess in accordance with chiral high-performance liquid chromatography studies. Initially, dipeptide was synthesized. The Boc-L-phenylalanine was treated with L-valine methyl ester to afford the dipeptide ester in 95% yield, which was saponificated with 1 M sodium hydroxide/tetrahydrofuran to provide the dipeptide [26, 27] in 94% yield. Further, the formation of tripeptide occurred in two steps involving removal of Boc group from (S)-(-)-N-(t-butoxycarbonyl)-1-(2-thiazolyl)-2-methylpropylamine to afford the amine hydrochloride. The coupling of amine hydrochloride with dipeptide in the presence of ClCOOi-Bu/N-methylmorpholine provided tripeptide in 54% yield [28]. Similarly, removal of Boc group from tripeptide with acetyl chloride in methanol afforded amine hydrochloride easily in almost quantitative yield, which was utilized for next step without further purification. Finally, double alkylation of amine with 4 eq. prenyl bromide in dimethylfuran at 70 °C for 2 h provided virenamide A in 91% yield, on the other hand, monoalkylation of amine hydrochloride with 2 eq. prenyl bromide in dimethylfuran at rt for 5 h easily afforded virenamide D in 79% yield [29].

Rink linker, developed for the formation of carboxamides, was utilized to afford a traceless pathway to thiazoles (Scheme 3.3) [30]. The acids were coupled to Rink linker resin, and the amide was transformed to thioamide utilizing LR. The cleavage utilizing haloketones afforded thiazoles [31].

The hydrolysis of nitrile with hydrogen peroxide/potassium carbonate/water afforded amide (93% yield), which was consequently transformed to thioamide (83% yield) utilizing LR. The cyclocondensation of thioamide with phenacyl bromide afforded thiazole in 84% yield. The reaction of lithiated thiazole with alkyl halides allowed the monoalkylation of methylene group to provide the final derivatives in 76–83% yield (Scheme 3.4) [32–34].

The radical cyclizations frequently involve addition of a carbon-centered radical to an unsaturation, often a C–C double or triple bond, but occasionally also a carbon heteroatom multiple bond [35–41]. In contrast, cyclization through intramolecular hemolytic substitution at carbon is hardly observed [42]. This is possibly due to challenges in forming C–C bonds with this kind of approach. The heteroatoms are well suited for this kind of chemistry, and researchers have taken benefit of homolytic intramolecular substitution for preparing new Se-based antioxidants [43, 44]. The hemolytic substitution could also be beneficial for the synthesis of thiazolines—again utilizing an aziridine as a starting compound. In the proposed formation of 4,5-dihydro-4-benzyl-2-phenyl thiazole (Scheme 3.5), it was believed that it would

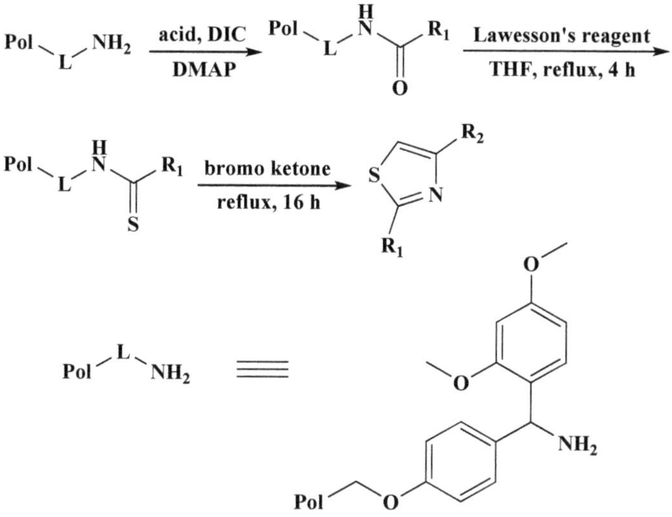

3.2 Synthesis of Thiazoles

Scheme 3.3 Synthesis of thiazoles

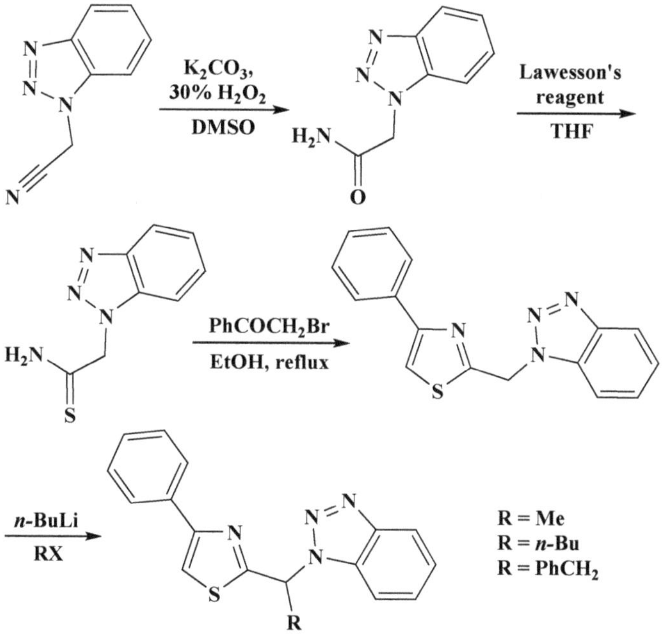

Scheme 3.4 Synthesis of thiazoles

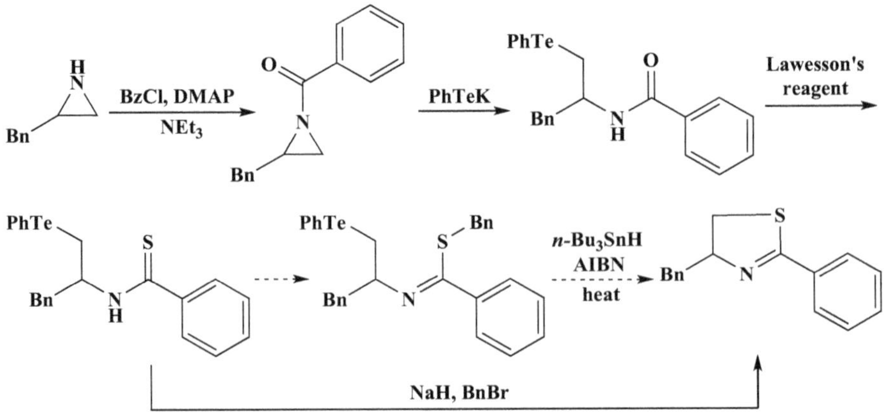

Scheme 3.5 Synthesis of 4,5-dihydro-4-benzyl-2-phenyl thiazole

be wise to utilize a telluride as a radical precursor. The reaction of a Sn radical with Te is much faster than its reaction with Se or S. The benzoylation worked well [45]. The benzoylated aziridine is activated and can thus be ring-opened with nucleophiles. The heating of Ph_2Te_2 in tetrahydrofuran with potassium hydride caused the reduction of Te–Te bond with the formation of potassium benzenetellurolate [46]. An aziridine ring-opening took place effectively to provide an amido telluride in 80% yield when added to KTePh in tetrahydrofuran/1,3-dimethyl-2-imidazolidinone at rt. The LR is particularly efficient for transforming amides to thioamides [9]. The telluride was thus refluxed with LR in tetrahydrofuran and the product passed through a silica column. The thioamide was isolated in 86% yield. The thioamide was deprotonated with sodium hydride and BnBr to install a good radical leaving group on sulfur. Strangely enough, no alkylation was found at sulfur. The product was a 1:1 mixture of unreacted starting compound and thiazoline. It appeared that BnBr reacted at Te and converted it into a good leaving group. Further, ring-closure took place through an ionic mechanism (Scheme 3.6). The idea was to add Sn radicals to a thioamide, thus transferring the radical to aziridine nitrogen. The aziridine would then ring-open rapidly to afford the more stable secondary radical. This radical in turn can underwent intramolecular homolytic substitution at sulfur for the synthesis of thiazoline.

The amide group was reacted with LR to synthesize the thioamide in 70% yield. The Cbz-protected L-proline thioamide [47, 48] cyclized with bromoketones using pyridine in EtOH (Scheme 3.7) to synthesize various thiazole ligands. In the case of a ligand with phenyl substituent, the reaction was initially carried out with phenacyl bromide (R = phenyl) under reflux conditions as reported by Pichota et al. [49]. Although the required product was formed in 63% yield, some racemization took place which provided cyclized product in only 90% enantiomeric excess. The product was formed in 65% yield and more than 99% enantiomeric excess when the reaction was performed at rt. Utilizing the similar process, the ligand substrates

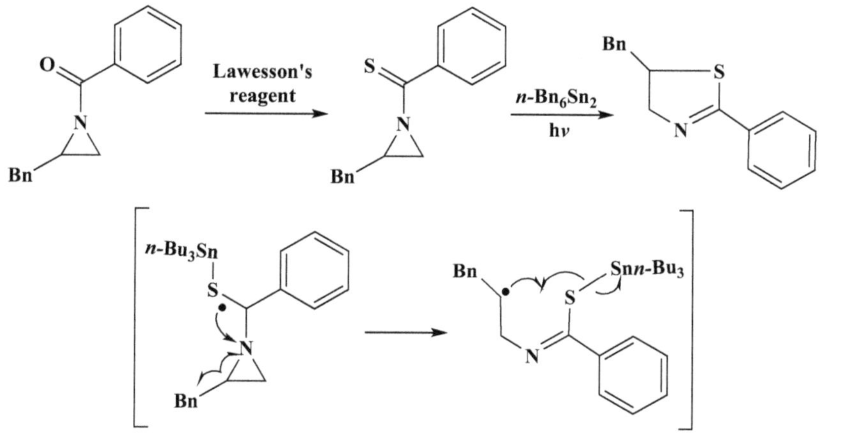

3.2 Synthesis of Thiazoles

Scheme 3.6 Synthesis of thiazoline

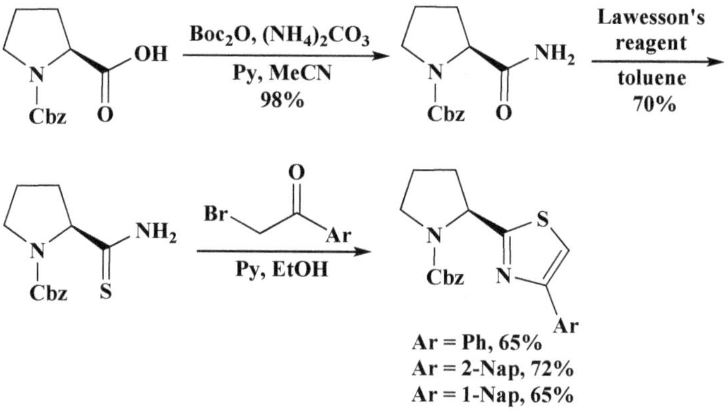

Scheme 3.7 Synthesis of thiazolopyrrolidines

containing bulky 2- and 1-naphthyl substituents were obtained in 72 and 65% yield and enantiomeric purity > 99% enantiomeric excess.

New pyrazolothiazole derivatives were obtained when thioamide was reacted with different α-haloketones in EtOH. The pyrazolothiazoles were synthesized from N-Boc-4-piperidone in five steps. The β-diketo ester was prepared from N-Boc-4 piperidone utilizing LiHMDS and diethyl oxalate [50]; further, it was condensed with CH_3NHNH_2 in EtOH to afford the pyrazole carboxylic acid ethyl ester [51], which upon reaction with liquid NH_3 in THF provided amide [52]. The amide was transformed into Boc-protected thioamide [53] utilizing LR in toluene. The Boc-protected thioamide was reacted with different α-bromo ketones under reflux conditions in EtOH to provide the Boc-cleaved pyrazolothiazoles (Scheme 3.8) [54, 55].

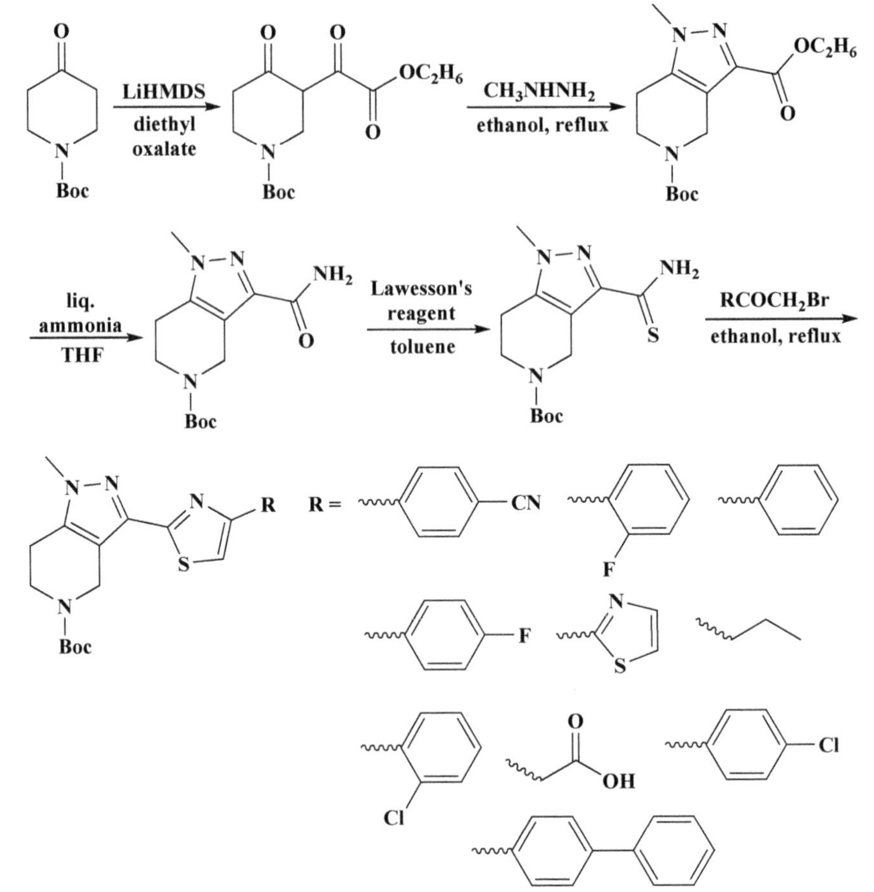

Scheme 3.8 Synthesis of pyrazolothiazoles

The indole-3-carbothioamides were synthesized from indoles via amides [56, 57]. The 7-chloro-1H-pyrrolo[2,3-c]pyridine was obtained when 2-chloro-3-nitropyridine was reacted with vinylmagnesium bromide under nitrogen atmosphere in tetrahydrofuran [58]. The 7-chloro-1H-pyrrolo[2,3-c]pyridine was transformed into N-methylated derivative, and both pyrrolo[2,3-c]pyridines were converted into 3-bromoacetylpyrrolo[2,3-c]pyridines in excellent yields (88–90%) through general acylating process [59]. The indolocarbothioamides were reacted with 3-bromoacetylpyrrolo[2,3-c]pyridines to afford the indolylthiazolylpyrrolo[2,3-c]pyridines in yields ranging from good to excellent (65–98%) (Scheme 3.9). The subsequent deprotection of N-t-butylcarboxylates utilizing TFA in dichloromethane under reflux, after neutralization with NaHCO$_3$, provided thiazoles in yields ranging from good to excellent (62–99%) [60, 61].

A modification of Hantzsch synthesis was utilized to synthesize the thiazole following the process reported by Schmidt et al. [62] (Scheme 3.10). The reaction of

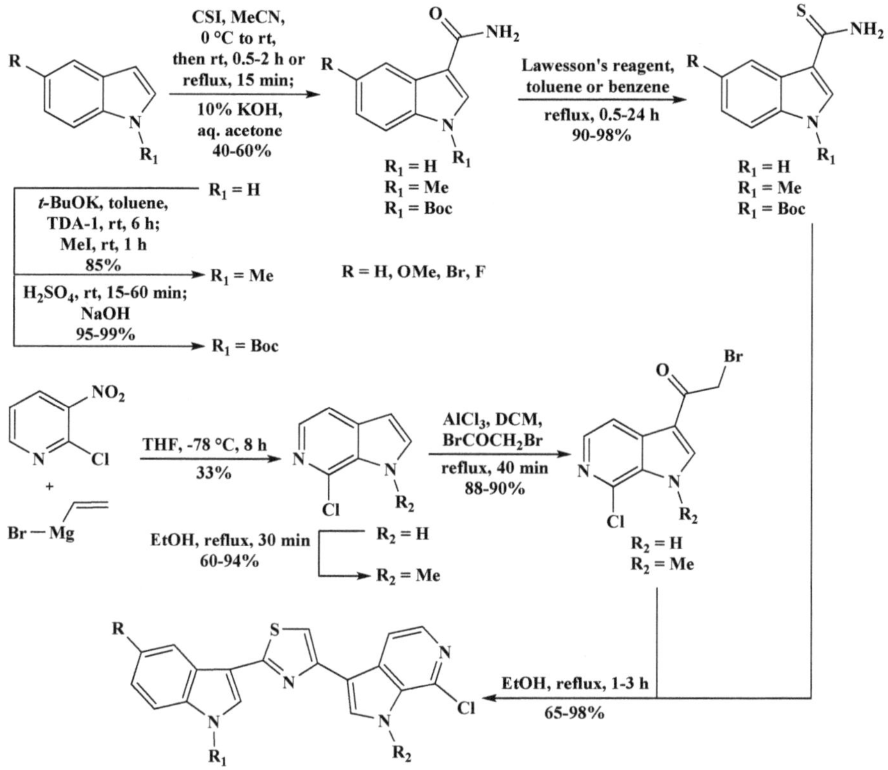

3.2 Synthesis of Thiazoles

Scheme 3.9 Synthesis of indolylthiazolylpyrrolo[2,3-c]pyridines

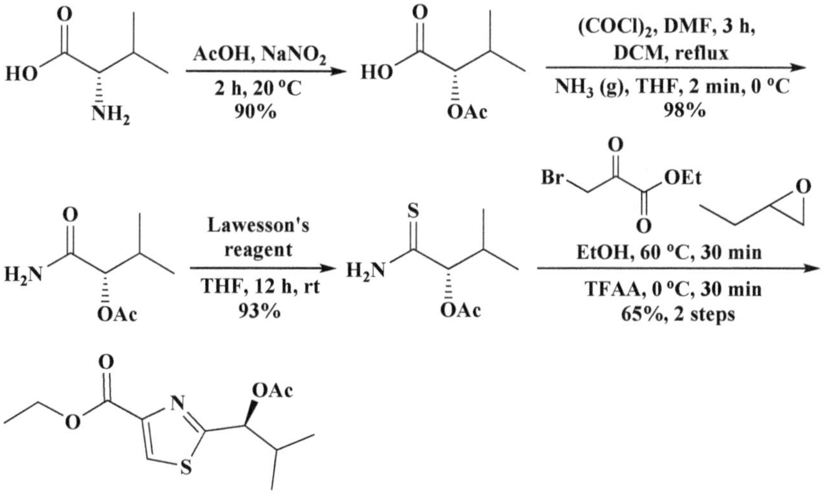

Scheme 3.10 Synthesis of thiazole

thioamide, easily prepared from (S)-valine, with ethyl bromopyruvate and after that dehydration of intermediate upon treatment with TFAA (trifluoroacetic anhydride) at 0 °C provided thiazole as white solid ($[\alpha]_D^{25} = -37$ (c 0.90, chloroform); $[\alpha]_D^{25} = -38.6$ (c 1.09, chloroform)) [63].

The thiazole ethylketones were synthesized in moderate yields by heating thioamides, formed when amides were reacted with LR in dry tetrahydrofuran, with 3-chloropentane-2,4-dione in absolute EtOH. The methyl ketones were gently heated with aminoguanidine hydrochloride in LiCl as a catalyst to provide the hydrazinecarboximidamides (Scheme 3.11) [64].

The second required fragment for the thiazoline-thiazole domain, thiazoline subunit, was prepared as described in Scheme 3.12. The N-Fmoc-L-threonine was reacted with D-serine methyl ester hydrochloride using hydroxybenzotriazole, 1-ethyl-3-(3-dimethylaminopropyl)carbodiimide and diisopropylethylamine to afford the dipeptide in 93% yield. The protection of free hydroxyls of dipeptide as triethylsilane ethers (triethylchlorosilane, imidazole, 81% yield) was followed by exposure of amide derivative to LR in refluxing benzene providing thioamide selectively in 83% yield. Further, the removal of N-Fmoc group (diethylamine, 83% yield) from this intermediate revealed free amine. In parallel, L-threonine derivative [65] was transformed to free amine through the action of trifluoroacetic acid, and then to azide, through a Cu-mediated (copper sulfate pentahydrate) diazo transfer reaction with trifluoromethanesulfonyl azide [66], in 84% total yield for the two steps. The resulting azido ester (the azide group serving as a masking device for the eventually needed amino group) was saponified, in quantitative yield, to its

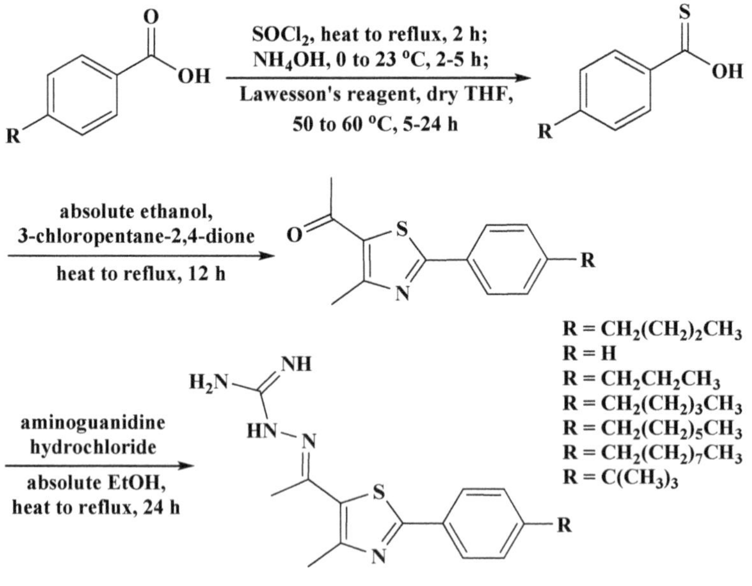

Scheme 3.11 Synthesis of thiazoles

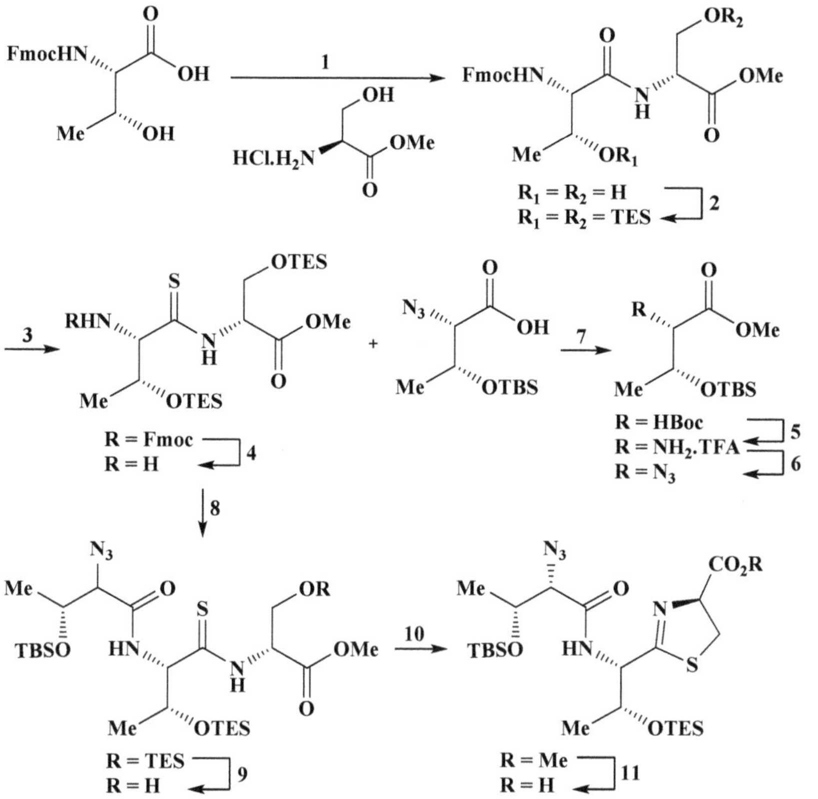

3.2 Synthesis of Thiazoles

Reagents and conditions: (1) 1 eq. D-Ser-OMe.HCl, 2 eq. i-Pr$_2$NEt, 1.2 eq. HOBt, 1.2 eq. EDC, CH$_2$Cl$_2$, 0 °C, 1 h; then 25 °C, 2 h, 93%, (2) 2.2 eq. TESCl, 3 eq. imidazole, DMF, 0 °C, 30 min; then 25 °C, 12 h, 81%, (3) 0.55 eq. Lawesson's reagent, benzene, reflux, 3 h, 83%, (4) 6.5 eq. Et$_2$NH, DMF, 0 °C, 30 min; then 25 °C, 30 min, 83%, (5) TFA/CH$_2$Cl$_2$ (1:1), 0 °C, 1.5 h (6) 3 eq. TfN$_3$, 4 eq. Et$_3$N, 0.05 eq. CuSO$_4$.5H$_2$O, MeOH/H$_2$O/CH$_2$Cl$_2$ (3.3:1:1), 25 °C, 1.5 h, 84%, two steps, (7) 3 eq. Me$_3$SnOH, 1,2-DCE, 80 °C, 3 h, 100%, (8) 1.1 eq. HATU, 1.1 eq. HOAt, 2 eq. i-Pr$_2$NEt, DMF, -20 °C, 20 min; then 0 °C, 20 min, 78%, (9) THF/AcOH/H$_2$O (10:3.3:1), 25 °C, 18 h, 60%, (10) 1.2 eq. DAST, CH$_2$Cl$_2$, -78 °C, 30 min, 88%, (11) 3 eq. Me$_3$SnOH, 1,2-dichloroethane, 80 °C, 1.5 h, 100%.

Scheme 3.12 Synthesis of thiazolines

carboxylic acid counterpart through the mild action of trimethyltin hydroxide, conditions that did not cause any epimerization at the azide group containing center. Two easily accessible building blocks were coupled together by the action of HATU, HOAt, and i-Pr$_2$Net to provide the tripeptide (78% yield) from which primary alcohol-bound triethylsilyl group was selectively removed upon exposure to acetic acid/tetrahydrofuran/water (10:3.3:1) at ambient temperature (60% yield, plus 17% recovered starting compound). An activation of hydroxy thioamide with DAST (diethylaminosulfur trifluoride) in dichloromethane at −78 °C afforded thiazoline in 88% yield, which was further reacted with trimethyltin hydroxide in dichloroethane at 80 °C to provide the carboxylic acid in quantitative yield, remarkably suffering no significant epimerization at any of its vulnerable centers. The selectivity and mildness of this trimethyltin hydroxide-based approach for hydrolyzing esters were indeed remarkable for thiazoline, which was highly sensitive and prone to epimerization at no less than three of its stereogenic sites [67–72].

The thiazole methyl ester was prepared via oxidation of thiazoline to thiazole. Since the amino acid-derived thiazolines have tendency for epimerization under either acidic or basic conditions, therefore, any cyclodehydration reaction leading to thiazoline and its further oxidation reaction to thiazole must be performed under near neutral conditions. Depending on earlier experience in the formation of amino acid-derived thiazoles and thiazolines [73], it was decided to use the Wipf process for the synthesis of Boc-L-proline-derived thiazole (Scheme 3.13). The coupling of L-serine and L-Boc-proline methyl ester provided dipeptide in 89% yield. The OH group in dipeptide was protected as its *t*-butyldimethylsilyl ether providing dipeptide, which was easily transformed into thioamide with LR [9]. Subsequent elimination of silicon-protecting group and treatment with Burgess reagent afforded thiazoline. An oxidation of thiazoline with active γ-manganese dioxide synthesized thiazole derivative in 44% total yield [74]. Following removal of the Boc group in thiazole derivative with trifluoroacetic acid in CH_2Cl_2, this residue was further condensed with Teoc-L-phenylalanine using hydroxybenzotriazole and 1-ethyl-3-(3-dimethylaminopropyl)carbodiimide to afford the tripeptide in 84% yield [75].

An appropriate activation of hydroxy group of a serine residue may also induce thiazoline synthesis. The Wipf thiazoline synthesis calls for cyclization of a thioamide derivative of serine using Burgess reagent. The utilization of other dehydration approaches (4-toluenesulfonyl chloride/triethylamine, thionyl chloride, and Mitsunobu conditions) led to an extensive epimerization of α-carbon of the thioamide [76, 77]. This approach needed a protection–deprotection sequence of serine hydroxy group, but it provided high overall yields and holds wide scope in terms of a thioamide segment (Scheme 3.14). Again, thiazolines aromatized smoothly upon treatment with manganese dioxide. This approach has been adapted for the formation of oxazoles, thiazines, and oxazines [78].

The formation of acid began from L-threonine. The oxazolidinone was obtained when free amino acid was reacted with triphosgene in dioxane [79–81] and subsequent esterification. The latter reaction was observed to proceed spontaneously when a methanolic solution of acid was allowed to stand overnight in small amount of 4-dimethylaminopyridine at rt. It was not clear whether esterification took place under autocatalysis conditions or whether it was promoted by traces of acidic contaminants or acyl chloride in acid. The ammonolysis of ester took place easily on dissolution into a methanolic solution of anhydrous ammonium (g) at rt with catalytic amount of 4-dimethylaminopyridine. The selective thionation of amide using oxazolidinone was completed with LR in refluxing benzene. An application of higher boiling solvents, like toluene or the customary xylenes, promoted the variable degrees of thionation of

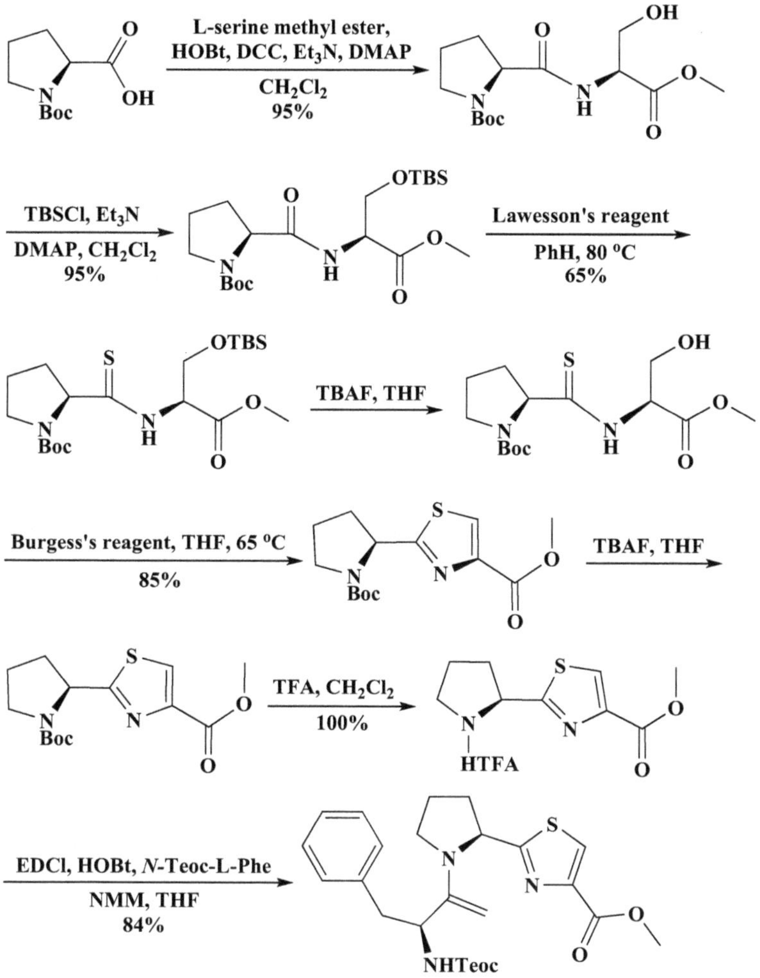

3.2 Synthesis of Thiazoles 75

Scheme 3.13 Synthesis of thiazole

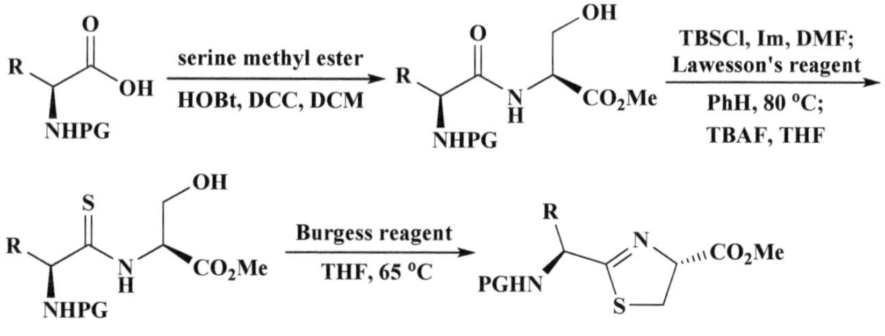

Scheme 3.14 Synthesis of dihydrothiazoles

oxazolidinone as well. Then, reaction of resultant compound with EBP in refluxing EtOH afforded ester, a common building block of the tripeptide functionality and the pyridine core, from L-threonine in a very satisfactory yield (75%) (Scheme 3.15).

A peptide approach synthesized heterocyclic compounds in the smallest number of steps (vs. a cross-coupling approach-theoretical calculation) and afforded high yields. The synthesis started with the construction of dipeptide (Scheme 3.16). This dipeptide was synthesized in 90% yield by coupling free amine H_2N-Ser(Bn)-OMe and free acid Boc-Thr(Ot-Bu)-OH utilizing TBTU and DIPEA in anhydrous dichloromethane. The removal of benzyl ether-protecting group was completed through hydrogenolysis employing palladium/carbon (10%) as a catalyst. A two-step process involving an intramolecular cyclization utilizing fluorinating agent diethylaminosulfur trifluoride (DAST) and potassium carbonate afforded oxazoline intermediate, which was oxidized utilizing bromochloroform and DBU to synthesize the ester in 85% total yield over three steps. The hydrolysis of ester with lithium hydroxide and subsequent coupling between the free acid and free amine H_2N-Ser(Bn)-OMe was carried out with TBTU and DIPEA. The hydrogenolysis of synthesized compound (in 95% yield) and subsequent transformation of the free serine into an oxazole utilizing DAST/BrCCl$_3$ afforded dioxazole (85% yield over three steps). The synthesis of thioamide was carried out with NH_4OH in MeOH followed by transformation of amide into thioamide utilizing LR (77% yield over two steps). A base-induced Hantzsch thiazole synthesis [23] was conducted with an excess of ethyl bromopyruvate and potassium bicarbonate to provide the intermediate thiazoline, which was dehydrated utilizing trifluoroacetic anhydride (TFAA) and pyridine

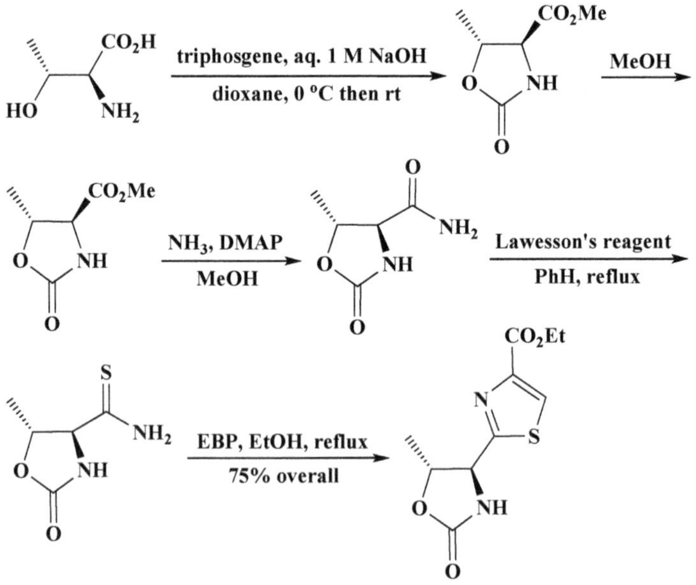

Scheme 3.15 Synthesis of oxazolidinothiazole

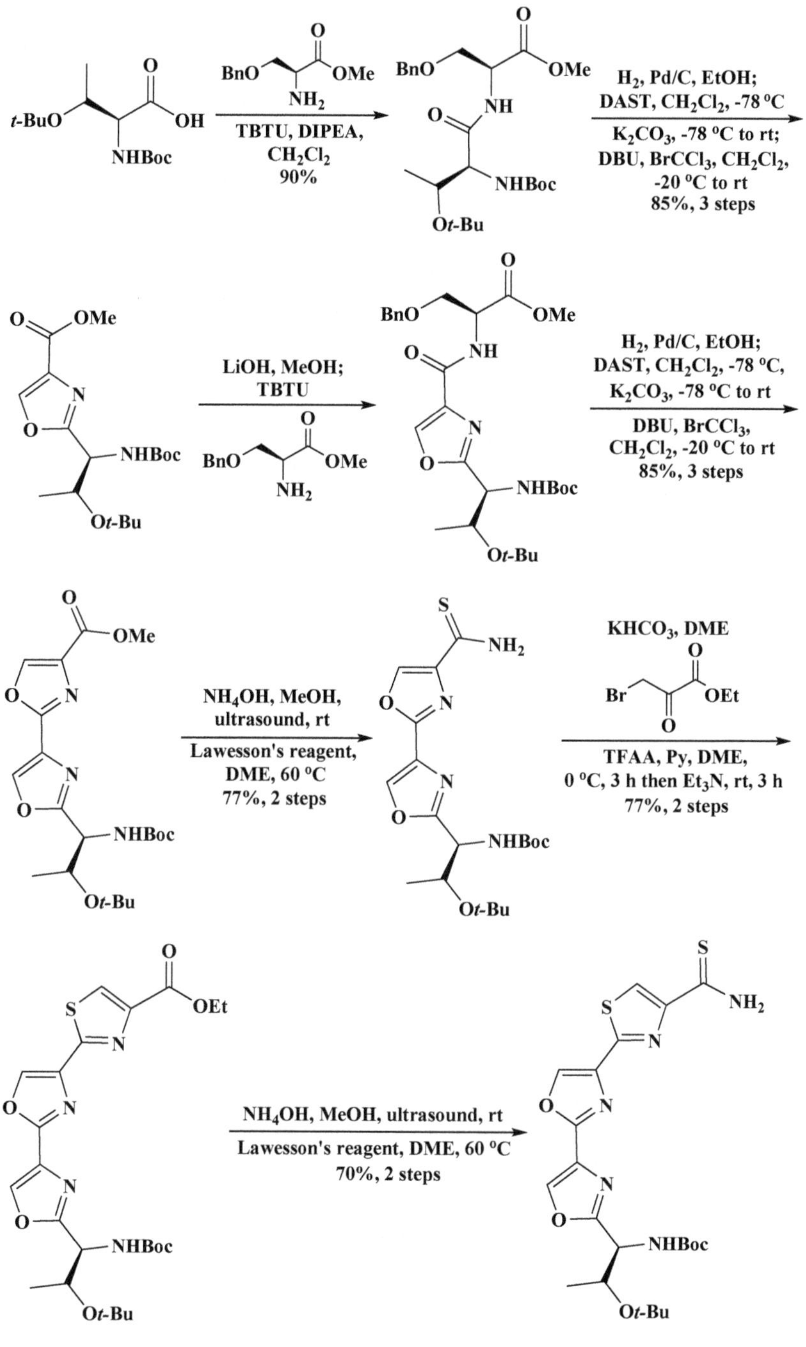

3.2 Synthesis of Thiazoles

Scheme 3.16 Synthesis of dioxazolothiazole

to afford the dioxazole–thiazole (77% yield over two steps). The dioxazole–thiazole thioamide was synthesized utilizing NH$_4$OH in MeOH and after that reaction with LR (70% yield over two steps) [82].

The formation of Ustat A analogue is shown in Scheme 3.17. Starting with the protection of free acid Boc-Thr(O*t*-Bu)-OH utilizing trimethylsilyl diazomethane (TMSD) in MeOH and further transformation into thioamide with NH$_4$OH and LR resulted in the synthesis of thioamide (60% yield over three steps). The Hantzsch thiazole synthesis took place by reacting thioamide, ethyl bromopyruvate, and potassium bicarbonate to afford the intermediate thiazoline, where subsequent dehydration was assisted with trifluoroacetic acid and pyridine at 0 °C to afford the thiazole (78% over two steps). The synthesis of thiazole thioamide was completed by a two-step process involving NH$_4$OH and LR (53% yield, two steps). The next two thiazole functionalities were installed by repeating the Hantzsch thiazole synthesis procedure, whereby the thioamide was reacted with ethyl bromopyruvate and potassium bicarbonate. Dehydration utilizing pyridine and TFAA and further reaction with sodium ethoxide in ethanol afforded dithiazole (97% yield over three steps). This method was repeated on dithiazole to synthesize the thioamide (73% yield over two steps) and further trithiazole (87% yield over three steps). The transformation of trithiazole into a thioamide utilizing NH$_4$OH and LR afforded Ustat A analogue (73% yield over two steps) [82].

The bis-thiazole was prepared by broadly utilized Hantzsch's approach. The amide was synthesized from acid (Scheme 3.18) using 2,2,2-trichloroethyl chloroformate/aqueous NH$_3$ in moderate yield. Further thionation of amide with LR afforded thioamide in good yield. Then, Hantzsch's reaction utilizing ethyl bromopyruvate provided bis-thiazole [83].

3.3 Synthesis of Benzothiazoles

The employment of TBAA and cyclization of thioformanilides by Dess–Martin periodinane afforded benzothiazole ring system [84] providing a new quinolone–benzothiazole hybrid molecule (Scheme 3.19). This synthesis was started by heating a mixture of *p*-toluidine and *t*-butylacetoacetate in dry xylene for a period of 5 min to provide a 1,3-dicarbonyl compound in 95% yield. The 1,3-dicarbonyl compound on reaction with 2-amino-5-chlorobenzophenone and ceric ammonium nitrate (10 mol%) in the presence of methanol at room temperature for 45 min afforded quinoline amide in 90% yield. The quinoline amide was further reacted with LR in dry toluene for a period of 1–2 h to provide the quinoline thioamide in 80% yield. The quinoline thioamide was finally cyclized by treating it with Dess–Martin periodinane in DCM solvent at rt for 15 min to provide the quinolone–benzothiazole hybrid molecule in 85% yield after column purification [85].

The nitration of *p*-toluic acid utilizing NH$_4$NO$_3$ with conc. sulfuric acid in dichloromethane at 0 °C afforded white solid 4-methyl-3-nitrobenzoic acid in 88% yield. The 4-methyl-3-nitrobenzoic acid was transformed to its acid chloride with

3.3 Synthesis of Benzothiazoles

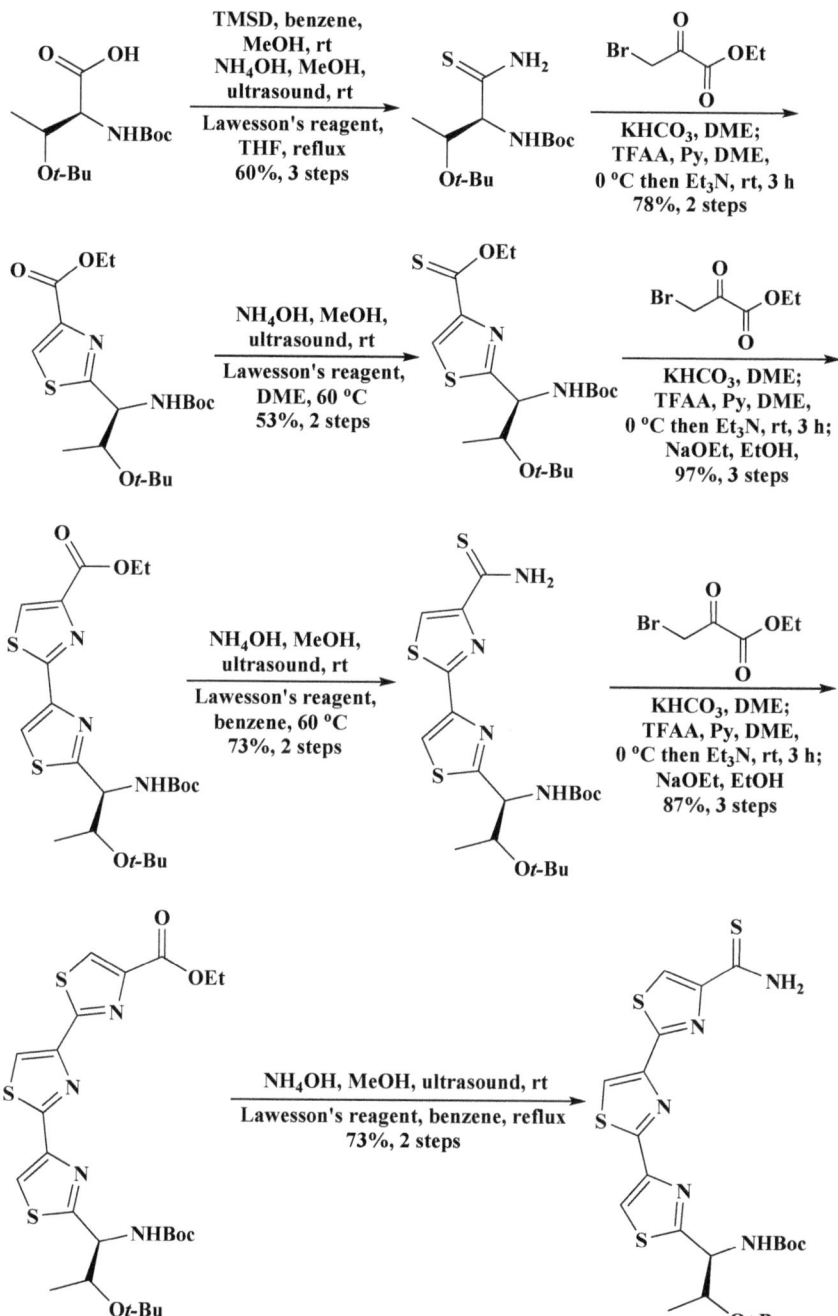

Scheme 3.17 Synthesis of tris-thiazole

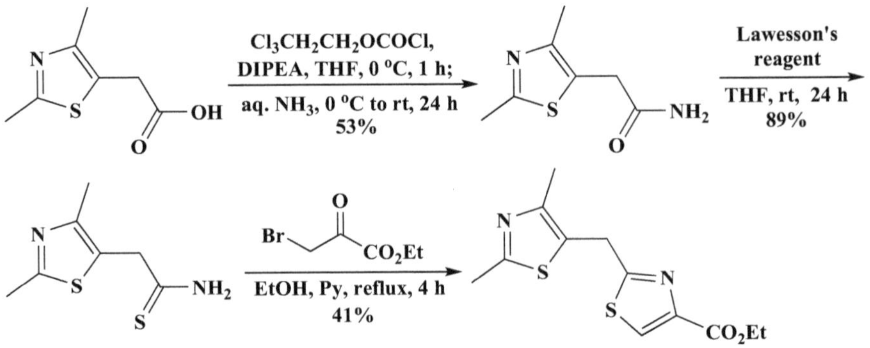

Scheme 3.18 Synthesis of bis-thiazole

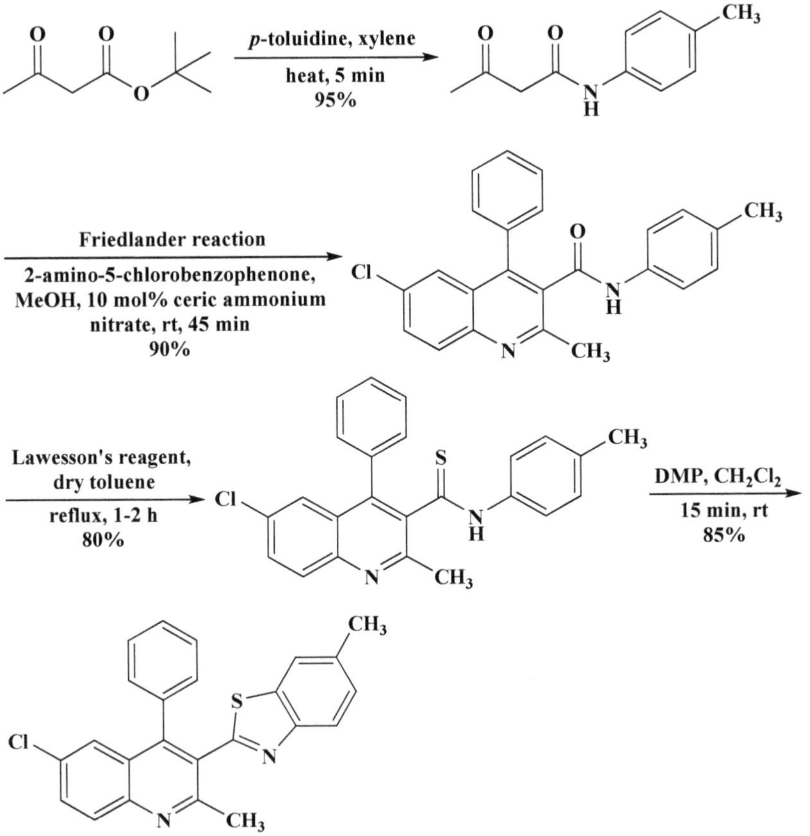

Scheme 3.19 Synthesis of quinolinobenzothiazole

3.3 Synthesis of Benzothiazoles

thionyl chloride followed by condensation with easily accessible *p*-anisidine in dry dichloromethane utilizing triethylamine to afford the light brown crystals of amide in 85% yield. The amide was reacted with LR in dry toluene under reflux conditions to afford the thioamide as pale yellow crystals in 8.5% yield [86]. An intermolecular free-radical cyclization of thioamide utilizing $K_3[Fe(CN)_6]$ in aqueous sodium hydroxide and EtOH at 100 °C for 120 min provided thiazole as a pale yellow solid in 88% yield (Scheme 3.20).

Different substituted anilines were reacted with KSCN in glacial CH_3COOH to afford the 2-substituted benzothiazoles. The 2-aryl-substituted benzothiazoles could be prepared by the reaction of substituted anilines with nitrobenzoyl chloride in pyridine under reflux and then reaction with LR and further cyclization of intermediate utilizing $K_3[Fe(CN)_6]$ (Scheme 3.21) [87–90].

The pyridinylbenzothiazole was synthesized utilizing Pd complexes. The reaction of 4-*t*-butylpicolinic acid, *N*,*N*′-dicyclohexylcarbodiimide, 4-*t*-butylaniline, 4-propanolamine, and dichloromethane gave 2-*t*-butylpyridine-2-carboxylic acid (4-*t*-butylphenyl)-amide, which on reaction with LR afforded carbothionic acid and finally cyclized to benzothiazole in the presence of $K_3[Fe(CN)_6]$ (Scheme 3.22) [91].

Wang et al. [92] prepared 4-fluorinated 2-phenylbenzothiazoles in multistep procedure including oxidation, benzylation, acid chloride synthesis, etc. (Scheme 3.23).

Wang et al. [92] prepared 4-fluorinated 2-phenylbenzothiazoles. The benzylation of 3-hydroxy-4-methoxybenzaldehyde through the protection of phenolic

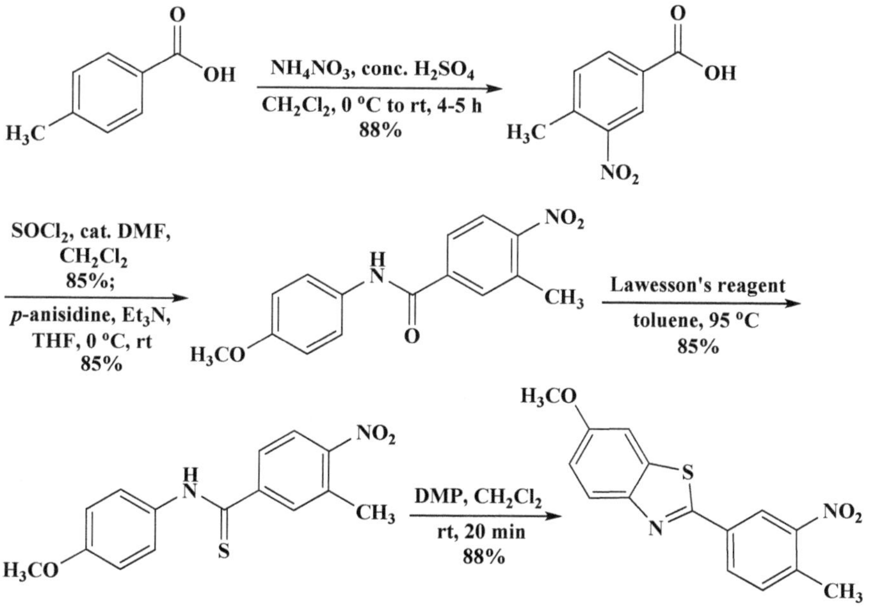

Scheme 3.20 Synthesis of benzothiazole

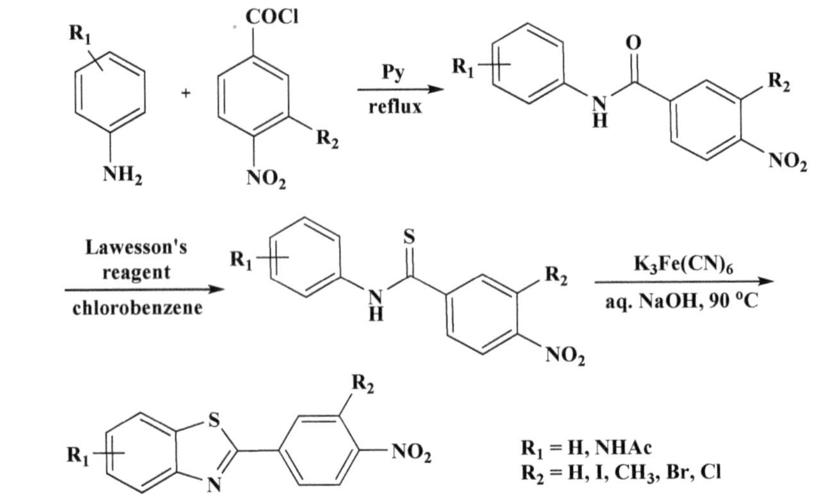

Scheme 3.21 Synthesis of 2-substituted benzothiazoles

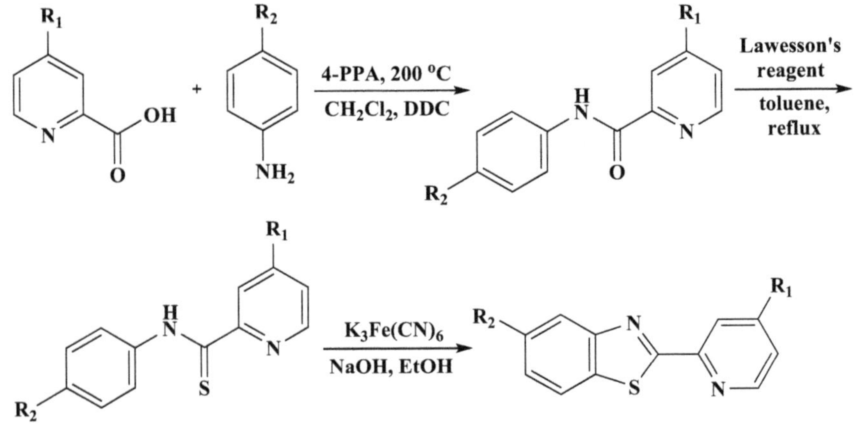

Scheme 3.22 Synthesis of pyridinylbenzothiazoles

OH group with benzyl bromide afforded 3-benzyloxy-4-methoxybenzaldehyde. Then, oxidation of 3-benzyloxy-4-methoxybenzaldehyde with sodium chlorite gave 3-benzloxy-4-methoxybenzoic acid, which was reacted with thionyl chloride to afford the 3-benzloxy-4-methoxybenzoyl chloride. The N-(2-fluorophenyl)-3,4-dimethoxybenzamide and 2-fluorobenzamides N-(2-fluorophenyl)-3-benzyloxy-4-methoxybenzamide were synthesized through the condensation of 3-benzloxy-4-methoxybenzoyl chloride or commercially accessible starting compound 3,4-dimethoxybenzoyl chloride with 2-fluoroaniline. The benzamides were transformed to their thiobenzamides N-(2-fluorophenyl)-3,4-dimethoxythiobenzamide

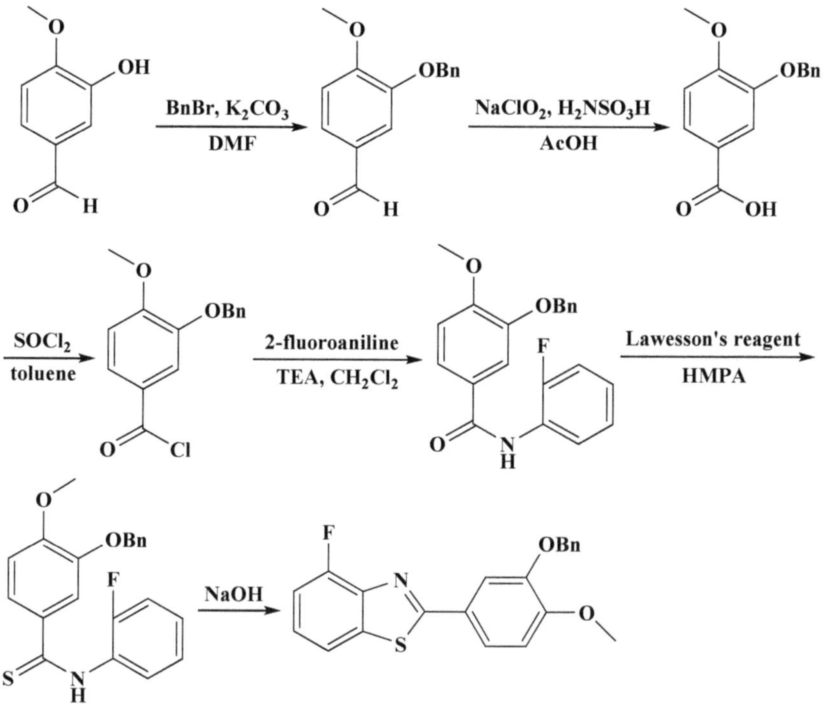

3.3 Synthesis of Benzothiazoles

Scheme 3.23 Synthesis of 4-fluorinated 2-phenylbenzothiazoles

and N-(2-fluorophenyl)-3-benzloxy-4-methoxythiobenzamide with LR in hexamethylphosphoramide (HMPA). The cyclization of thiobenzamides through a modified approach of Jacobson thioanilide radical cyclization utilizing $K_3[Fe(CN)_6]$ and aqueous NaOH gave 4-fluorobenzothiazoles through 4-fluoro-2-(3-benzloxy-4-methoxyphenyl)benzothiazole (Scheme 3.24) [91].

Serdons et al. [93, 94] described the formation of benzothiazole. The o-anisidine was reacted with p-nitrobenzoyl chloride to synthesize the N-2′-methoxyphenyl-4-nitrobenzamide. The amide was further transformed to thiobenzamide with LR (2,4-bis(4-methoxyphenyl)-1,3-dithia-2,4-diphosphetane-2,4-disulfide), which is a beneficial thiation reagent to replace the carbonyl oxygen atoms of amides, esters, and ketones with sulfur. The thiobenzamide was cyclized to 2-(4′-nitrophenyl)-benzothiazole in the presence of $K_3[Fe(CN)_6]$ (Scheme 3.25) [91].

The reaction of a series of heteroaromatic aldehydes with 2-aminothiophenol afforded 2-hetarylbenzothiazoles (Scheme 3.26). The 2-hetarylbenzothiazoles were studied in complex formation reactions using copper(II) and cobalt(II) perchlorate. According to X-ray data, coordination compounds have similar geometry to the geometry of the active site of superoxide dismutase. However, the resulting complex compounds had a very low solubility in H_2O. The benzothiazole ring must enter different hydrophilic substituents to synthesize the low molecular weight analogues

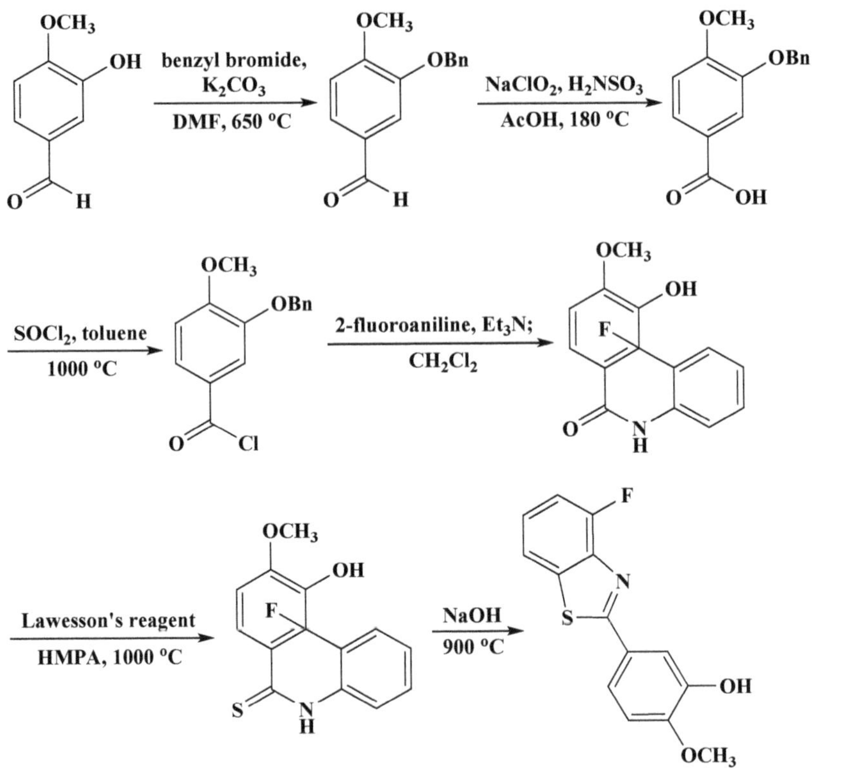

Scheme 3.24 Synthesis of 4-fluorobenzothiazole

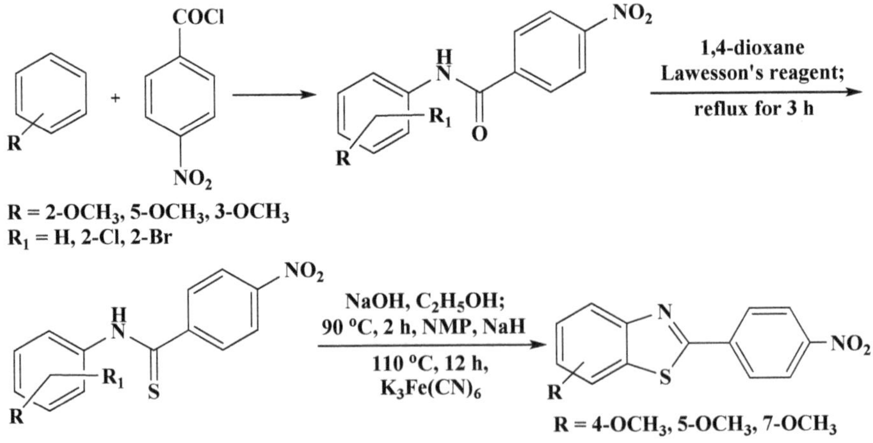

R = 2-OCH₃, 5-OCH₃, 3-OCH₃
R₁ = H, 2-Cl, 2-Br

R = 4-OCH₃, 5-OCH₃, 7-OCH₃

Scheme 3.25 Synthesis of 2-(4'-nitrophenyl)-benzothiazoles

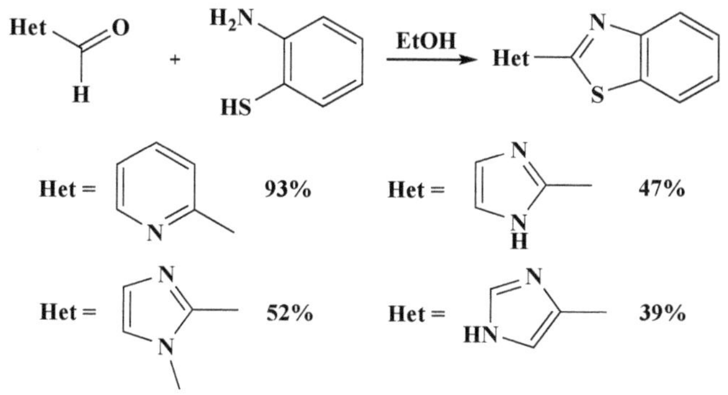

3.3 Synthesis of Benzothiazoles

Scheme 3.26 Synthesis of 2-hetarylbenzothiazoles

of SOD having good solubility in H_2O. Two methods have been proposed for the synthesis of 6-methoxy-2-pyridin-2-yl-1,3-benzothiazole (Scheme 3.27) [95].

The dienophile was obtained from easily accessible 2,5-dimethoxyaniline as shown in Scheme 3.28. The N-benzoylation of 2,5-dimethoxyaniline and subsequent thionation utilizing LR synthesized thiobenzamide. The reaction with sodium hydroxide and $K_3[Fe(CN)_6]$ in accordance to the approach of Jacobson [96] provided benzothiazole in 80% yield, and oxidative demethylation with CAN afforded dienophile [9, 97–99].

The reaction of 4-fluoroaniline with 4-nitrobenzoyl chloride in pyridine afforded amide. The amide was transformed to its thio derivative thioamide utilizing LR under reflux in toluene. Finally, the thio derivative was cyclized to thiazole in the presence of $K_3[Fe(CN)_6]$ by Jacobson's approach [96] followed by reduction and then coupling with bromopentanoyl chloride synthesized final compound (Scheme 3.29) [99].

Ten novel compounds (2-arylbenzothiazole) were prepared successfully in good yields through thioamide and 2,3-dichloro-5,6-dicyano-1,4-benzoquinone (DDQ) following the reaction sequence shown in Scheme 3.30. An intramolecular cyclization of thioformanilides occurred with 2,3-dichloro-5,6-dicyano-1,4-benzoquinone (DDQ) without metal catalyst in a stirred solution of thioamide in CH_2Cl_2 at rt. DDQ is a well-known oxidizing agent and has been proved to be a versatile reagent for different organic conversions involving the deprotection of functional groups, cleavage of linker molecules from solid supports, incorporation of unsaturation, and potential uses for the formation of C–C and C-heteroatom bonds. The thioamide compound can exist as thioiminol A, which reacted with 2,3-dichloro-5,6-dicyano-1,4-benzoquinone (DDQ) to generate the sulfanyl radical. Subsequently, 1,5-homolytic radical cyclization followed by aromatization of radical intermediate afforded 2-arylbenzothiazole [100].

The thiobenzamide was easily obtained in 50% yield from commercially accessible 2,4,5-trimethoxybenzaldehyde (Scheme 3.31). An addition of 2,4,5-trimethoxybenzaldehyde to a cold (−5 °C) solution of 50% aqueous HNO_3 resulted in *ipso* nitration [101, 102] to synthesize the nitrobenzene. Subsequent reduction and

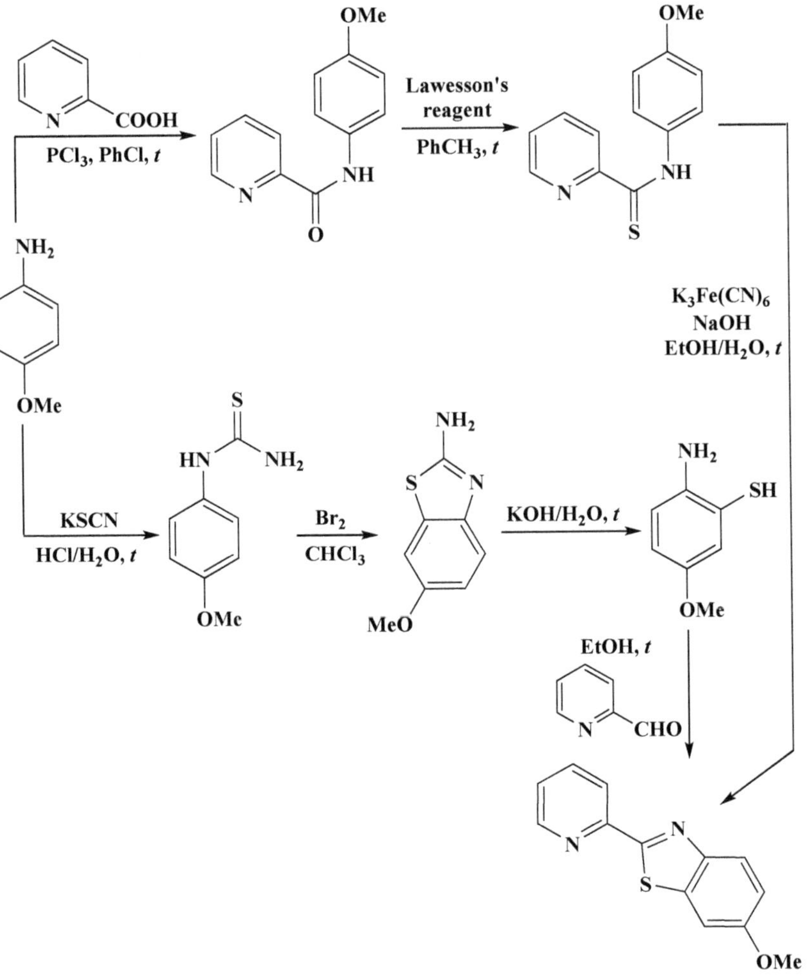

Scheme 3.27 Synthesis of 6-methoxy-2-pyridin-2-yl-1,3-benzothiazole

after that condensation with benzoyl chloride and thionation utilizing LR synthesized thiobenzamide. The reaction of thiobenzamide with 1.5 M sodium hydroxide followed by 20% potassium ferricyanide at rt for 1 d, however, did not synthesize the desired benzothiazole, but rather, afforded benzothiazole [103], the product of *ipso* substitution of OMe group. Not only an unexpected product was formed, cyclization also took place in low yield (15%) [104].

The AIBN-induced cyclization of aminothiobenzamides took place in good yield with the replacement of *o*-OMe group, on the other hand, attempts at a similar reaction utilizing thiobenzamides failed, and only starting compound was recovered (thiobenzamides were synthesized from known benzamide which was itself easily accessible by Schotten–Baumann reaction of *o*-anisidine). The AIBN-induced cyclization of

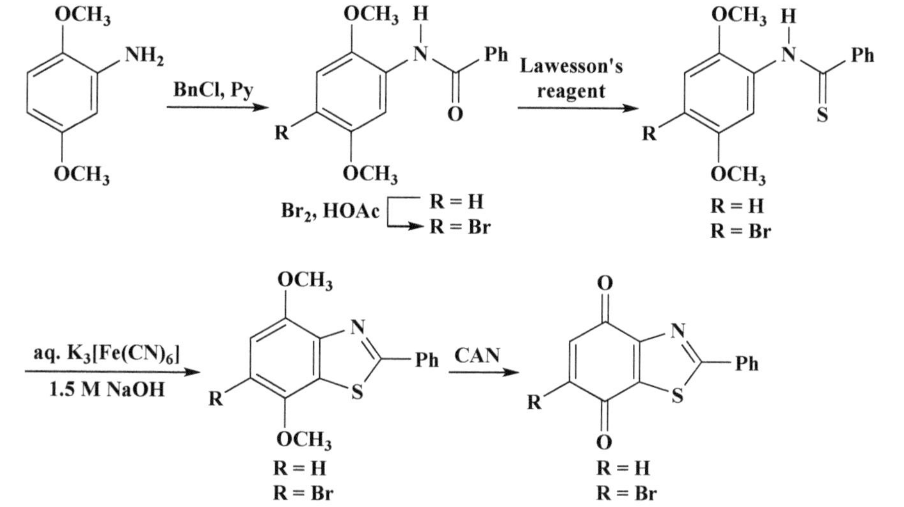

3.3 Synthesis of Benzothiazoles

Scheme 3.28 Synthesis of benzothiazolediones

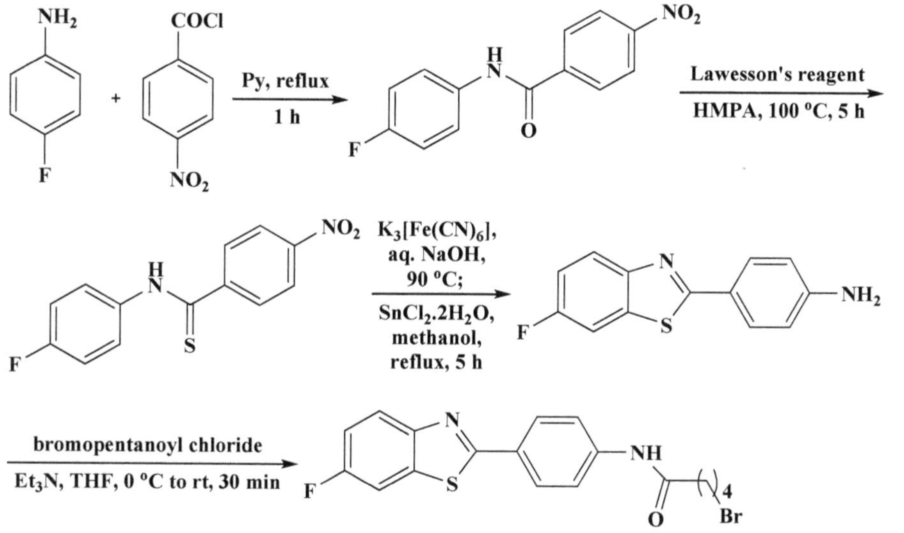

Scheme 3.29 Synthesis of benzothiazole

o-methoxythiobenzamides needed a benzene ring with high electron density. The AIBN-induced cyclization completed with the *ipso* substitution of OMe group when there are two or more electron-releasing groups on the primary ring. There was no reaction when there was only one ERG on the primary ring or two ERGs and one EWG. The Jacobson cyclization [96] of *o*-methoxythiobenzamides having one

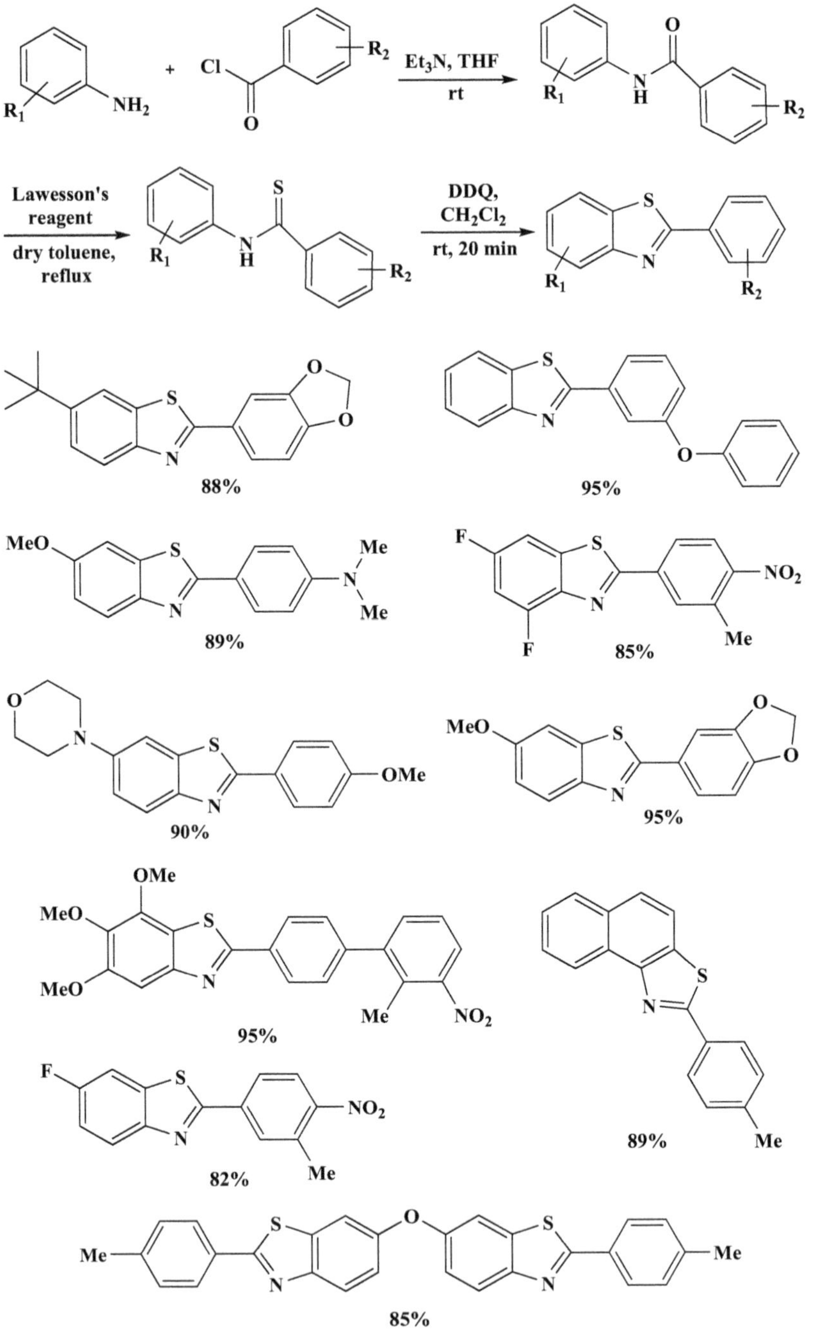

Scheme 3.30 Synthesis of 2-arylbenzothiazoles

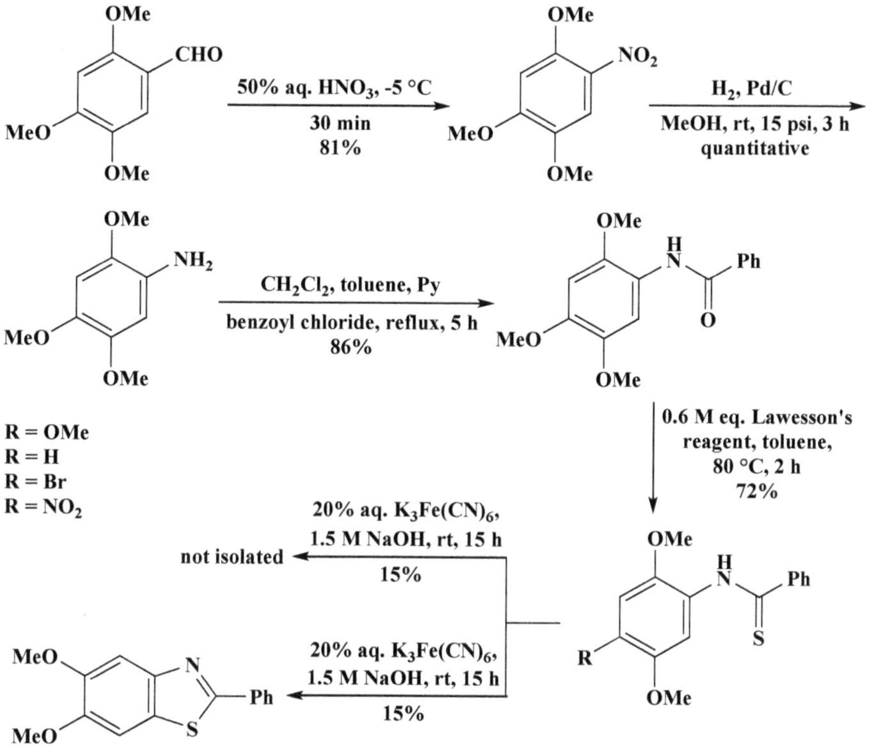

3.3 Synthesis of Benzothiazoles

Scheme 3.31 Synthesis of 5,6-dimethoxy-2-phenylbenzothiazole

or two ERGs on the primary ring took place with the replacement of *o*-hydrogen. When there were three ERGs on the primary ring, however, cyclization took place in very low yield, and with the *ipso* substitution of *o*-OMe group (Scheme 3.32) [99, 103–105].

An alternate method (Scheme 3.33) was developed to construct the benzothiazole core in the PIB compound [106]. Basically, commercially accessible 4-methoxyaniline was reacted with *p*-nitrobenzoyl chloride in pyridine. The benzamides were then transformed into thiobenzamide by reaction with LR [2,4-bis(4-methoxyphenyl)-1,3-dithia-2,4-diphosphetane 2,3-disulfide] as a thiation reagent [9]. The thiobenzamide was cyclized to its aryl benzothiazole through a Jacobson synthesis [96] employing oxidizing agent $K_3[Fe(CN)_6]$ in aqueous NaOH. The *O*-methyl group in aryl benzothiazole was substituted by more acid labile protective group (MOM), followed by reduction of NO_2 group to an amine with tin(II) chloride. This time-consuming protective group chemistry was required to avoid the *O*-methylation when MeI was utilized as a labeling agent. The KOH proved to be superior to K_2CO_3 for promoting the *N*-methylation of protected substrates amino benzothiazole. At the end of the synthesis, PIB was obtained in low RCY (15%) [107].

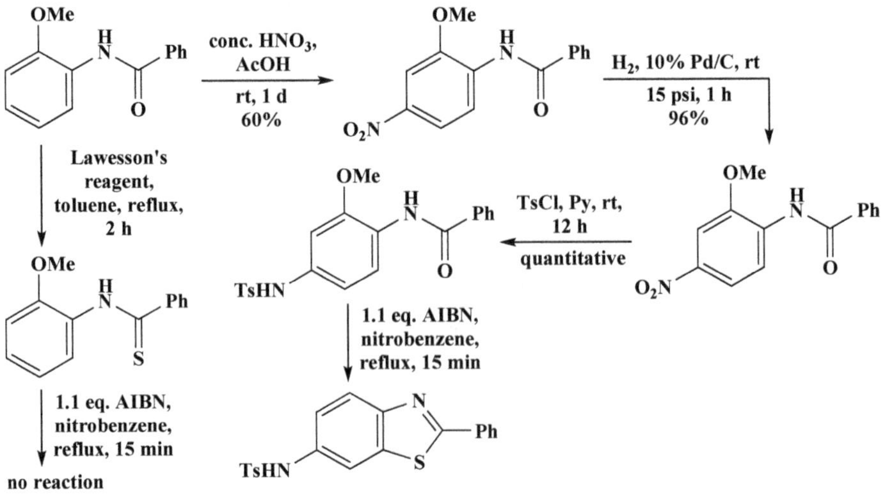

Scheme 3.32 Synthesis of benzothiazole

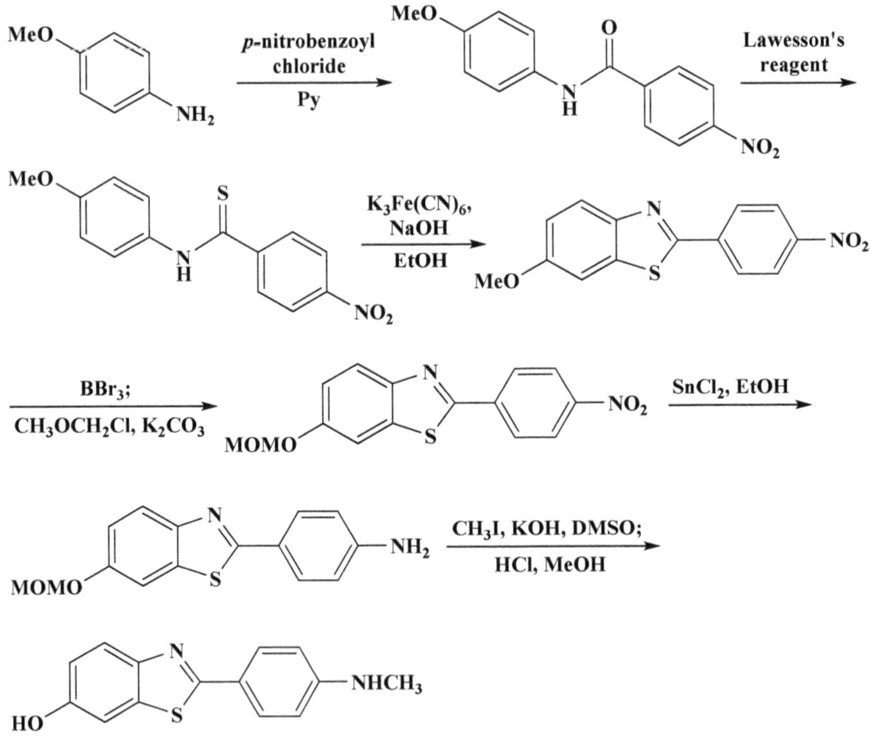

Scheme 3.33 Synthesis of 2-(4-(methylamino)phenyl)benzothiazol-6-ol

3.3 Synthesis of Benzothiazoles

It is well established that the progress of cascade reactions for the efficient formation of small molecules is an important pursuit in combinatorial chemistry in terms of operational simplicity and assembly efficiency [108–132]. The benzamide could be smoothly converted into benzothioamide using LR [133–135]. In the meantime, the elaboration of heterocyclic compounds via Cu-mediated coupling reactions is well developed [136–147]. The ease of access of benzothioamide from benzamide and the development of Cu-mediated cross-coupling reactions prompted examination of the cascade one-pot reaction starting from 2-haloaniline. The 2-haloaniline was treated with acid chloride to provide the N-(2-iodophenyl)benzamide, which was further transformed to N-(2-iodophenyl)-benzothioamide with LR. Subsequently, benzothiazole was obtained by Cu(I)-mediated intramolecular carbon–sulfur coupling of N-(2-iodophenyl)-benzothioamide (Scheme 3.34) [148].

A series of Schiff's base of many benzothiazole derivatives was prepared. The p-nitro benzothiazole carboxylic acid was prepared by Jacobson synthesis (Scheme 3.35) [96]. It was further reduced to p-amino benzothiazole carboxylic acid with NH_4Cl and Fe metal. The resulting product was then condensed with different aromatic or heterocyclic aldehydes utilizing conc. H_2SO_4 as a catalyst and EtOH as a solvent to afford various Schiff bases [149].

A parent benzothiazole molecule was synthesized by Jacobson's approach [96]. The benzothiazole was reacted with different aromatic aldehydes to afford the Schiff bases followed by esterification of carboxyl group utilizing different alcohols. In Scheme 3.36, the parent benzothiazole molecule was constructed by a Jacobson's approach [96] utilizing LR, the obtained product was utilized for the formation of different benzothiazole-6-carboxylate derivatives [150].

All the benzothiazole derivatives synthesized by this method have considerable anti-inflammatory properties. A series of new benzothiazole derivatives was prepared, which are used as sulfa drugs. The benzothiazole derivatives like substituted 2-benzylbenzo[d]thiazole-6-sulfonamides were obtained from substituted benzyl bromide (Scheme 3.37) [151].

The 4-iodoaniline was nearly quantitatively transformed to N-(4-iodophenyl)acetamide with acetic acid anhydride. The replacement of oxygen by sulfur, to afford the thioacetamide, was investigated utilizing two diverse approaches, either by reaction with LR under MWI (70% yield) [134] or by heating under reflux with phosphorus pentasulfide and aluminum oxide as a solid support (80% yield) [152]. The latter approach was found to be more efficient, not only because of the higher yields but also due to the cheap reagent, simple reaction conditions, and a cleaner product. The reaction with LR resulted in the synthesis of side-products derived from the reagent itself, which could not be removed smoothly. The 6-iodo-2-methylbenzothiazole was synthesized from thioacetamide by Jacobson's cyclization [96] utilizing $K_3[Fe(CN)_6]$ in an aqueous solution of sodium hydroxide (60% yield). The synthetic pathway to 6-iodo-2-methylbenzothiazole given here offered slight enhanced yield and allowed the synthesis on a larger scale. Finally, the last step involved a MW-assisted oxidation of 2-methyl group of 6-iodo-2-methylbenzothiazole with SeO_2. It was observed that 6-iodo-2-methylbenzothiazole was largely immune to the action of selenium dioxide in general solvents like

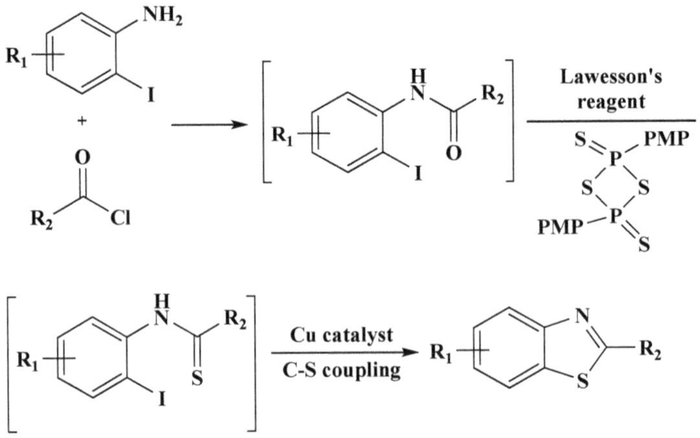

entry	R_1	R_2	yield (%)
1	H	C_6H_5	87
2	H	4-MeOC$_6$H$_4$	80
3	H	3-MeC$_6$H$_4$	87
4	H	2-MeOC$_6$H$_4$	82
5	H	3,5-(CF$_3$)$_2$C$_6$H$_3$	75
6	H	2-furyl	81
7	H	Me	66
8	4-CH$_3$	C_6H_5	93
9	4-CH$_3$	4-MeOC$_6$H$_4$	83
10	4-CH$_3$	3-MeC$_6$H$_4$	91
11	4-CH$_3$	3,5-(CF$_3$)$_2$C$_6$H$_3$	90
12	4-CH$_3$	2-furyl	90
13	4-F	C_6H_5	92
14	4-F	4-MeOC$_6$H$_4$	95
15	4-F	3-MeC$_6$H$_4$	91
16	4-F	3,5-(CF$_3$)$_2$C$_6$H$_3$	75
17	4-F	2-furyl	84
18	4-CF$_3$	C_6H_5	87
19	4-CF$_3$	4-MeOC$_6$H$_4$	93
20	4-CF$_3$	3,5-(CF$_3$)$_2$C$_6$H$_3$	77

Scheme 3.34 Synthesis of benzothiazoles

EtOH, dioxane, or a mixture of dioxane/H$_2$O under conventional heating or MWI. However, oxidation by selenium dioxide was made successful with a combination of glacial CH$_3$COOH as a solvent and MWI to provide the 6-iodobenzothiazole-2-carbaldehyde in 55% yield and a short period of time (20 min). This method was very beneficial and employed to other 2-methyl-substituted benzothiazoles. Nevertheless, the methyl oxidation of similar 6-(N,N-dimethylamino)-2-methylbenzothiazole [153] with selenium dioxide took place only in nonpolar dioxane with a slightly lower yield (25%). The 2-bromobenzothiazole-6-carbaldehyde, a counterpart of 6-iodobenzothiazole-2-carbaldehyde, was prepared as shown in Scheme 3.38 [154, 155].

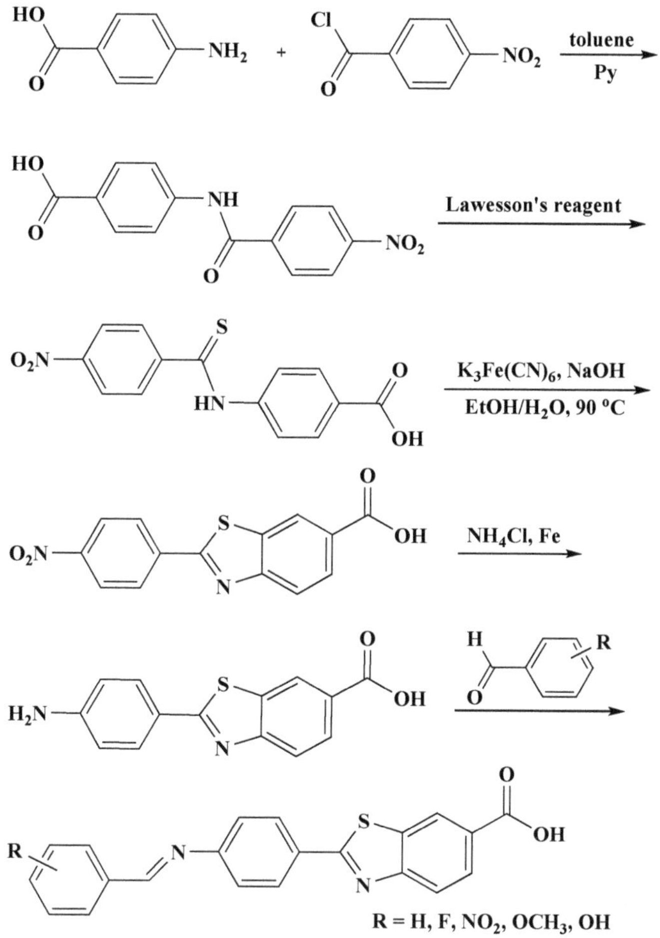

3.3 Synthesis of Benzothiazoles

Scheme 3.35 Synthesis of benzothiazoles

Three compounds were prepared by this route (Scheme 3.39) [156]. During the formation of 4-hydroxy-2-(4′-aminophenyl)-1,3-benzothiazole, ring-closure of benzothiazole not afforded two isomers and additional separation was not needed. The *o*-anisidine was reacted with *p*-nitrobenzoyl chloride to give the *N*-2′-methoxyphenyl-4-nitrobenzamide. The amide was further transformed to thiobenzamide utilizing LR (2,4-bis(4-methoxyphenyl)-1,3-dithia-2,4-diphosphetane-2,4-disulfide), which is a beneficial thiation reagent to substitute the oxygen atoms of carbonyl of ketones, amides, and esters with sulfur. The thiobenzamide was cyclized to 2-(4′-nitrophenyl)-benzothiazole using $K_3[Fe(CN)_6]$. The NO_2 group was reduced to an amine group with tin chloride, and then methyl ether was demethylated utilizing boron tribromide in CH_2Cl_2 at 70 °C to afford

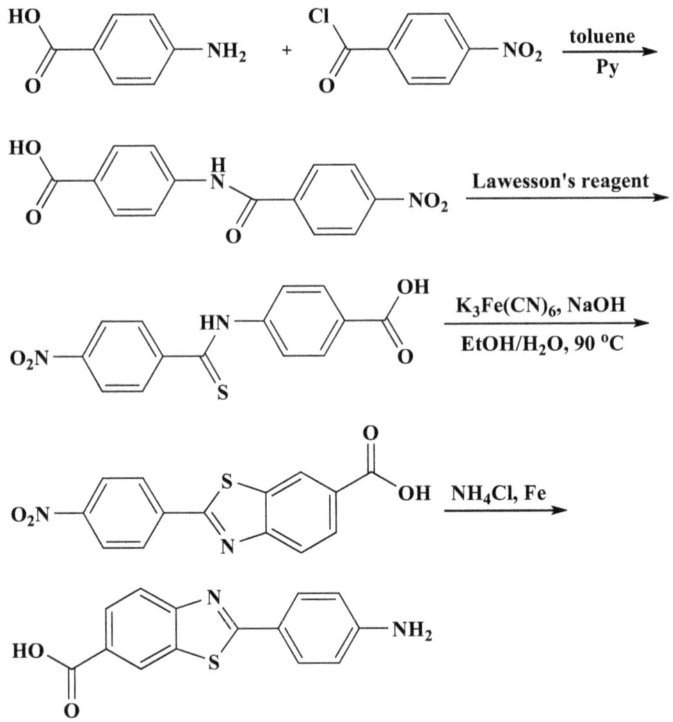

Scheme 3.36 Synthesis of 2-(4-aminophenyl)benzothiazole-6-carboxylic acid

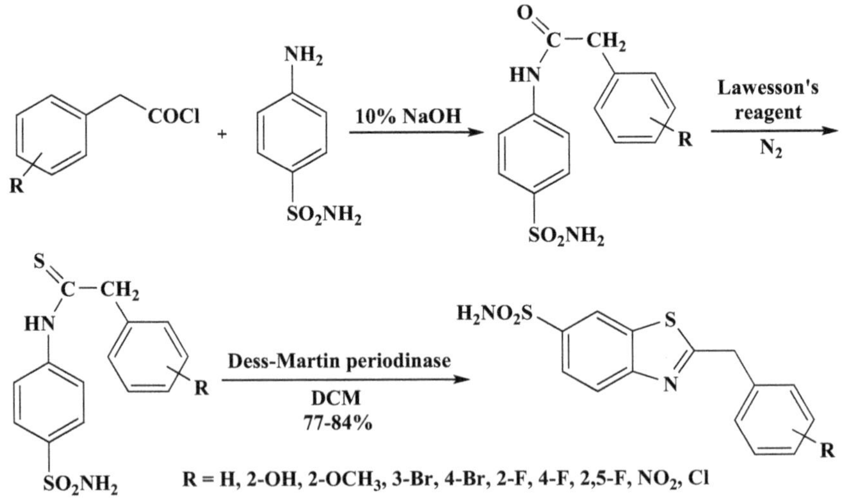

Scheme 3.37 Synthesis of benzothiazoles

R = H, 2-OH, 2-OCH$_3$, 3-Br, 4-Br, 2-F, 4-F, 2,5-F, NO$_2$, Cl

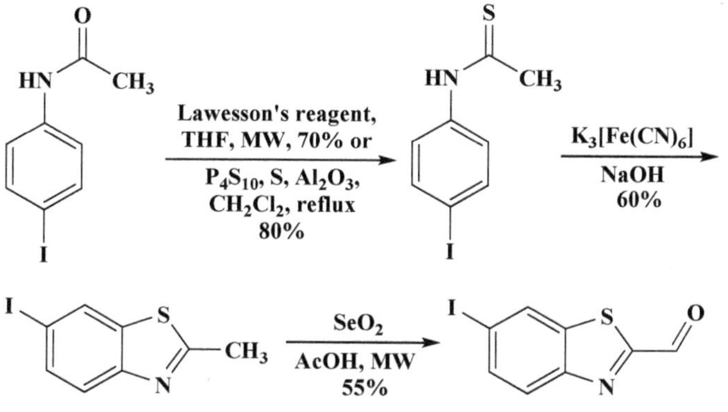

3.3 Synthesis of Benzothiazoles

Scheme 3.38 Synthesis of 2-iodobenzothiazole-6-carbaldehyde

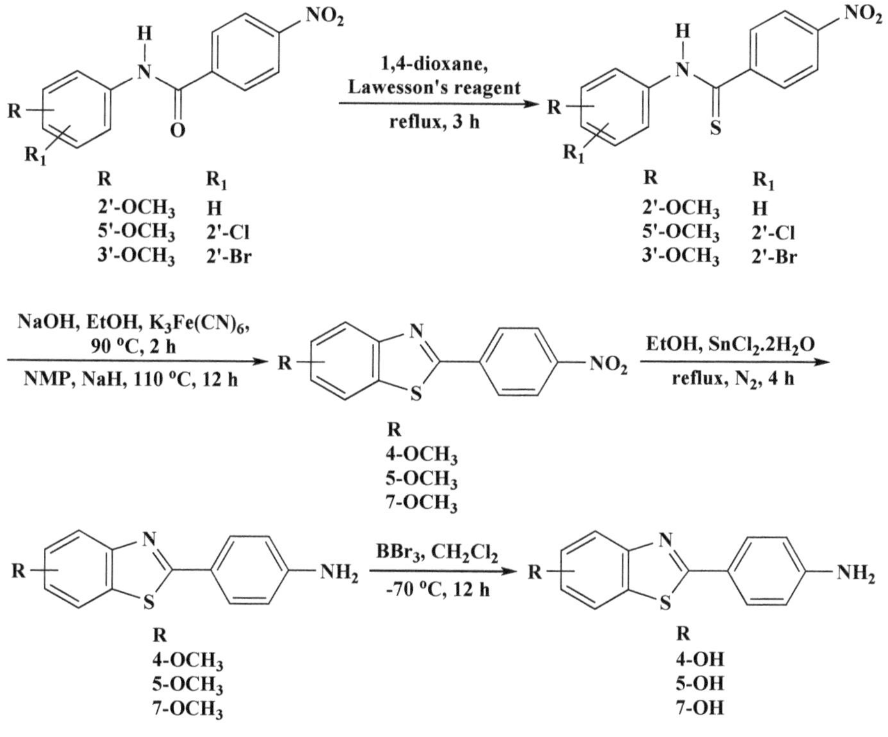

Scheme 3.39 Synthesis of 5-hydroxy-2-(4'-aminophenyl)-1,3-benzothiazoles

the 4-hydroxy-2-(4'-aminophenyl)-1,3-benzothiazole. For the formation of 7-hydroxy-2-(4'-aminophenyl)-1,3-benzothiazole and 5-hydroxy-2-(4'-aminophenyl)-1,3-benzothiazole, ring-closure led to two isomers by a method reported by Hutchinson et al. [157]. When a halogen atom (usually Cl or Br) was present in the position where the ring-closure took place and sodium hydride or sodium methoxide was utilized in combination with NMP (N-methyl-2-pyrrolidone) as a solvent, the ring-closure was specific for the intended position. The starting compound 2-bromo-3-aminoanisole acted as a substrate for the formation of 7-hydroxy-2-(4'-aminophenyl)-1,3-benzothiazole [158]. However, the commercially accessible HCl salt of 6-chloro-manisidine was utilized for the formation of 5-hydroxy-2-(4'-aminophenyl)-1,3-benzothiazole in a much higher yield [159].

The 4-, 5-, 6-, and 7-hydroxy derivatives of DF 203 were prepared by a pathway described in Scheme 3.40 [160]. An interaction of suitable anisidine with 3-methyl-4-nitrobenzoyl chloride in pyridine afforded different MeO-substituted nitrobenzanilides, which were transformed to thiobenzanilides employing LR in HMPA (hexamethylphosphoramide). The Jacobson cyclization [96] utilizing $K_3[Fe(CN)_6]$ in aqueous NaOH afforded MeO-substituted nitrobenzothiazoles. Whereas the 2- and 4-methoxythiobenzanilides afforded only a single benzothiazole product, in the case of 3-methoxythiobenzanilide a mixture of 5- and 7-substituted nitrobenzothiazoles was obtained; these isomers were separated by column chromatography. The reduction of nitrobenzothiazoles to their arylamines took place employing $SnCl_2.2H_2O$

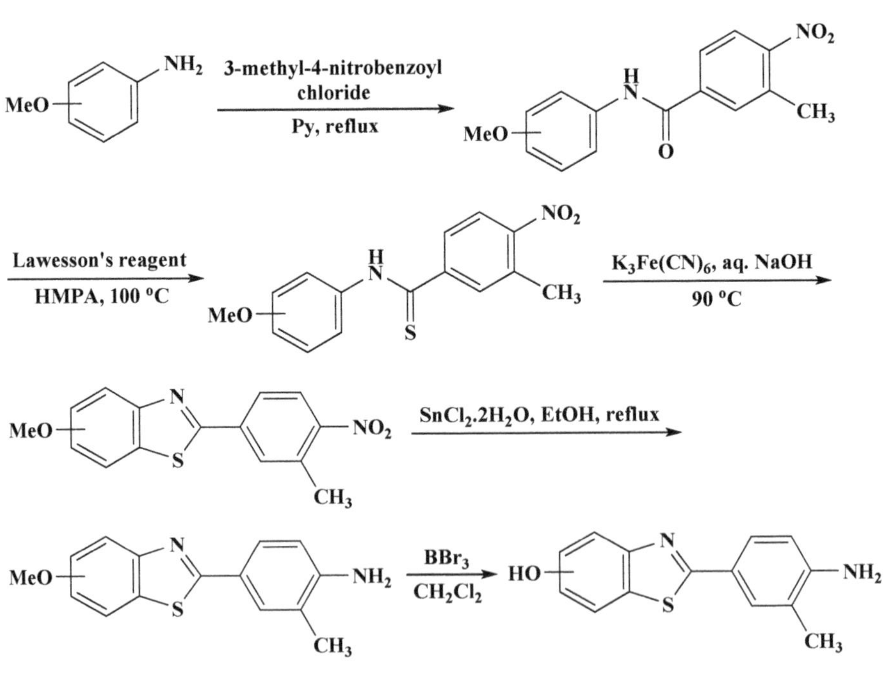

Scheme 3.40 Synthesis of hydroxylated 2-(4-aminophenyl)benzothiazole

3.3 Synthesis of Benzothiazoles

in refluxing EtOH. Finally, demethylation of methoxyarylamines was completed with excess of BBr$_3$ in CH$_2$Cl$_2$ to afford the required OH derivatives of DF 203. Otherwise, demethylation utilizing BBr$_3$ could take place at nitrophenyl stage to afford the phenols and after that reduction of NO$_2$ groups afforded hydroxylated 2-(4-aminophenyl)benzothiazoles; however, this was observed to be a less efficient pathway for desired compounds [161].

The formation of 4-fluoro-, 6-fluoro-, 4,5-difluoro-, 4,6-difluoro-, and 5,7-difluoro-DF 203 involving Jacobson's cyclization [96] as a key step is described in Scheme 3.41 [161, 162]. The reaction of suitable fluorinated aniline with 3-methyl-4-nitrobenzoyl chloride afforded a fluorinated benzanilide, which was transformed to thiobenzanilide utilizing LR. The Jacobson cyclization [96] of thiobenzanilide to 2-(4-nitrophenyl)benzothiazole and then NO$_2$ group reduction afforded desired fluorinated 2-(4-aminophenyl)benzothiazole in good yield. Although beneficial for the formation of a variety of substituted benzothiazoles, the Jacobsen's cyclization [96] suffered from one specific regioselectivity disadvantage in some cases. For instance, the cyclization of 3-fluoro- or 3,4-difluoro-substituted thiobenzanilides afforded a mixture of regioisomeric fluorinated benzothiazole products (5-fluoro and 7-fluoro-benzothiazoles in 10:1 ratio from 3-fluorothiobenzanilide; 5,6-difluoro- and 6,7-difluorobenzothiazoles in 2:1 ratio from 3,4-difluorothiobenzanilide), because of the accessibility of two cyclization sites *ortho* to nitrogen which result in two different products [157].

To probe the CYP1A1-induced mode of action of 5F 203 in vivo and for clinical imaging analysis, 5F 203 was chosen for radio-labeling with the positron emission tomography (PET) isotope F-18. It was hypothesized that prior induction of CYP1A1 with 5F 203 should result in the retention of [^{18}F]5F 203 at the tumor site by synthesis of reactive intermediates binding to cellular macromolecules [163]. The

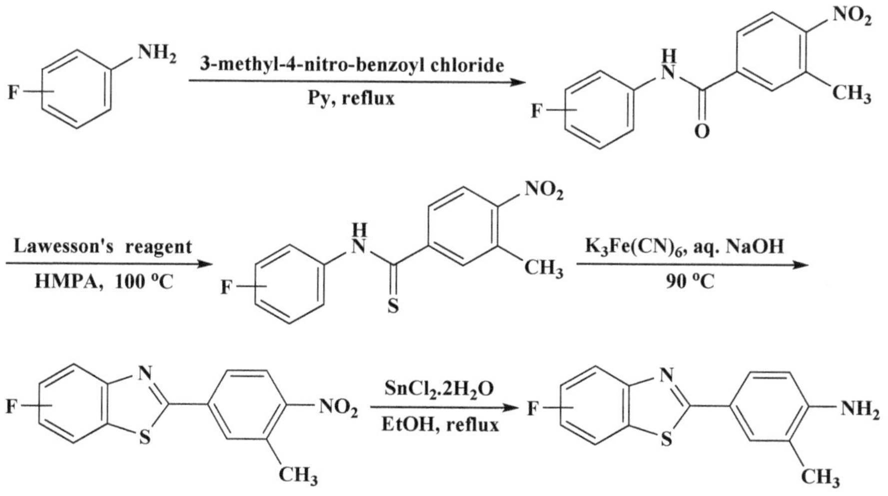

Scheme 3.41 Synthesis of fluorinated benzothiazole

formation of [^{18}F]5F 203 is given in Scheme 3.42, and included the Pd-catalyzed stannylation of 2-(4-amino-3-methylphenyl)-5-bromobenzothiazole and after that amine group protection to afford the 18F 203 substrate molecule. The ^{18}F-label was installed utilizing ^{18}F-F and after that rapid deprotection and isolation of the PET probe molecule [161].

The formation of 2-(4'-aminophenyl)-7-hydroxy-1,3-benzothiazole (and of 2-(4'-aminophenyl)-5-hydroxy-1,3-benzothiazole) caused surprising challenges when the route reported by Shi and coworkers [156] was utilized for similar molecules (Scheme 3.43). The reaction of *m*-anisidine with *p*-nitrobenzoyl chloride followed by transformation of amide to a thioamide took place efficiently. However, subsequently

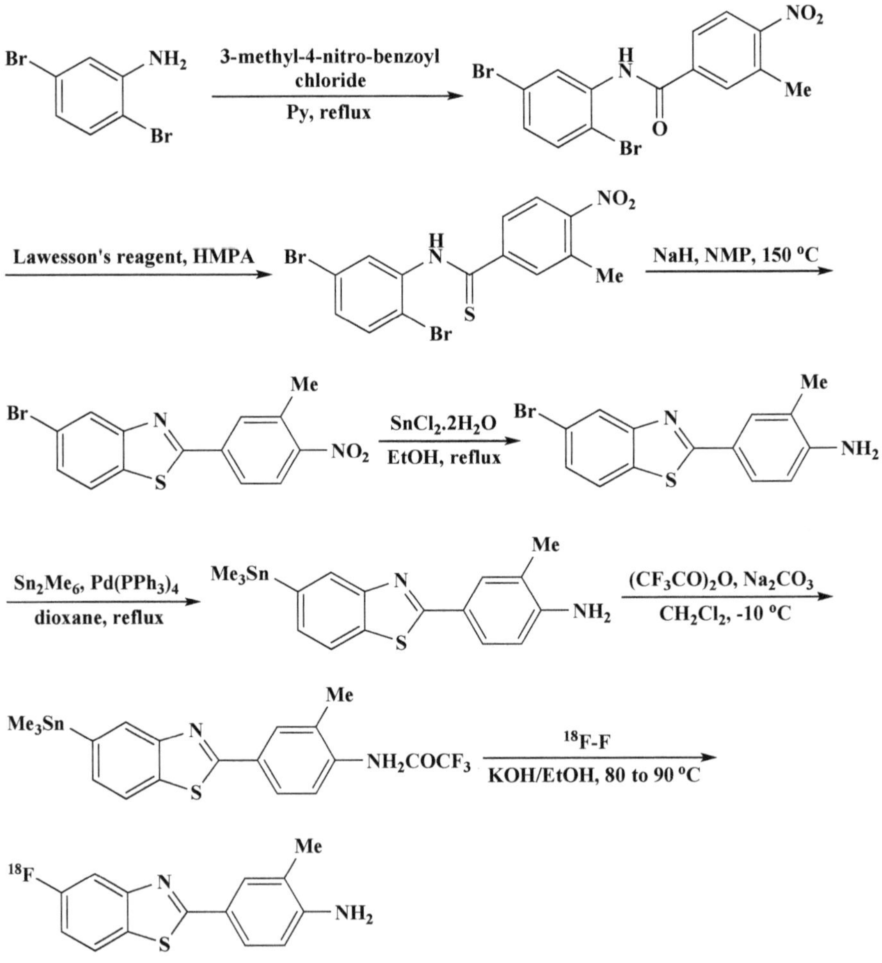

Scheme 3.42 Synthesis of fluorinated benzothiazole

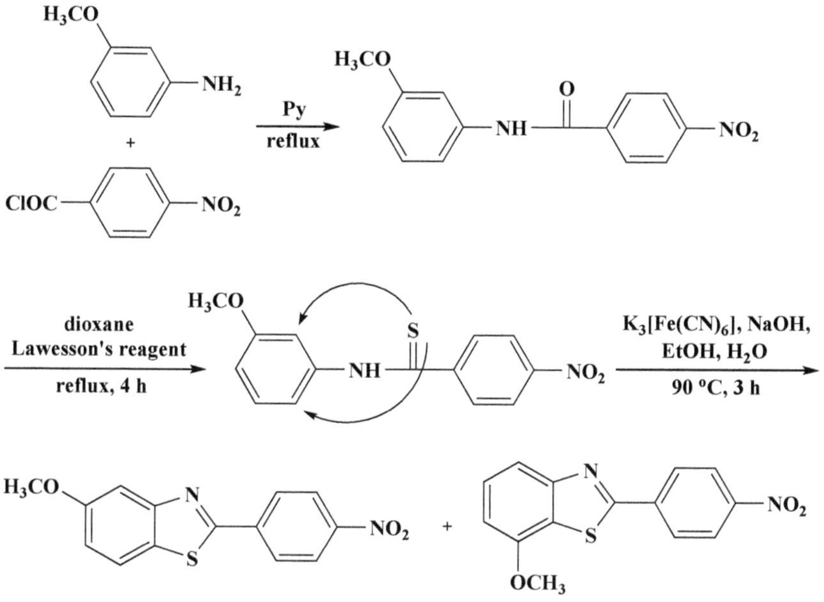

3.3 Synthesis of Benzothiazoles

Scheme 3.43 Synthesis of 1,3-benzothiazoles

ring-closure in the presence of $K_3[Fe(CN)_6]$ afforded a mixture of two phenyl-benzothiazole isomers with the MeO substituent in the 7-position or 5-position, respectively.

The starting material 2-bromo-3-aminoanisole was synthesized according a method reported for the formation of 2-(4'-aminophenyl)-7-hydroxy-1,3-benzothiazole. The 2-bromo-3-aminoanisole was treated with p-nitrobenzoyl chloride to synthesize an amide (57% yield), followed by transformation of the amide to a thioamide in the presence of LR. This process afforded thioamide, and the carbonyl oxygen of amides, esters, and ketones was reacted with sodium hydride in N-methylpyrrolidinone (an inert solvent with a high boiling point) to afford the benzothiazole in 63% yield. The reduction of NO_2 group with stannous chloride in boiling EtOH afforded amine derivative in 62% yield. The removal of O-methyl group with BBr_3 in CH_2Cl_2 afforded low yields, since the deprotected mixture had to be purified with MPLC. The successive reaction steps for the formation of final compound are described in Scheme 3.44 [158].

The formation of 2-(4'-aminophenyl)-5-hydroxy-1,3-benzothiazole was achieved. However, the commercially accessible hydrochloric salt of 6-chloro-m-anisidine was utilized as starting product. This HCl salt was treated with p-nitrobenzoyl chloride to synthesize the amide in a much higher yield (78%). The following steps involved the conversion into a thioamide with LR, ring-closure with sodium methoxide in N-methylpyrrolidinone (5%), reduction of NO_2 group with $SnCl_2$ (77%), and elimination of O-methyl group with boron tribromide (76% yield) (Scheme 3.45) [164].

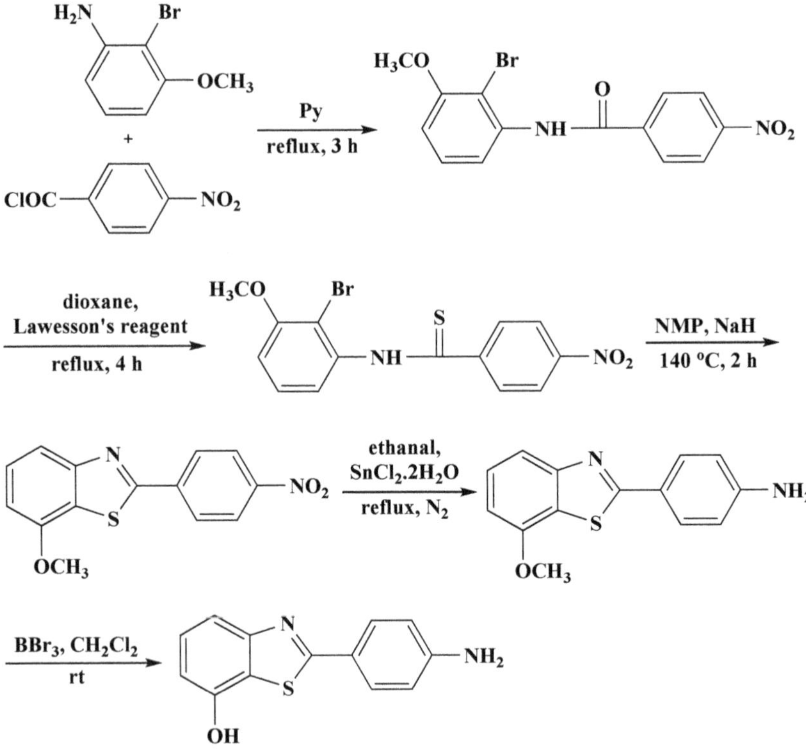

Scheme 3.44 Synthesis of 2-(4'-aminophenyl)-7-hydroxy-1,3-benzothiazole

3.4 Synthesis of Fused Thiazoles

The 5*H*-thiazolo[2,3-*d*][1,5]benzothiazepines were synthesized as shown in Scheme 3.46. The 1,5-benzothiazepin-4-ones, synthesized utilizing earlier described methodologies [165, 166], were converted into 1,5-benzothiazepine-4-thiones by reacting with LR in refluxing toluene. Successively, a bromoacetaldehyde diethyl acetal was added to a solution of 1,5-benzothiazepine-4-thiones in butan-2-one and H_2O [167]. The obtained mixtures were heated under reflux for different duration of time to provide the thiazolo[2,3-*d*][1,5]benzothiazepinium bromides, which upon reaction with NaOH in 50% MeOH-H_2O provided 5*H*-thiazolo[2,3-*d*][1,5]benzothiazepines [168]. A secondary 2-styrylbenzothiazole was synthesized via a thiazepine ring contraction and hydrogen sulfide elimination according to a method reported by Kaupp and Matthies [169] when a 2-phenyl group was present in the thione substrate.

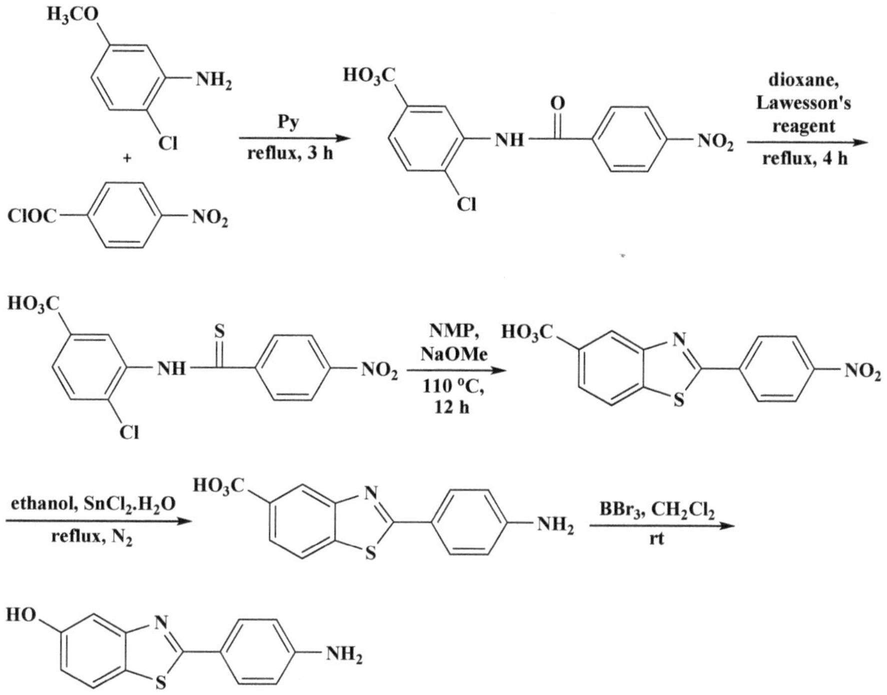

3.4 Synthesis of Fused Thiazoles

Scheme 3.45 Synthesis of 2-(4'-aminophenyl)-5-hydroxy-1,3-benzothiazole

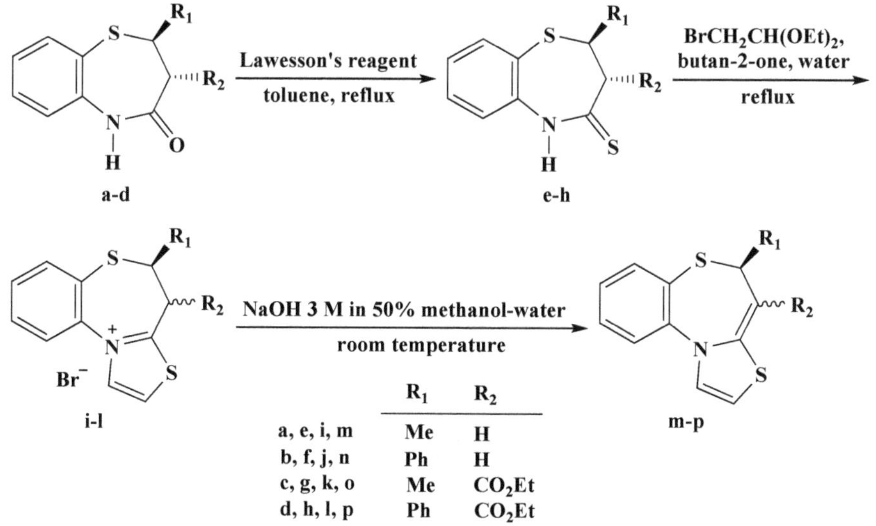

	R_1	R_2
a, e, i, m	Me	H
b, f, j, n	Ph	H
c, g, k, o	Me	CO_2Et
d, h, l, p	Ph	CO_2Et

Scheme 3.46 Synthesis of 5H-thiazolo[2,3-d][1,5]benzothiazepines

3.5 Synthesis of Fused Benzothiazoles

Three amines were individually acylated with all or some of acetic, propionic, and *i*-butyric/*n*-butyric anhydrides (and pyridine) and benzoylated utilizing benzoyl chloride-Et$_3$N to afford the *N*-(5-isoquinolinyl/quinolinyl)amides and *N*-(6-benzothiazolyl)amides. These were then easily thionated to thioamides by refluxing with LR in C$_6$H$_6$ [9, 170]. The thioamide (R = Me) was cyclized with bromine in CH$_3$COOH or in MeCN and then dethionation regenerated *N*-(5-isoquinolinyl/quinolinyl)amide (R = Me). The desired cyclization of thioamide (R = Me) to 2-methylthiazolo[4,5-*f*]isoquinoline was completed by Jacobson reaction [96] using aqueous alkaline K$_3$[Fe(CN)$_6$] at rt. Due to this success, each of thioamides was cyclized to 2-alkyl/phenyl derivatives of thiazolo[4,5-*f*]isoquinolines, thiazolo[4,5-*f*]quinolines (Scheme 3.47), and benzo[1,2-*d*:4,3-*d'*]bis-thiazoles in excellent yields (Scheme 3.48) [105, 171, 172].

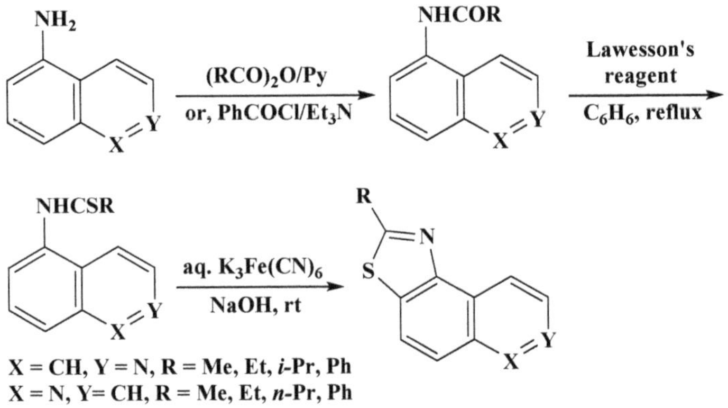

Scheme 3.47 Synthesis of 2-alkyl/phenyl derivatives of thiazolo[4,5-*f*]isoquinolines and thiazolo[4,5-*f*]quinolines

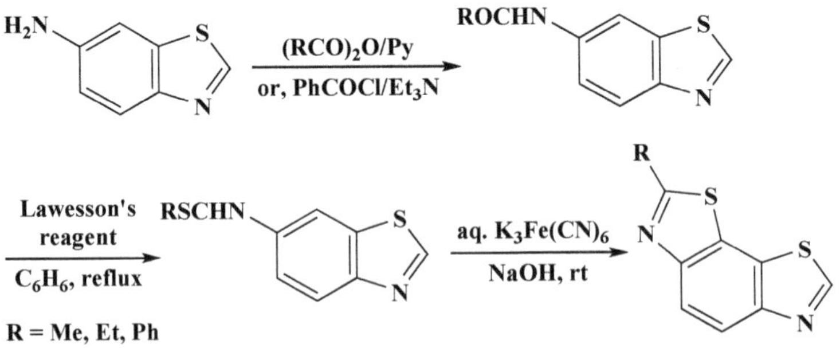

Scheme 3.48 Synthesis of 2-alkyl/phenyl derivatives of benzo[1,2-*d*:4,3-*d'*]bis-thiazoles

3.5 Synthesis of Fused Benzothiazoles

The thioamide quinolines were cyclized either with bromine in refluxing $CHCl_3$ (Hugerschoff reaction) or bromine in MeCN at rt to 2-alkylamino/2-anilinothiazolo[4,5]quinolines and 2-alkylthiothiazolo[4,5]quinolines. The thioamidoquinolines were cyclized with aqueous alkaline ferricyanide (Jacobson reaction [96]) to 2-alkyl/phenylthiazolo[4,5]quinolines (Scheme 3.49) [172].

An intramolecular Jacobson cyclization [96] of N-phenothiazinyl-benzothioamide/analogues utilizing less expensive and ecologically favorable catalyst iron(III) was developed for the preparation of a series of novel phenothiazine derivatives with fused thiazole unit. The first step of synthetic path employed was the regioselective amination of halogeno-10-alkylphenothiazine substrate, Pd-catalyzed amination appeared to be a suitable pathway for the synthesis of 2-amino-10-alkylphenothiazines [173]. The Cu-catalyzed coupling of aryl halides

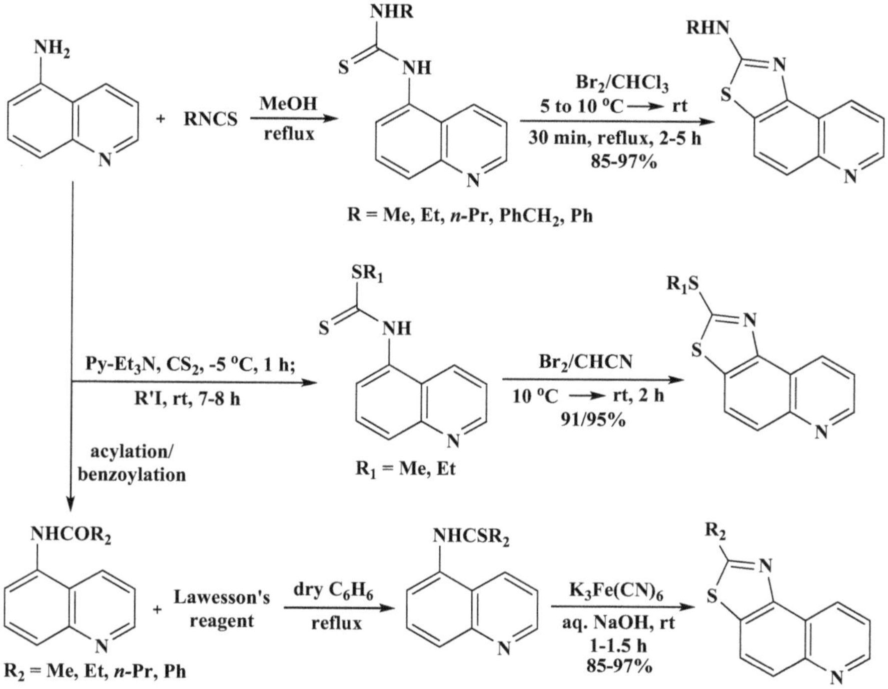

Scheme 3.49 Synthesis of 2-alkyl/phenylthiazolo[4,5]quinolines

with aqueous NH$_3$ was alternatively utilized for the MW-assisted amination of bromophenothiazine to afford the 3-amino-10-methyl-10H-phenothiazine in very good yield after 2 h irradiation. The amino phenothiazines were functionalized to aromatic amides utilizing substituted aromatic acid chlorides for driving the equilibrium toward product synthesis. The 2,4-bis(4-methoxyphenyl)-1,3-dithiaphosphetane-2,4-disulfide, i.e., LR has been utilized for the transformation of oxygen functionalities into thio analogues in moderate yield. The thiazolephenothiazines, along with a small amount of by-product, were obtained in moderate yields after Fe-mediated carbon–hydrogen functionalization/carbon–sulfur bond formation under mild conditions (Schemes 3.50 and 3.51) [174].

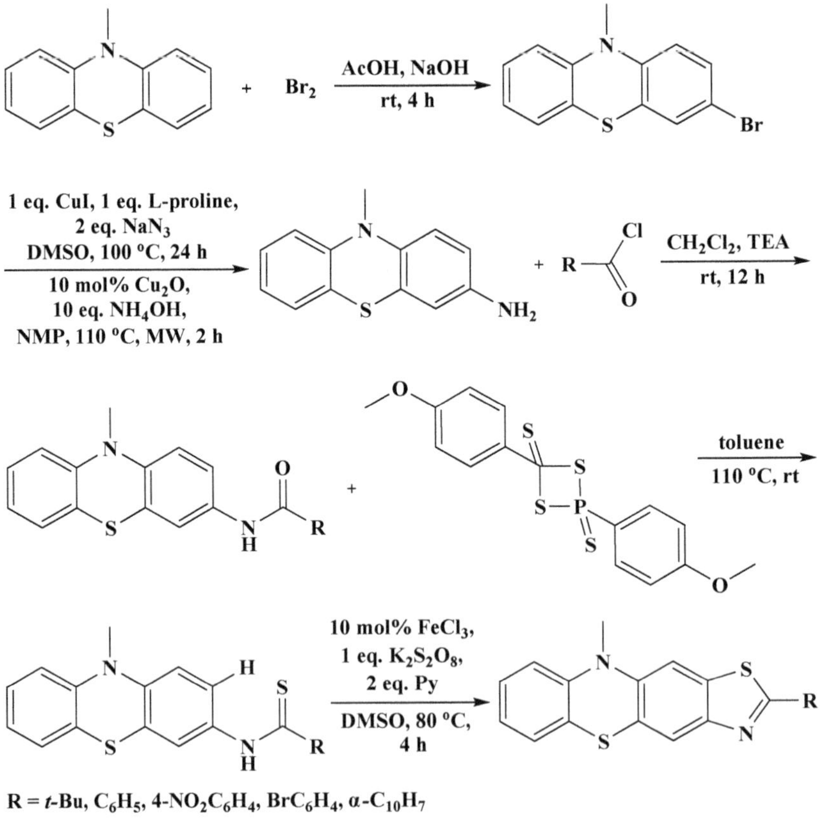

R = t-Bu, C$_6$H$_5$, 4-NO$_2$C$_6$H$_4$, BrC$_6$H$_4$, α-C$_{10}$H$_7$

Scheme 3.50 Synthesis of thiazolophenothiazines

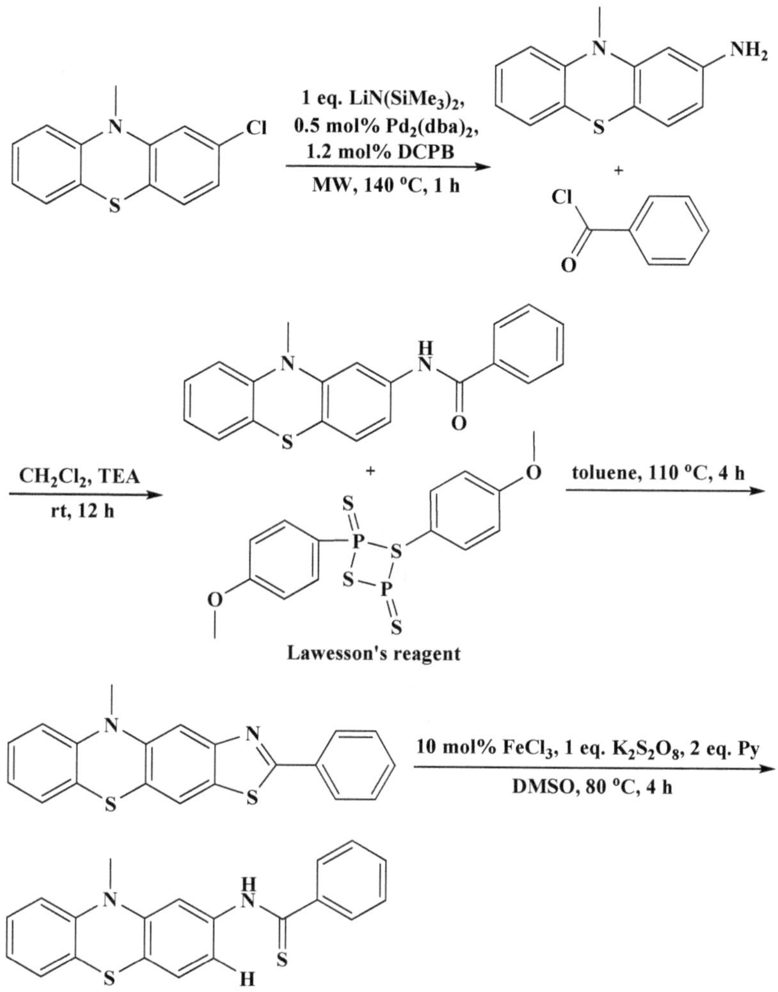

3.5 Synthesis of Fused Benzothiazoles

Scheme 3.51 Synthesis of phenothiazines

References

1. (a) N. Kaur. 2015. Environmentally benign synthesis of five-membered 1,3-*N*,*N*-heterocycles by microwave irradiation. Synth. Commun. 45: 909-943. (b) N. Kaur and D. Kishore. 2014. Microwave-assisted synthesis of seven- and higher-membered *O*-heterocycles. Synth. Commun. 44: 2739–2755. (c) N. Kaur, Y. Verma, P. Grewal, N. Ahlawat, P. Bhardwaj and N.K. Jangid. 2020. Photochemical C-N bond forming reactions for the synthesis of five-membered fused *N*-heterocycles. Synth. Commun. 50: 1286–1334. (d) N. Kaur. 2017. Gold catalysts in the synthesis of five-membered *N*-heterocycles. Curr. Organocatal. 4: 122–154. (e) N. Kaur. 2014. Microwave-assisted synthesis of five-membered *O*,*N*-heterocycles. Synth. Commun. 44: 3509–3537. (f) N. Kaur and D. Kishore. 2014. Microwave-assisted synthesis of six-membered *O*-heterocycles. Synth. Commun. 44: 3047–3081.

2. (a) N. Kaur. 2015. Synthesis of five-membered *N*,*N*,*N*- and *N*,*N*,*N*,*N*-heterocyclic compounds: Applications of microwaves. Synth. Commun. 45: 1711–1742. (b) N. Kaur. 2015. Six-membered heterocycles with three and four *N*-heteroatoms: Microwave-assisted synthesis. Synth. Commun. 45: 151–172. (c) N. Kaur and D. Kishore. 2014. Microwave-assisted synthesis of six-membered *S*-heterocycles. Synth. Commun. 44: 2615–2644. (d) N. Kaur. 2019. Synthesis of three-membered and four-membered heterocycles with the assistance of photochemical reactions. J. Heterocycl. Chem. 56: 1141–1167. (e) N. Kaur, P. Grewal, P. Bhardwaj, M. Devi and Y. Verma. 2019. Nickel-catalyzed synthesis of five-membered heterocycles. Synth. Commun. 49: 1543–1577. (f) N. Kaur. 2013. A new approach to anti-HIV chemotherapy devised by linking the vital fragments of active RT inhibitors such as etravirine to the molecular framework of anti-HIV prone privileged nucleus of 1,4-benzodiazepine as possible substitute to 'HAART'. Int. J. Pharm. Biol. Sci. 4: 309–317. (g) N. Kaur. 2018. Copper catalysts in the synthesis of five-membered *N*-polyheterocycles. Curr. Org. Synth. 15: 940-971.
3. (a) N. Kaur. 2015. Applications of microwaves in the synthesis of polycyclic six-membered *N*,*N*-heterocycles. Synth. Commun. 45: 1599–1631. (b) N. Kaur. 2018. Photochemical irradiation: Seven- and higher-membered *O*-heterocycles. Synth. Commun. 48: 2935–2964. (c) N. Kaur. 2015. Recent impact of microwave-assisted synthesis on benzo derivatives of five-membered *N*-heterocycles. Synth. Commun. 45: 539–568. (d) N. Kaur and D. Kishore. 2014. Microwave-assisted synthesis of seven- and higher-membered *N*-heterocycles. Synth. Commun. 44: 2577–2614. (e) N. Kaur. 2014. Microwave-assisted synthesis of five-membered *S*-heterocycles. J. Iran. Chem. Soc. 11: 523–564. (f) A. Anand, N. Kaur and D. Kishore. 2014. An efficient one pot protocol to the annulation of face 'd' of benzazepinone ring with pyrazole, isoxazole and pyrimidine nucleus through the corresponding oxoketene dithioacetal derivative. Adv. Chem. 1–5. (g) N. Kaur and D. Kishore. 2013. Metal and non-metal based catalysts for oxidation of organic compounds. Catal. Surv. Asia 17: 20–42.
4. T.J. Curphey. 2002. Thionation with the reagent combination of phosphorus pentasulfide and hexamethyldisiloxane. J. Org. Chem. 67: 6461–6473.
5. T.J. Curphey. 2002. Thionation of esters and lactones with the reagent combination of phosphorus pentasulfide and hexamethyldisiloxane. Tetrahedron Lett. 43: 371–373.
6. T.J. Curphey. 2000. A superior procedure for the conversion of 3-oxoesters to 3*H*-1,2-dithiole-3-thiones. Tetrahedron Lett. 41: 9963-9966.
7. M. Szostak and J. Aube. 2009. Synthesis and rearrangement of a bridged thioamide. Chem. Commun. 46: 7122–7124.
8. E. Campaigne. 1964. Thiones and thials. Chem. Rev. 39: 1–77.
9. M.P. Cava and M.I. Levinson. 1985. Thionation reactions of Lawesson's reagents. Tetrahedron 41: 5061–5087.
10. R.A. Cherkasov, G.A. Kutyrev and A.N. Pudovik. 1985. Tetrahedron report number 186. Tetrahedron 41: 2567–2624.
11. J. Nagaoka. 1995. Lawesson's reagent. J. Synth. Org. Chem. Jpn. 53: 1138–1140.
12. D. Brillon. 1992. Recent developments in the area of thionation methods and related synthetic applications. Sulfur Rep. 12: 297–332.
13. M. Jesberger, T.P. Davis and L. Barner. 2003. Applications of Lawesson's reagent in organic and organometallic syntheses. Synthesis 13: 1929–1958.
14. H.Z. Lecher, R.A. Greenwood, K.C. Whitehouse and T.H. Chao. 1956. The phosphonation of aromatic compounds with phosphorus pentasulfide. J. Am. Chem. Soc. 78: 518–522.
15. R. Schiess. 2013. Total synthesis of cyclopropyl-epothilone B analogs and studies towards the total synthesis of michaolide E. PhD Thesis, ETH Zurich.
16. C.J. Moody and J.C.A. Hunt. 1999. Synthesis of virenamide B, a cytotoxic thiazole-containing peptide. J. Org. Chem. 64: 8715–8717.
17. M.C. Bagley, R.T. Buck, S.L. Hind and C.J. Moody. 1998. Synthesis of functionalised oxazoles and bis-oxazoles. J. Chem. Soc. Perkin Trans. 1 3: 591–600.
18. H.Z. Lecher, R.A. Greenwood, K.C. Whitehouse and T. Chao. 1956. The phosphonation of aromatic compounds with phosphorus pentasulfide. J. Am. Chem. Soc. 78: 5018–5022.

19. V.V. Sureshbabu, S.A. Naik and G. Nagendra. 2009. Synthesis of Boc-amino tetrazoles derived from α-amino acids. Synth. Commun. 39: 395–406.
20. P. Tavecchia, P. Gentili and M. Kurz. 1995. Degradation studies of antibiotic MDL 62,879 (GE2270A) and revision of the structure. Tetrahedron 51: 4867–4869.
21. H.M. Müller, O. Delgado and T. Bach. 2007. Total synthesis of the thiazolyl peptide GE2270 A. Angew. Chem. Int. Ed. 46: 4771–4774.
22. M.W. Bredenkamp, C.W. Holzapfel and W.J. van Zyl. 1990. The chiral synthesis of thiazole amino acid enantiomers. Synth. Commun. 20: 2235–2249.
23. E. Aguilar and A.I. Meyers. 1994. Reinvestigation of a modified Hantzsch thiazole synthesis. Tetrahedron Lett. 35: 2473–2476.
24. E.A. Merritt and M.C. Bagley. 2007. Holzapfel-Meyers-Nicolaou modification of the Hantzsch thiazole synthesis. Synthesis 22: 3535–3541.
25. Y. Hamada, K. Hayashi and T. Shioiri. 1991. Efficient stereoselective synthesis of dolastatin 10, an antineoplastic peptide from a sea hare. Tetrahedron Lett. 32: 931-934.
26. S. Ray, M.G.B. Drew and A. Banerjee. 2006. The role of terminal tyrosine residues in the formation of tripeptide nanotubes: A crystallographic insight. Tetrahedron 62: 7274–7283.
27. Y.Q. Fu, B. Xu, X. Zou, C. Ma, X. Yang, K. Mou, G. Fu, Y. Lü and P. Xu. 2007. Design and synthesis of a novel class of furan-based molecules as potential 20S proteasome inhibitors. Bioorg. Med. Chem. Lett. 17: 1102–1106.
28. Z.Y. Chen and T. Ye. 2006. The first total synthesis of aeruginosamide. New J. Chem. 30: 518–520.
29. H. Gan, Z. Chen, Z. Fang and K. Guo. 2013. Concise and efficient total syntheses of virenamides A and D. J. Adv. Chem. 4: 488–493.
30. J.-F. Pons, Q. Mishir, A. Nouvet and F. Brookfield. 2000. Thiazole formation via traceless cleavage of rink resin. Tetrahedron Lett. 41: 4965–4968.
31. V. Krchnak and M.W. Holladay. 2002. Solid phase heterocyclic chemistry. Chem. Rev. 102: 61-92.
32. A.R. Katritzky, S. Rachwal, K.C. Caster, F. Mahni, K.W. Law and O. Rubio. 1987. The chemistry of N-substituted benzotriazoles. Part 1. 1-(Chloromethyl)benzotriazole. J. Chem. Soc. Perkin Trans. 1 0: 781–789.
33. A.R. Katritzky, J. Chen and Z. Yang. 1995. 1-(Cyanomethyl)benzotriazole as a convenient precursor for the synthesis of 2-substituted thiazoles. J. Org. Chem. 60: 5638–5642.
34. A.R. Katritzky and S. Rachwal. 2010. Synthesis of heterocycles mediated by benzotriazole. 1. Monocyclic systems. Chem. Rev. 110: 1564–1610.
35. J.J.C. Grove and C.W. Holzapfel. 1997. Samarium(II) iodide mediated transformations of carbohydrate derived iodo oxime ethers into stereodefined alkoxy aminocyclopentanes. Tetrahedron Lett. 38: 7429–7432.
36. W.R. Bowman, P.T. Stephenson, N.K. Terrett and A.R. Young. 1995. Radical cyclisations of imines and hydrazones. Tetrahedron 51: 7959–7980.
37. C.F. Sturino and A.G. Fallis. 1994. Samarium(II) iodide induced radical cyclizations of halo- and carbonylhydrazones. J. Am. Chem. Soc. 116: 7447–7448.
38. P.A. Bartlett, K.L. McLaren and P.C. Ting. 1988. Radical cyclization of oxime ethers. J. Am. Chem. Soc. 110: 1633–1634.
39. P. Devin, L. Fensterbank and M. Malacria. 1999. Tin-free radical chemistry: Intramolecular addition of alkyl radicals to aldehydes and ketones. Tetrahedron Lett. 40: 5511–5514.
40. W.-T. Jiaang, H.-C. Lin, K.-H. Tang, L.-B. Chang and Y.-M. Tsai. 1999. The study of the kinetics of intramolecular radical cyclizations of acylsilanes via the intramolecular competition method. J. Org. Chem. 64: 618–628.
41. S.-Y. Chang, W.-T. Jiaang, C.-D. Cherng, K.-H. Tang, C.-H. Huang and Y.-M. Tsai. 1997. The scope and limitations of intramolecular radical cyclizations of acylsilanes with alkyl, aryl, and vinyl radicals. J. Org. Chem. 62: 9089-9098.
42. J.C. Walton. 1998. Homolytic substitution: A molecular ménage à trois. Acc. Chem. Res. 31: 99–107.

43. N. Al-Maharik, L. Engman, J. Malmström and C.H. Schiesser. 2001. Intramolecular homolytic substitution at selenium: Synthesis of novel selenium-containing vitamin E analogues. J. Org. Chem. 66: 6286–6290.
44. L. Engman, M.J. Laws, J. Malmström, C.H. Schiesser and L.M. Zugaro. 1999. Toward novel antioxidants: Preparation of dihydrotellurophenes and selenophenes by alkyltelluride-mediated tandem $S_{RN}1$/Shi reactions. J. Org. Chem. 64: 6764–6770.
45. G.S. Bates and M.A. Varelas. 1980. A mild, general preparation of N-acyl aziridines and 2-substituted 4(S)-benzyloxazolines. Can. J. Chem. 58: 2562–2566.
46. A. Krief, M. Trabelsi and W. Dumont. 1992. Syntheses of alkali selenolates from diorganic diselenides and alkali metal hydrides: Scope and limitations. Synthesis 10: 933-935.
47. E. Owusu-Ansah, A.C. Durow, J.R. Harding, A.C. Jordan, S.J. O'Connell and C.L. Willis. 2011. Synthesis of dysideaproline E using organocatalysis. Org. Biomol. Chem. 9: 265–272.
48. J. Lloyd, H.J. Finlay, W. Vacarro, T. Hyunh, A. Kover, R. Bhandaru, L. Yan, K. Atwal, M.L. Conder, T. Jenkins-West, H. Shi, C. Huang, D. Li, H. Sun and P. Levesque. 2010. Pyrrolidine amides of pyrazolodihydropyrimidines as potent and selective KV1.5 blockers. Bioorg. Med. Chem. Lett. 20: 1436–1439.
49. A. Pichota, J. Duraiswamy, Z. Yin, T. Keller and M. Schreiber. 2007. Pdf inhibitors. PCT Int. Appl.
50. D.W. Arthur and Y. Lester. 1922. Amide formation from esters of secondary alkyl malonic acids. J. Am. Chem. Soc. 44: 1564–1567.
51. B.Z. Alexey and A. Hans. 2006. Enantioswitchable catalysts for the asymmetric transfer hydrogenation of aryl alkyl ketones. Org. Lett. 8: 5129–5132.
52. J. Rudolph, L. Chen, D. Majumdar, W.H. Bullock, M. Burns, T. Claus, F.E.D. Cruz, M. Daly, F.J. Ehrgott, J.S. Johnson, J.N. Livingston, R.W. Schoenleber, J. Shapiro, L. Yang, M. Tsutsumi and X. Ma. 2007. Indanylacetic acid derivatives carrying 4-thiazolyl-phenoxy tail groups, a new class of potent PPAR α/γ/δ pan agonists: Synthesis, structure-activity relationship, and in vivo efficacy. J. Med. Chem. 50: 984–1000.
53. R. Skov, R. Smyth, A.R. Larsen, N.M. Frimodt and G. Kahlmeter. 2005. Evaluation of cefoxitin 5 and 10 µg discs for the detection of methicillin resistance in *Staphylococci*. J. Antimicrob. Chemother. 55: 157–161.
54. B.A. Arthington, M. Motley, D.W. Warnock and C.J. Morrison. 2000. Comparative evaluation of PASCO and national committee for clinical laboratory standards M27-A broth microdilution methods for antifungal drug susceptibility testing of yeasts. J. Clin. Microbiol. 38: 2254–2260.
55. K. Sivagurunathan, S.R.M. Kamil and S.S. Shafi. 2013. Efficient synthesis of novel pyrazolo thiazole derivatives and its antifungal activity studies. J. Pharm. Res. 2: 1–3.
56. A. Carbone, M. Pennati, P. Barraja, A. Montalbano, B. Parrino, V. Spanò, A. Lopergolo, S. Sbarra, V. Doldi, N. Zaffaroni, G. Cirrincione and P. Diana. 2014. Synthesis and antiproliferative activity of substituted 3[2-(1H-indol-3-yl)-1,3-thiazol-4-yl]-1H-pyrrolo[3,2-b]pyridines, marine alkaloid nortopsentin analogues. Curr. Med. Chem. 21: 1654–1666.
57. A. Carbone, M. Pennati, B. Parrino, A. Lopergolo, P. Barraja, A. Montalbano, V. Spanò, S. Sbarra, V. Doldi, M. de Cesare, G. Cirrincione, P. Diana and N. Zaffaroni. 2013. Novel 1H-pyrrolo[2,3-b]pyridine derivative nortopsentin analogues: Synthesis and antitumor activity in peritoneal mesothelioma experimental models. J. Med. Chem. 56: 7060–7072.
58. C. Ganser, E. Lauermann, A. Maderer, T. Stauder, J.P. Kramb, S. Plutizki, T. Kindler, M. Moehler and G. Dannhardt. 2012. Novel 3-azaindolyl-4-arylmaleimides exhibiting potent antiangiogenic efficacy, protein kinase inhibition, and antiproliferative activity. J. Med. Chem. 55: 9531-9540.
59. P. Diana, A. Carbone, P. Barraja, A. Montalbano, B. Parrino, A. Lopergolo, M. Pennati, N. Zaffaroni and G. Cirrincione. 2011. Synthesis and antitumor activity of 3-(2-phenyl-1,3-thiazol-4-yl)-1H-indoles and 3-(2-phenyl-1,3-thiazol-4-yl)-1H-7-azaindoles. ChemMedChem 6: 1300–1309.

60. B. Parrino, A. Carbone, G. DiVita, C. Ciancimino, A. Attanzio, V. Spanò, A. Montalbano, P. Barraja, L. Tesoriere, M.A. Livrea, P. Diana and G. Cirrincione. 2015. 3-[4-(1H-Indol-3-yl)-1,3-thiazol-2-yl]-1H-pyrrolo[2,3-b]pyridines, nortopsentin analogues with antiproliferative activity. Mar. Drugs 13: 1901–1924.
61. A. Carbone, B. Parrino, G. DiVita, A. Attanzio, V. Spano, A. Montalbano, P. Barraja, L. Tesoriere, M.A. Livrea, P. Diana and G. Cirrincione. 2015. Synthesis and antiproliferative activity of thiazolyl-bis-pyrrolo[2,3-b]pyridines and indolyl-thiazolyl-pyrrolo[2,3-c]pyridines, nortopsentin analogues. Mar. Drugs 13: 460–492.
62. U. Schmidt, P. Gleich, H. Griesser and R. Utz. 1986. Amino acids and peptides; 581. Synthesis of optically active 2-(1-hydroxyalkyl)-thiazole-4-carboxylic acids and 2-(1-aminoalkyl)-thiazole-4-carboxylic acids. Synthesis 12: 992-998.
63. R.V. Pirovani, G.A. Brito, R.C. Barcelos and R.A. Pilli. 2015. Enantioselective total synthesis of (+)-lyngbyabellin M. Mar. Drugs 13: 3309–3324.
64. H. Mohammad, A.S. Mayhoub, A. Ghafoor, M. Soofi, R.A. Alajlouni, M. Cushman and M.N. Seleem. 2014. Discovery and characterization of potent thiazoles versus methicillin- and vancomycin-resistant *Staphylococcus aureus*. J. Med. Chem. 57: 1609–1615.
65. P.T. Nyffeler, C.-H. Liang, K.M. Koeller and C.-H. Wong. 2002. The chemistry of amine-azide interconversion: Catalytic diazotransfer and regioselective azide reduction. J. Am. Chem. Soc. 124: 10773–10778.
66. P.B. Alper, S.-C. Hung and C.-H. Wong. 1996. Metal catalyzed diazo transfer for the synthesis of azides from amines. Tetrahedron Lett. 37: 6029–6032.
67. K.C. Nicolaou, A.A. Estrada, M. Zak, S.H. Lee and B.S. Safina. 2005. A mild and selective method for the hydrolysis of esters with trimethyltin hydroxide. Angew. Chem. Int. Ed. 44: 1378–1382.
68. R.L.E. Furlan, E.G. Mata and O.A. Mascaretti. 1998. Efficient, non-acidolytic method for the selective cleavage of N-Boc amino acid and peptide phenacyl esters linked to a polystyrene resin. J. Chem. Soc. Perkin Trans. 1 2: 355–358.
69. R.L.E. Furlan, E.G. Mata, O.A. Mascaretti, C. Pena and M.P. Coba. 1998. Selective detachment of Boc-protected amino acids and peptides from merrifield, PAM and Wang resins by trimethyltin hydroxide. Tetrahedron 54: 13023–13034.
70. R.L.E. Furlan and O.A. Mascaretti. 1997. Esterification, transesterification and deesterification mediated by organotin oxides, hydroxides and alkoxides. Aldrichimica Acta 30: 55–69.
71. R.L.E. Furlan, E.G. Mata and O.A. Mascaretti. 1996. Cleavage of carboxylic esters effected by organotin oxides and hydroxides under classical heating and microwave irradiation. A comparative study. Tetrahedron Lett. 37: 5229–5232.
72. K.C. Nicolaou, B.S. Safina, M. Zak, S.H. Lee, M. Nevalainen, M. Bella, A.A. Estrada, C. Funke, F.J. Zecri and S. Bulat. 2005. Total synthesis of thiostrepton. Retrosynthetic analysis and construction of key building blocks. J. Am. Chem. Soc. 127: 11159–11175.
73. C.D.J. Boden, G. Pattenden and T. Ye. 1995. The synthesis of optically active thiazoline and thiazole derived peptides from N-protected α-amino acids. Synlett 5: 417–419.
74. H. Ishihara and K. Shimura. 1988. Further evidence for the presence of a thiazoline ring in the isoleucylcysteine dipeptide intermediate in bacitracin biosynthesis. FEBS Lett. 226: 319–323.
75. Z. Chen, J. Deng and T. Ye. 2003. Total synthesis of *cis,cis*-ceratospongamide. ARKIVOC (vii): 268–285.
76. P. Wipf and C.P. Miller. 1994. Synthesis of peptide thiazolines from β-hydroxythioamides. An investigation of racemization in cyclodehydration protocols. Tetrahedron Lett. 35: 5397–5400.
77. P. Wipf and S. Venkatraman. 1996. An improved protocol for azole synthesis with PEG-supported Burgess reagent. Tetrahedron Lett. 37: 4659–4662.
78. P. Wipf and G.B. Hayes. 1998. Synthesis of oxazines and thiazines by cyclodehydration of hydroxy amides and thioamides. Tetrahedron 54: 6987–6998.
79. M.A. Ciufolini and Y.-C. Shen. 1997. Studies toward thiostrepton antibiotics: Assembly of the central pyridine-thiazole cluster of micrococcins. J. Org. Chem. 62: 3804–3805.

80. E. Falb, A. Nudelmann and A. Hassner. 1993. A convenient synthesis of chiral oxazolidin-2-ones and thiazolidin-2-ones and an improved preparation of triphosgene. Synth. Commun. 23: 2839–2844.
81. L. Cotarca, A. Delogu and V. Šunjic. 1996. Bis(trichloromethyl) carbonate in organic synthesis. Synthesis 5: 553–576.
82. C.-M. Pan, C.-C. Lin, S.J. Kim, R.P. Sellers and S.R. McAlpine. 2012. Progress toward the synthesis of urukthapelstatin A and two analogues. Tetrahedron Lett. 53: 4065–4069.
83. L. Landeira, Y. Imbriago, G. Serra, E. Manta, J. Saldana and L. Scarone. 2013. Synthesis and anthelmintic evaluation of [2,5']-bis-heterocycles as bengazole analogs. Rev. Latinoamer. Quím. 41: 38–49.
84. D.S. Bose and M. Idrees. 2006. Hypervalent iodine mediated intramolecular cyclization of thioformanilides: Expeditious approach to 2-substituted benzothiazoles. J. Org. Chem. 71: 8261–8263.
85. D.S. Bose, M. Idrees, N.M. Jakka and J.V. Rao. 2010. Diversity-oriented synthesis of quinolines via Friedländer annulation reaction under mild catalytic conditions. J. Comb. Chem. 12: 100–110.
86. B. Srinivas, N.S. Devi, G.K. Sreenivasulu and R. Parameshwar. 2015. Synthesis and characterization of quinazolinobenzodiazepine-benzothiazole hybrid derivatives. Der Pharma Chem. 7: 251–256.
87. K.J. Malik, F.V. Manvi, B.K. Nanjwade and S. Singh. 2009. Synthesis and screening of some new 2-amino substituted benzothiazole derivatives for antifungal activity. Drug Invent. Today 1: 32–34.
88. W. Hu, Y. Chen, C. Liao, H. Yu, Y. Tsai, S. Huang, F. Tsai, H. Shen, L. Chang and J. Wang. 2010. Synthesis, and biological evaluation of 2-(4-aminophenyl)benzothiazole derivatives as photosensitizing agents. Bioorg. Med. Chem. 18: 6197–6207.
89. S.L. Khokra, K. Arora, H. Mehta, A. Aggarwal and M. Yadav. 2011. Common methods to synthesize benzothiazole derivatives and their medicinal significance: A review. Int. J. Pharm. Sci. Res. 2: 1356–1377.
90. S. Jaiswal, A.P. Mishra and A. Srivastava. 2012. The different kinds of reaction involved in synthesis of 2-substituted benzothiazole and its derivatives. Res. J. Pharm. Biol. Chem. Sci. 3: 631–641.
91. S.Z.K. Hisamoddin, S. Priyanka, S.P. Yogesh and P. Nilam. 2014. Benzothiazole the molecule of diverse biological activities. Pharma. Sci. Monitor. 5: 207–225.
92. M. Wang, M. Gao, B.H. Mock, K.D. Miller, G.W. Sledge, G.D. Hutchins and Q.-H. Zheng. 2006. Synthesis of carbon-11 labeled fluorinated 2-arylbenzothiazoles as novel potential PET cancer imaging agents. Bioorg. Med. Chem. 14: 8599–8607.
93. K. Serdons, K. van Laere, P. Janssen, H. Kung, G. Bormans and A. Verbruggen. 2009. Synthesis and evaluation of three ^{18}F-labeled aminophenylbenzothiazoles as amyloid imaging agents. J. Med. Chem. 52: 7090–7102.
94. K. Serdons, C. Terwinghe, P. Vermaelen, K. van Laere, H. Kung, L. Mortelmans, G. Bormans and A. Verbruggen. 2009. Synthesis and evaluation of ^{18}F-labeled 2-phenylbenzothiazoles as positron emission tomography imaging agents for amyloid plaques in Alzheimer's disease. J. Med. Chem. 52: 1428–1437.
95. E.S. Barskaya, N. Abdullaeva, I. Yudin, A. Drobyazko, E.K. Beloglazkina, A. Maguga and N.V. Zyk. 2014. 2-(Pyridin-2-yl)-1,3-benzothiazoles: Alternative synthetic approaches and study of complexation reactions. 3rd International Conference of Organic Chemistry (ICOC-2014) "Organic synthesis - driving force of life development", Javakhishvili Tbilisi State University.
96. P. Jacobson. 1886. Ueber bildung von anhydroverbindungen des orthoamidophenylmercaptans aus thioaniliden. Ber. 19: 1067–1077.
97. N. Castagnoli, P. Jacob, P.S. Callery and A.T. Shulgin. 1976. A convenient synthesis of quinones from hydroquinone dimethyl ethers. Oxidative demethylation with ceric ammonium nitrate. J. Org. Chem. 41: 3627–3629.

98. I. Thomsen, K. Clausen, S. Scheibye and S.O. Lawesson. 1984. Thiation with 2,4-bis(4-methoxyphenyl)-1,3,2,4-dithiaphosphetane 2,4-disulfide: N-Methylthiopyrrolidone. Org. Synth. 62: 158–164.
99. M.A. Lyon, S. Lawrence, D.J. Williams and Y.A. Jackson. 1999. Synthesis and structure verification of an analogue of kuanoniamine A. J. Chem. Soc. Perkin Trans. 1 4: 437–442.
100. A.L.V.R. Reddy, S.V.L. Reddy, C.N. Reddy, G. Chandrasekhar and T.B. Reddy. 2014. Synthesis and antibacterial activity studies of 2-aryl benzothiazoles and its derivatives. World J. Pharm. Pharma. Sci. 3: 1335–1343.
101. R.B. Moodie and K. Schofield. 1976. Ipso attack in aromatic nitration. Acc. Chem. Res. 9: 287–292.
102. J.G. Traynham. 1983. Aromatic substitution reactions: When you've said *ortho*, *meta*, and *para* you haven't said it all. J. Chem. Educ. 60: 937-941.
103. M.F.G. Stevens, C.J. McCall, P. Lelieveld, P. Alexander and A. Richter. 1994. Structural studies on bioactive compounds. 23. Synthesis of polyhydroxylated 2-phenylbenzothiazoles and a comparison of their cytotoxicities and pharmacological properties with genistein and quercetin. J. Med. Chem. 37: 1689–1695.
104. N.K. Downer and Y.A. Jackson. 2004. Synthesis of benzothiazoles via *ipso* substitution of *ortho*-methoxythiobenzamides. Org. Biomol. Chem. 2: 3039–3043.
105. Y.A. Jackson, M.A. Lyon, N. Townsend, K. Bellabe and F. Soltanik. 2000. Reactions of some N-(2,5-dimethoxyaryl)thiobenzamides: En route to an analogue of kuanoniamine A. J. Chem. Soc. Perkin Trans. 1 2: 205–210.
106. C.A. Mathis, Y.M. Wang, D.P. Holt, G.F. Huang, M.L. Debnath and W.E. Klunk. 2003. Synthesis and evaluation of ^{11}C-labeled 6-substituted 2-arylbenzothiazoles as amyloid imaging agents. J. Med. Chem. 46: 2740–2754.
107. G.R. Morais, A. Paulo and I. Santos. 2012. A synthetic overview of radiolabeled compounds for β-amyloid targeting. Eur. J. Org. Chem. 7: 1279–1293.
108. J. Montgomery. 2004. Nickel-catalyzed reductive cyclizations and couplings. Angew. Chem. Int. Ed. 43: 3890–3908.
109. E. Negishi, C. Coperet, S. Ma, S.Y. Liou and F. Liu. 1996. Cyclic carbopalladation. A versatile synthetic methodology for the construction of cyclic organic compounds. Chem. Rev. 96: 365–394.
110. L.F. Tietze. 1996. Domino reactions in organic synthesis. Chem. Rev. 96: 115–136.
111. R. Grigg and V. Sridharan. 1999. Palladium catalysed cascade cyclisation-anion capture, relay switches and molecular queues. J. Organomet. Chem. 576: 65–87.
112. T. Miura and M. Murakami. 2007. Formation of carbocycles through sequential carborhodation triggered by addition of organoborons. Chem. Commun. 3: 217–224.
113. K. Agapiou, D.F. Cauble and M.J. Krische. 2004. Copper-catalyzed tandem conjugate addition-electrophilic trapping: Ketones, esters, and nitriles as terminal electrophiles. J. Am. Chem. Soc. 126: 4528–4529.
114. K. Subburaj and J. Montgomery. 2003. A new catalytic conjugate addition/aldol strategy that avoids preformed metalated nucleophiles. J. Am. Chem. Soc. 125: 11210–11211.
115. H.-C. Guo and J.-A. Ma. 2006. Catalytic asymmetric tandem transformations triggered by conjugate additions. Angew. Chem. Int. Ed. 45: 354–366.
116. S.E. Denmark and A. Thorarensen. 1996. Tandem [4+2]/[3+2] cycloadditions of nitroalkenes. Chem. Rev. 96: 137–166.
117. J.A. Porco, F.J. Schoenen, T.J. Stout, J. Clardy and S.L. Schreiber. 1990. Transannular Diels-Alder route to systems related to dynemicin A. J. Am. Chem. Soc. 112: 7410–7411.
118. G.A. Molander and C.R. Harris. 1996. Sequenced reactions with samarium(II) iodide. Tandem nucleophilic acyl substitution/ketyl-olefin coupling reactions. J. Am. Chem. Soc. 118: 4059–4071.
119. C. Chen, M.E. Layton, S.M. Sheehan and M.D. Shair. 2000. Synthesis of (+)-CP-263,114. J. Am. Chem. Soc. 122: 7424–7425.
120. F. Shi, X. Li, Y. Xia, L. Zhang and Z.-X. Yu. 2007. DFT study of the mechanisms of in water Au(I)-catalyzed tandem [3,3]-rearrangement/Nazarov reaction/[1,2]-hydrogen shift of enynyl

acetates: A proton-transport catalysis strategy in the water-catalyzed [1,2]-hydrogen shift. J. Am. Chem. Soc. 129: 15503–15512.
121. S.W. Youn, J.-Y. Song and D.I. Jung. 2008. Rhodium-catalyzed tandem conjugate addition-Mannich cyclization reaction: Straightforward access to fully substituted tetrahydroquinolines. J. Org. Chem. 73: 5658–5661.
122. K.-G. Ji, X.-Z. Shu, J. Chen, S.-C. Zhao, Z.-J. Zheng, L. Lu, X.-Y. Liu and Y.-M. Liang. 2008. $PtCl_2$-Catalyzed tandem triple migration reaction toward (Z)-1,5-dien-2-yl esters. Org. Lett. 10: 3919–3921.
123. Z. Chen, X. Yang and J. Wu. 2009. AgOTf-Catalyzed tandem reaction of N'-(2-alkynylbenzylidene)hydrazide with alkyne. Chem. Commun. 23: 3469–3471.
124. Q. Ding, Z. Wang and J. Wu. 2009. Tandem electrophilic cyclization-[3+2] cycloaddition-rearrangement reactions of 2-alkynylbenzaldoxime, DMAD, and Br_2. J. Org. Chem. 74: 921-924.
125. K. Gao and J. Wu. 2008. $Sc(OTf)_3$-Catalyzed or t-BuOK promoted tandem reaction of 2-(2-(alkynyl)benzylidene)malonate with indole. Org. Lett. 10: 2251–2254.
126. Q. Ding and J. Wu. 2008. Access to functionalized isoquinoline N-oxides via sequential electrophilic cyclization/cross-coupling reactions. Adv. Synth. Catal. 350: 1850–1854.
127. Q. Ding and J. Wu. 2008. A facile route to 2,4-dihydro-1H-benzo[d][1,3]thiazines via silver-catalyzed tandem addition-cyclization reactions. J. Comb. Chem. 10: 541–545.
128. L. Zhang and J. Wu. 2007. Friedländer synthesis of quinolines using a Lewis acid-surfactant-combined catalyst in water. Adv. Synth. Catal. 349: 1047–1051.
129. Z. Wang, R. Fan and J. Wu. 2007. Palladium-catalyzed regioselective cross-coupling reactions of 3-bromo-4-tosyloxyquinolin-2(1H)-one with arylboronic acids. A facile and convenient route to 3,4-disubstituted quinolin-2(1H)-ones. Adv. Synth. Catal. 349: 1943–1948.
130. Q. Ding, Y. Ye, R. Fan and J. Wu. 2007. Selective synthesis of 2,3-disubstituted-2H-isoindol-1-ylphosphonate and 2,3-disubstituted-1,2-dihydroisoquinolin-1-ylphosphonate via metal-tuned reaction of α-amino (2-alkynylphenyl)methylphosphonate. J. Org. Chem. 72: 5439–5442.
131. L. Zhang, T. Meng, R. Fan and J. Wu. 2007. General and efficient route for the synthesis of 3,4-disubstituted coumarins via Pd-catalyzed site-selective cross-coupling reactions. J. Org. Chem. 72: 7279–7286.
132. Z. Wang, B. Wang and J. Wu. 2007. Diversity-oriented synthesis of functionalized quinolin-2(1H)-ones via Pd-catalyzed site-selective cross-coupling reactions. J. Comb. Chem. 9: 811–817.
133. D. Bernardi, L.A. Ba and G. Kirsch. 2007. One-pot preparation of 2-(alkyl)arylbenzothiazoles from the corresponding o-halobenzanilides. Synlett 13: 2121–2123.
134. S.V. Rajender and K. Dalip. 1999. Microwave-accelerated solvent-free synthesis of thioketones, thiolactones, thioamides, thionoesters, and thioflavonoids. Org. Lett. 1: 697–700.
135. G. Evindar and R.A. Batey. 2006. Parallel synthesis of a library of benzoxazoles and benzothiazoles using ligand-accelerated copper-catalyzed cyclizations of ortho-halobenzanilides. J. Org. Chem. 71: 1802–1805.
136. D. Ma and Q. Cai. 2008. Copper/amino acid catalyzed cross-couplings of aryl and vinyl halides with nucleophiles. Acc. Chem. Res. 41: 1450–1460.
137. B. Zou, Q. Yuan and D. Ma. 2007. Synthesis of 1,2-disubstituted benzimidazoles by a Cu-catalyzed cascade aryl amination/condensation process. Angew. Chem. Int. Ed. 46: 2598–2601.
138. F. Liu and D. Ma. 2007. Assembly of conjugated enynes and substituted indoles via CuI/amino acid-catalyzed coupling of 1-alkynes with vinyl iodides and 2-bromotrifluoroacetanilides. J. Org. Chem. 72: 4844–4850.
139. Y. Chen, X. Xie and D. Ma. 2007. Facile access to polysubstituted indoles via a cascade Cu-catalyzed arylation-condensation process. J. Org. Chem. 72: 9329-9334.
140. Y. Chen, Y. Wang, Z. Sun and D. Ma. 2008. Elaboration of 2-(trifluoromethyl)indoles via a cascade coupling/condensation/deacylation process. Org. Lett. 10: 625–628.

141. Q. Yuan and D. Ma. 2008. A one-pot coupling/hydrolysis/condensation process to pyrrolo[1,2-*a*]quinoxaline. J. Org. Chem. 73: 5159–5162.
142. R. Martin, R.M. Rodriguez and S.L. Buchwald. 2006. Domino Cu-catalyzed C-N coupling/hydroamidation: A highly efficient synthesis of nitrogen heterocycles. Angew. Chem. Int. Ed. 45: 7079–7082.
143. H. Miyamoto, Y. Okawa, A. Nakazaki and S. Kobayashi. 2006. Highly diastereoselective one-pot synthesis of spirocyclic oxindoles through intramolecular Ullmann coupling and Claisen rearrangement. Angew. Chem. Int. Ed. 45: 2274–2277.
144. A. Coste, M. Toumi, K. Wright, V. Razafimahaleo, F. Couty, J. Marrot and G. Evano. 2008. Copper-catalyzed cyclization of iodo-tryptophans: A straightforward synthesis of pyrroloindoles. Org. Lett. 10: 3841–3844.
145. X. Liu, H. Fu, Y. Jiang and Y. Zhao. 2009. A simple and efficient approach to quinazolinones under mild copper-catalyzed conditions. Angew. Chem. Int. Ed. 48: 348–351.
146. X. Lv, Y. Liu, W. Qian and W.L. Bao. 2008. Copper(I)-catalyzed one-pot synthesis of 2-iminobenzo-1,3-oxathioles from *ortho*-iodophenols and isothiocyanates. Adv. Synth. Catal. 350: 2507–2512.
147. X. Lv and W.L. Bao. 2007. A β-keto ester as a novel, efficient, and versatile ligand for copper(I)-catalyzed C-N, C-O, and C-S coupling reactions. J. Org. Chem. 72: 3863–3867.
148. Q. Ding, X.-G. Huang and J. Wu. 2009. Facile synthesis of benzothiazoles via cascade reactions of 2-iodoanilines, acid chlorides and Lawesson's reagent. J. Comb. Chem. 11: 1047–1049.
149. P.P. Prabhu, S. Pande and C.S. Shastry. 2011. Synthesis and biological evaluation of Schiff's bases of some new benzothiazole derivatives as antimicrobial agents. Int. J. ChemTech Res. 3: 185–191.
150. P.P. Prabhu, C.S. Shastry, S.S. Pande and T.P. Selvam. 2011. Design, synthesis, characterization and biological evaluation of benzothiazole-6-carboxylate derivatives. Res. Pharm. 1: 6–12.
151. R. Mahtab, A. Srivastava, N. Gupta, S.K. Kushwaha and A. Tripathi. 2014. Synthesis of novel 2-benzylbenzo[*d*]thiazole-6-sulfonamide derivatives as potential antiinflammatory agent. J. Chem. Pharm. Sci. 7: 34–38.
152. V. Polshettiwar and M.P. Kaushik. 2006. Alumina encapsulated phosphorus pentasulfide (P_4S_{10}/Al_2O_3) mediated efficient thionation of long chain amides. Tetrahedron Lett. 47: 2315–2317.
153. P. Hrobarik, I. Sigmundova and P. Zahradnik. 2005. Preparation of novel push-pull benzothiazole derivatives with reverse polarity: Compounds with potential non-linear optic application. Synthesis 4: 600–604.
154. G. Bold, S. Wojeik, G. Caravatti, R. Lindauer, C. Stierlin, J. Gertsch, M. Wartmann and K.H. Altmann. 2006. Structure-activity relationships in side-chain-modified epothilone analogues - how important is the position of the nitrogen atom? ChemMedChem 1: 37–40.
155. P. Hrobarik, V. Hrobarikova, I. Sigmundova, P. Zahradník, M. Fakis, I. Polyzos and P. Persephonis. 2011. Benzothiazoles with tunable electron-withdrawing strength and reverse polarity: A route to triphenylamine-based chromophores with enhanced two-photon absorption. J. Org. Chem. 76: 8726–8736.
156. D.-F. Shi, T.D. Bradshaw, S. Wrigley, C.J. McCall, P. Lelieveld, I. Fichtner and M.F.G. Stevens. 1996. Antitumor benzothiazoles. 3.1 Synthesis of 2-(4-aminophenyl)benzothiazoles and evaluation of their activities against breast cancer cell lines in vitro and in vivo. J. Med. Chem. 39: 3375–3384.
157. I. Hutchinson, M.F.G. Stevens and A.D. Westwell. 2000. The regiospecific synthesis of 5- and 7-monosubstituted and 5,6-disubstituted 2-arylbenzothiazoles. Tetrahedron Lett. 41: 425–428.
158. M.E. Krolski, A.F. Renaldo, D.E. Rudisill and J.K. Stille. 1988. Palladium-catalyzed coupling of 2-bromoanilines with vinylstannanes. A regiocontrolled synthesis of substituted indoles. J. Org. Chem. 53: 1170–1176.

159. K. Serdons, T. Verduyckt, D. Vanderghinste, P. Borghgraef, J. Cleynhens, F. van Leuven, H. Kung, G. Bormans and A. Verbruggen. 2009. ^{11}C-Labelled PIB analogues as potential tracer agents for in vivo imaging of amyloid β in Alzheimer's disease. Eur. J. Med. Chem. 44: 1415–1426.
160. E. Kashiyama, I. Hutchinson, M.-S. Chua, S.F. Stinson, L.R. Phillips, G. Kaur, E.A. Sausville, T.D. Bradshaw, A.D. Westwell and M.F.G. Stevens. 1999. Antitumor benzothiazoles. 8.1 Synthesis, metabolic formation, and biological properties of the C- and N-oxidation products of antitumor 2-(4-aminophenyl)benzothiazoles. J. Med. Chem. 42: 4172–4184.
161. T.D. Bradshaw and A.D. Westwell. 2004. The development of the antitumour benzothiazole prodrug, phortress, as a clinical candidate. Curr. Med. Chem. 11: 1241–1253.
162. I. Hutchinson, M.-S. Chua, H.L. Browne, V. Trapani, T.D. Bradshaw, A.D. Westwell and M.F.G. Stevens. 2001. Antitumor benzothiazoles. 14.1 Synthesis and in vitro biological properties of fluorinated 2-(4-aminophenyl)benzothiazoles. J. Med. Chem. 44: 1446–1455.
163. G.D. Brown, H. Wilson, A.D. Westwell, I. Hutchinson, M.F.G. Stevens, P.M. Price, E. Aboagye, S.K. Luthra and F. Brady. 2001. Radiolabelling of the potential anti-cancer agent, 5-fluoro-2-(4-amino-3-methylphenyl)benzothiazole (5F203) with fluorine-18. J. Label. Compd. Radiopharm. 44: S374-S375.
164. T. Verduyckt. 2004. Radiolabelled derivatives of thioflavin T for visualization of fibrillar brain amyloid. PhD Thesis, Katholieke Universiteit Leuven.
165. J. Krapcho, E.R. Spitzmiller and C.F. Turk. 1963. Substituted 2,3-dihydro-1,5-benzothiazepin-4(5H)-ones and 3,4-dihydro-2-phenyl-(2H)-1,6-benzothiazocin-5(6H)-ones. J. Med. Chem. 6: 544–546.
166. B. Letois, J.C. Lancelot, C. Saturnino, M. Robba and P. de Caprariis. 1992. Synthesis of trans-thienyl-3-ethoxycarbonyl-2,3-dihydro[1,5]benzothiazepin-4(5H)-ones. J. Heterocycl. Chem. 29: 1769–1772.
167. V. Ambrogi, G. Grandolini, L. Perioli, L. Giusti, A. Lucacchini and C. Martini. 1993. Studies on annelated 1,4-benzothiazines and 1,5-benzothiazepines. VII. Synthesis and inhibition of benzodiazepine receptor binding of some 4,5-dihydro-tetrazolo[5,1-d]-1,5-benzothiazepines and 5-phenyl-s-triazolo[3,4-d]-1,5-benzothiazepines. Farmaco 48: 665–676.
168. G. Bruno, A. Chimirri, R. Gitto, S. Grasso, F. Nicolo, R. Scopelliti and M. Zappala. 1997. Synthesis and structural characteristics of novel 5H-thiazolo[2,3-d][1,5]benzothiazepine derivatives. J. Chem. Soc. Perkin Trans. 1 15: 2211–2216.
169. G. Kaupp and D. Matthies. 1987. Komplexe eliminierungen und solvolysen von N-acyl-1,5-benzothiazepin-4-onen. Chem. Ber. 120: 1741–1745.
170. B.S. Pedersen and S.-O. Lawesson. 1979. Studies on organophosphorus compounds - XXVIII. Tetrahedron 35: 2433–2437.
171. S.R.H. Edge. 1923. XVIII - Benzbisthiazoles. Part II. J. Chem. Soc. Trans. 123: 153–156.
172. M. Chakrabarty, A. Mukherji, S. Karmakar, R. Mukherjee, K. Nagai, A. Geronikaki and P. Eleni. 2010. An expedient synthesis of novel 2-substituted thiazolo[4,5-f]isoquinolines/quinolines and benzo[1,2-d:4,3-d']bisthiazoles and their potential as inhibitors of COX-1 and COX-2. ARKIVOC (xi): 265–290.
173. L.I. Găină, L.N. Mătărângă-Popa, E. Gál, P. Boar, P. Lönnecke, E. Hey-Hawkins, C. Bischin, R. Silaghi-Dumitrescu, I. Lupan, C. Cristea and L. Silaghi-Dumitrescu. 2013. Microwave-assisted catalytic amination of phenothiazine; reliable access to phenothiazine analogues of Tröger's base. Eur. J. Org. Chem. 24: 5500–5508.
174. H. Wang, L. Wang, J. Shang, X. Li, H. Wang, J. Guia and A. Lei. 2012. Fe-Catalysed oxidative C-H functionalization/C-S bond formation. Chem. Commun. 48: 76–78.

Chapter 4
Thiadiazole Synthesis

4.1 Introduction

The heterocyclic compounds are center of focus in the field of medicinal research due to their valuable medicinal properties. The heterocyclic compounds are abundant in nature and have attracted more attention because their structural subunits are present in many natural products like antibiotics, vitamins, hormones, etc. The heterocyclic compounds are known as the key section of organic chemistry. They have widespread importance not only biologically and industrially but also in the development of human society [1–3].

In the last few years, the formation of substituted thiadiazolines [4, 5] and related compounds has attracted attention as these compounds constitute the structural scaffolds of many naturally occurring alkanoids which show a variety of medicinal and industrial applications. The technical applications of these compounds include photographic materials, optically active liquid crystals, and dyes. The thiadiazolines, thiadiazoles, and oxadiazolines show a number of biological properties, and they serve as anthelmintics [6, 7], antihypertensive [8], antitumor [9–11], analgesic, anticancer, anti-inflammatory, antibacterial [12–16], and tyrosinase inhibitory agents [17]. The macrocycles bearing thiadiazoline and thiadiazole subunits have exciting host guest complexation features [18] and show antibacterial activities as well [19–22].

Previous to the work in thionation chemistry reported by Lawesson et al. [23, 24] in 1978, cyclization reactions were generally performed employing P_4S_{10} under reflux conditions. The LR is usually known to have benefits over P_4S_{10} in terms of shorter reaction times, lesser equivalents of the reagent, and superior reproducibility [25, 26]. Although the formation of LR was published by Lecher [27] in 1956, it was the systematic explorations of the compound by Lawesson et al. [23, 24] that led to it becoming arguably the most important agent in thionation chemistry. It is commercially accessible and prepared by heating anisole and phosphorus pentasulfide for many hours, affording about 80% yield.

4.2 Synthesis of 1,2,3-Thiadiazoles

The 5-ethyl-4-ethoxycarbonyl-1,2,3-thiadiazole was prepared from diazoketone through the thioketone intermediates employing LR (Scheme 4.1) [28].

The formation of alkyl-substituted pyrrolothiadiazoles (R = Bn) and (R = Me) started from itaconic acid or its dimethylester (Scheme 4.2). Cyclization toward the pyrrolidine system was optimized based on a method reported by Wu and Feldkamp [29]. Subsequently, lactams were converted into thiolactams employing LR [30]. Because the work-up conditions for reactions with this reagent were problematic, sometimes, Kugelrohr distillation was a very smooth way to purify the products. The thiolactams afforded hydrazono cyclization precursors cleanly by treating with ethyl carbazate on refluxing in tetrahydrofuran using mercury(II) acetate. The reaction progress decreased considerably without this reagent. The equilibrium concentrations were shifted by precipitation of mercury(II) sulfide. The actual Hurd-Mori cyclization toward alkyl-thiadiazoles afforded fully aromatized products with both substrates. However, it turned out to be unsatisfactory with respect to the yields of pyrrolothiadiazoles 25% (R = Bn) and 15% (R = Me), respectively, even under

Scheme 4.1 Synthesis of 5-ethyl-4-ethoxycarbonyl-1,2,3-thiadiazole

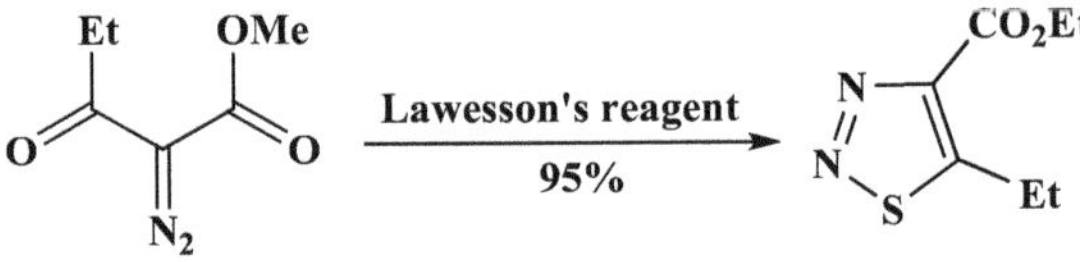

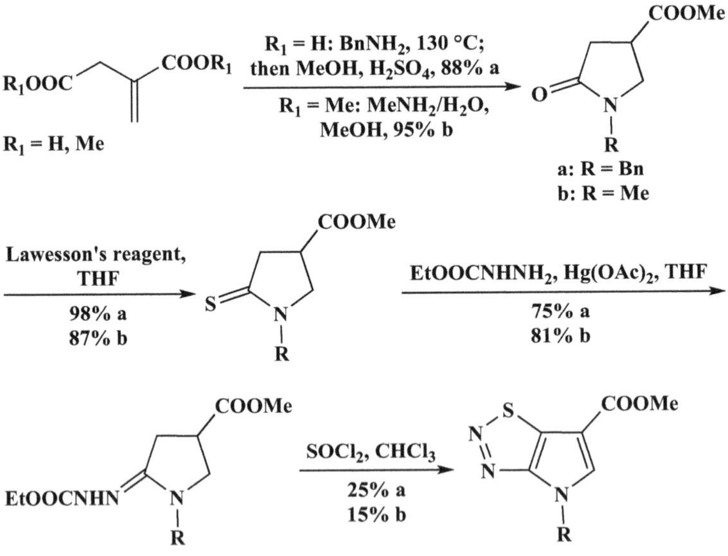

Scheme 4.2 Synthesis of alkyl-substituted pyrrolothiadiazoles

4.2 Synthesis of 1,2,3-Thiadiazoles

optimized reaction conditions. The pyrrolidino precursors needed harsher conditions (CHCl₃ at reflux) to show rational reaction rates. However, at these temperatures, the stability of starting compound and/or of intermediate products turns out to be a limiting factor. In all optimized experiments, noteworthy decomposition was found by darkening the reaction mixtures and/or the precipitation of insoluble material. Subsequently, other option to overcome the poor cyclization yields was by the replacement of protecting group at the pyrrolidine system with EWG [31].

The 5-oxopyrrolidin-3-carboxylate methylester was synthesized as shown in Scheme 4.3. The transformation to thiolactam was completed in excellent yields employing LR. An incorporation of carbamate protecting group afforded thiopyrrole in only moderate yields, as complete conversion not occurred. Almost quantitative conversion was reported based on the consumed starting compounds. The reaction with ethyl carbazate not needed mercury additive and afforded iminopyrrole in excellent yield. Different from the alkyl substrates, Hurd-Mori reaction with iminopyrrole needed cooling and took place easily to provide the fully aromatized thiadiazole. No significant side-products were formed and aromatized thiadiazole was isolated in 94% yield after easy recrystallization. Final deprotection of pyrrolo nitrogen was completed by the reaction of a methanolic solution of aromatized thiadiazole with silica gel to provide the final compound quantitatively [31–34].

The 2-diazo-1,3-diketones, prepared from HFA[Asp(Cl)] and diazoethyl acetate, were treated with LR to afford the β-(1,2,3-thiadiazol-5-yl)alaninates (Scheme 4.4) [35]. The diazoketones of HFA(Ida) were reacted in the similar manner [36, 37].

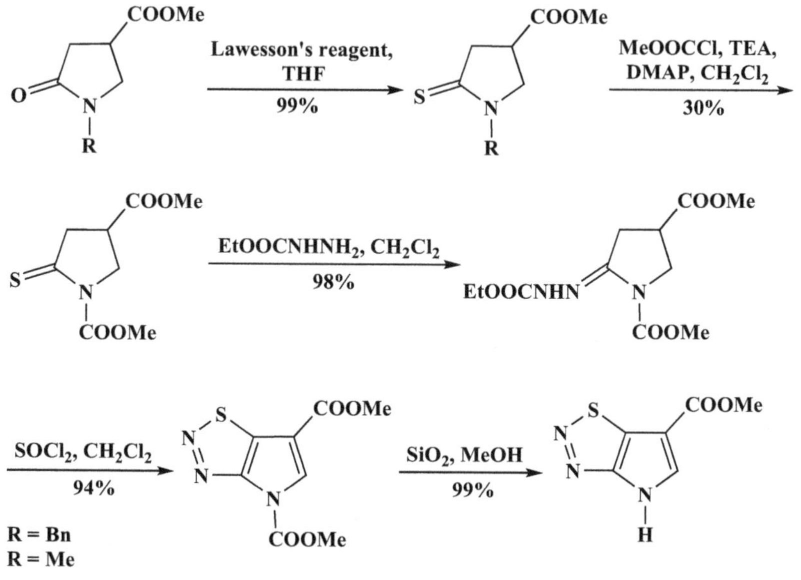

Scheme 4.3 Synthesis of pyrrolothiadiazole

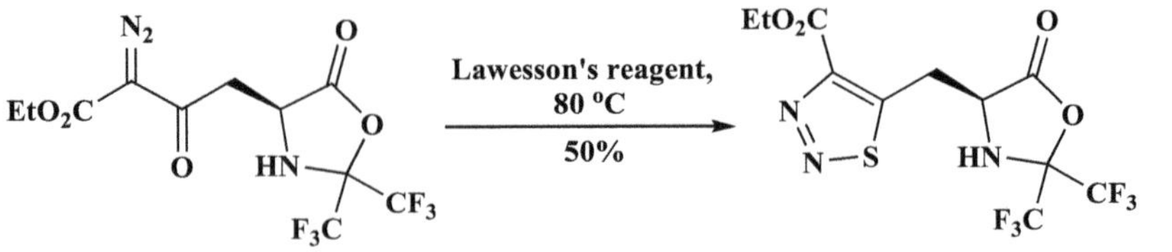

Scheme 4.4 Synthesis of β-(1,2,3-thiadiazol-5-yl)alaninates

Scheme 4.5 Synthesis of 1,2,3-thiadiazole

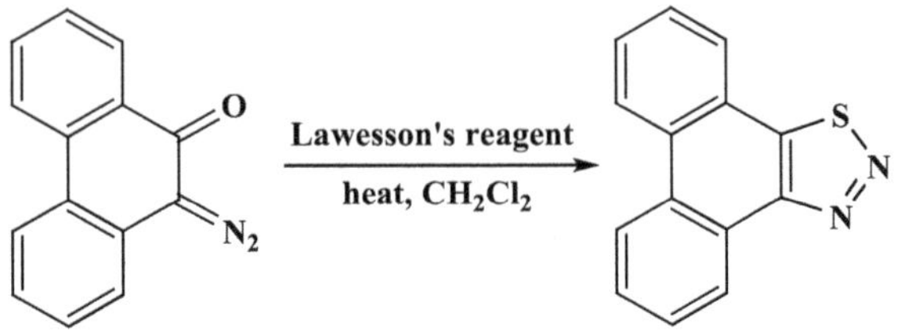

Scheme 4.6 Synthesis of 1,2,3-thiadiazoles

R_1 = allyl, Me, Bn, t-Bu, CH$_2$CH$_2$SiMe$_3$
R_2 = Me, Et, cyclopentyl, t-Bu

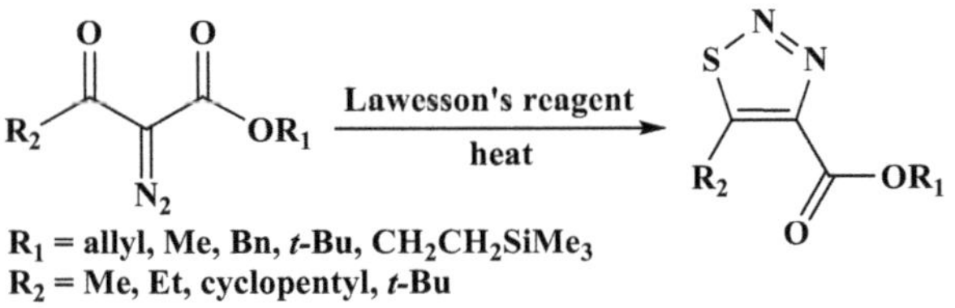

The 1,2,3-thiadiazoles were prepared by reacting α-diazoketones with LR (Schemes 4.5 and 4.6). This reaction worked particularly well for α-diazoketones where the ketone and the diazo group were held *cis* [38]. When the size of R_1 was increased, the reaction for the synthesis of thiadiazole needed harsh conditions [28]. No thiadiazole synthesis was reported for azobenzil, which would be reasonable as the molecule was not likely to be in the *cis*-arrangement [39].

4.3 Synthesis of 1,2,4-Thiadiazoles

The Lawesson's reagent transformed 1,2-oxazole into a 1,2,4-thiadiazole. This conversion proceeded through a thioamide intermediate, which underwent rearrangement (Scheme 4.7) [39, 40].

The rational design, formation, and anticancer activity of some nortopsentin analogues bearing 1,2,4-thiadiazole as main heterocyclic spacers instead of parent imidazole ring were described. A facile and high yielding synthesis of bis(indolyl)-1,2,4-thiadiazoles occurred employing a relatively benign iodobenzene diacetate (IBD) reagent. A fast formation of bis(indolyl)-1,2,4-thiadiazoles included the oxidative dimerization of indolyl thioamides employing IBD at rt (Scheme 4.8) [41–43].

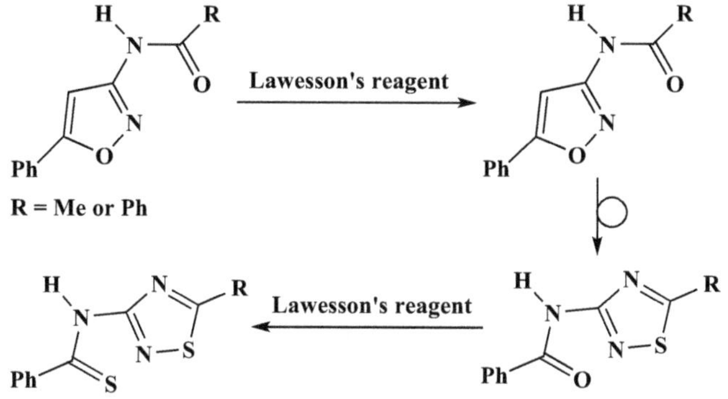

4.3 Synthesis of 1,2,4-Thiadiazoles

Scheme 4.7 Synthesis of 1,2,4-thiadiazoles

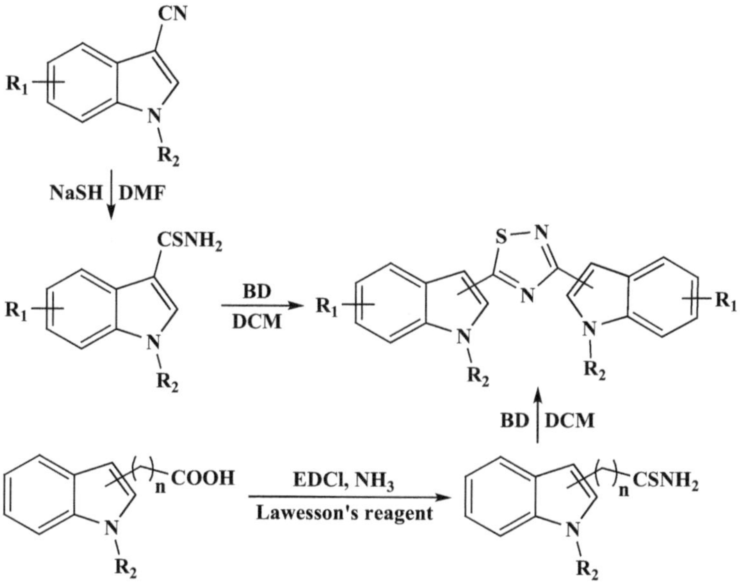

Scheme 4.8 Synthesis of bis(indolyl)-1,2,4-thiadiazoles

The 5-(4′-methoxy)phenyl-1,2,4-thiadiazole was synthesized by this synthetic route using 4-methoxythiobenzamide as a stating compound in place of thiobenzamide. The 4-methoxythiobenzamide was obtained by thionation of 4-methoxybenzamide with LR in refluxing THF (Scheme 4.9) [44].

The 5-(4′-cyano)phenyl-1,2,4-thiadiazole was prepared from 4-cyanothiobenzamide in this synthetic pathway. The 4-cyanothiobenzamide is not commercially accessible and was synthesized by thionation of 4-cyanobenzamide

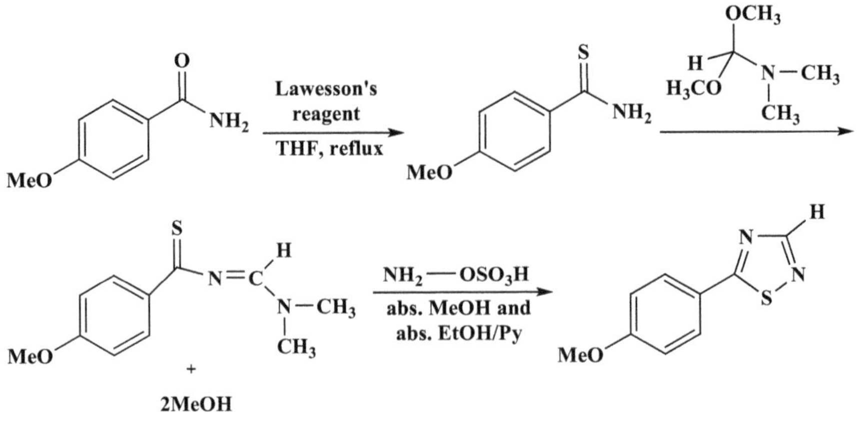

Scheme 4.9 Synthesis of 5-(4'-methoxy)phenyl-1,2,4-thiadiazole

with LR. The 4-cyanobenzamide was prepared from 4-cyanobenzoic acid by the reaction of 4-cyanobenzoyl chloride with aqueous NH_3 (Scheme 4.10) [45].

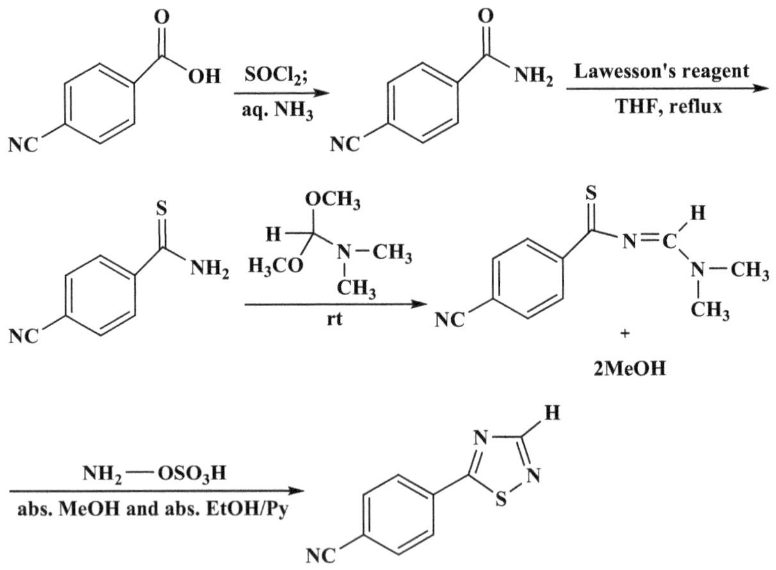

Scheme 4.10 Synthesis of 5-(4'-cyano)phenyl-1,2,4-thiadiazole

4.4 Synthesis of 1,3,4-Thiadiazoles

Many 2,5-diaryl-1,3,4-thiadiazoles were synthesized in 74–91% yield employing LR under MWI without solvent in 3–8 min [46]. For instance, 2-phenyl-5-*m*-tolyl-1,3,4-thiadiazole was obtained in 91% yield from of a mixture of diamide and LR (Scheme 4.11) [47].

Dalip and coworkers [48] prepared a series of 5-(3-indolyl)-2-substituted 1,3,4-thiadiazoles. The *N,N'*-diacylhydrazines, prepared by reaction of indole-3-carboxylic acid with aryl or heteroaryl hydrazides, were reacted with LR to afford the indolyl-1,3,4-thiadiazoles in good yield. The indolyl-1,3,4-thiadiazole with 4-benzyloxy-3-methoxyphenyl and 5-bromo indolyl substituents is the most active in suppressing the growth of cancer cells. A mixture of indole-3-carboxylic acid, 1-ethyl-3-(3-dimethylaminopropyl)carbodiimide hydrochloride, and 1-hydroxybenzotriazole in dry THF was stirred at rt for 15 min. Suitable arylhydrazide was added into this reaction mixture with continued stirring at rt for 6 h. Then, the solid 1,2-diacylhydrazine was filtered off. A mixture of 1,2-diacylhydrazines and LR in THF was refluxed at 80 °C for 5 h (Scheme 4.12) [43, 49, 50].

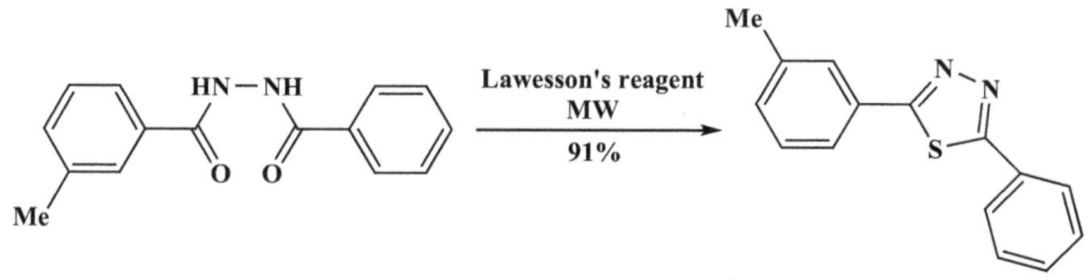

Scheme 4.11 Synthesis of 2-phenyl-5-*m*-tolyl-1,3,4-thiadiazole

R = H, Br
R_1 = C_6H_5, 2,3-di-OCH$_3$C$_6$H$_3$

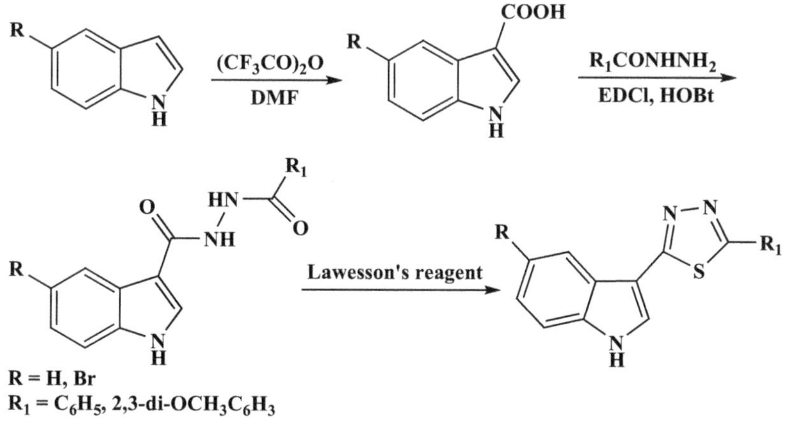

Scheme 4.12 Synthesis of 5-(3-indolyl)-2-substituted 1,3,4-thiadiazoles

The 1,3,4-thiadiazoles were synthesized by the reaction of diacylhydrazines with a sulfur source. The reaction included thionation of CO groups and after that cyclization with removal of hydrogen sulfide. The P_2S_5 is generally employed for this cyclization but needs more reaction time and excess reagent, which mostly affords less yields and side-products (Scheme 4.13). The utilization of LR afforded higher yields and cleaner reactions. This cyclization took place under MW and solvent-free conditions to provide the 1,3,4-thiadiazoles in high yields and less reaction times [49].

Kaleta and coworkers [51] synthesized novel thiadiazole derivatives utilizing fluorous LR (Scheme 4.14).

The thionation of amides, 1,4-diketones, N-(2-oxoalkyl)amides, and N,N'-acylhydrazines with fluorous LR produced thioamides, thiophenes, 1,3-thiazoles, and 1,3,4-thiadiazoles in high yields (Scheme 4.15). In most of the cases, the final product was obtained by simple filtration [51, 52].

The most popular approach for the formation of this family of compounds included cyclization and dehydration of thiohydrazides or other substrates with S–C–N–N–C–S functionality [53–55]. Generally, $POCl_3$ or H_2SO_4 has been utilized in these reactions. Another pathway for 1,3,4-thiadiazole ring formation was through

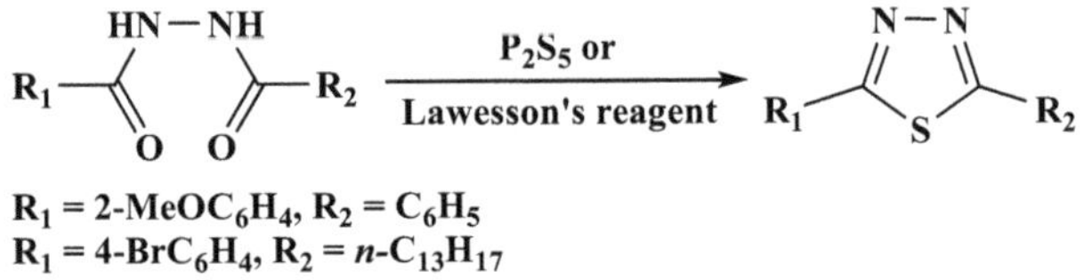

$R_1 = 2\text{-MeOC}_6H_4$, $R_2 = C_6H_5$
$R_1 = 4\text{-BrC}_6H_4$, $R_2 = n\text{-}C_{13}H_{17}$

Scheme 4.13 Synthesis of 1,3,4-thiadiazoles

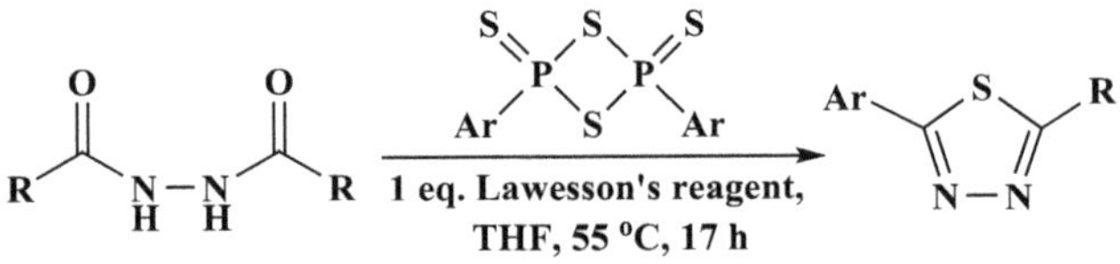

Scheme 4.14 Synthesis of thiadiazoles

Scheme 4.15 Synthesis of thioamides, thiophenes, 1,3-thiazoles, and 1,3,4-thiadiazoles

X, Y = NH, CH, OMe

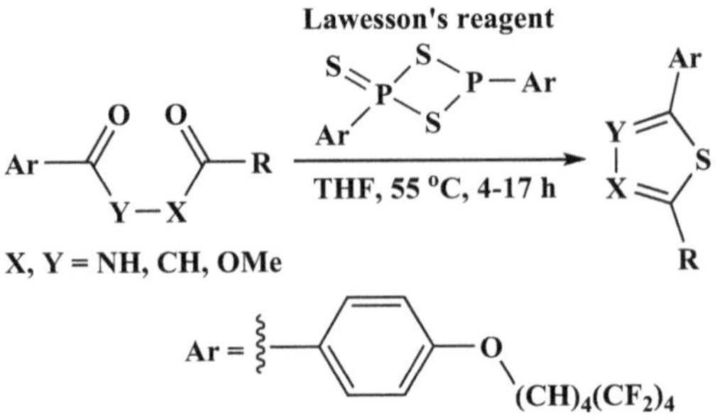

4.4 Synthesis of 1,3,4-Thiadiazoles

the exchange of oxygen atom in 1,3,4-oxadiazole with sulfur utilizing P_4S_{10} or $SC(NH_2)_2$ [56]. Lawesson has reported the thionation of 1,2-diacylhydrazines with 2,4-bis(4-methoxyphenyl)-1,2,3,4-dithiadiphosphetane (Lawesson's reagent), and after that spontaneous cyclization and dehydrosulfurization as an improved approach of thiadiazole ring synthesis. Disappointingly, this analysis only dealt with a narrow class of compounds, mainly dialkyl thiadiazoles. The present approach is a beneficial pathway for the formation of different types of substituted 1,3,4-thiadiazoles (Scheme 4.16). Their formation by another approach was problematic and the product yields were low. This approach was particularly beneficial for the formation of bis-thiadiazoles (Scheme 4.17) and polymers bearing 1,3,4-thiadiazole unit

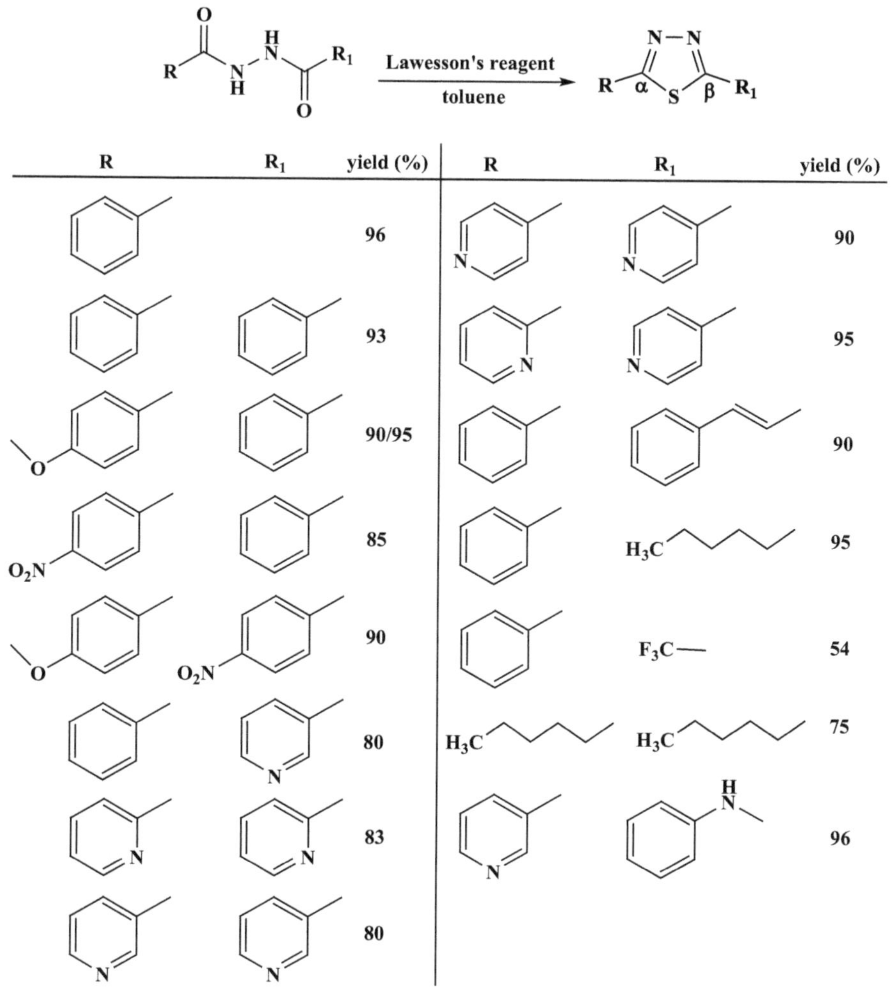

Scheme 4.16 Synthesis of 1,3,4-thiadiazoles

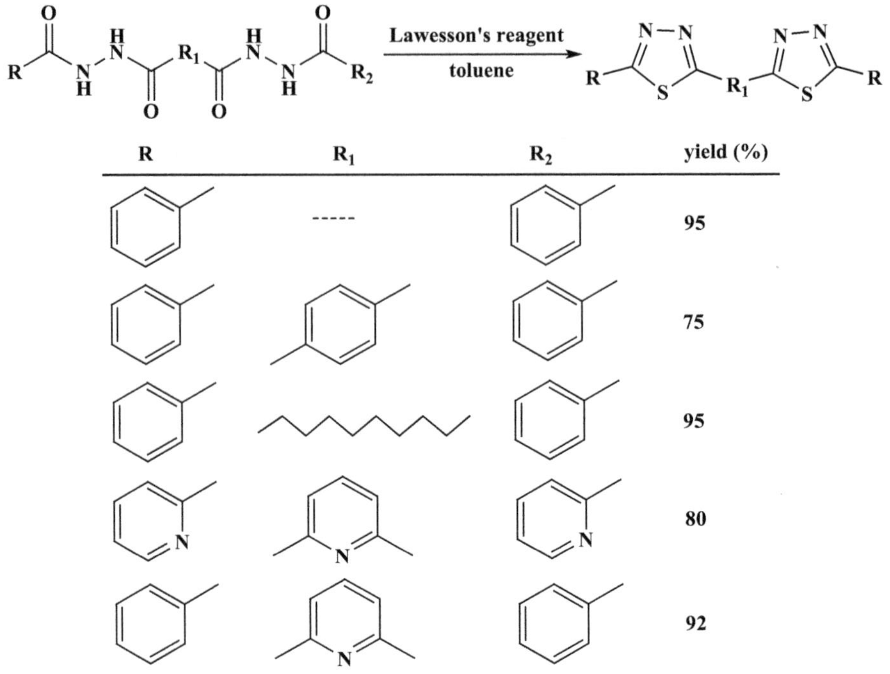

Scheme 4.17 Synthesis of bis-thiadiazoles

(Scheme 4.18), which are interesting complexing agents and may be utilized as liquid crystals and conducting polymers. The mild reaction conditions allowed the synthesis of few sensitive compounds in good yield and purity. No by-product was obtained thus making the isolation and purification of 1,3,4-thiadiazoles easy and

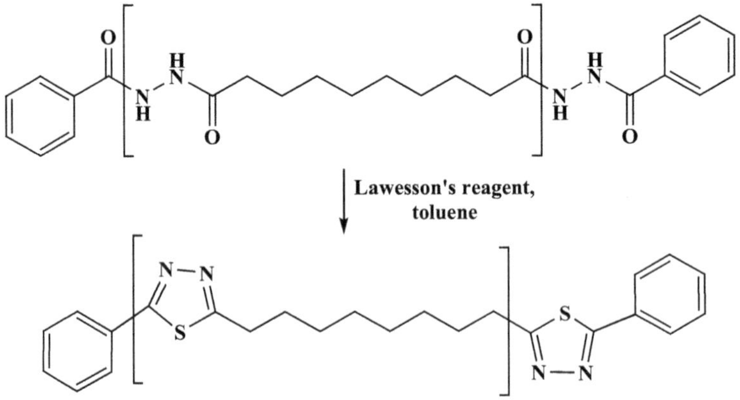

Scheme 4.18 Synthesis of polymers bearing 1,3,4-thiadiazole unit

4.4 Synthesis of 1,3,4-Thiadiazoles

straightforward from the reaction mixture. The mechanism of this reaction was not known with certainty but perhaps included thionation of both CO groups and after that simultaneous cyclization with the loss of H_2S [57, 58].

The lack of structural variation was obviously related to the ease of formation of these units and the limitations of synthetic approach in the literature. The cyclization of diacylhydrazides utilizing a source of sulfur like P_2S_5 or LR afforded 1,3,4-thiadiazoles substituted at the 2- and 5-positions with alkyl, cycloalkyl, and aryl units (Scheme 4.19) [59, 60].

The 1,3,4-thiadiazoles are relatively common in the liquid crystal literature although the variety of structural modifications that have been studied are actually very less. The majority of systems possess 1,3,4-thiadiazole nucleus substituted at the 2- and 5-positions by a combination of aryl and alkyl/cycloalkyl units or aryl units (Scheme 4.20) [61–67]. Classically, these compounds are synthesized through the sulfurization of suitably substituted N,N'-diacyldiazanes which are in turn prepared by reacting hydrazides with acid chlorides. Previous reports employed P_2S_5 as the sulfurization reagent although LR soon emerged as the most reproducible and reliable reagent [59, 68, 69].

The esters were synthesized by standard Williamson ether synthesis [70]. The hydrazides were prepared by treatment of esters with NH_2NH_2 in C_2H_5OH [71]. The formation of ethyl oxalyl diazanes was completed by treatment of hydrazides with ethyl oxalyl chloride in tetrahydrofuran and TEA. The cyclization of ethyl oxalyl diazanes employing LR at rt afforded purified 1,3,4-thiadiazole-2-carboxylate esters in excellent yields. The saponification of 1,3,4-thiadiazole-2-carboxylate esters afforded sodium salts. These sodium salts were unstable when subjected to high vacuum for prolonged time and easily decarboxylate once all traces of H_2O have been removed by drying method. The transformation of sodium salts to acid chlorides

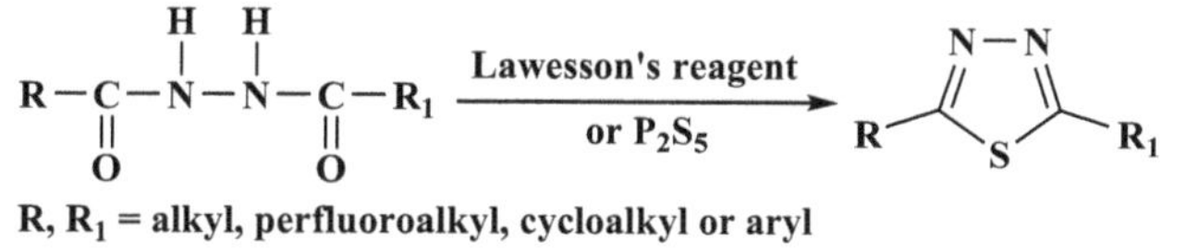

Scheme 4.19 Synthesis of 1,3,4-thiadiazoles

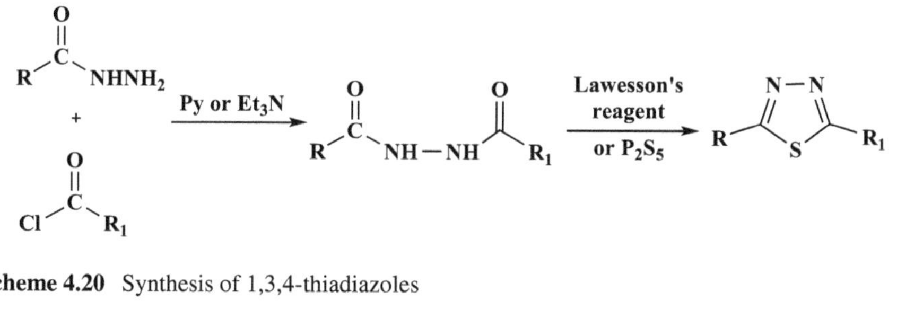

Scheme 4.20 Synthesis of 1,3,4-thiadiazoles

followed by in situ esterification afforded desired compounds (Schemes 4.21 and 4.22) [60].

The potential scope of this approach was evident when the methodology was extended to one-step formation of 1,3,4-thiadiazoles from acid hydrazides. The 4-toluic acid and 3-fluorophenyl hydrazide were treated with 2 eq. phosphorus pentasulfide in T$_3$P under reaction conditions of entry 2. After 4 h of heating at 80 °C, product (entry 1) was formed in 92% yield along with a small amount of 1,3,4-oxadiazole (3%) as by-product. After some optimization reactions, it was observed that application of T$_3$P (1.2 eq. 50% solution in ethyl acetate) and 1.5 eq. phosphorus pentasulfide was optimum to promote the reaction completion. However, the formation of side-product 1,3,4-oxadiazole was not avoided under any of the conditions. The outcome was same when LR was utilized instead of phosphorus pentasulfide. However, the reaction was unsuccessful in providing the thiadiazole when T$_3$P in dimethylformamide was employed. On further examination, it was observed that the thionation reagent (phosphorus pentasulfide or LR) was completely consumed by

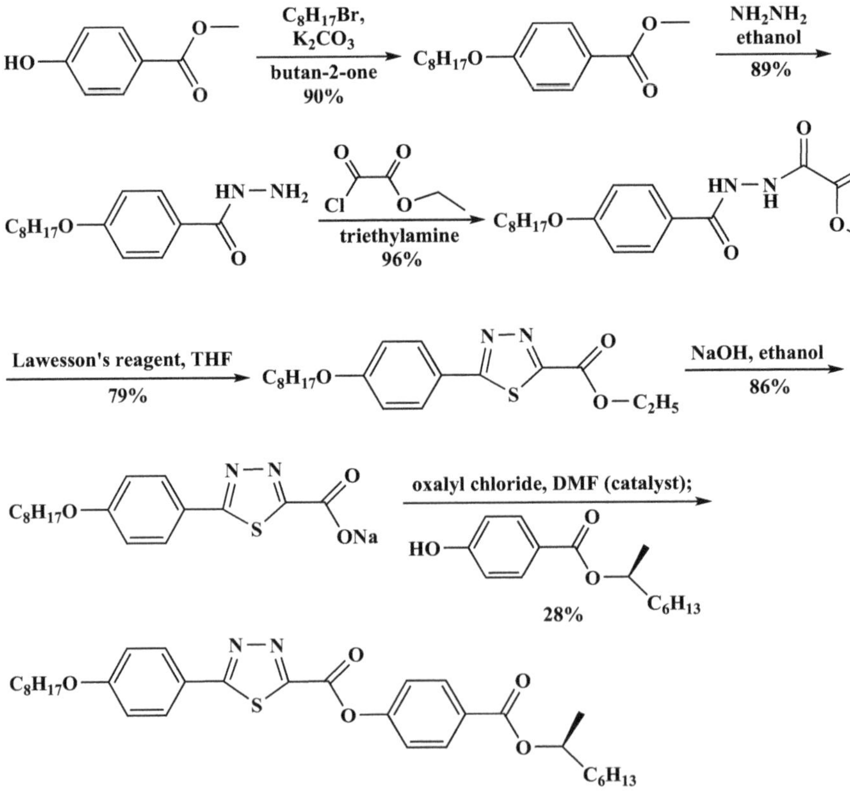

Scheme 4.21 Synthesis of 1,3,4-thiadiazole-2-carboxylate esters

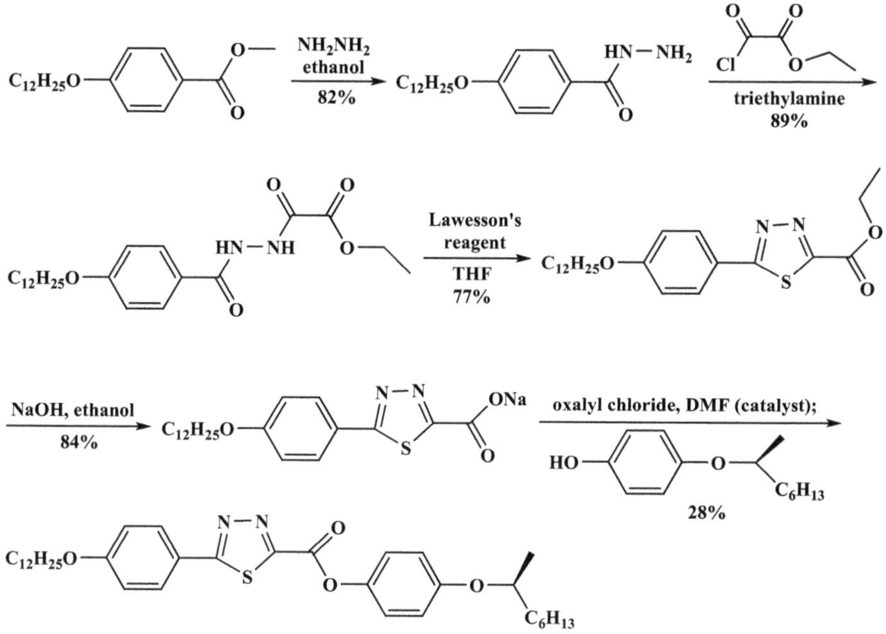

4.4 Synthesis of 1,3,4-Thiadiazoles

Scheme 4.22 Synthesis of 1,3,4-thiadiazole-2-carboxylate esters

dimethylformamide (reaction medium) to afford the *N,N*-dimethylthioformamide, thus unavailable for the synthesis of thiadiazole (Scheme 4.23) [72, 73].

The diacyldiazane derivative was cyclized to 1,3,4-thiadiazole-2-carboxylate ester employing an appropriate sulfurization agent. The chemoselective sulfurization of an

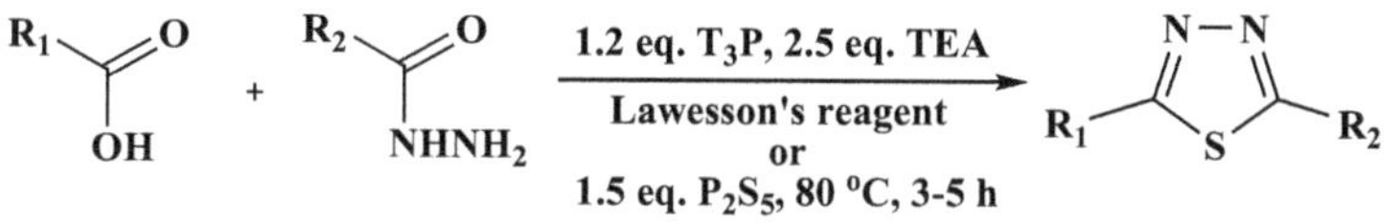

entry	R_1	R_2	yield (%)
1	4-Me-Ph	3-F-Ph	92
2	4-CN-Ph	5-Br-pyridin-3-yl	86
3	cyclopentyl	3-Me-4-NO_2-Ph	93
4	cyclobutyl	*t*-Bu	87
5	2-Me-propyl	3-F-Ph	85
6	4-NO_2-Ph	*t*-Bu	90
7	5-Br-furan-2-yl	Me	94
8	*N*-Boc-piperidin-4-yl	3-Me-4-NO_2-Ph	87
9	4-CF_3-β-styryl	H	88
10	4-Cl-2-Me-Ph	Me	93

Scheme 4.23 Synthesis of thiadiazoles

amide carbonyl with an ester carbonyl was described by Scheibye and coworkers [74] employing LR. The main diazane was targeted as an appropriate 1,3,4-thiadiazole substrate (Scheme 4.24). A standard Williamson etherification was employed for the synthesis of ether from phenol [75]. The benzohydrazide was synthesized by the reaction of a large excess of hydrazine hydrate with ester [76]. The ethyl oxalyl chloride was reacted with benzohydrazide to afford the tricarbonyl intermediate. These compounds were highly polar, difficult to purify by chromatography, and were utilized crude in the consequent cyclization step. The sulfurization and ring-closure of tricarbonyl intermediate occurred utilizing LR in anhydrous tetrahydrofuran [77]. An application of fresh LR was very valuable because older reagent

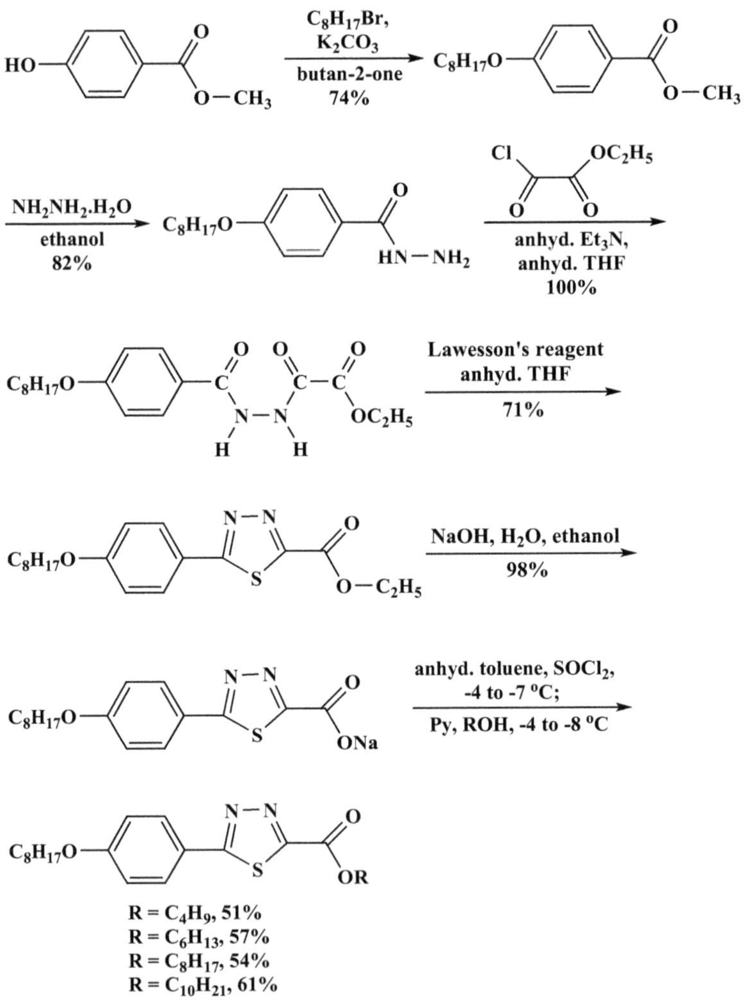

Scheme 4.24 Synthesis of 1,3,4-thiadiazoles

4.4 Synthesis of 1,3,4-Thiadiazoles

afforded less yields (in few cases a 25% reduction in yield was reported). Careful monitoring of the reaction was also important as prolonged reaction times resulted in the formation of significant amounts of unknown side-products. At this point the ester was hydrolyzed to carboxylic acid derivative and subsequently esterified with suitable alcohols [78]. Unfortunately, the 1,3,4-thiadiazole-2-carboxylic acid was proved to be unstable in solution and underwent spontaneous decarboxylation (the solid carboxylic acids also underwent decarboxylation over many days). An esterification of 5-(4-methoxyphenyl)-1,3,4-thiadiazole-2-carboxylic acid with p-cresol employing N,N'-dicyclohexylcarbodiimide/4-dimethylaminopyridinr afforded ester in unsatisfactory yield (25%) with the remainder of the material being lost to decarboxylation. The decarboxylation has been seen earlier in 5-amino-1,3,4-thiadiazole-2-carboxylic acids [79–82]. Therefore, the sodium salt of carboxylic acid (a stable compound unless heated in solution on which decarboxylation was reported) was isolated. A direct transformation of carboxylic acid into acid chloride derivative followed by in situ esterification with desired alcohols was expected to provide the ester adducts. This reaction was proved to be highly sensitive to temperature and competitive decarboxylation was again found unless the reaction temperature was carefully maintained at 26–28 °C. For instance, reaction temperatures of 210–217 °C afforded decarboxylation side-product in 20% yield while temperatures of 24–27 °C resulted in 5–17% decarboxylation. The reaction temperatures of 26–28 °C afforded only 0–4% decarboxylation and yields of the purified final esterification products were good (51–61%) [69].

The aryl methyl ester on treatment with hydrazine monohydrate was transformed into hydrazide. The hydrazide was reacted with methyl oxalyl chloride and TEA to afford the diacyl hydrazide intermediate [83], which was finally cyclized to 1,3,4-thiadiazole upon reaction with LR (Scheme 4.25).

The amino-substituted thiadiazoles are very popular building blocks because of their versatility. In contrast to many examples of the application of aryl, amino-, and halo-substituted 1,3,4-thiadiazoles, a few examples of efficient formation of 2-carboxy-1,3,4-thiadiazole compounds are reported in the literature. One example of a methyl ester-substituted 1,3,4-thiadiazole was described by Garfunkle and coworkers

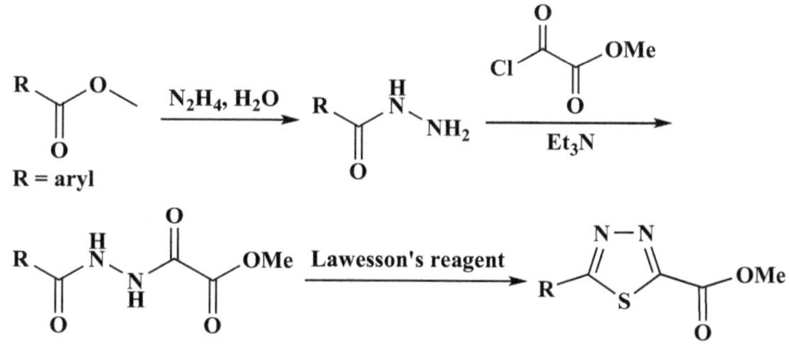

Scheme 4.25 Synthesis of 1,3,4-thiadiazoles

[83] as part of the formation of inhibitors of fatty acid amide hydrolase (FAAH). The family of inhibitors included compounds bearing different five-membered heterocycles such as 1,3,4-thiadiazoles. The hydrazide compounds were treated with methyl oxalyl chloride to provide the tricarbonyl compounds, which were then cyclized with LR to afford the 2-carboxy-1,3,4-thiadiazoles. The Grignard or organolithium reagents were then employed to transform the ester group into a ketone and provided final products, 1,3,4-thiadiazoles. Although the ester functionality was not preserved in the desired compounds and experimental details were less, the methodology showed the application of 2-carboxy ester substitution in an intermediate step (Scheme 4.26) [69, 84].

The reaction of 1,4-diamide, bearing an ester functional group, with P_4S_{10} in pyridine at 80 °C afforded thiadiazole with decreased yield, almost 30%, and no product was formed when more than 0.01 mol of starting compound was employed (Scheme 4.27) [68]. In contrast, successful conversion to thiadiazole was achieved when 1,4-diamide was first reacted with Lawesson's reagent in a refluxing solvent

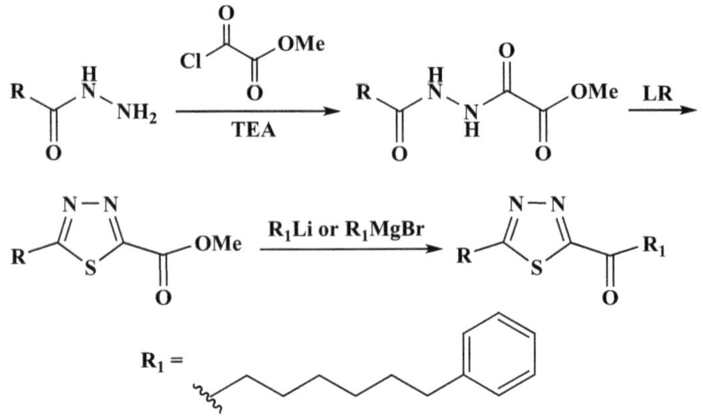

Scheme 4.26 Synthesis of 1,3,4-thiadiazoles

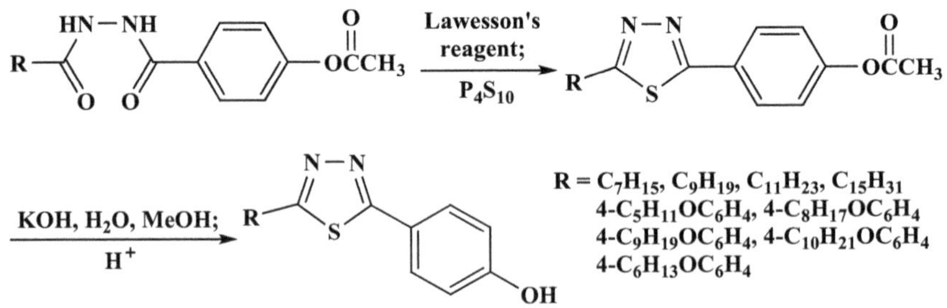

Scheme 4.27 Synthesis of thiadiazoles

4.4 Synthesis of 1,3,4-Thiadiazoles

mixture of toluene (dry)/pyridine (dry) for 4 h and further reacted with P_4S_{10} in pyridine for 4 h at 80 °C. The crude product was hydrolyzed to hydroxyphenylthiadiazole in 52–90% yield [85].

The thionation–cyclization of 1,2-diacylhydrazines to 1,3,4-thiadiazoles occurred by reacting with LR under MW heating without solvent (Scheme 4.28) [47].

The 16,17-unsaturated N,N'-diacylhydrazines were reacted with LR to afford the 17β-1′,3′,4′-thiadiazoles, but the efforts for the formation of amino-substituted analogue were not successful (Scheme 4.29). Although the S-containing reagent was employed in excess, synthesis of 17β-1′,3′,4′-oxadiazoles was also found in few cases. The deacetylation of 3β-acetates in basic media provided thiadiazole [86, 87].

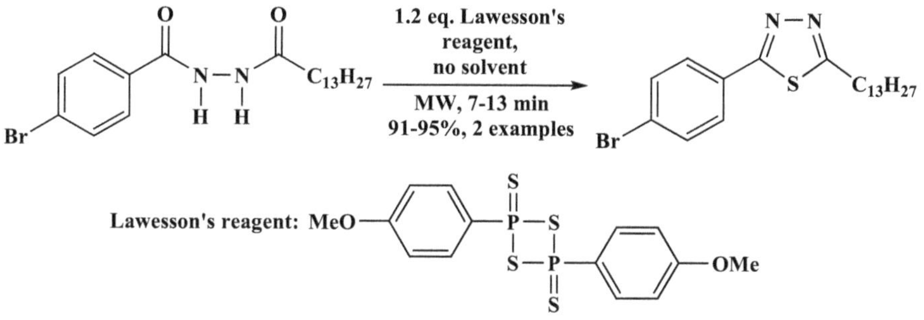

Scheme 4.28 Synthesis of 1,3,4-thiadiazole

R = Ph, 3-pyridyl, 4-pyridyl, 2-furyl, CH_3, NH_2

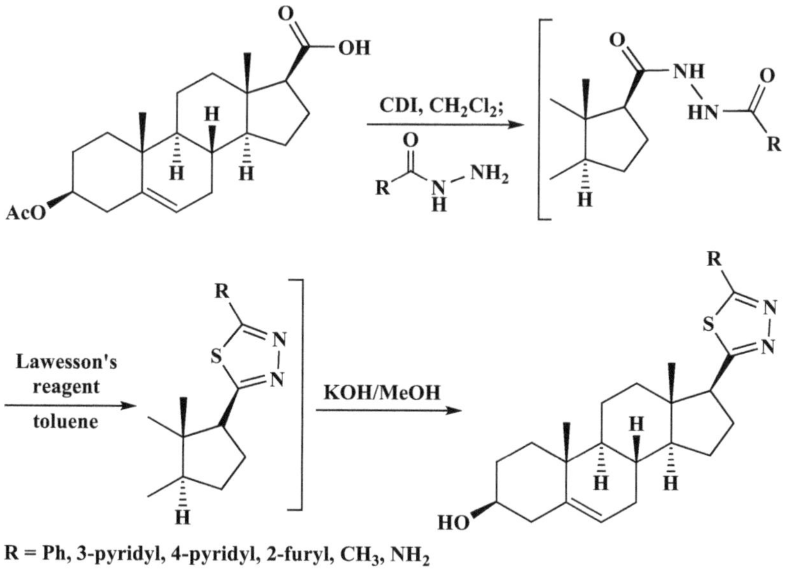

Scheme 4.29 Synthesis of 17β-1′,3′,4′-thiadiazoles

A thiazole bearing *C*-glucoside was prepared as described in Scheme 4.30. The reaction of acid and a hydrazide, like furan-2-carbohydrazide, employing 1-ethyl-3-(3-dimethylaminopropyl)carbodiimide, HOBt, and NMM in *N,N*-dimethylformamide (DMF) afforded acylhydrazide, which was cyclized utilizing LR to produce the thiadiazole in 81% over two steps. The deprotection of four benzyl groups employing trimethyliodosilane [88] followed by peracetylation utilizing Ac$_2$O and filtration by either column chromatography or recrystallization produced tetraacetate. The tetraacetate was hydrolyzed with sodium methoxide in CH$_3$OH to afford the *C*-glucoside in 31% yield over three steps [89].

A classical reaction of acid with hydrazide and then cyclization generated acylhydrazide, which was reacted with trimethylsilyltrifluoromethanesulfonate and Ac$_2$O to afford the tetraacetate. The peracetylated compound was hydrolyzed to afford the thiadiazole derivative in 10% yield over four steps. This alternative deprotection

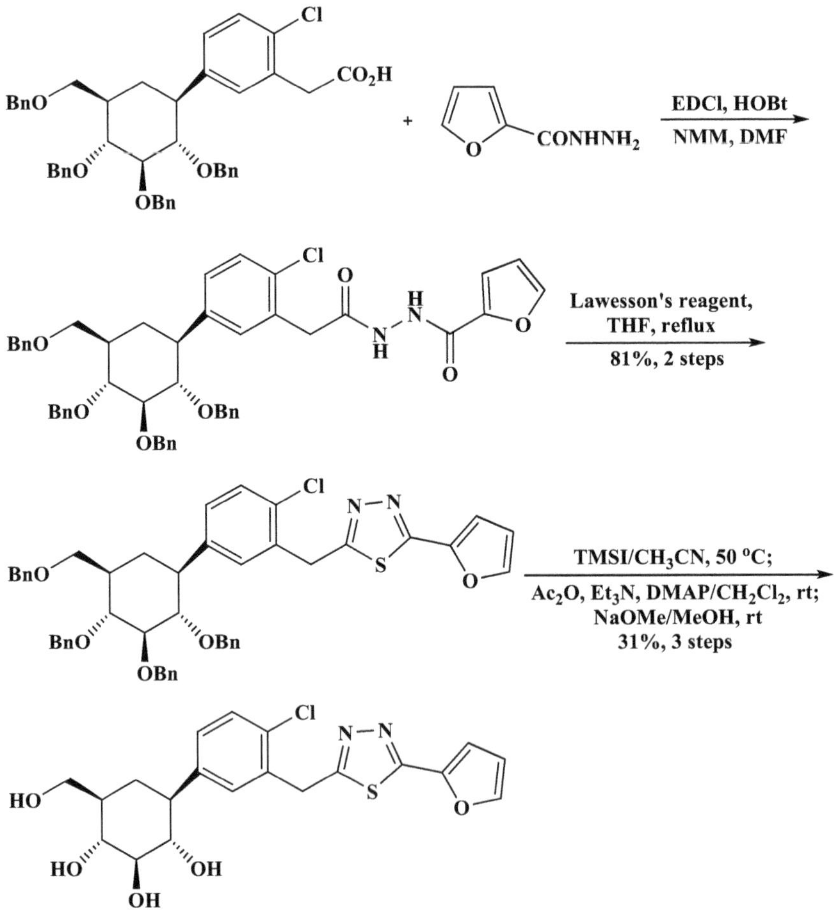

Scheme 4.30 Synthesis of thiazole bearing C-glucoside

4.4 Synthesis of 1,3,4-Thiadiazoles

approach was frequently used, particularly in the case of unsatisfactory reactions with trimethylsilyl iodide (Scheme 4.31) [89, 90].

A traditional Robinson–Gabriel cyclization of 2-acylamino carbonyl compounds was modified for the formation of oxazole. This cyclization used Burgess reagent (including its PEG-supported analogue) as a mild dehydrating reagent under single-mode MW heating conditions. The most problematic cyclization of unstable 2-acylamino aldehydes also occurred easily under these conditions. The multi-substituted oxazoles were synthesized from simple ketone and primary amide building blocks in a MW-assisted one-pot procedure [91]. The ketones were made to react with hypervalent iodine(III) sulfonate [hydroxyl-(2,4-dinitrobenzene)-sulfonyloxy)iodo]benzene (HDNIB) under MW and solvent-free conditions to produce the intermediate [(2,4-dinitrobenzene)sulfonyl]oxy carbonyl compound. The [(2,4-dinitrobenzene)sulfonyl]oxycarbonyl compound was condensed with acetamide or benzamide under MW heating conditions (Scheme 4.32) [92].

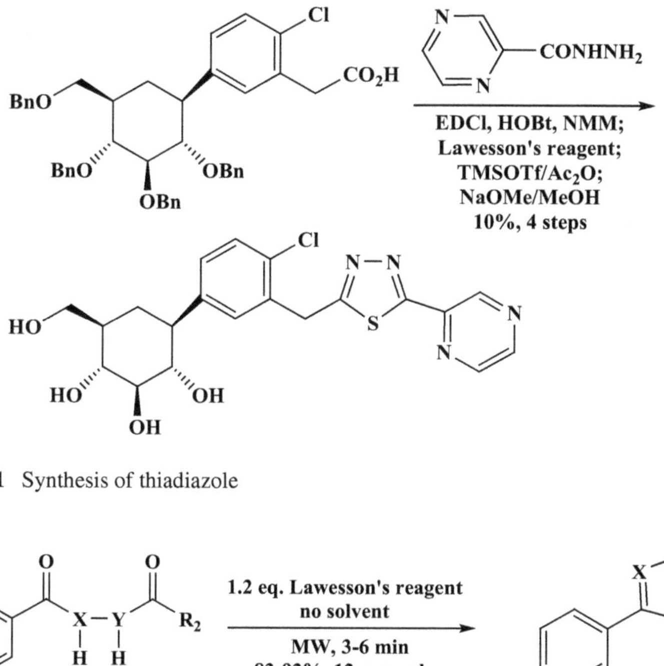

Scheme 4.31 Synthesis of thiadiazole

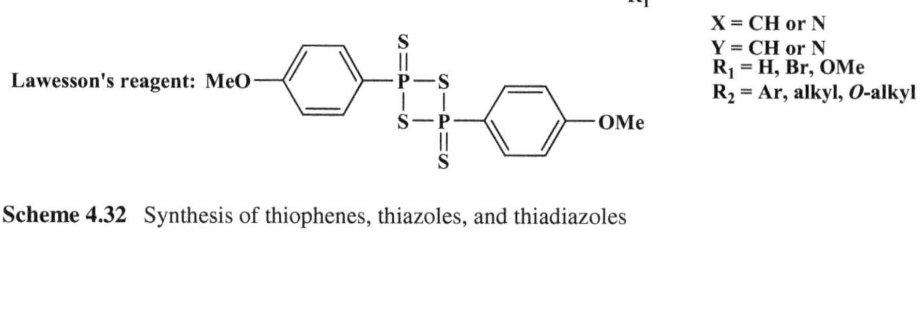

Scheme 4.32 Synthesis of thiophenes, thiazoles, and thiadiazoles

A MW-assisted approach offered potentially beneficial and efficient methods for the synthesis of 1,3,4-thiadiazoles. The researchers [47] discovered a class of thiophenes, 1,3-thiazoles, and 1,3,4-thiadiazoles through the cyclization of different dicarbonyl compounds. These compounds were mixed with LR in the absence of solvent and irradiated in a conventional MW oven. All but two of twelve different compounds were obtained in less reaction times and yields in excess of 83%. Han and coworkers [93] also described that MW-irradiated Lawesson's cyclizations for the synthesis of 2,5-diaryl-1,3,4-thiadiazoles were very efficient. The reaction mixture was removed immediately at this point to avoid the decomposition of products. After silica column purification, the 2,5-diaryl-1,3,4-thiadiazole products were obtained in high yield (80–91%) (Scheme 4.33). The benefits of MW-assisted reactions were very less reaction times, no need of solvent, and no requirement of anhydrous hydrocarbon solvents. However, the reactions could be difficult to scale up because of the probability of local superheating of the reaction mixture, although many compounds were formed on a multigram scale without significant decrease in yields.

The most common ring synthesizing method for 5-membered sulfur heterocyclic compounds in the literature involved the cyclization of 1,4-dicarbonyl species utilizing a thionating agent. Although the methods to afford these dicarbonyl compounds were highly diverse, analogous conditions were used to prepare the 1,3,4-thiadiazole, 1,3-thiazole, and thiophene systems from such compounds, based on the identity of 3- and 4-position bridging atoms (Scheme 4.34). A suggested mechanism included a ring-closure of 1,4-dicarbonyl compounds in which thionation of both carbonyl groups was the first step, and after that in situ cyclization and removal of hydrogen sulfide [23–27].

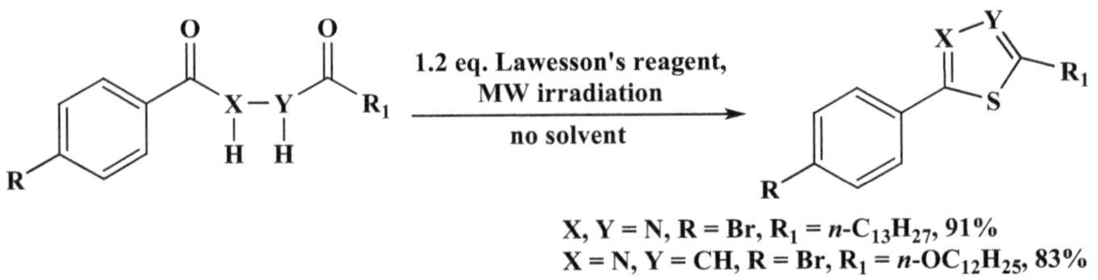

X, Y = N, R = Br, R_1 = n-$C_{13}H_{27}$, 91%
X = N, Y = CH, R = Br, R_1 = n-$OC_{12}H_{25}$, 83%

Scheme 4.33 Synthesis of 2,5-disubstituted 1,3,4-thiadiazoles

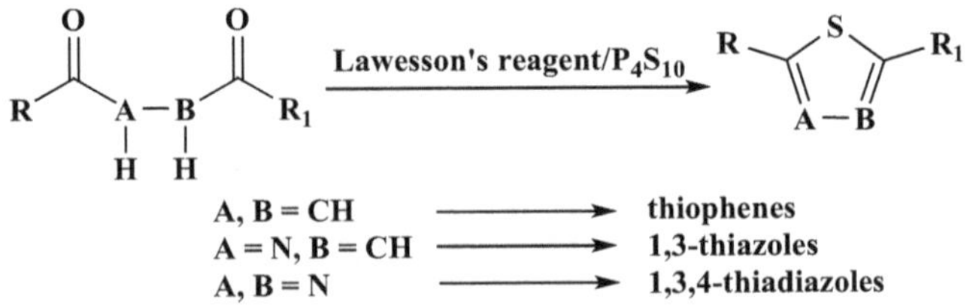

A, B = CH → thiophenes
A = N, B = CH → 1,3-thiazoles
A, B = N → 1,3,4-thiadiazoles

Scheme 4.34 Synthesis of 1,3,4-thiadiazoles, 1,3-thiazoles, and thiophenes

4.4 Synthesis of 1,3,4-Thiadiazoles

The substituted benzoic acids were transformed into esters by reacting them with $SOCl_2$ and CH_3OH, followed by reflux for 12 h. The esters were reacted with hydrazine hydrate in CH_3OH by refluxing at 60 °C to provide its hydrazide. The hydrazides were dissolved in dry THF and reacted with chloroacetyl chloride and TEA, followed by reflux to provide their N-chloroacetyl-N'-aroyl hydrazine. The N-chloroacetyl-N'-aroyl hydrazines were converted into thiadiazoles by reacting with LR in p-xylene at 140 °C (Scheme 4.35) [94].

Lee and coworkers [95] discovered 2-(4-((1H-1,2,4-triazol-1-yl)methyl)-5-(4-bromophenyl)-1-(2-chlorophenyl)-1H-pyrazol-3-yl)-5-t-butyl-1,3,4-thiadiazole (GCC-2680) as an effective, selective, and orally effective cannabinoid-1 receptor antagonist. The thiadiazoles were synthesized [96–102] through (i) treatment of carboxylic acid with a hydrazide compound and coupling reagents (1-ethyl-3-(3-dimethylaminopropyl)carbodiimide, 4-dimethylaminopyridine) and (ii) thionation–cyclization of product utilizing LR [47]. Alternatively, the acylhydrazine intermediate was also prepared by the reaction of hydrazide with acid-assisted coupling reagents like 4-dimethylaminopyridine, 1-ethyl-3-(3-dimethylaminopropyl)carbodiimide or 1-ethyl-3-(3-dimethylaminopropyl)carbodiimide, hydroxybenzotriazole, and N-methylmorpholine. For this sequence, the required hydrazide was synthesized by reacting ester with NH_2NH_2 in refluxing ethanol (Scheme 4.36) [103].

The thiophenes (X–Y = CH) were synthesized by LR-assisted cyclization of 1,4-dicarbonyl compounds under MWI without solvent [47]. The reaction was carried out by mixing two solid reagents in a glass tube kept inside a household MW apparatus and irradiating until the evolution of hydrogen sulfide stopped (Scheme 4.37) [104].

A MW-assisted condensation of NH_2OH with enaminoketones provided isoxazoles [105]. In a process similar to the Paal–Knorr thiophene formation, 2-aminoacyl

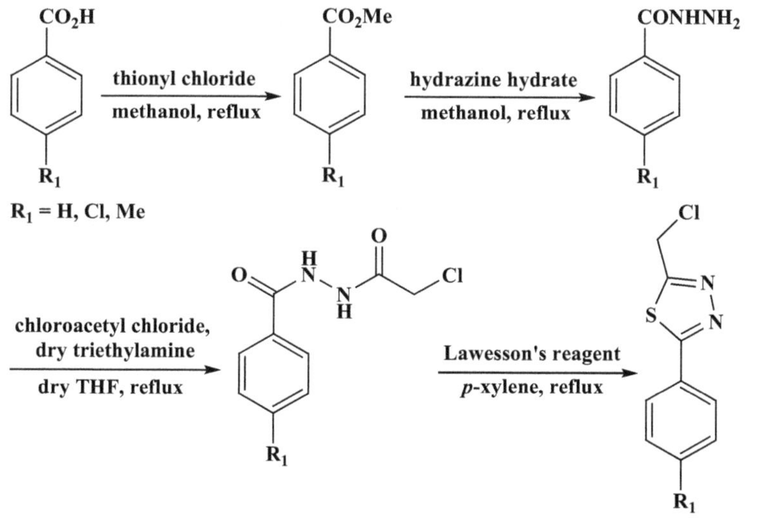

Scheme 4.35 Synthesis of thiadiazoles

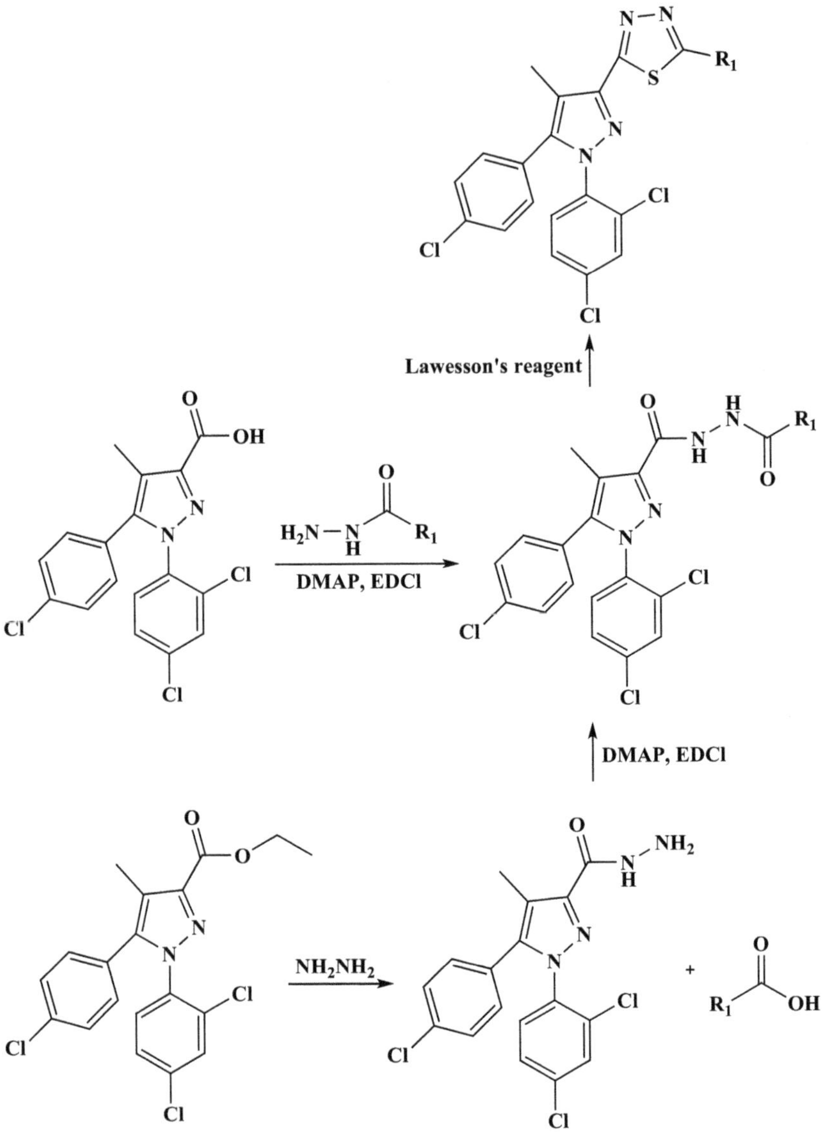

Scheme 4.36 Synthesis of thiadiazoles

carbonyl compounds were cyclized to thiazoles employing LR under MW heating (Scheme 4.38) [47].

The structure–activity relationship analysis on a series of diarylpyrazolyl thiadiazoles identified that cannabinoid-1 receptor antagonists have

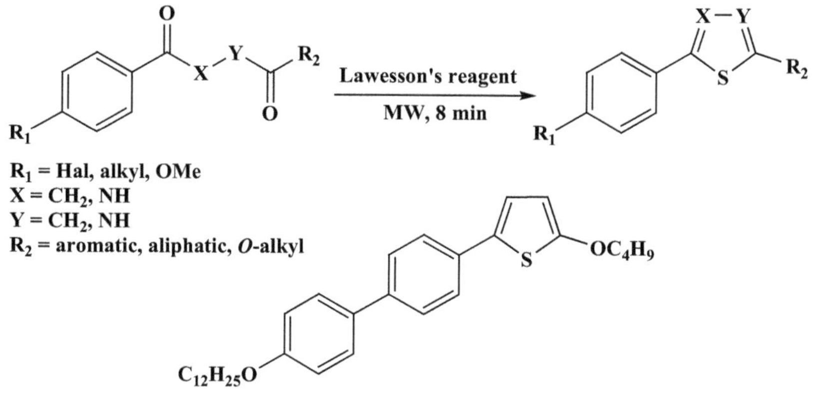

4.4 Synthesis of 1,3,4-Thiadiazoles

Scheme 4.37 Synthesis of thiophenes, thiazoles, and thiadiazoles

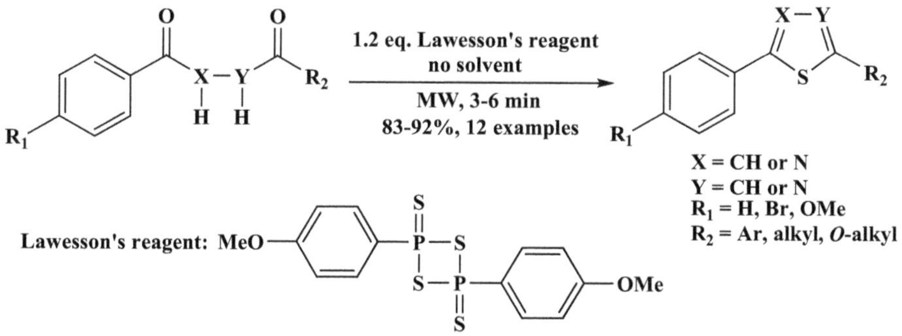

Scheme 4.38 Synthesis of thiophenes, thiazoles, and thiadiazoles

excellent selectivity and potency. Based on its exceptional in vivo efficiency in animal models and its favorable toxicological and pharmacokinetic profiles, 2-(4-((1H-1,2,4-triazol-1-yl)methyl)-5-(4-bromophenyl)-1-(2-chlorophenyl)-1H-pyrazol-3-yl)-5-t-butyl-1,3,4-thiadiazole was chosen as a preclinical applicant for curing obesity. The thiadiazole bearing 1,2,4-triazole was also prepared by a reaction sequence involving the key intermediate bromide. The bromide, prepared by treatment of pyrazole with N-bromosuccinimide in catalytic amounts of azobis-isobutyronitrile, was reacted with 1,2,4-triazole sodium derivatives to afford the ester. The hydrazinolysis of ester afforded hydrazide, which was reacted with an acid to afford the acyl hydrazide. The hydrolysis of ester and activation of acid followed by reaction with hydrazide in TEA provided acyl hydrazide. The thionation–cyclization was carried out employing LR under MWI to afford the thiadiazole (Scheme 4.39) [95, 103].

The thionation–cyclization of 1,2-diacylhydrazidines to 1,3,4-thiadiazoles was carried out with LR under MWI in a domestic microwave without solvent

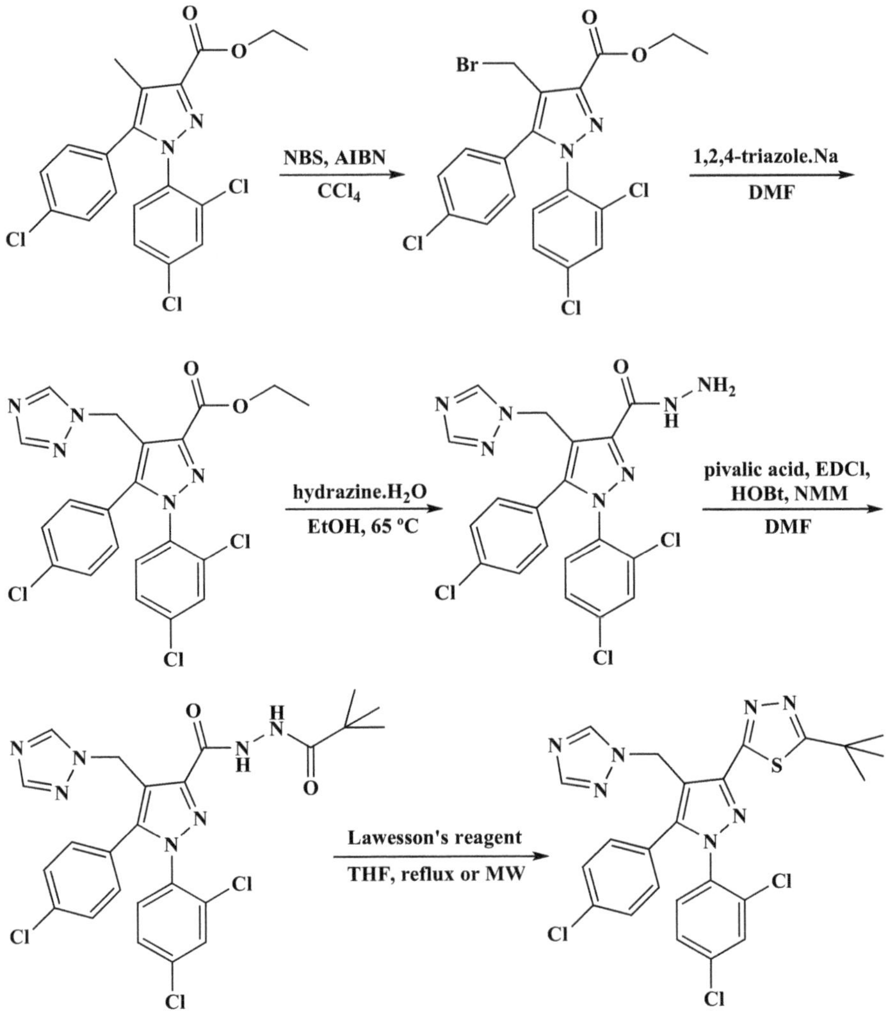

Scheme 4.39 Synthesis of thiadiazole bearing 1,2,4-triazole

(Scheme 4.40). This ring-closure approach was extended for the formation of different liquid crystals [47, 106].

The salt was found to be much more stable in comparison to free acid, and transformation to acid chloride in situ and after that treatment with different alcohols at rt provided first reasonably positive results (41% yield of aryl ester from carboxylate salt). Sybo et al. [69] then extended Bradley's work [60], determining that the major factor in controlling decarboxylation of the carboxylate salt was the temperature of reaction. The esterification reactions occurred with the best yields when a temperature range of −8 to −6 °C was maintained. Below this temperature range, the reaction not took place, whereas above this range the rate of decarboxylation

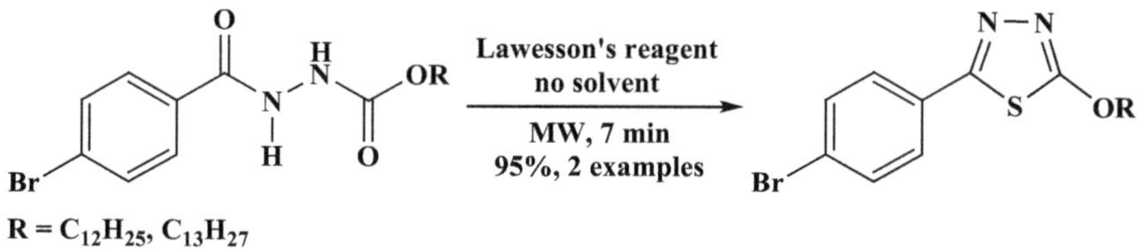

4.4 Synthesis of 1,3,4-Thiadiazoles

Scheme 4.40 Synthesis of 1,3,4-thiadiazoles

enhanced significantly. This caused the experimental procedure to be very complicated, as the addition of alcohols would cause the temperature of the solution to fluctuate. Together with a long reaction time to completely hydrolyze the ethyl ester, the reaction beginning after cyclization of the 1,3,4-thiadiazole ring proved to be quite demanding. Despite this, Sybo et al. [69] successfully prepared four liquid crystals (butyl, hexyl, octyl, and decyl esters) in moderate yield (51–61%, one-step) (Scheme 4.41).

More examples were reported employing late-stage cyclization strategies [107]. Despite this growing class in the literature, there persist relatively less examples of the method being employed to prepare the 1,3,4-thiadiazoles with substitution other than aromatic groups (Scheme 4.42). One of these examples was observed in work of Deokar and coworkers [108] as part of an analysis on the antimicrobial

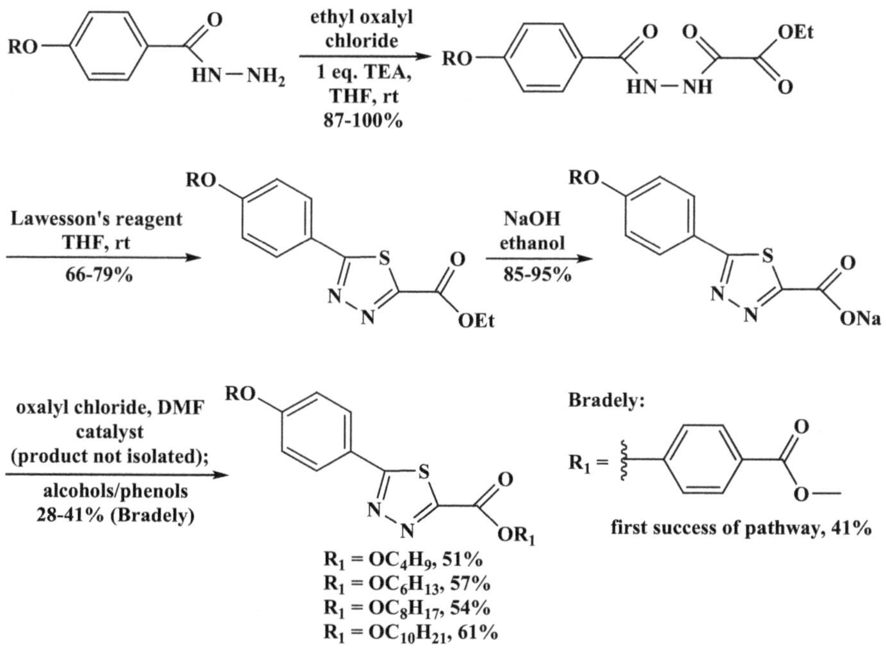

Scheme 4.41 Synthesis of 1,3,4-thiadiazoles

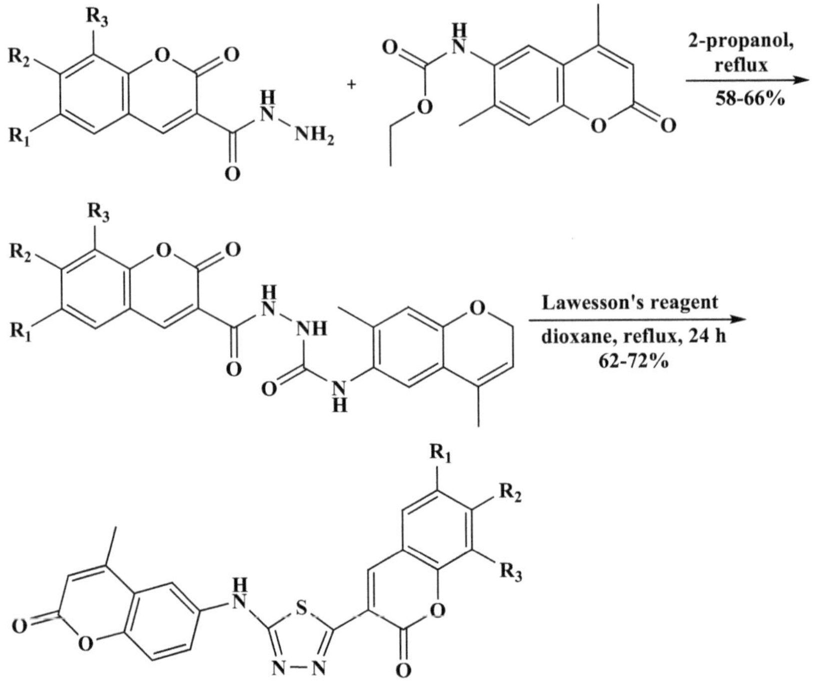

Scheme 4.42 Synthesis of 1,3,4-thiadiazoles

activity of some oxadiazole/thiadiazole compounds. The hydrazide was refluxed in 2-propanol with carbamate to provide the hydrazinecarboxamide, which was consequently cyclized employing LR in refluxing dioxane. The final product involved an amine linkage between the 2-position on the 1,3,4-thiadiazole heterocyclic and an aromatic functionality. However, many examples of amino-, halo-, nitrile-, and other substituted 1,3,4-thiadiazoles are observed within the context of ring modifying strategies.

Han and coworkers [109] employed Lawesson's cyclization approach to prepare the precursors to rod- and H-shaped liquid crystals. The final desired compounds had a 2,5-diphenyl-1,3,4-thiadiazole group with an ester-linked 4-decyloxyphenyl functionality. The main purpose of the analysis was to develop 1,3,4-thiadiazole-based liquid crystals with low-temperature mesophase ranges and to compare the liquid crystal behavior of dimers (H-shaped in this case) with that of their monomers. The 4-substituted benzoic acids were treated with $SOCl_2$ to provide the 4-substituted benzoyl chlorides. The 4-methoxybenzoic hydrazide was further treated with acyl chlorides to afford the 1,4-dicarbonyl compounds, which were further cyclized employing LR. The 1,3,4-thiadiazole precursors were further deprotected to phenols and reacted with 4-decyloxybenzoyl chloride to afford the final targets. One concept

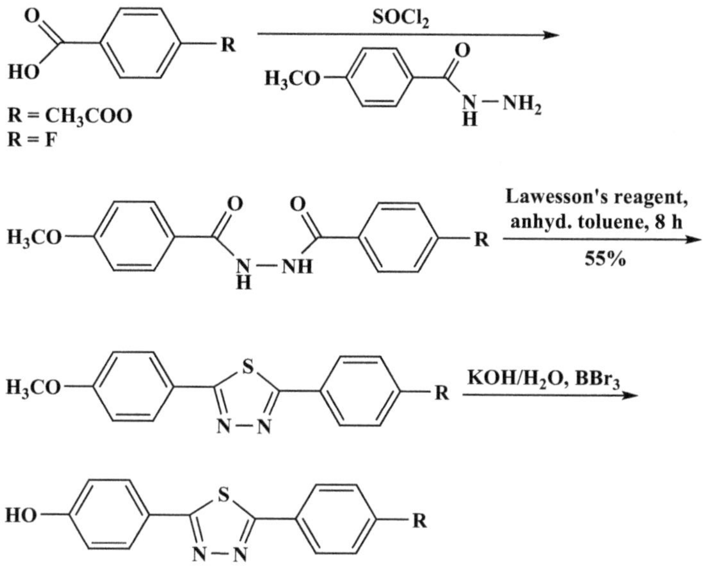

4.4 Synthesis of 1,3,4-Thiadiazoles

Scheme 4.43 Synthesis of 1,3,4-thiadiazoles

that can be seen in this synthetic methodology is another general method to 1,4-dicarbonyl compounds before cyclization: the synthesis of acyl chloride intermediate in situ (Scheme 4.43).

References

1. (a) N. Kaur. 2015. Palladium-catalyzed approach to the synthesis of S-heterocycles. Catal. Rev. 57: 478–564. (b) N. Kaur. 2019. Synthetic routes to seven and higher membered S-heterocycles by use of metal and nonmetal catalyzed reactions. Phosphorus, Sulfur, Silicon Relat. Elem. 194: 186–209. (c) N. Kaur. 2018. Synthesis of six- and seven-membered and larger heterocycles using Au and Ag catalysts. Inorg. Nano Met. Chem. 48: 541–568. (d) N. Kaur. 2019. Synthesis of five-membered heterocycles containing nitrogen heteroatom under ultrasonic irradiation. Mini Rev. Org. Chem. 16: 481–503. (e) N. Kaur, P. Bhardwaj, M. Devi, Y. Verma and P. Grewal. 2019. Photochemical reactions in five and six-membered polyheterocycles synthesis. Synth. Commun. 49: 2281–2318. (f) N. Kaur. 2019. Application of silver-promoted reactions in the synthesis of five-membered O-heterocycles. Synth. Commun. 49: 743–789.
2. (a) N. Kaur. 2015. Palladium catalysts: Synthesis of five-membered N-heterocycles fused with other heterocycles. Catal. Rev. 57: 1–78. (b) N. Kaur. 2019. Synthesis of seven and higher-membered heterocycles using ruthenium catalysts. Synth. Commun. 49: 617–661. (c) N. Kaur. 2018. Ionic liquid promoted eco-friendly and efficient synthesis of six-membered N-polyheterocycles. Curr. Org. Synth. 15: 1124–1146. (d) N. Kaur. 2019. Copper catalyzed synthesis of seven and higher-membered heterocycles. Synth. Commun. 49: 879–916. (e) N. Kaur. 2019. Ionic liquid assisted synthesis of S-heterocycles. Phosphorus, Sulfur, Silicon

Relat. Elem. 194: 165–185. (f) K. Rao, R. Tyagi, N. Kaur and D. Kishore. 2013. An expedient protocol to the synthesis of benzo(b)furans by palladium induced heterocyclization of corresponding 2-allylphenols containing electron rich and electron capturing substituents in the arene ring. J. Chem. 1–5. (g) N. Kaur, R. Tyagi and D. Kishore. 2016. Expedient protocols for the installation of 1,5-benzoazepino based privileged templates on 2-position of 1,4-benzodiazepine through a phenoxyl spacer. J. Heterocycl. Chem. 53: 643–646.
3. (a) N. Kaur. 2015. Role of microwaves in the synthesis of fused five-membered heterocycles with three N-heteroatoms. Synth. Commun. 45: 403–431. (b) N. Kaur. 2019. Palladium acetate and phosphine assisted synthesis of five-membered N-heterocycles. Synth. Commun. 49: 483–514. (c) N. Kaur. 2019. Ionic liquid: An efficient and recyclable medium for the synthesis of fused six-membered oxygen heterocycles. Synth. Commun. 49: 1679–1707. (d) N. Kaur. 2018. Ultrasound-assisted green synthesis of five-membered O- and S-heterocycles. Synth. Commun. 48: 1715–1738. (e) N. Kaur, R. Tyagi, M. Srivastava and D. Kishore. 2014. Application of dimethylaminomethylene ketone in heterocycles synthesis: Synthesis of 2-(isoxazolo, pyrazolo and pyrimido) substituted analogs of 1,4-benzodiazepin-5-carboxamides linked through an oxyphenyl bridge. J. Heterocycl. Chem. 51: E50–E54. (f) R. Tyagi, N. Kaur, B. Singh and D. Kishore. 2014. A novel synthetic protocol for the heteroannulation of oxocarbazole and oxoazacarbazole derivatives through corresponding oxoketene dithioacetals. J. Heterocycl. Chem. 51: 18–23.
4. M.H. Shih and C.L. Wu. 2005. Efficient syntheses of thiadiazoline and thiadiazole derivatives by the cyclization of 3-aryl-4-formylsydnone thiosemicarbazones with acetic anhydride and ferric chloride. Tetrahedron 61: 10917–10925.
5. V.A. Ogurtsov, O.A. Rakitin, C.W. Rees and A.A. Smolentsev. 2005. Synthesis of 1,3,4-thiadiazolines from 1,2-dithiole-3-thiones. Mendeleev Commun. 15: 55–56.
6. X. Xiong, L.X. Zhang, A.J. Zhang and X.S. Lu. 2002. Synthesis of 6-aryl-3-(D-gluco-pentitol-1-yl)-7H-1,2,4-triazolo[3,4-b][1,3,4]thiadiazines. Synth. Commun. 32: 3455–3459.
7. X.C. Wang, L. Zheng, Z.J. Quang and D.J. Xu. 2003. Solvent-free synthesis of 2-furyl-5-aryloxyacetylamido-1,3,4-thiadiazoles under microwave irradiation. Synth. Commun. 33: 2891–2897.
8. C.-H. Oh, H.-W. Cho, D. Baek and J.-H. Cho. 2002. Synthesis and antibacterial activity of 1β-methyl-2-(5-substituted thiazolo pyrrolidin-3-ylthio)carbapenem derivatives. Eur. J. Med. Chem. 37: 743–754.
9. M.A. Ilies, D. Vullo, J. Pastorek, A. Scozzafava, M. Ilies, M.T. Caproiu, S. Pastorokova and C.T. Supuran. 2003. Carbonic anhydrase inhibitors. Inhibition of tumor-associated isozyme IX by halogenosulfanilamide and halogenophenylaminobenzolamide derivatives. J. Med. Chem. 46: 2187–2196.
10. S. Masereel, R.F. Abbate, A. Scozzafava and C.T. Supuran. 2002. Carbonic anhydrase inhibitors: Anticonvulsant sulfonamides incorporating valproyl and other lipophilic moieties. J. Med. Chem. 45: 312–320.
11. C.T. Supuran and A. Scozzafava. 2000. Carbonic anhydrase inhibitors: Aromatic sulfonamides and disulfonamides act as efficient tumor growth inhibitors. J. Enzym. Inhib. 15: 597–610.
12. M. Amir and K. Shikha. 2004. Synthesis and anti-inflammatory, analgesic, ulcerogenic and lipid peroxidation activities of some new 2-[(2,6-dichloroanilino) phenyl]acetic acid derivatives. Eur. J. Med. Chem. 39: 535–545.
13. N. Demirbas, S.A. Karaoglu, A. Demirbas and K. Sancak. 2004. Synthesis and antimicrobial activities of some new 1-(5-phenylamino[1,3,4]thiadiazol-2-yl)methyl-5-oxo[1,2,4]triazole and 1-(4-phenyl-5-thioxo[1,2,4]triazol-3-yl)methyl-5-oxo[1,2,4]triazole derivatives. Eur. J. Med. Chem. 39: 793–804.
14. E. Palaska, G. Sahin, P. Kelicen, N.T. Durlu and G. Altinok. 2002. Synthesis and anti-inflammatory activity of 1-acylthiosemicarbazides, 1,3,4-oxadiazoles, 1,3,4-thiadiazoles and 1,2,4-triazole-3-thiones. Farmaco 57: 101–107.
15. B.S. Holla, K.N. Poorjary, B.S. Rao and M.K. Shivananda. 2002. New bis-aminomercaptotriazoles and bis-triazolothiadiazoles as possible anticancer agents. Eur. J. Med. Chem. 37: 511–517.

16. L.F. Awad and E.S.H. El-Ashry. 1998. Synthesis and conformational analysis of seco C-nucleosides and their diseco double-headed analogues of the 1,2,4-triazole, 1,2,4-triazolo[3,4-b]1,3,4-thiadiazole. Carbohydr. Res. 312: 9–22.
17. M.T.H. Khan, M.I. Choudary, K.M. Khan, M. Rani and A. Rahman. 2005. Structure-activity relationships of tyrosinase inhibitory combinatorial library of 2,5-disubstituted-1,3,4-oxadiazole analogues. Bioorg. Med. Chem. 13: 3385–3395.
18. J.S. Bradshaw, P. Huzzthy, C.W. McDaniel, C.Y. Zhu, N.K. Dalley and R.M. Izatt. 1990. Enantiomeric recognition of organic ammonium salts by chiral dialkyl-, dialkenyl-, and tetramethyl-substituted pyridino-18-crown-6 and tetramethyl-substituted bispyridino-18-crown-6 ligands: Comparison of temperature-dependent proton NMR and empirical force field techniques. J. Org. Chem. 55: 3129–3137.
19. C.P. Collier, G. Mattersteig, E.W. Wong, Y. Luo, K. Beverly, J. Sampaio, F.M. Raymo, J.F. Stoddart and J.R. Heath. 2000. A [2]catenane-based solid state electronically reconfigurable switch. Science 289: 1172–1175.
20. A.J. Elizabeth, E.S.A. Eric, A.M. Osgood, M.R. Lisa and B.B. Lawrence. 2003. Antibacterial activity of quinolone-macrocycle conjugates. Bioorg. Med. Chem. 13: 1635–1638.
21. L. Jian, L. Tington and W. Yongmei. 2008. Synthesis, structure and biological activity of pseudopeptidic macrolides based on an amino alcohol. Eur. J. Med. Chem. 43: 19–24.
22. S.T. Huang, H.S. Kuo, C.L. Hsiao and Y.L. Lin. 2002. Efficient synthesis of 'redox-switched' naphthoquinone thiol-crown ethers and their biological activity evaluation. Bioorg. Med. Chem. 10: 1947–1952.
23. S. Scheibye, B.S. Pdersen and S.O. Lawesson. 1978. Studies on organophosphorus compounds. XXII. The dimer of p-methoxyphenylthionophosphine sulfide as a thiation reagent. A new route to O-substituted thioesters and dithioesters. Bull. Soc. Chim. Belg. 87: 293–297.
24. S. Scheibye, B.S. Pedersen and S.O. Lawesson. 1978. Studies on organophosphorus compounds. XXI. The dimer of p-methoxyphenylthionophosphine sulfide as thiation reagent. A new route to thiocarboxamides. Bull. Soc. Chim. Belg. 87: 229–238.
25. T. Ozturk, E. Ertas and O. Mert. 2007. Use of Lawesson's reagent in organic synthesis. Chem. Rev. 107: 5210–5278.
26. M.I. Levinson and M.P. Cava. 1985. Thionation reactions of Lawesson's reagents. Tetrahedron 41: 5061–5087.
27. H.Z. Lecher, R.A. Greenwood, K.C. Whitehouse and T.H. Chao. 1956. The phosphonation of aromatic compounds with phosphorus pentasulfide. J. Am. Chem. Soc. 78: 518–522.
28. M. Caron. 1986. Convenient preparation of 5-alkyl-4-carbalkoxy-1,2,3-thiadiazoles. J. Org. Chem. 51: 4075–4077.
29. Y.H. Wu and R.F. Feldkamp. 1961. Pyrrolidines. I. 1-Substituted 3-pyrrolidinylmethyl alcohols and chlorides. J. Org. Chem. 26: 1519–1524.
30. B. Yde, N.M. Yousif, U. Pedersen, I. Thomsen and S.O. Lawesson. 1984. Studies on organophosphorus compounds XLVII. Preparation of thiated synthons of amides, lactams and imides by use of some new P,S-containing reagents. Tetrahedron 40: 2047–2052.
31. P. Stanetty, M. Turner and M.D. Mihovilovic. 2005. Synthesis of pyrrolo[2,3-d][1,2,3]thiadiazole-6-carboxylates via the Hurd-Mori reaction. Investigating the effect of the N-protecting group on the cyclization. Molecules 10: 367–375.
32. W. Bartmann, G. Beck, J. Knolle and R.H. Rupp. 1982. Synthese von $(\pm)$-(E)-2-(1-thia-4-äthoxycarbonylbutyl)-4-(3-hydroxy-1-octenyl)-Δ^1-pyrrolin und seiner analogen. Tetrahedron Lett. 23: 2947–2950.
33. E. Arvanitis, M. Motevalli and P.B. Wyatt. 1996. Enantioselective synthesis of (S)-2-(aminomethyl)butanedioic acid using chiral β-alanine α-enolate equivalents. Tetrahedron Lett. 37: 4277–4280.
34. O. Tsuge, S. Kanemasa, A. Hatada and K. Matsuda. 1986. Synthetic versatility of N-(silylmethyl)imines: Water-induced generation of N-protonated azomethine ylides of nonstabilized type and fluoride-induced generation of 2-azaallyl anions. Bull. Chem. Soc. Jpn. 59: 2537–2545.

35. K. Burger, M. Rudolph, H. Neuhauser and M. Gold. 1992. Easy access to amino acids with a diazo function in the side chain starting from aspartic acid. Synthesis 11: 1150–1156.
36. K. Burger, H. Neuhauser and A. Worku. 1993. Hexafluoraceton als schutzgruppen- und aktivierungsreagenz in der aminosäure- und peptidchemie N-substituierte glycinderivate aus iminodiessigsäure/application of hexafluoroacetone as protecting and activating reagent in amino acid and peptide chemistry N-substituted glycine derivatives from iminodiacetic acid. Z. Naturforsch. B 48: 107–120.
37. J. Spengler, C. Bottcher, F. Albericio and K. Burger. 2006. Hexafluoroacetone as protecting and activating reagent: New routes to amino, hydroxy, and mercapto acids and their application for peptide and glyco- and depsipeptide modification. Chem. Rev. 106: 4728–4746.
38. M.I. Levinson and M.P. Cava. 1982. The synthesis of acenaphtho[1,2-c][1,2,3]thiadiazole and phenanthro[9,10-c][1,2,3]thiadiazole. Heterocycles 19: 241–243.
39. M.S.J. Foreman and J.D. Woollins. 2000. Organo-P-S and P-Se heterocycles. J. Chem. Soc. Dalton Trans. 10: 1533–1543.
40. S. Buscemi and N. Vivona. 1994. Molecular rearrangements in heterocyclic synthesis. A generalized synthesis of 1,2,4-thiadiazoles from 3-acylamino-1-oxa-2-azoles. Heterocycles 38: 2423–2432.
41. D. Kumar, N.M. Kumar, K.-H. Chang, R. Gupta and K. Shah. 2011. Synthesis and in vitro anticancer activity of 3,5-bis(indolyl)-1,2,4-thiadiazoles. Bioorg. Med. Chem. Lett. 21: 5897–5900.
42. D. Kumar, V. Arun, N.M. Kumar, G. Acosta, B. Noel and K. Shah. 2012. A facile synthesis of novel bis-(indolyl)-1,3,4-oxadiazoles as potent cytotoxic agents. ChemMedChem 7: 1915–1920.
43. N.M. Kumar and D. Kumar. 2013. Recent developments on synthetic indoles as potent anticancer agents. Chem. Biol. Interfaces 3: 276–303.
44. J.W. Pavlik and P. Tongcharoensirikul. 2000. Photochemistry of 3- and 5-phenylisothiazoles. Competing phototransposition pathways. J. Org. Chem. 65: 3626–3632.
45. J.W. Pavlik, C. Changtong and V. Tsefrikas. 2003. Photochemistry of phenyl-substituted 1,2,4-thiadiazoles. ^{15}N-Labeling studies. J. Org. Chem. 68: 4855–4861.
46. H.-M. Huang, H.-T. Yu, P.-L. Chen, J. Han and J.-B. Meng. 2004. Synthesis of 2,5-diaryl-1,3,4-thiadiazoles via Lawesson's reagent under microwave irradiation. Chin. J. Org. Chem. 24: 502–505.
47. A.A. Kiryanov, P. Sampson and A.J. Seed. 2001. Synthesis of 2-alkoxy-substituted thiophenes, 1,3-thiazoles, and related S-heterocycles via Lawesson's reagent-mediated cyclization under microwave irradiation: Applications for liquid crystal synthesis. J. Org. Chem. 66: 7925–7929.
48. K. Dalip, N.M. Kumar, H.C. Kuei and C. Kavita. 2010. Synthesis and anticancer activity of 5-(3-indolyl)-1,3,4-thiadiazoles. Eur. J. Med. Chem. 45: 4664–4668.
49. R. Yadav, D. Yadav and S.K. Paliwal. 2012. Novel biphenyl imidazo[2,1-b][1,3,4]-thiadiazole - a versatile scaffold. Int. J. Pharm. Sci. 2: 20–37.
50. B.F. Abdel-Wahab and H.A. Mohamed. 2012. Synthetic access to benzazolyl (pyrazoles, thiazoles, or triazoles). Turk. J. Chem. 36: 805–826.
51. Z. Kaleta, B.T. Makowski, T. Soos and R. Dembinski. 2006. Thionation using fluorous Lawesson's reagent. Org. Lett. 8: 1625–1628.
52. R.T. Loto, C.A. Loto and A.P.I. Popoola. 2012. Corrosion inhibition of thiourea and thiadiazole derivatives: A review. J. Mater. Environ. Sci. 3: 885–894.
53. H.N. Dogan, A. Duran, S. Rollas, G. Sener, M.K. Uysalb and D. Güilenc. 2002. Synthesis of new 2,5-disubstituted-1,3,4-thiadiazoles and preliminary evaluation of anticonvulsant and antimicrobial activities. Bioorg. Med. Chem. 10: 2893–2898.
54. M. Kritsanida, A. Mouroutsou, P. Marakos, N. Pouli, S. Papakonstantinou-Gaufalias, C. Pannecouque, M. Witvrouw and E. de Clercq. 2002. Synthesis and antiviral activity evaluation of some new 6-substituted 3-(1-adamantyl)-1,2,4-triazolo[3,4-b][1,3,4]thiadiazoles. Farmaco 57: 253–257.
55. M. Sato, T. Kamita, K. Nakadera and K.I. Mukaida. 1995. Thermotropic liquid-crystalline polymers having five-membered heterocycles as mesogens-2. Homo- and copolymers

composed of 1,3- or 1,4-phenylenebis(1,3,4-thiadiazole 2,5-diyl) unit. Eur. Polym. J. 31: 395–400.
56. N. Linganna and R. Lokanantha. 1998. Transformation of 1,3,4-oxadiazoles to 1,3,4-thiadiazoles using thiourea. Synth. Commun. 28: 4611–4617.
57. I. Thomsen, K. Clausen, S. Scheibye and S.O. Lawesson. 1984. Thiation with 2,4-bis(4-methoxyphenyl)-1,3,2,4-dithiaphosphetane 2,4-disulfide: N-Methylthiopyrrolidone. Org. Synth. 62: 158–164.
58. B. Gierczyk and M. Zalas. 2005. Synthesis of substituted 1,3,4-thiadiazoles using Lawesson's reagent. Org. Prep. Proced. 37: 213–222.
59. M. Jesberger, T.P. Davis and L. Barner. 2003. Applications of Lawesson's reagent in organic and organometallic syntheses. Synthesis 13: 1929–1958.
60. P. Bradley, P. Sampson and A.J. Seed. 2005. Preliminary communication: The synthesis of new mesogenic 1,3,4-thiadiazole-2-carboxylate esters via a novel ring-closure. Liq. Cryst. Today 14: 15–18.
61. R. Lunkwitz, C. Tschierske, A. Langhoff, F. Giebelmann and P. Zugenmaier. 1997. Axial chiral allenylacetates as novel ferroelectric liquid crystals. J. Mater. Chem. 7: 1713–1721.
62. C. Tschierske, H. Zaschke, H. Kresse, A. Madicke, D. Demus, D. Girdziunaite and G.Y. Bak. 1990. Liquid crystalline thiadiazole derivatives IV. New liquid crystalline materials with broad smectic C ranges. Mol. Cryst. Liq. Cryst. 191: 223–230.
63. S. Sugita, S. Toda, T. Yamashita and T. Teraji. 1993. Ferroelectric liquid crystals having various cores. Effect of core structure on physical properties. Bull. Chem. Soc. Jpn. 66: 568–572.
64. S.-I. Sugita, S. Toda and T. Teraji. 1993. Synthesis and mesomorphic properties of ferroelectric liquid crystals bearing 2-phenyl-1,3,4-thiadiazole rings. Mol. Cryst. Liq. Cryst. 237: 33–38.
65. T. Geelhaar. 1988. Ferroelectric mixtures and their physico-chemical properties. Ferroelectrics 85: 329–349.
66. W. Schafer, U. Rosenfeld, H. Zaschke, H. Stettin and H. Kresse. 1989. Kristallin-flüssige 1,3,4-thiadiazole. II [1]. 1,3,4-Thiadiazole mit cyclohexanstrukturfragmenten. J. Prakt. Chem. 331: 631–636.
67. K. Dimitrowa, J. Hauschild, H. Zaschke and H. Schubert. 1980. Kristallin-flüssige 1,3,4-thiadiazole. I. Biphenyl- und terphenylanaloge 1,3,4-thiadiazole. J. Prakt. Chem. 322: 933–944.
68. C. Tschierske and D. Girdziunaite. 1991. Synthese von 1,3,4-thiadiazol-derivaten. J. Prakt. Chem. 333: 135–137.
69. B. Sybo, P. Bradley, A. Grubb, S. Miller, K.J.W. Proctor, L. Clowes, M.R. Lawrie, P. Sampson and A.J. Seed. 2007. 1,3,4-Thiadiazole-2-carboxylate esters: New synthetic methodology for the preparation of an elusive family of self-organizing materials. J. Mater. Chem. 17: 3406–3411.
70. H.H. Freedman and R.A. Dubois. 1975. An improved Williamson ether synthesis using phase transfer catalysis. Tetrahedron Lett. 16: 3251–3254.
71. S.R. Sandler and W. Karo. 1983. Hydrazine derivatives, hydrazones, and hydrazides. H.H. Wasserman (Ed.). Organic functional group preparations. Academic Press, New York, 434–465.
72. S.R. Varma and D. Kumar. 1999. Microwave-accelerated solvent-free synthesis of thioketones, thiolactones, thioamides, thionoesters, and thioflavonoids. Org. Lett. 1: 697–700.
73. J.K. Augustine, V. Vairaperumal, S. Narasimhan, P. Alagarsamy and A. Radhakrishnan. 2009. Propylphosphonic anhydride (T_3P®): An efficient reagent for the one-pot synthesis of 1,2,4-oxadiazoles, 1,3,4-oxadiazoles, and 1,3,4-thiadiazoles. Tetrahedron 65: 9989–9996.
74. S. Scheibye, A.A. El-Barbary, S.-O. Lawesson, H. Fritz and G. Rihs. 1982. Studies on organophosphorus compounds - XLI. Tetrahedron 38: 3753–3760.
75. A.A. Kiryanov, P. Sampson and A.J. Seed. 2001. Synthesis and mesomorphic properties of 1,1-difluoroalkyl-substituted biphenylthienyl and terphenyl liquid crystals. A comparative study of mesomorphic behavior relative to alkyl, alkoxy and alkanoyl analogs. J. Mater. Chem. 11: 3068–3077.

76. P.A.S. Smith. 1947. Organic reactions. R. Adams (Ed.). John Wiley & Sons Inc. New York, 3: 366–369.
77. V.M. Sonpatki, M.R. Herbert, L.M. Sandvoss and A.J. Seed. 2001. Troublesome alkoxythiophenes. A highly efficient synthesis via cyclization of γ-keto esters. J. Org. Chem. 66: 7283–7286.
78. A. Hassner and V. Alexanian. 1978. Direct room temperature esterification of carboxylic acids. Tetrahedron Lett. 19: 4475–4478.
79. E.S.H. El-Ashry, M.A.M. Nassr, Y. El-Kilany and A. Mousaad. 1986. Synthesis and reactions of 2-(p-chloroanilino)-5-(D-galacto-1,2,3,4,5-pentahydroxypentyl)-1,3,4-thiadiazole. J. Prakt. Chem. 328: 1–6.
80. D. Spinelli, R. Noto, G. Consiglio and F. Buccheri. 1977. Studies on the decarboxylation reactions. Note II. Kinetic study of the decarboxylation reaction of 5-amino-1,3,4-thiadiazole-2-carboxylic acid (I) to 2-amino-1,3,4-thiadiazole (III). J. Heterocycl. Chem. 14: 309–311.
81. R. Noto, G. Werber, F. Buccheri and C. Arnone. 1987. Studies on decarboxylation reactions. Part 6. Kinetic study of decarboxylation of 5-amino-1,3,4-oxadiazole-2-carboxylic acid and its N-phenyl derivatives at high hydrochloric acid concentrations. J. Heterocycl. Chem. 24: 1457–1459.
82. R. Noto, M. Ciofalo, F. Buccheri, G. Werber and D. Spinelli. 1991. Studies on decarboxylation reactions. Part 7. Kinetic study of the decarboxylation of 2-amino- and 2-phenylaminothiazole-5-carboxylic acids. J. Chem. Soc. Perkin Trans. 2 349-352.
83. J. Garfunkle, C. Ezzili, T.J. Rayl, D.G. Hochstatter, I. Hwang and D.L. Boger. 2008. Optimization of the central heterocycle of α-ketoheterocycle inhibitors of fatty acid amide hydrolase. J. Med. Chem. 51: 4392–4403.
84. A. Seed. 2007. Synthesis of self-organizing mesogenic materials containing a sulfur-based five-membered heterocyclic core. Chem. Soc. Rev. 36: 2046–2069.
85. T. Ozturk, E. Ertas and O. Mert. 2010. A Berzelius reagent, phosphorus decasulfide (P_4S_{10}), in organic syntheses. Chem. Rev. 110: 3419–3478.
86. D. Kovács, Z. Kádár, G. Mótyán, G. Schneider, J. Wölfling, I. Zupkó and E. Frank. 2012. Synthesis, characterization and biological evaluation of some novel 17-isoxazoles in the estrone series. Steroids 77: 1075–1085.
87. D. Kovács, J. Wölfling, N. Szabó, M. Szécsi, I. Kovács, I. Zupkó and E. Frank. 2013. An efficient approach to novel 17-5'-(1',2',4')-oxadiazolyl androstenes via the cyclodehydration of cytotoxic O-steroidacylamidoximes, and an evaluation of their inhibitory action on 17α-hydroxylase/C17,20-lyase. Eur. J. Med. Chem. 70: 649–660.
88. M.E. Jung and M.A. Lyster. 1997. Quantitative dealkylation of alkyl ethers via treatment with trimethylsilyl iodide. A new method for ether hydrolysis. J. Org. Chem. 42: 3761–3764.
89. J. Lee, S.-H. Lee, H.J. Seo, E.-J. Son, S.H. Lee, M.E. Jung, M.W. Lee, H.-K. Han, J. Kim, J. Kang and J. Lee. 2010. Novel C-aryl glucoside SGLT2 inhibitors as potential antidiabetic agents: 1,3,4-Thiadiazolylmethylphenyl glucoside congeners. Bioorg. Med. Chem. 18: 2178–2194.
90. P. Angibeaud and J.-P. Utile. 1991. Cyclodextrin Chemistry; Part I. Application of a regioselective acetolysis method for benzyl ethers. Synthesis 9: 737–738.
91. J.C. Lee, H.J. Choi and Y.C. Lee. 2003. Efficient synthesis of multi-substituted oxazoles under solvent-free microwave irradiation. Tetrahedron Lett. 44: 123–125.
92. C.T. Brain and J.M. Paul. 1999. Rapid synthesis of oxazoles under microwave conditions. Synlett 10: 1642–1644.
93. J. Han, X. Chang, X. Wang, L. Zhu, M. Pang and J. Meng. 2009. Microwave-assisted synthesis and liquid crystal properties of 1,3,4-thiadiazole-based liquid crystals. Liq. Cryst. 36: 157–163.
94. V.S. Murthy, B.K. Manuprasad and S. Shashi. 2012. Synthesis and antimicrobial activity of novel (3,5-dichloro-4-((5-aryl-1,3,4-thiadiazol-2-yl)methoxy)phenyl) aryl methanones. J. Appl. Pharm. Sci. 2: 172–176.
95. J. Lee, H.J. Seo, S.H. Lee, J. Kim, M.E. Jung, S.-H. Lee, K.-S. Song, J. Lee, S.Y. Kang, M.J. Kim, M.-S. Kim, E.-J. Son, M.W. Lee and H.-K. Han. 2010. Discovery

of 2-(4-((1H-1,2,4-triazol-1-yl)methyl)-5-(4-bromophenyl)-1-(2-chlorophenyl)-1H-pyrazol-3-yl)-5-*tert*-butyl-1,3,4-thiadiazole (GCC2680) as a potent, selective and orally efficacious cannabinoid-1 receptor antagonist. Bioorg. Med. Chem. 18: 6377–6388.

96. L.S. Lin, T.J. Lanza, J.P. Jewell, P. Liu, S.K. Shah, H. Qi, X. Tong, J. Wang, S. Xu, T.M. Fong, C.-P. Shen, J. Lao, J. Chen, L.P. Shearman, D.S. Stribling, K. Rosko, A. Strack, D.J. Marsh, Y. Feng, S. Kumar, K. Samuel, W. Yin, L. van der Ploeg, M.T. Goulet and W.K. Hagman. 2006. Discovery of N-[(1S,2S)-3-(4-chlorophenyl)-2-(3-cyanophenyl)-1-methylpropyl]-2-methyl-2-{[5-(trifluoromethyl)pyridin-2-yl]oxy}propanamide (MK-0364), a novel, acyclic cannabinoid-1 receptor inverse agonist for the treatment of obesity. J. Med. Chem. 49: 7584–7587.

97. C.-Y. Chen, L.F. Frey, S. Shultz, D.J. Wallace, K. Marcantonio, J.F. Payack, E. Vazquez, S.A. Springfield, G. Zhou, P. Liu, G.R. Kieczykowski, A.M. Chen, B.D. Phenix, U. Singh, J. Strine, B. Izzo and S.W. Krska. 2007. Catalytic, enantioselective synthesis of taranabant, a novel, acyclic cannabinoid-1 receptor inverse agonist for the treatment of obesity. Org. Process Res. Dev. 11: 616–623.

98. L.S. Lin, S. Ha, R.G. Ball, N.N. Tsou, L.A. Castonguay, G.A. Doss, T.M. Fong, C.P. Shen, J.C. Xiao, M.T. Goulet and W.K. Hagmann. 2008. Conformational analysis and receptor docking of N-[(1S,2S)-3-(4-chlorophenyl)-2-(3-cyanophenyl)-1-methylpropyl]-2-methyl-2-{[5-(trifluoromethyl)pyridin-2-yl]oxy}propanamide (Taranabant, MK-0364), a novel, acyclic cannabinoid-1 receptor inverse agonist. J. Med. Chem. 51: 2108–2114.

99. D.A. Griffith, J.R. Hadcock, S.C. Black, P.A. Iredale, P.A. Carpino, P. DaSilva Jardine, R. Day, J. DiBrino, R.L. Dow, M.S. Landis, R.E. O'Connor and D.O. Scott. 2009. Discovery of 1-[9-(4-chlorophenyl)-8-(2-chlorophenyl)-9H-purin-6-yl]-4-ethylaminopiperidine-4-carboxylic acid amide hydrochloride (CP-945,598), a novel, potent, and selective cannabinoid type 1 receptor antagonist. J. Med. Chem. 52: 234–237.

100. Barth, P. Casellas, C. Congy, S. Martinez and M. Ridaldi. 1995. Pyrazole-3-carboxamide derivatives, process for their preparation and pharmaceutical compositions in which they are present. US Patent 5,462,960.

101. R. Lan, Q. Liu, P. Fan, S. Lin, S.R. Frnando, D. McCallion and R. Pertwee. 1999. Makriyannis. Structure-activity relationships of pyrazole derivatives as cannabinoid receptor antagonists. J. Med. Chem. 42: 769–776.

102. R. Katoch-Rouse, O.A. Pavlova, T. Caulder, A.F. Hoffman, A.G. Mukhin and A.G. Horti. 2003. Synthesis, structure-activity relationship, and evaluation of SR141716 analogues: Development of central cannabinoid receptor ligands with lower lipophilicity. J. Med. Chem. 46: 642–645.

103. M.R. Mahmoud and M.F. Ismail. 2014. Recent developments in chemistry of 1,3,4-thiadiazoles. J. Adv. Chem. 10: 2823–2853.

104. M. Rodriquez and M. Taddei. 2006. Synthesis of heterocycles via microwave-assisted cycloadditions and cyclocondensations. Top. Heterocycl. Chem. 1: 213–266.

105. V. Molteni, M.M. Hamilton, L. Mao, C.M. Crane, A.P. Termin and D.M. Wilson. 2002. Aqueous one-pot synthesis of pyrazoles, pyrimidines and isoxazoles promoted by microwave irradiation. Synthesis 12: 1669–1674.

106. T. Besson and V. Thiery. 2006. Microwave-assisted synthesis of sulfur and nitrogen-containing heterocycles. Top. Heterocycl. Chem. 1: 59–78.

107. T.S. Hu, K.T. Lin, C.C. Mu, H.M. Kuo, M.C. Chen and C.K. Lai. 2014. Polar effect in columnar unsymmetric 1,3,4-oxa(thia)diazoles. Tetrahedron 70: 9204–9213.

108. H. Deokar, J. Chaskar and A. Chaskar. 2014. Synthesis and antimicrobial activity evaluation of novel oxadiazino thiadiazino-indole and oxadiazole/thiadiazole derivatives of 2-oxo-2H-benzopyran. J. Heterocycl. Chem. 51: 719–725.

109. J. Han, Q. Wang, J. Wu and L.R. Zhu. 2015. Synthesis and liquid crystalline property of H-shaped 1,3,4-thiadiazole dimers. Liq. Cryst. 42: 127–133.

Chapter 5
Five-Membered *S*-Heterocycle Synthesis

5.1 Introduction

Today analogues of heterocyclic compounds and their derivatives have become strong interest in pharmaceutical research field due to their valuable biological and pharmacological activities. Heterocycles are abundant in nature and are important because their structural subunits are present in numerous natural products like vitamins, hormones, and antibiotics [1–4]. A large number of pharmaceutical compounds belong to a major class of sulfur-containing heterocyclic compounds. The versatile synthetic use and biological action of these heterocyclic compounds have motivated the pharmacologist to plan, design, and execute new methodologies for the synthesis of novel drugs. The thiophene compounds and their derivatives are an important class of heterocyclic compounds, particularly, 2-amino-substituted thiophene compounds have a wide range of biological activities like antibacterial, antifungal, analgesic, anti-inflammatory, antioxidant, and antitumor and also local anesthetic action. The five-membered *S*-containing heterocyclic compounds are important synthetic intermediates and have found a diversity of uses in medical, agricultural, and material chemistry. The reagents like P_2S_5 or Lawesson's reagent serve as sulfurizing agents and also as dehydrating agents, allowing a reaction route that could result in the synthesis of *S*-heterocyclic compounds [5–8].

Lawesson reagent is a general sulfuration reagent which can convert CO group of aldehydes, ketones, amides, or esters to thiocarbonyl species. Numerous new thionating reagents have been prepared and utilized for the formation of organosulfur compounds in the past years. Curphey's thionating reagent (P_4S_{10}/hexamethyldisiloxane) is known to be one of the best substitute for Lawesson's reagent. Likewise, P_4S_{10} in combination with numerous supports like alumina and silica displays good thionating character with enhancement in the yield. The application of polymer-supported and in situ produced thionating reagent is another development in this area. The use of ionic liquids and solid supports to develop an efficient and ecologically benign "green" thionating reagent is yet to be explored. However, there has been a widespread worldwide progress in this area, this field of chemistry is yet in the early stages, therefore, both basic research and practical work are necessary in the development of thionation procedures for the formation of organosulfur compounds. The LR has remained the most important reagent in thionation chemistry and was followed by P_4S_{10}. Usually, LR has benefits over P_4S_{10} in terms of the need for excess P_4S_{10} and reduced yields [9–11].

5.2 Synthesis of Thiophenes

Kiryanov et al. [12] examined the formation of liquid crystals bearing five-membered sulfur-containing heterocyclic compounds having thiophene derivatives while searching for potential applicants for ferroelectric display applications. They reported that high yields of thiophenes were observed when LR-assisted cyclization of numerous 1,4-dicarbonyl compounds was carried out under MWI. Little or no by-products were obtained, and reaction times were very less (3–13 min) in these variations of classical Paal–Knorr thiophene synthesis, which was conducted without solvent in conventional MW oven. Ongoing with the trend for solvent-free reactions, a MW alternative of classical Paal–Knorr thiophene synthesis has been described [13]. The thionation–cyclization of a variety of 1,4-dicarbonyl compounds with solid LR delivered thiophene compounds rapidly in high yield and with minimal purification in comparison to the equivalent solution-phase reactions (Scheme 5.1).

A combination of Friedel–Crafts and Lawesson reaction afforded 5-N,N-dialkylamino-2,2′-bithiophenes from N,N-dialkylamino-4-(2′-thienyl)-4-oxobutanamides (Scheme 5.2) [14]. The 5-pyrrolidino-2,2′-bithiophene was prepared from new pyrrolidino-4-(2′-thienyl)-4-oxobutanamide by same synthetic pathway for comparing the influence of electronic nature of 5-N,N-dialkylamino groups on the optical properties of phenylazobithiophenes. A direct amidation of 4-oxo-(2′-thienyl)butanoic acid [15–17] with pyrrolidine took place via N,N'-dicyclohexylcarbodiimide-butyl alcohol-assisted reaction. Amide was found as a

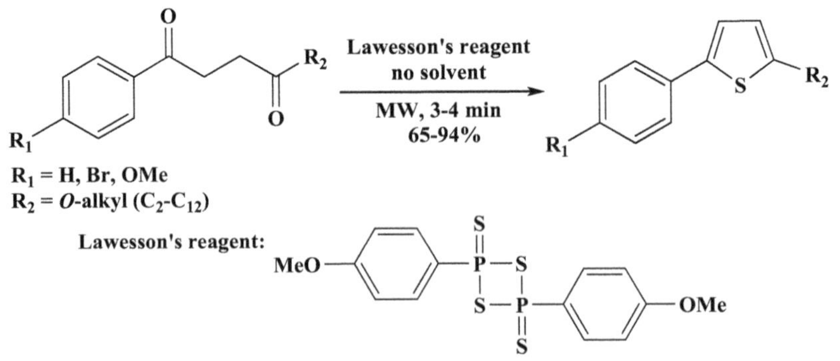

Scheme 5.1 Synthesis of thiophenes

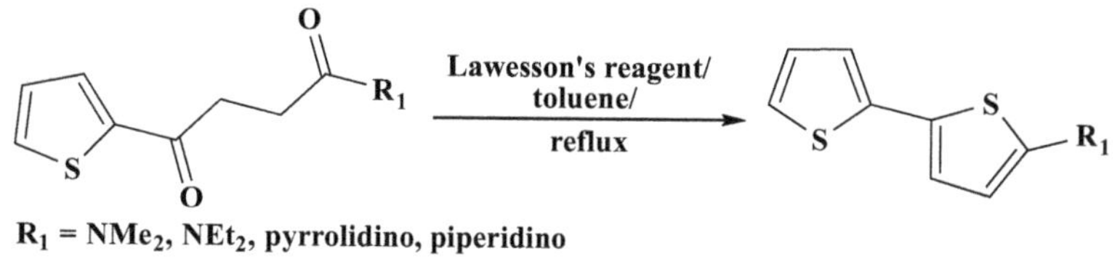

5.2 Synthesis of Thiophenes

Scheme 5.2 Synthesis of bis-thiophenes

R_1 = NMe$_2$, NEt$_2$, pyrrolidino, piperidino

colorless solid in good yield (80%). The reaction of amide with an equimolar amount of Lawesson's reagent in toluene afforded bithiophene in 47% yield. The formation of 5-pyrrolidino-2,2'-bithiophene has been described by Effenberger et al. [18] via two different procedures. A mixture of two compounds was obtained in 43% yield by a palladium-catalyzed coupling reaction through an organotin intermediate. These two compounds were formed in 70:30 ratio. The 5-pyrrolidino-2,2'-bithiophene and phenyl-5-pyrrolidinothiophene could not be separated neither by chromatography nor by recrystallization method. Thus, an another pathway for the formation of 5-pyrrolidino-2,2'-bithiophene was via lithiation of 2,2'-bithiophene followed by reaction with sulfur to afford the 5-mercapto-2,2'-bithiophene in 40% yield, subsequent reaction with pyrrolidine allowed the formation of 5-pyrrolidino-2,2'-bithiophene in 25–37% yield. As compared to Effenberger's methods [18], the 5-pyrrolidino-2,2'-bithiophene was formed in higher yield via combination of Friedel–Crafts and Lawesson reactions from economically feasible and commercially accessible reagents utilizing simple workup processes which allowed the good yield synthesis and easy isolation of this derivative.

The effect of additives like Lewis or Brønsted acids, bases, and molecular sieves as H_2O absorbers was analyzed. It was observed that a very easy process involving LR in hot toluene or benzene afforded the highest yield of bis-thiophenes. The scope of this reaction was examined under these conditions through the utilization of diverse substituents R (Scheme 5.3). The best results were observed utilizing electron-neutral aryl units (entries 3 and 4) or heteroaromatic residues (entries 7 and 8) while aliphatic residues were found to be less suitable for this reaction providing poor or moderate yields. This pathway offered simplistic access to tetracyclic heteroarenes in which furans, thiophenes, or pyrroles were connected through 2,2'- and 3,3'-linkages. Such compounds might find interesting uses with respect to organic electronic materials [19–21].

Paal–Knorr reaction that follows this Stetter reaction will result in aromatization and the synthesis of a short oligomer. Similar to the formation of trimers, symmetric pentamers were obtained if a symmetric bifunctional Michael acceptor was utilized in a Stetter reaction with a monofunctional aldehyde. No result was observed when 2,5-thiophenedicarboxaldehyde was utilized in combination with regular Michael acceptor (Schemes 5.4, 5.5, and 5.6) [22].

The 1,4-diketones could be cyclized when heated with hydrogen sulfide and HCl [23] or P-S compounds like P_2S_5 or Lawesson's reagent (Scheme 5.7). This reaction

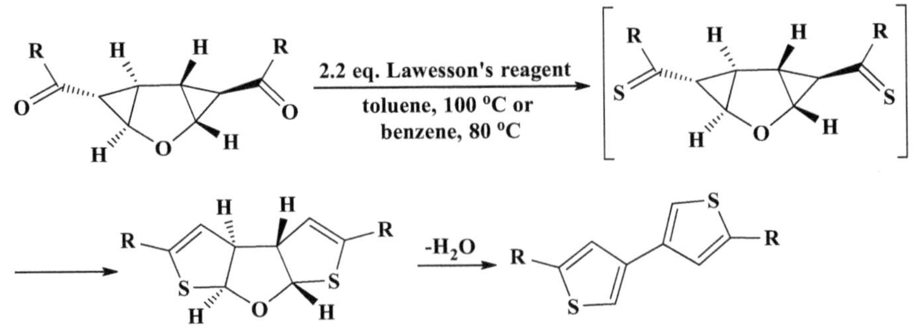

Scheme 5.3 Synthesis of bis-thiophenes

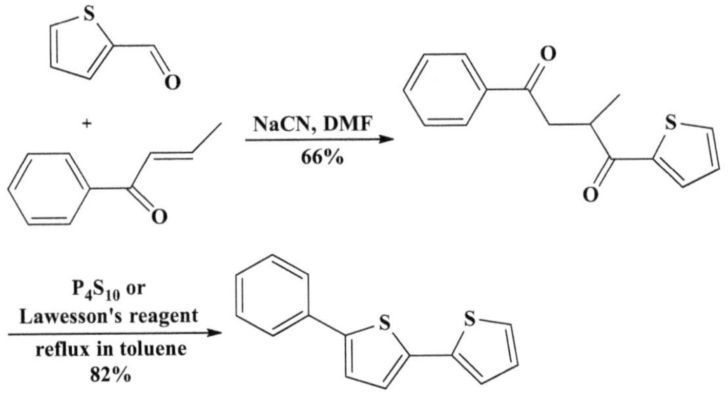

entry	R	yield (%)
1	Me	33
2	n-Pr	19
3	Ph	51
4	C_6H_4Me	45
5	C_6H_4OMe	traces
6	2-furyl	41
7	2-thienyl	56
8	2-pyrryl	29

Scheme 5.4 Synthesis of bis-thiophene

allowed the introduction of side-chains on the heterocyclic compounds obtained in the reaction [22].

Sonpatki et al. [24] reported that 2-alkoxythiophenes were synthesized by the reaction of 1,4-dicarbonyl species with LR (Scheme 5.8); however, this method needed an aryl substituent at C5-position of the 2-alkoxythiophene. Alternatively, identical methods with other sulfurization reagents afforded low yield/reproducibility or problematic mixtures of products [25, 26].

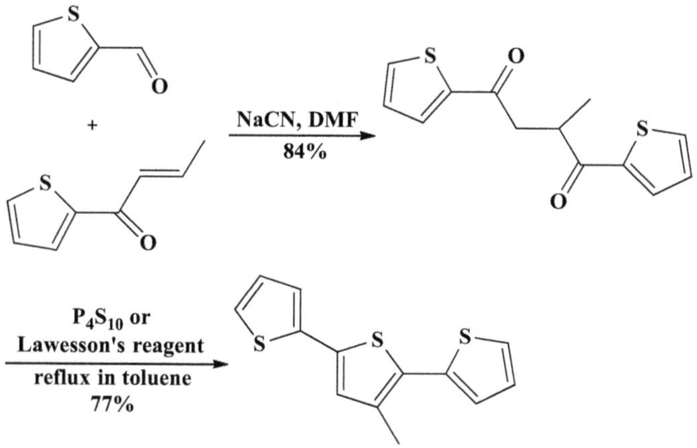

5.2 Synthesis of Thiophenes

Scheme 5.5 Synthesis of tris-thiophene

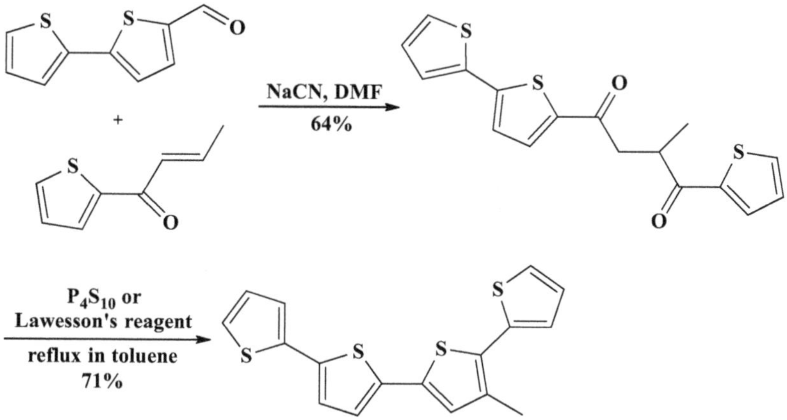

Scheme 5.6 Synthesis of tetrakis-thiophene

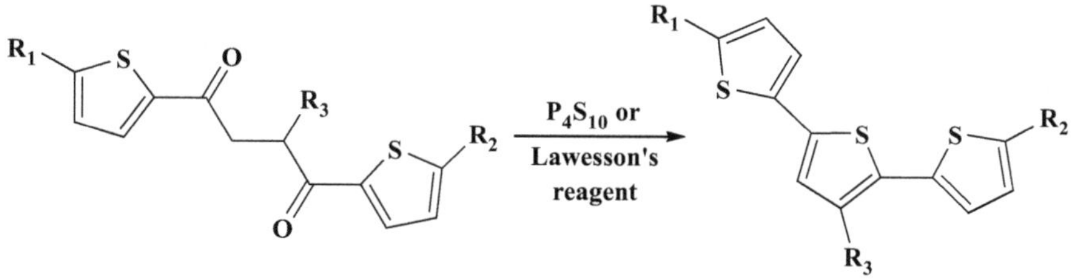

Scheme 5.7 Synthesis of tris-thiophenes

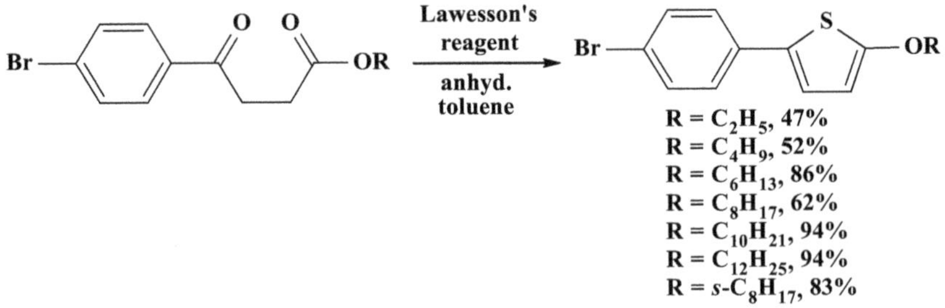

R = C$_2$H$_5$, 47%
R = C$_4$H$_9$, 52%
R = C$_6$H$_{13}$, 86%
R = C$_8$H$_{17}$, 62%
R = C$_{10}$H$_{21}$, 94%
R = C$_{12}$H$_{25}$, 94%
R = s-C$_8$H$_{17}$, 83%

Scheme 5.8 Synthesis of 2-alkoxythiophenes

The indole derivatives were transformed into indole-3-carboxaldehydes via Vilsmeier–Haack reaction with POCl$_3$ and DMF. The indole-3-carboxaldehydes were protected by benzensulfonyl species at NH to afford the N-benzensulfonyl-indole-3-carboxaldehyde. The valuable and versatile intermediates 1,4-bis(indolyl)diketones were generated for the formation of 2,5-bis(3-indolyl)thiophenes. The Stetter reaction of electron-deficient indole aldehyde and divinyl sulfone utilizing CH$_3$COONa and thiazolium chloride as catalyst under reflux in EtOH afforded 1,4-diketones. The diketones were transformed into bis(indolyl)thiophenes in the presence of LR under reflux in toluene. Further, hydrolysis of bis(indolyl)thiophenes with potassium hydroxide in refluxing EtOH afforded 2,5-bis(3-indolyl)thiophenes (Scheme 5.9) [27].

A simple one-pot reaction of hydroxy amides with Lawesson's reagent gave heterocyclic compounds like tetrahydrothiophene-2-imines and tetrahydrothiophene-2-thione (Scheme 5.10) [28].

This protocol can be employed for the formation of a tetraarylated thiophene (Scheme 5.11). The [3 + 2]-cycloaddition reaction of ene and ketone and a subsequent S–O exchange reaction with LR afforded dihydrothiophene [28, 29]. The oxidation of dihydrothiophene with 2,3-dichloro-5,6-dicyano-1,4-benzoquinone afforded thiophene without its regioisomer in a 48% yield after 3 steps. The triarylthiophene was synthesized by bromination of thiophene with N-bromosuccinimide (95%) followed by Suzuki–Miyaura coupling (81%). The triarylthiophene was hydrolyzed and after that decarboxylative coupling with p-CF$_3$C$_6$H$_4$I afforded tetraarylthiophene in 66% yield (two steps) [30].

The LR has been employed for the formation of 2,5-thiopheneophane but instead furan was obtained [31]. A related thiophene formation occurred by the treatment of epoxycarbonyls with LR using tosylic acid (Ts = toluene-p-sulfonyl) (Scheme 5.12) [32, 33].

An application of LR allowed the cyclization to thiophene derivatives in place of furan (Scheme 5.13). The epoxide was converted into ethyl 4-(chloromethyl)thiophene-2-acetate when reacted with LR in small amount of p-toluenesulfonic acid [32, 34]. The p-toluenesulfonic acid must be added only after complete stirring with LR to avoid the cyclization to furan ring. The furan was obtained in some extent along with thiophene if p-toluenesulfonic acid was added

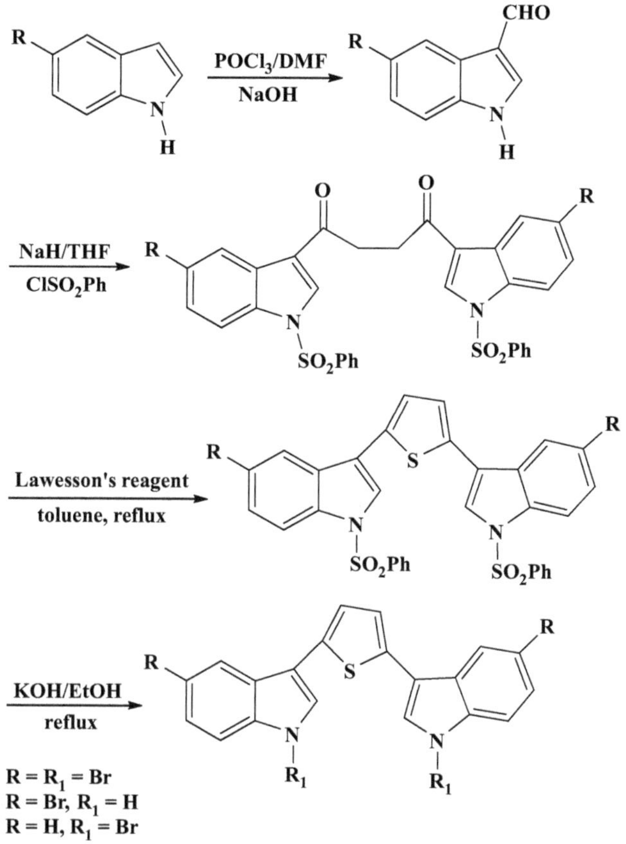

5.2 Synthesis of Thiophenes

Scheme 5.9 Synthesis of 2,5-bis(3-indolyl)thiophenes

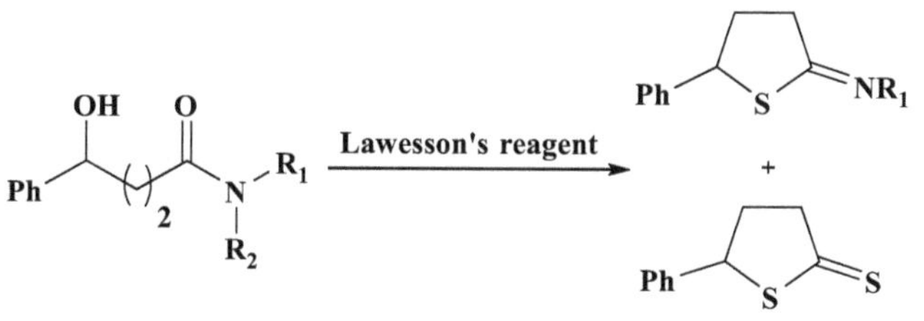

Scheme 5.10 Synthesis of tetrahydrothiophene-2-imines and tetrahydrothiophene-2-thione

before the LR. The ethyl 4-(iodomethyl)thiophene-2-acetate was prepared by transhalogenation of chloride with sodium iodide in $(CH_3)_2CO$. The ethyl esters of 4-alkylthiophene-2-acetic acid were also synthesized in the similar way from the reaction of iodide with Grignard reagents using Li_2CuCl_4 in catalytic amounts [35].

Scheme 5.11 Synthesis of tetraarylthiophene

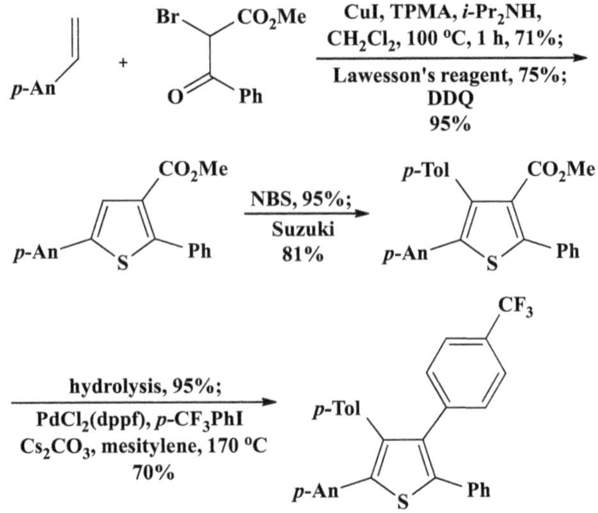

Scheme 5.12 Synthesis of thiophenes

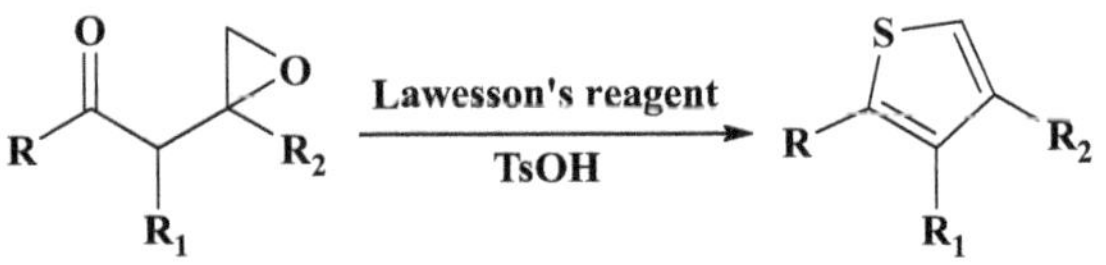

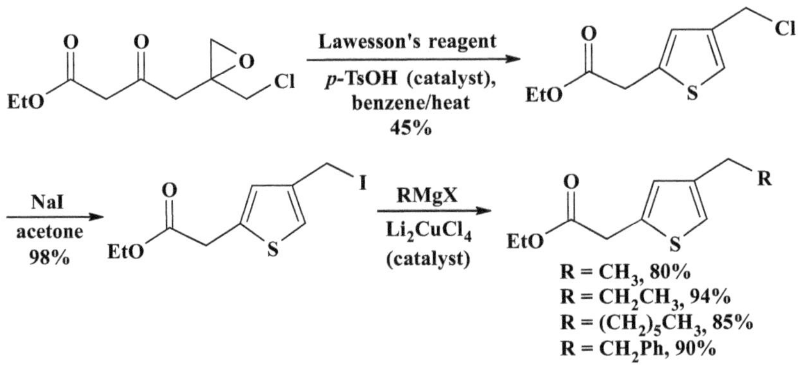

Scheme 5.13 Synthesis of ethyl esters of 4-alkylthiophene-2-acetic acid

The nickel-catalyzed reductive coupling approach to pyrrole synthesis provided an efficient route to furans and thiophenes from the common dicarbonyl intermediate (Scheme 5.14). The furans with the same broad substitution scope can be synthesized under acidic and microwave heating conditions as with pyrroles. Similarly, thiophenes can be prepared from 1,4-dicarbonyl compounds by simple heating with LR [36].

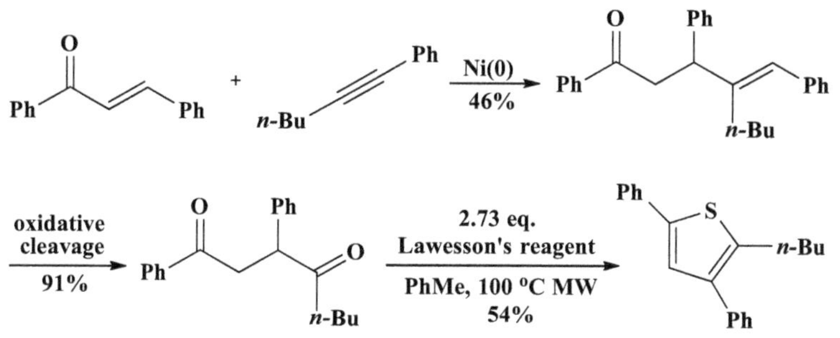

5.2 Synthesis of Thiophenes

Scheme 5.14 Synthesis of thiophene

The cyclization of 1,4-diketones afforded furans, pyrroles, and thiophenes. The furans were prepared by acid-catalyzed ring-closure [37], and thiophenes were synthesized by ring-closure with phosphorus pentasulfide or LR [38]. The unsubstituted and N-substituted pyrroles were synthesized by the reaction of a 1,4-diketone with NH_3, ammonium carbonate [39], or ammonium acetate and primary amines [40], respectively (Scheme 5.15).

The coincidence of solvent in the oxidative deacetylation, and the Paal–Knorr synthesis allowed one-pot synthetic conversion of easily accessible 1,5-dicarbonyl compound into furan, thiophene, and pyrrole (Scheme 5.16). The oxidative deacetylation of starting ketones utilizing Mn(III)/Co(II) catalysts in acetic acid at 25 °C for 3 h and after that reaction with 1 eq. H_2SO_4 at 90 °C for 1 h generated furan in 54% yield. Then, reaction with 1.5 eq. LR at 90 °C for 1 h and with 10 eq. NH_4OAc at

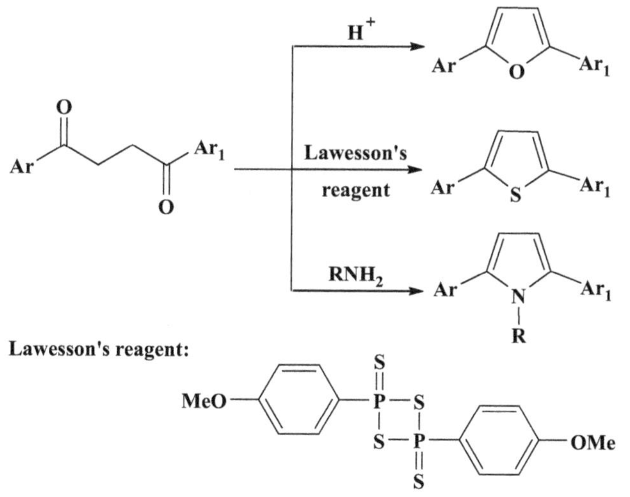

Scheme 5.15 Synthesis of furans, thiophenes, and pyrroles

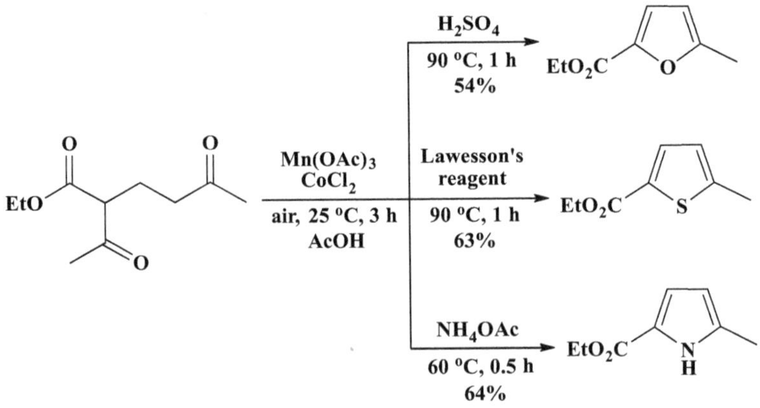

Scheme 5.16 Synthesis of furan, thiophene, and pyrrole

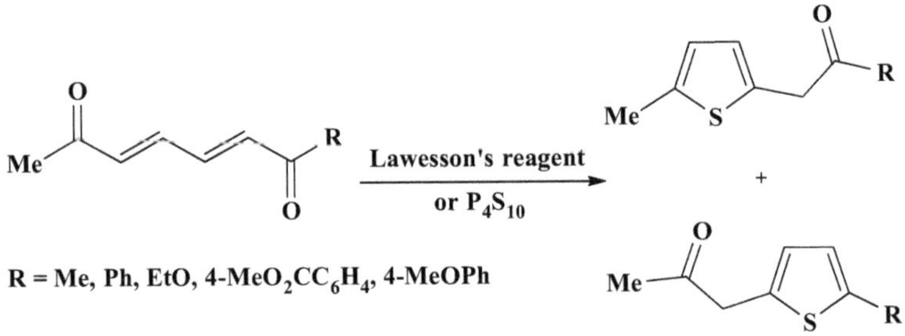

Scheme 5.17 Synthesis of 2,5-disubstituted thiophenes

60 °C for 0.5 h afforded thiophene in 63% yield and pyrrole in 64% yield, respectively [41].

The unsaturated diketones were reacted with LR to afford a mixture of two possible isomers. The product ratio changes when the reaction was carried out with BF$_3$ [42]. The 1,6-dioxo compounds, bearing 2,4-diene functionalities, afforded 2,5-disubstituted thiophenes on reaction with phosphorus pentasulfide (Scheme 5.17). The mechanism included thionation of one of the CO groups followed by Michael-type addition, which afforded 2,5-disubstituted thiophenes. An another proposed mechanism consisted of an addition of sulfur to 4-C unit between the two carbonyl groups [33, 43].

5.3 Synthesis of Benzothiophenes

The thiophenes were synthesized through ring-closure of 1,4-dithioketone (Schemes 5.18 and 5.19) [44]. The reactions of more substituted diketones provided furans in

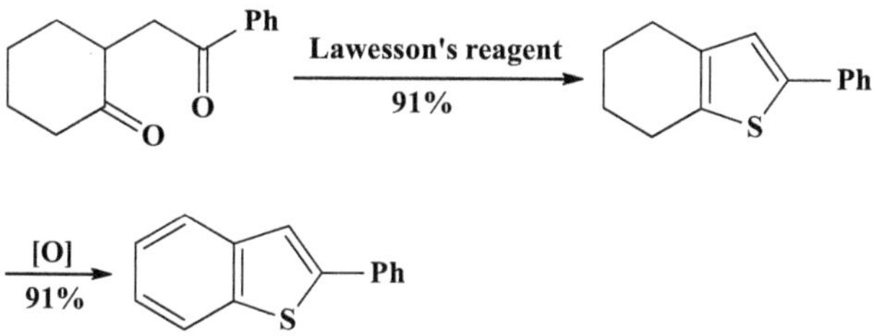

5.3 Synthesis of Benzothiophenes

Scheme 5.18 Synthesis of 2-phenylthiophene

Scheme 5.19 Synthesis of 2-phenylcyclopentathiophene

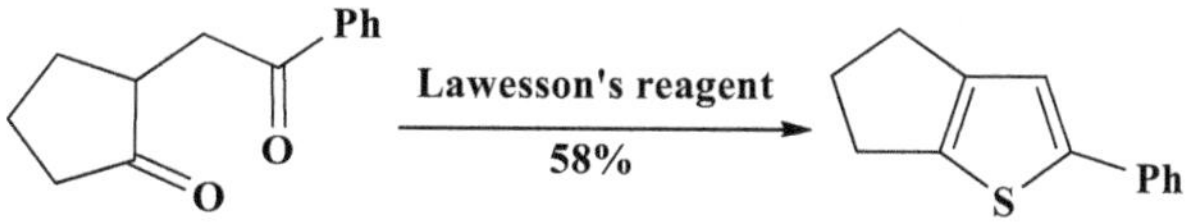

significant yields, also the presence of EDGs on aromatic groups at 1- and 4-positions enhanced the yield of furans while EWGs in these positions reduced their yield [45]. A substitute mild method for the synthesis of thiophenes from 1,4-diketones was the reaction with a Sn/S/B system [34, 46, 47].

The cross-conjugated enaminones were prepared by this reaction [48]. The subsequent reaction of enaminones with LR [49, 50] in CH$_2$Cl$_2$ at 0 °C resulted in the formation of enaminothiones (in situ), quenching of which with α-bromoketones/ethyl bromoacetate afforded thiophene framework in good yields (59–79%). The probable reaction mechanism suggested for the synthesis of thiophene involved the initial nucleophilic attack of the sulfur atom of enaminothiones at the methylene bromide carbon of α-bromoketones/ethyl bromoacetate to afford the intermediate. This intermediate was deprotonated in situ with triethylamine. The ring-closure occurred spontaneously by the removal of a dimethylamine molecule from the intermediate to afford the thiophenes (Scheme 5.20) [51].

The reaction of 2-acylbenzamides with Lawesson's reagent afforded numerous products depending on the groups attached to starting material (Scheme 5.21) [9, 52].

The benzothiophene was prepared in moderate to good yield from commercially accessible rhodamine B base via "one-pot synthesis" or a one-step reaction (Scheme 5.22) [53].

The 3-phenyl-1(3H)-isobenzothiophene-1-thione was prepared in one-step. The 2-benzoylbenzoic acid was thionated in the presence of LR to afford the 3-phenyl-1(3H)isobenzothiophene-1-thione directly (Scheme 5.23) [54].

The 3-methyl-1(3H)-isobenzothiophene-1-thione was synthesized in a similar manner to that of the phenyl-substituted case. The reaction of 2-acetylbenzoic acid with 1 eq. LR under reflux in toluene afforded 3-methyl-1(3H)isobenzothiophene-1-thione (Scheme 5.24) [54].

R_2 = -OCH_2CH_3, phenyl, 4-methoxyphenyl, 4-methylphenyl
Ar = phenyl, 4-methoxyphenyl, 4-chlorophenyl

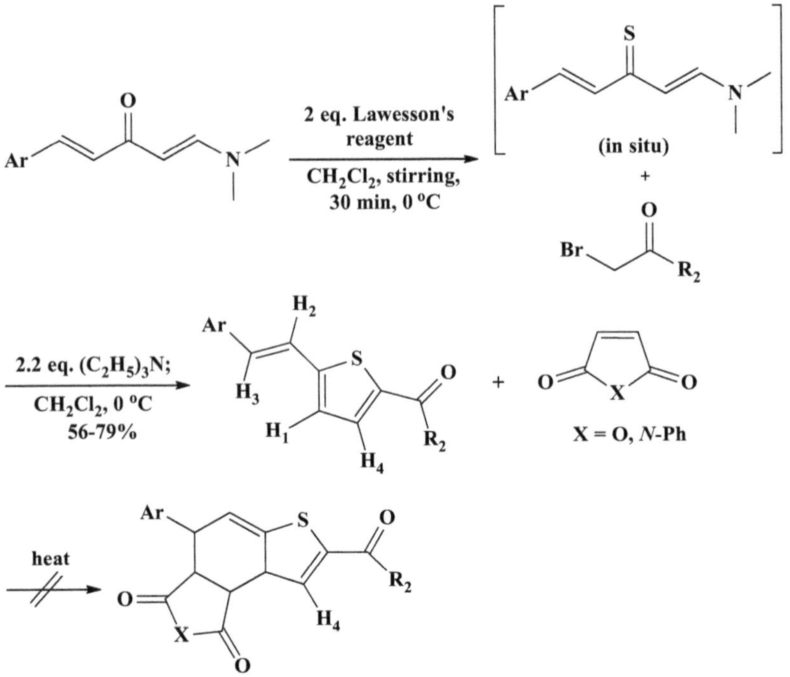

Scheme 5.20 Synthesis of thiophenes

Scheme 5.21 Synthesis of benzothiophene-1-thiones

R = H, $PhCH_2$
R_1 = Et, $PhCH_2$
R_2 = Ph, $PhCH_2$

rhodamine B base

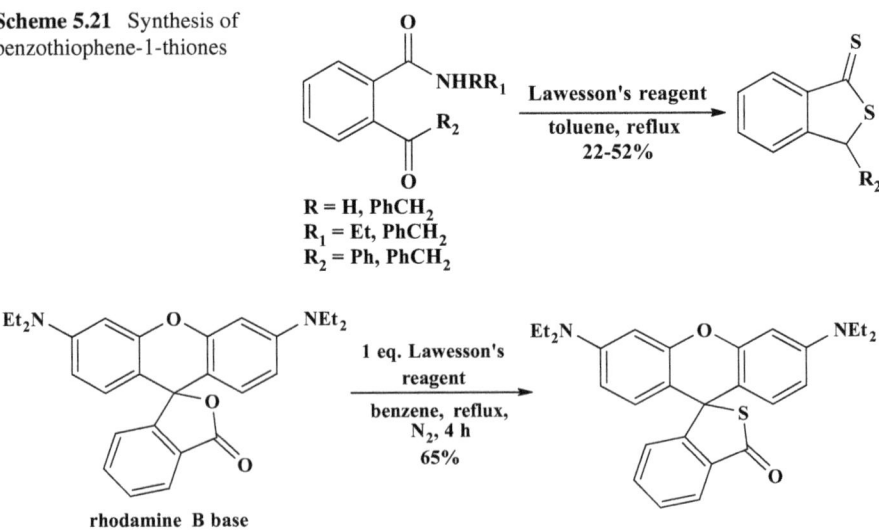

Scheme 5.22 Synthesis of benzothiophene

5.3 Synthesis of Benzothiophenes

Scheme 5.23 Synthesis of 3-phenyl-1(3H)-isobenzothiophene-1-thione

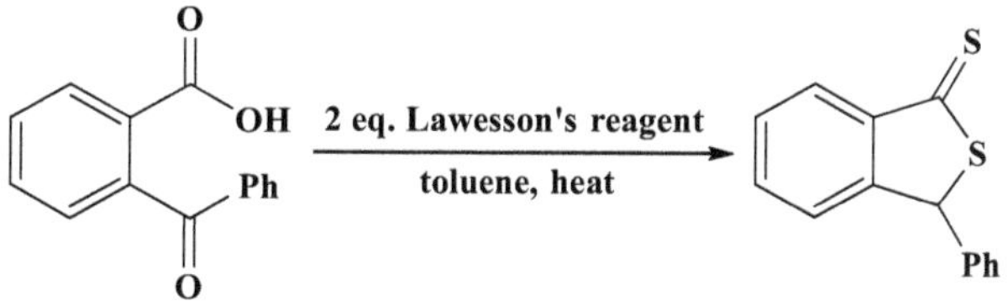

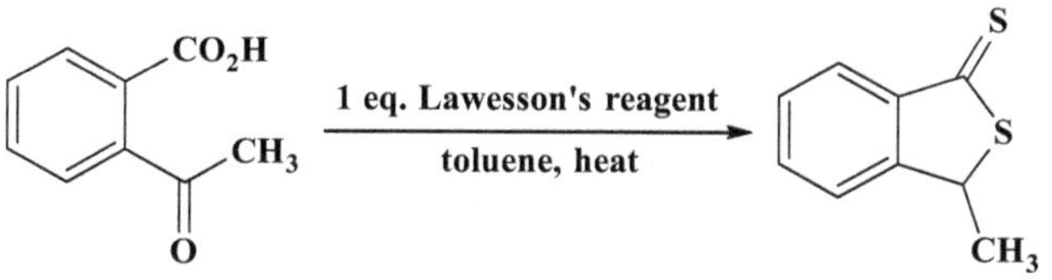

Scheme 5.24 Synthesis of 3-methyl-1(3H)-isobenzothiophene-1-thione

Instead of 1,4-diketones, γ-butyrolactones can be used as precursors. A Grignard reagent led to ring-opening of lactone. Treatment with phosphorus–sulfur compounds will result in thiophene formation (Scheme 5.25) [55].

Clement and Mohanakrishnan [56] reported the formation of symmetrical and unsymmetrical naphth-annelated thienyl heterocyclic compounds through thionation of hydroxyketones/diketones with LR (Scheme 5.26).

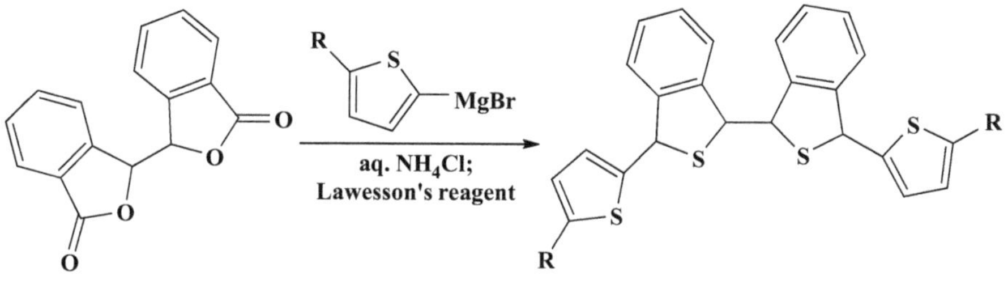

Scheme 5.25 Synthesis of 3,3'-di(thiophen-2-yl)-1,1',3,3'-tetrahydro-1,1'-bibenzothiophenes

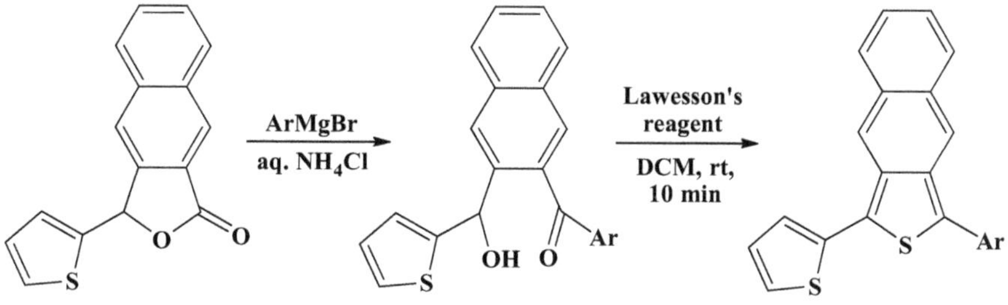

Scheme 5.26 Synthesis of naphthalene-annelated thiophenes

5.4 Synthesis of Fused Thiophenes

The synthesis of thieno[2,3-*b*]indole from 1-ethylisatin and $C_6H_5COCH_3$ has been investigated in details (Scheme 5.27). The traditional method [57] (path C) leading to aldol adduct (which was then utilized without purification), its dehydration into intermediate, reduction of latter to indolin-2-ones, and finally cyclization of indolin-2-ones with LR resulted in the synthesis of thieno[2,3-*b*]indole in 25% yield. It was observed that the reaction of intermediate with LR in toluene solution under reflux for 1 h afforded thieno[2,3-*b*]indole; however, the best yield reached to only 57% (path B). The LR served as a source of H_2S to reduce the C=C double bond in the intermediate, and secondly, as a thiation agent to obtain the thieno[2,3-*b*]indole by means of Paal–Knorr reaction. The aldol adduct has also been reacted with LR in toluene to afford the thieno[2,3-*b*]indole in low yield (10%) through the intermediacy of intermediate, depending on the starting isatin (path A). Also, the one-pot formation (path D) of thieno[2,3-*b*]indole occurred via reaction of isatin with (phenacylidene)triphenylphosphorane and subsequent cyclization of intermediate based on the path B. The yield of thieno[2,3-*b*]indole obtained by one-pot process was proved to

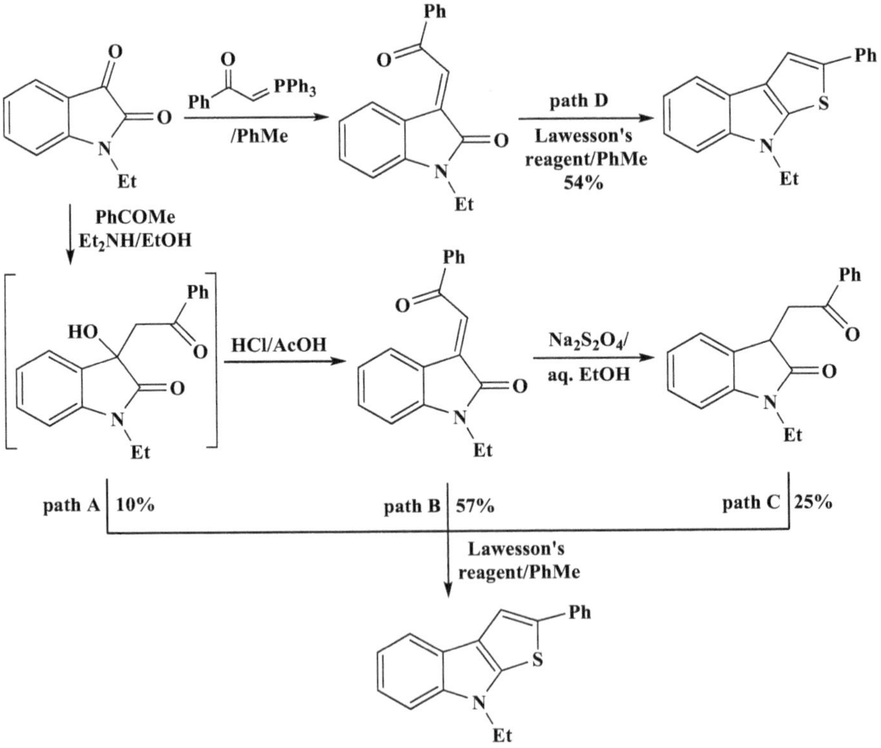

Scheme 5.27 Synthesis of thieno[2,3-*b*]indole

be close to that of thieno[2,3-*b*]indole derived from the path B. Path D needed more expensive phosphorane derivative, which was synthesized by prefunctionalization of PhCOCH$_3$, and this methodology was regarded as an alternative synthetic pathway just in some specific cases. Therefore, two-step approach to transform the isatins into thieno[2,3-*b*]indoles through the intermediacy of intermediate (path B) has been chosen as the most convenient and efficient one. A series of thieno[2,3-*b*]indoles having both electron-rich and electron-deficient (hetero)aromatic fragments at C-2 has been prepared in good to moderate yields through the two-step synthetic process (path B) from isatin and acetylated (hetero)arenes [58].

Dehydration of aldol-type adducts into 3-(2-oxo-2-(hetero)arylethylidene)indolin-2-ones took place in CH$_3$COOH solution with the incorporation of HCl (method A) or in dichloromethane solution with an excess of thionyl chloride (method B), when 3-(2-oxo-2-(hetero)arylethylidene)indolin-2-ones could not be prepared by method A (Scheme 5.28). It should be observed that thieno[2,3-*b*]indoles having 4-CN or 2-NO$_2$ phenyl substituents at C-2 have been synthesized in high yields from suitable indolin-2-ones by reaction with Lawesson's reagent under reaction conditions without displacement of CN or NO$_2$ groups [58].

Starting from *cis*-3,4-dibenzoyltricyclo[3.1.0.0^{2,6}]hexane [59–61] and tricyclo[3.1.0.0]hexanedione [62], there are many possibilities for the formation of the desired type of compound. However, only two reactions have thus far met with success: Those to the thiophenes in 20 and 9% yield from *cis*-3,4-dibenzoyltricyclo[3.1.0.0^{2,6}]hexane and LR (toluene, Et$_3$N, 22 °C, 7 d) and from tricyclo[3.1.0.0]hexanedione and bis-phenacylthioether (potassium hydroxide in MeOH, 20 °C, 18 h), respectively. The latter reaction afforded glycol product first, which was dehydrated with SOCl$_2$ in pyridine (0 °C, 1.5 h). The 3,4-diaminothiophene was transformed into thieno[3,4-*b*]quinoxaline (27%) (Scheme 5.29) [63].

The fused thiophene systems were prepared by the reaction of chloropropenylpyrazine with thiourea (Scheme 5.30). After deprotection, the yield was found to be 14% [43, 64].

The ring cyclization to thienocyclohexanone followed by ring-expansion of Schmidt-type occurred [65]. This route utilized Beckman rearrangement of cyclohexanone oxime with 5,5-dimethyl-1,3-cyclohexandione (dimedone) as a starting material (Scheme 5.31). The dimedone was reacted with 2-bromoacetophenone using EtONa in EtOH to furnish the tricarbonyl compound. The reaction of tricarbonyl compound with Lawesson's reagent gave tetrahydrobenzothiophene-4-ones. The tetrahydrobenzothiophene-4-ones were then treated with NH$_2$OH·HCl to afford the oxime. Rearrangement of oxime posed an interesting problem as alkyl (path A) and heteroaryl (path B) migrations were promising with ultimate synthesis of [3,2-*c*] or [3,2-*b*]thienoazepine. Generally, the aryl migration was preferred. Many cases were considered in which substantial alkyl migration also took place. The ring-expansion of oxime to afford the mixture of thienoazepines was affected with polyphosphoric acid (Beckmann conditions).

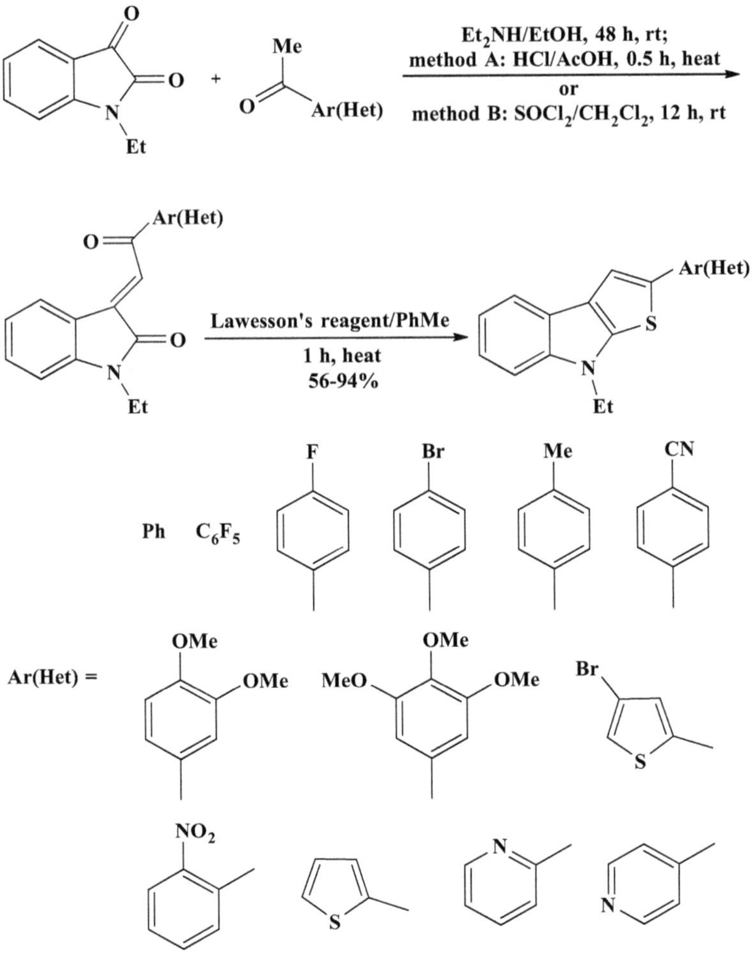

Scheme 5.28 Synthesis of thieno[2,3-b]indoles

Initial attempts to convert a mixture of four isomers of "southern half" into dithiocarbonyl compounds using LR in toluene at 100 °C led to two unexpected products, phosphaspirocycle and oxidative annulation product (Scheme 5.32) [66–68]. The formation of six-membered ring was followed by sulfur extrusion to afford the phosphaspirocycles. As the thioenamide moiety of phosphaspirocycles showed no reactivity toward amines, such as propylamine, this remarkable compound, which contributed to the exotic molecules related to the "southern half," was not further studied. In analogy, the thienopyrrole was isolated among other thionated products when a mixture of (E)- and (Z)-starting compound was heated to reflux in toluene in the presence of Lawesson's reagent (Scheme 5.33). Because all these unexpected side-products were formed at very high temperatures, the thionation of

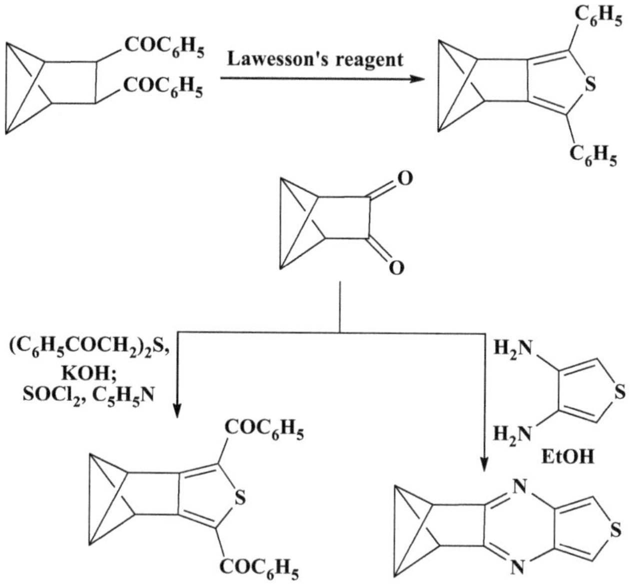

5.4 Synthesis of Fused Thiophenes

Scheme 5.29 Synthesis of thieno[3,4-b]quinoxaline

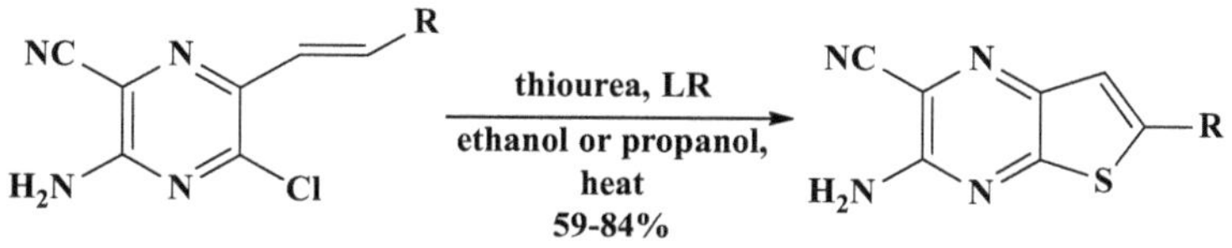

Scheme 5.30 Synthesis of thieno[2,3-b]pyrazines

R = Me, 2-thienyl, CH(CH$_3$)$_2$, C=CHPh, Ph, 4-CH$_2$OC$_6$H$_4$

four (E/Z)-isomers of "southern half" with Lawesson's reagent at lower temperatures was studied next.

5.5 Synthesis of Dithioles

The ketoesters were reacted with Lawesson's reagent and elemental sulfur in anhydrous toluene at 110 °C to give the dithiole-3-thiones in good yield (Scheme 5.34) [69].

Many diverse pathways have been employed to synthesize the DTTs. In many cases, elemental S or P$_2$S$_5$ was utilized to dehydrogenate and sulfurize an allylic methyl group to afford the desired products (Scheme 5.35) [70–74]. In addition,

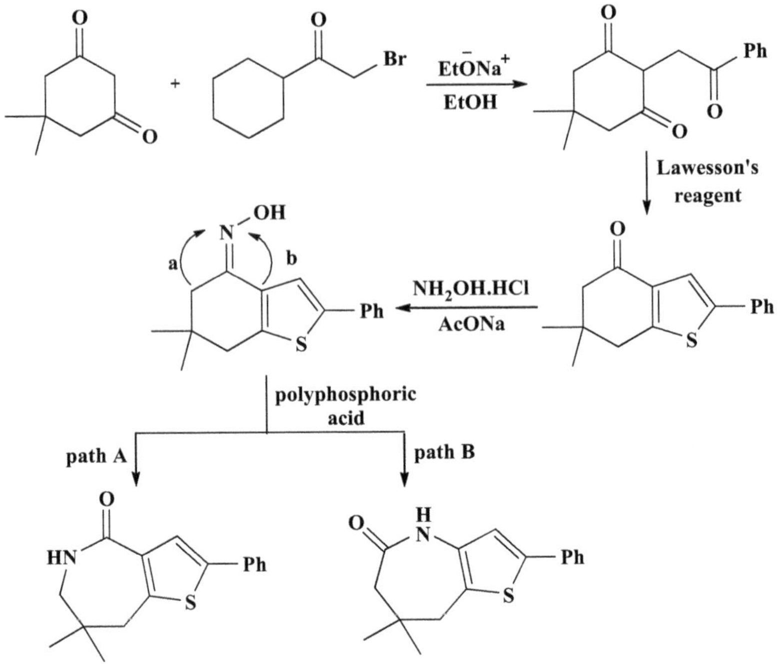

Scheme 5.31 Synthesis of [3,2-c] or [3,2-b]thienoazepine

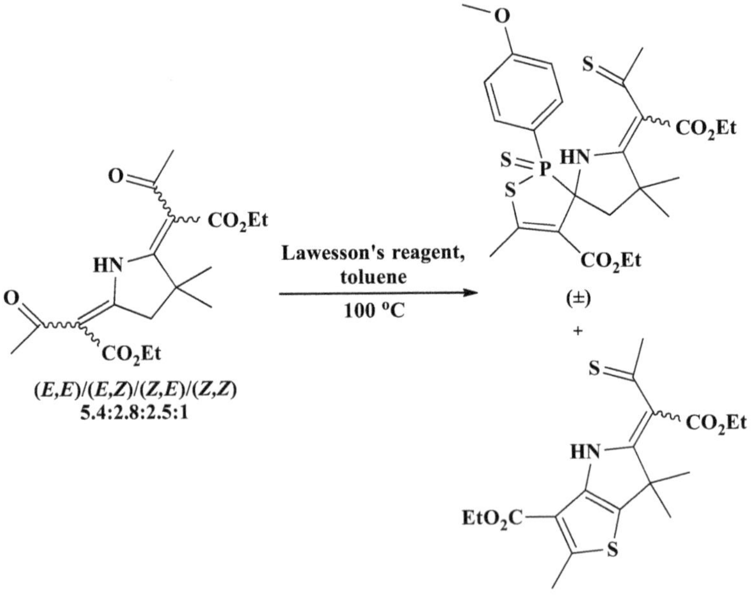

Scheme 5.32 Synthesis of phosphaspirocycle and thiophene

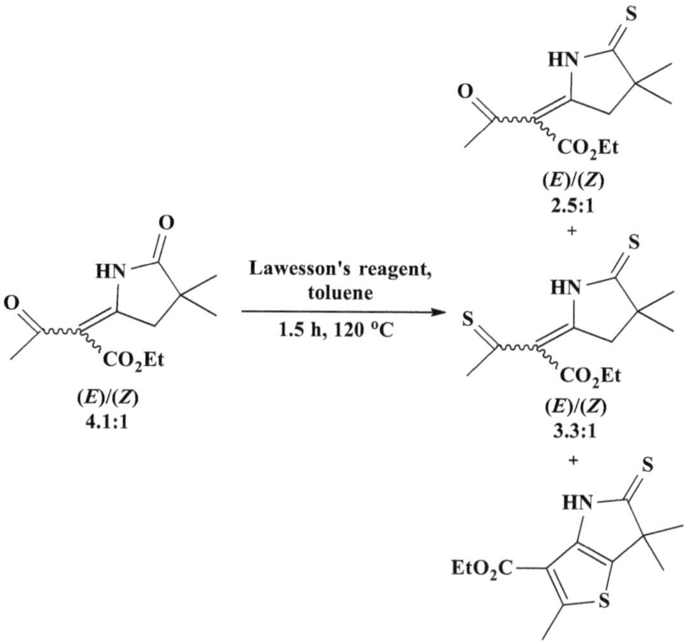

5.5 Synthesis of Dithioles

Scheme 5.33 Synthesis of pyrroles and thienopyrrole

Scheme 5.34 Synthesis of dithiole-3-thiones

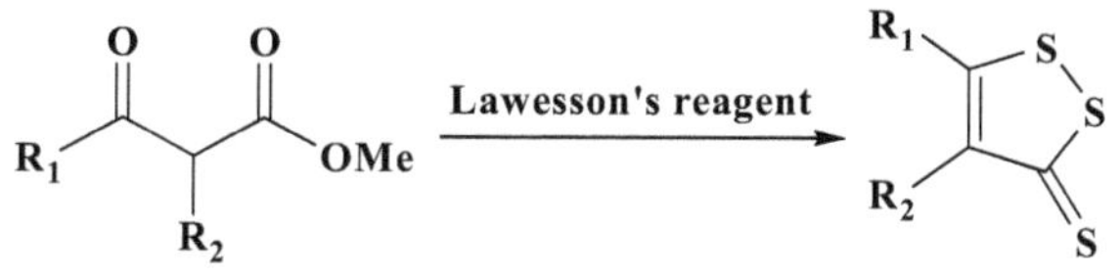

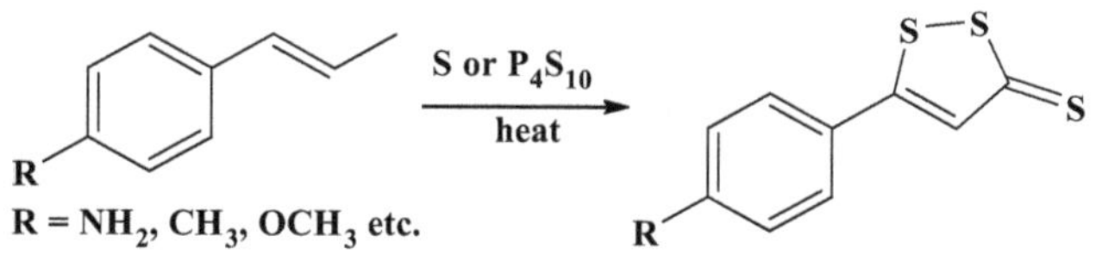

Scheme 5.35 Synthesis of dithiole-3-thiones

β-ketoesters were also reacted with Lawesson's reagent to synthesize the DTTs (Scheme 5.36) [75].

One-pot synthesis of 3H-1,2-dithiole-3-thiones has been reported using P_2S_5/S_8 in boiling xylene or Lawesson's reagent (LR)/S_8 in boiling dioxane and 2-mercaptobenzothiazole (MBT) in the presence of ZnO as a catalyst. The reaction was performed under N_2 atmosphere. Lawesson's reagent system led to cleaner reaction than those with P_2S_5/S_8 (Scheme 5.37) [76].

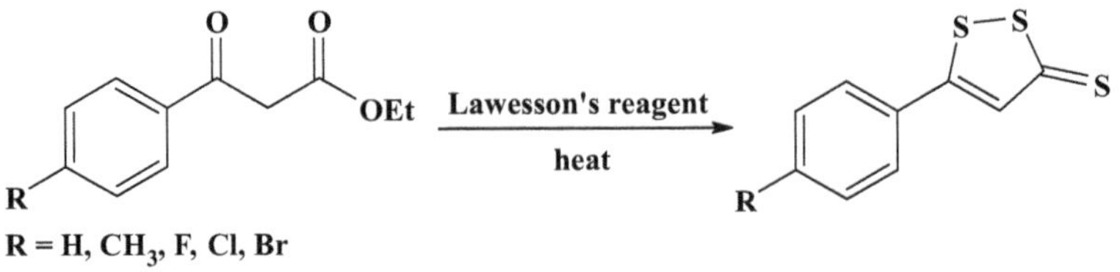

Scheme 5.36 Synthesis of dithiole-3-thiones

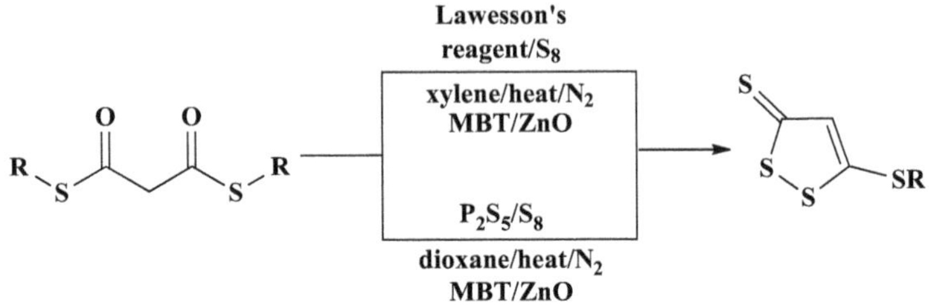

Scheme 5.37 Synthesis of 3H-1,2-dithiole-3-thiones

The reaction of ketoamides with Lawesson's reagent afforded a variety of diverse products [52]. A mixture of thioamide and sulfur heterocyclic compound was obtained from β-ketoamides (Scheme 5.38) [33].

Aimar et al. [76] described a one-pot formation of 3H-1,2-dithiole-3-thiones from dithiomalonic esters in the presence of LR and S in boiling xylene with 2-mercaptobenzothiazole and also zinc oxide as catalysts (Scheme 5.39) [77].

The synthesis of two dithiolethione rings fused to thienothiophene was completed by reaction of thienothiophenes with Lawesson's reagent in boiling xylene using S_8 (Scheme 5.40) [9, 78].

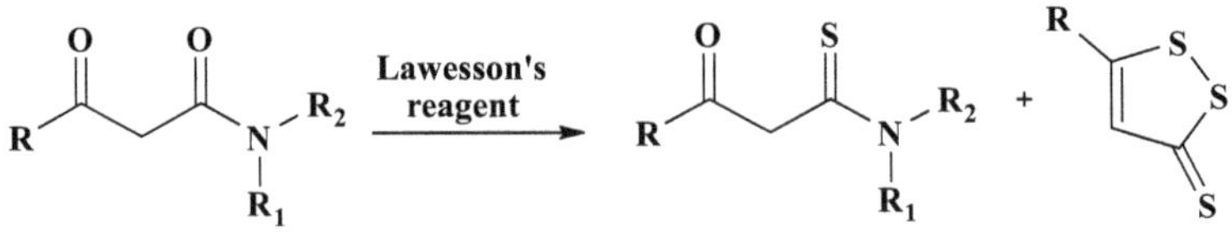

Scheme 5.38 Synthesis of thioamides and 1,2-dithiole-3-thiones

Scheme 5.39 Synthesis of 3H-1,2-dithiole-3-thiones

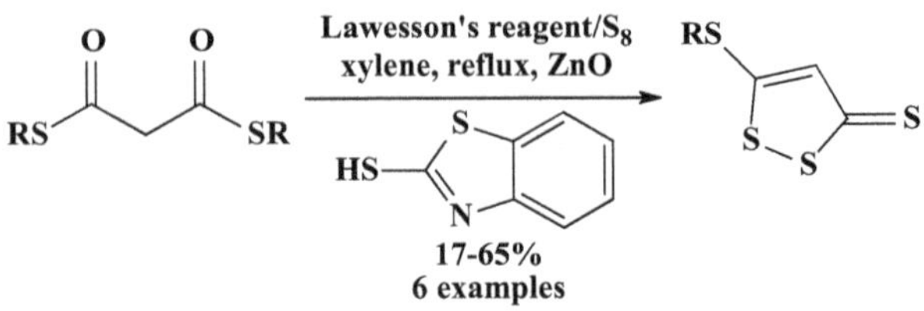

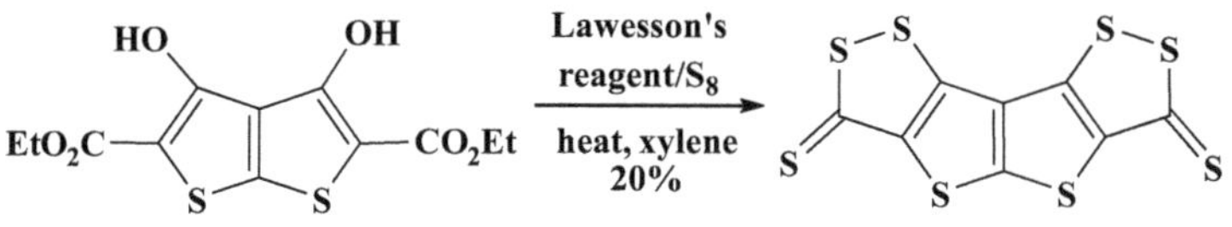

5.5 Synthesis of Dithioles

Scheme 5.40 Synthesis of dithiolothione thienothiophene

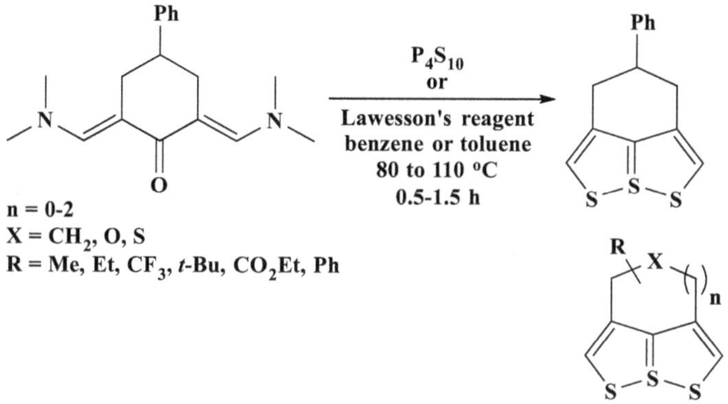

Scheme 5.41 Synthesis of bridged trithiapentalenes

The reaction of keto dienamine with P_4S_{10} resulted in the synthesis of bridged trithiapentalene. The reaction of dienamine, synthesized by the reaction of 4-phenylcyclohexanone with dimethylamino-*t*-butoxymethane (Bredereck's reagent), with P_4S_{10} (or LR) in refluxing benzene or toluene afforded trithiapentalene in 41% yield (Scheme 5.41) [79]. This pathway was extended to many bridged trithiapentalenes [43].

5.6 Synthesis of Trithioles

The reaction of selected α,β-unsaturated steroidal ketones with LR in dichloromethane and toluene under standard reaction conditions and with a combination of P_2S_5 and HMDO (P_4S_{10}/hexamethyldisiloxane) in 1,2-dichlorobenzene (*o*-DCB) under MWI was examined, and for this purpose, many cholestane, androstane, and also pregnane carbonyl derivatives were selected. Depending on the reagent and the solvent, nineteen new *S*-containing compounds including dithiones, α,β-unsaturated 3-thiones, dimer-sulfides, 1,2,4-trithiolanes, and phosphonotrithioates were obtained. Same reactions were carried out under milder conditions in dichloromethane as a solvent (refluxing for 45 min) for enhancing the yield of thioketones and the thioketones were formed in higher yield (28–70%) (Scheme 5.42). All unsaturated ketones afforded 1,2,4-trithiolanes (11–79%) as major product with

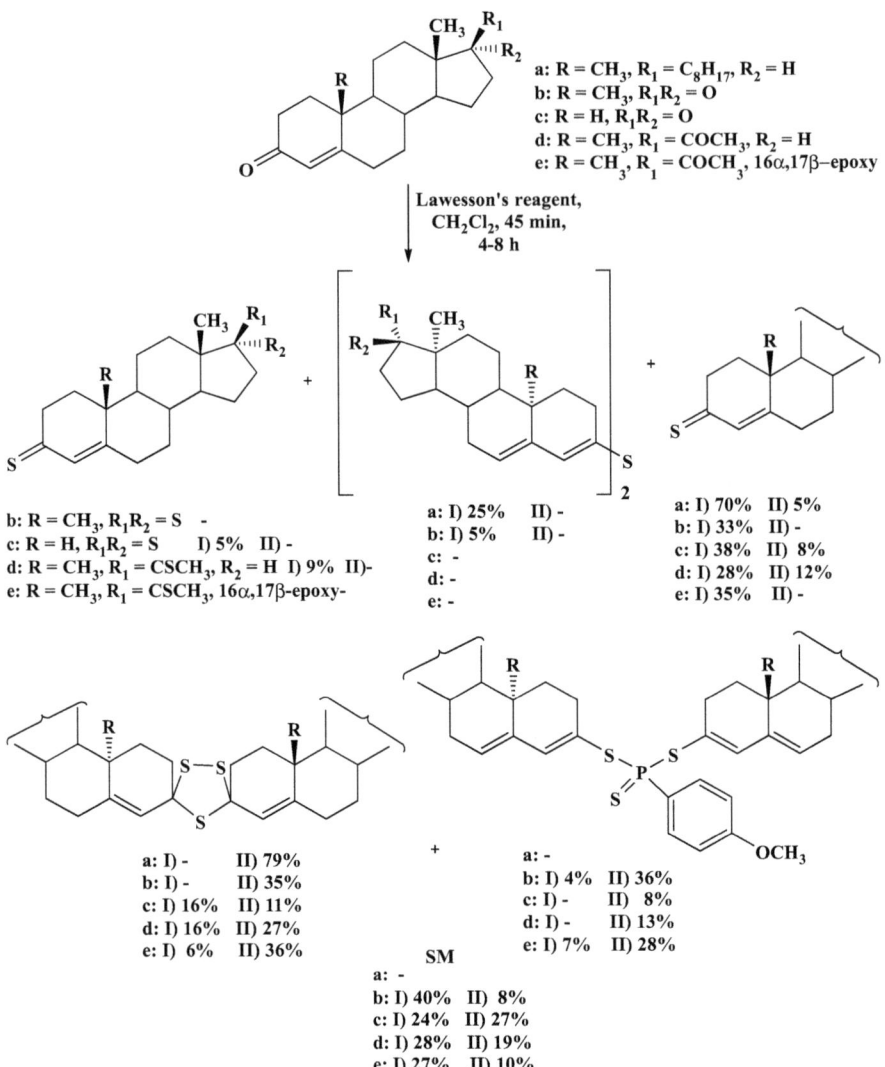

Scheme 5.42 Synthesis of trithiolanes

increase in the time of reaction in dichloromethane (reflux 4–8 h, depending on substrate). Besides, steroids also afforded (4-methoxyphenyl)phosphonotrithioates (8–36%) as a result of further reaction of firstly synthesized thioketones with Lawesson's reagent. In some cases, the thioketones were still present in the reaction mixture and isolated in very poor yield (5–12%) [80].

References

1. (a) N. Kaur. 2015. Benign approaches for the microwave-assisted synthesis of five-membered 1,2-N,N-heterocycles. J. Heterocycl. Chem. 52: 953–973. (b) N. Kaur and D. Kishore. 2014. Nitrogen-containing six-membered heterocycles: Solid-phase synthesis. Synth. Commun. 44: 1173–1211. (c) N. Kaur. 2015. Synthesis of fused five-membered N,N-heterocycles using microwave irradiation. Synth. Commun. 45: 1379–1410. (d) N. Kaur. 2017. Ionic liquids: Promising but challenging solvents for the synthesis of N-heterocycles. Mini Rev. Org. Chem. 14: 3–23.
2. (a) N. Kaur. 2014. Palladium-catalyzed approach to the synthesis of five-membered O-heterocycles. Inorg. Chem. Commun. 49: 86–119. (b) N. Kaur. 2017. Methods for metal and non-metal-catalyzed synthesis of six-membered oxygen-containing polyheterocycles. Curr. Org. Synth. 14: 531–556. (c) N. Kaur and D. Kishore. 2014. Solid-phase synthetic approach toward the synthesis of oxygen-containing heterocycles. Synth. Commun. 44: 1019–1042. (d) N. Kaur. 2014. Microwave-assisted synthesis of five-membered O-heterocycles. Synth. Commun. 44: 3483–3508. (e) N. Kaur and D. Kishore. 2014. Microwave-assisted synthesis of six-membered O,O-heterocycles. Synth. Commun. 44: 3082–3111. (f) N. Kaur. 2014. Microwave-assisted synthesis of five-membered O,N,N-heterocycles. Synth. Commun. 44: 3229–3247.
3. (a) N. Kaur. 2015. Metal catalysts: Applications in higher-membered N-heterocycles synthesis. J. Iran. Chem. Soc. 12: 9–45. (b) N. Kaur. 2015. Insight into microwave-assisted synthesis of benzo derivatives of five-membered N,N-heterocycles. Synth. Commun. 45: 1269–1300. (c) N. Kaur. 2017. Photochemical reactions: Synthesis of six-membered N-heterocycles. Curr. Org. Synth. 14: 972–998. (d) N. Kaur. 2014. Microwave-assisted synthesis of seven-membered S-heterocycles. Synth. Commun. 44: 3201–3228. (e) N. Kaur, P. Bhardwaj, M. Devi, Y. Verma, N. Ahlawat and P. Grewal. 2019. Ionic liquids in the synthesis of five-membered N,N-, N,N,N- and N,N,N,N-heterocycles. Curr. Org. Chem. 23: 1214–1238. (f) N. Kaur. 2015. Polycyclic six-membered N-heterocycles: Microwave-assisted synthesis. Synth. Commun. 45: 35–69.
4. (a) N. Kaur. 2015. Microwave-assisted synthesis: Fused five-membered N-heterocycles. Synth. Commun. 45: 789–823. (b) N. Kaur. 2015. Advances in microwave-assisted synthesis for five-membered N-heterocycles synthesis. Synth. Commun. 45: 432–457. (c) N. Kaur. 2015. Application of microwave-assisted synthesis in the synthesis of fused six-membered heterocycles with N-heteroatom. Synth. Commun. 45: 173–201. (d) N. Kaur. 2015. Microwave-assisted synthesis of fused polycyclic six-membered N-heterocycles. Synth. Commun. 45: 273–299. (e) N. Kaur. 2016. Metal catalysts for the formation of six-membered N-polyheterocycles. Synth. React. Inorg. Met. Org. Nano Met. Chem. 46: 983–1020.
5. B. Jiang, C. Yang and J. Wang. 2002. Enantioselective synthesis of marine indole alkaloid hamacanthin B. J. Org. Chem. 67: 1396–1398.
6. G. Bringmann, S. Tasler, H. Endress and J. Muhlbacher. 2001. En route to the first stereoselective synthesis of axially chiral biscarbazole alkaloids. Chem. Commun. 8: 761–762.
7. A.R. Katritzky, S. LeDoux and S. Nair. 2003. Benzannulation of 3-substituted pyrroles to indoles. J. Org. Chem. 68: 5728–5730.
8. R.B. Bedford and C.S. Cazin. 2002. A novel catalytic one-pot synthesis of carbazoles via consecutive amination and C-H activation. Chem. Commun. 20: 2310–2311.
9. T. Ozturk, E. Ertas and O. Mert. 2007. Use of Lawesson's reagent in organic syntheses. Chem. Rev. 107: 5210–5278.
10. M.I. Levinson and M.P. Cava. 1982. The synthesis of acenaphtho[1,2-c][1,2,3]thiadiazole and phenanthro[9,10-c][1,2,3]thiadiazole. Heterocycles 19: 241–243.
11. H.Z. Lecher, R.A. Greenwood, K.C. Whitehouse and T.H. Chao. 1956. The phosphonation of aromatic compounds with phosphorus pentasulfide. J. Am. Chem. Soc. 78: 518–522.
12. A.A. Kiryanov, P. Sampson and A.J. Seed. 2001. Synthesis of 2-alkoxy-substituted thiophenes, 1,3-thiazoles, and related S-heterocycles via Lawesson's reagent-mediated cyclization under microwave irradiation: Applications for liquid crystal synthesis. J. Org. Chem. 66: 7925–7929.

13. T. Besson and V. Thiery. 2006. Microwave-assisted synthesis of sulfur and nitrogen-containing heterocycles. Top. Heterocycl. Chem. 1: 59–78.
14. M.M.M. Raposo and G. Kirsch. 2001. A combination of Friedel-Crafts and Lawesson reactions to 5-substituted 2,2'-bithiophenes. Heterocycles 55: 1487–1498.
15. M.M.M. Raposo, A.M.B.A. Sampaio and G. Kirsch. 2005. Arylamino-thieno-oxobutanamides under Lawesson's conditions: Competition between thienylpyrrole and bithiophene formation. Synthesis 2: 199–210.
16. I.L. Goldfarb, B.P. Fabrichnyi and I.F. Shalavina. 1958. Synthesis of 3-(2-thienyl)alanine. Bull. Acad. Sci. USSR Div. Chem. Sci. 7: 89–91.
17. L. Fieser and R.G. Kennelly. 1935. A comparison of heterocyclic systems with benzene. 1IV. Thionaphthenequinones. J. Am. Chem. Soc. 57: 1611–1616.
18. F. Effenberger, F. Wurthner and F. Steybe. 1995. Synthesis and solvatochromic properties of donor-acceptor-substituted oligothiophenes. J. Org. Chem. 60: 2082–2091.
19. O. Gidron, Y. Diskin-Posner and M. Bendikov. 2010. α-Oligofurans. J. Am. Chem. Soc. 132: 2148–2150.
20. C.-Q.M. Mishra and P. Bauerle. 2009. Functional oligothiophenes: Molecular design for multidimensional nano architectures and their applications. Chem. Rev. 109: 1141–1276.
21. J. Kaschel, C.D. Schmidt, M. Mumby, D. Kratzert, D. Stalke and D.B. Werz. 2013. Donor-acceptor cyclopropanes with Lawesson's and Woollins' reagents: Formation of bisthiophenes and unprecedented cage-like molecules. Chem. Commun. 49: 4403–4405.
22. G.H. Degenhart. 2008. Synthesis of conjugated oligomers. PhD Thesis, Universiteit Leiden.
23. H.J. Kooreman and H. Wynberg. 1967. The chemistry of polythienyls. Part III: The synthesis of terthienyls. Recl. Trav. Chim. Pays-Bas 86: 37–55.
24. V.M. Sonpatki, M.R. Herbert, L.M. Sandvoss and A.J. Seed. 2001. Troublesome alkoxythiophenes. A highly efficient synthesis via cyclization of γ-keto esters. J. Org. Chem. 66: 7283–7286.
25. J. Brunet, D. Paquer and P. Rioult. 1977. Sulfurization of γ-oxo esters; study of the resulting sulfur compounds. Phosphorus, Sulfur, Silicon Relat. Elem. 3: 377–379.
26. K. Dimitrowa, J. Hauschild, H. Zaschke and H. Schubert. 1980. Kristallin-flüssige 1,3,4-thiadiazole. I. Biphenyl- und terphenyl analoge 1,3,4-thiadiazole. J. Prakt. Chem. 322: 933–944.
27. P. Diano, A. Carbone, P. Barraja, A. Montalbano, A. Martorana, G. Dattolo, O. Gia, L.D. Via and G. Cirrincione. 2007. Synthesis and antitumor properties of 2,5-bis(3'-indolyl)thiophenes: Analogues of marine alkaloid *Nortopsentin*. Bioorg. Med. Chem. Lett. 17: 2342–2346.
28. H.Z. Lecher, R.A. Greenwood, K.C. Whitehouse and T.H. Cho. 1956. The phosphonation of aromatic compounds with phosphorus pentasulfide. J. Am. Chem. Soc. 78: 5018–5022.
29. M.P. Cava and M.I. Levinson. 1985. Thionation reactions of Lawesson's reagents. Tetrahedron 41: 5061–5087.
30. S. Ishikawa, Y. Noda, M. Wada and T. Nishikata. 2015. A copper-catalyzed formal [3+2]-cycloaddition for the synthesis of all different aryl-substituted furans and thiophenes. J. Org. Chem. 80: 7555–7563.
31. F. Hadj-Abo, S. Bienz and M. Hesse. 1994. Synthesis of [10]heterophanes using a ring enlargement reaction. Tetrahedron 50: 8665–8672.
32. K.-T. Kang and J.U. Sun. 1995. A convenient synthesis of thiophenes from β,γ-epoxycarbonyl compounds. Synth. Commun. 25: 2647–2653.
33. M.S.J. Foreman and J.D. Woollins. 2000. Organo-P-S and P-Se heterocycles. J. Chem. Soc. Dalton Trans. 10: 1533–1543.
34. E.K. Kim, K.U. Lee, B.Y. Cho, Y.B. Kim and K.-T. Kang. 2001. Regioselective synthesis of 2,4-disubstituted thiophenes. Liq. Cryst. 28: 339–345.
35. K.-T. Kang, Y.B. Hwang, M.Y. Kim, S.K. Lee and J.G. Lee. 2002. Synthesis of 4-alkylfuran-2-acetates and 4-alkylthiophene-2-acetates using 2-(chloromethyl)-3-(tributylstannyl)propene. Bull. Korean Chem. Soc. 23: 1333–1336.
36. B.B. Thompson. 2012. The reductive couplings of enones and alkynes and application to heterocycle synthesis. PhD Thesis, University of Michigan.

37. A. Smith. 1890. XLV - Contributions from the chemical laboratory of the University of Edinburgh. No. II. On desylacetophenone. J. Chem. Soc. 57: 643–652.
38. H. Wynberg and J. Metselaar. 1984. A convenient route to polythiophenes. Synth. Commun. 14: 1–9.
39. D.M. Young and C.F.H. Allen. 2003. 2,5-Dimethylpyrrole. Org. Synth. 16: 25–25.
40. H. Stetter and M. Schreckenberg. 1974. A new method for adding aldehydes to activated double bonds, IV. Adding aromatic and heterocyclic aldehydes to α,β-unsaturated ketones. Chem. Rep. 107: 2453–2458.
41. Y. Ju, D. Miao, R. Yu and S. Koo. 2015. Tandem catalytic oxidative deacetylation of acetoacetic esters and heteroaromatic cyclizations. Org. Biomol. Chem. 13: 2588–2599.
42. C.W. Ong, C.M. Chen and L.F. Wang. 1998. Convenient synthesis of 2,5-disubstituted thiophene from 1,6-dioxo-2,4-diene. Tetrahedron Lett. 39: 9191–9192.
43. T. Ozturk, E. Ertas and O. Mert. 2010. A Berzelius reagent, phosphorus decasulfide (P_4S_{10}), in organic syntheses. Chem. Rev. 110: 3419–3478.
44. D.R. Shridhar, M. Jogibhukta, P.S. Rao and V.K. Handa. 1982. An improved method for the preparation of 2,5-disubstituted thiophenes. Synthesis 12: 1061–1062.
45. F. Freeman, D.S.H.L. Kim and E. Rodriguez. 1992. Preparation of 1,4-diketones and their reactions with bis(trialkyltin) or bis(triphenyltin)sulfide-boron trichloride. J. Org. Chem. 57: 1722–1727.
46. D.R. Shridhar, M. Jogibhukta, P.S. Rao and V.K. Handa. 1984. An improved method for the synthesis of 2-thioxo-2,3-dihydro-1,3-benzoxazoles [2(3H)-benzoxazolethiones]. Synthesis 11: 936–937.
47. T. Nishhio, N. Okuda and C. Kashima. 1988. A facile route to 4,5,6,7-tetrahydrothiophenes and benzothiophenes. J. Heterocycl. Chem. 25: 1437–1438.
48. P. Singh, P. Sharma, K. Bisetty and M.P. Mahajan. 2009. Cycloaddition reactions of cross-conjugated enaminones. Tetrahedron 65: 8478–8485.
49. S. Scheibye, J. Kristensen and S.-O. Lawesson. 1979. Studies on organophosphorus compounds - XXVII. Tetrahedron 35: 1339–1343.
50. G. Minetto, L.F. Raveglia, A. Sega and M. Taddei. 2005. Microwave-assisted Paal-Knorr reaction - three-step regiocontrolled synthesis of polysubstituted furans, pyrroles and thiophenes. Eur. J. Org. Chem. 24: 5277–5288.
51. P. Singh, P. Sharma, K. Bisetty and M.P. Mahajan. 2011. Synthesis and docking studies of thiophene scaffolds in COX-2. ARKIVOC (x): 55–70.
52. T. Nishio. 1998. Sulfur-containing heterocycles derived by reaction of ω-keto amides with Lawesson's reagent. Helv. Chim. Acta 81: 1207–1214.
53. X.-Q. Zhan, Z.-H. Qian, H. Zheng, B.-Y. Su, Z. Lan and J.-G. Xu. 2008. Rhodamine thiospirolactone. Highly selective and sensitive reversible sensing of Hg(II). Chem. Commun. 16: 1859–1861.
54. T. Nishio. 1995. Photoreactions of 1,3-dihydroisobenzofuran-1-thiones and 1,3-dihydro-2-benzothiophene-1-thiones with alkenes. J. Chem. Soc. Perkin Trans. 1 5: 561–568.
55. A.K. Monankrishnan, P. Amaladass and J.A. Clement. 2007. Synthesis of end-blocked thienyl oligomers incorporating benzo[c]thiophene. Tetrahedron Lett. 48: 779–784.
56. J.A. Clement and A.K. Mohanakrishnan. 2010. Synthesis and characterization of naphth-annelated thiophene analogs. Tetrahedron 66: 2340–2350.
57. J. Levy, D. Royer, J. Guilhem, M. Cesario and C. Pascard. 1987. Benzothieno[2,3-b]indole and pyridothiéno[2,3-b]indole. Bull. Soc. Chim. Fr. 1: 193–198.
58. R.A. Irgashev, A.A. Karmatsky, G.L. Rusinov and V.N. Charushin. 2015. A new and convenient synthetic way to 2-substituted thieno[2,3-b]indoles. Beilstein J. Org. Chem. 11: 1000–1007.
59. M. Christl and S. Freund. 1985. Substituierte benzobenzvalene und diazabenzobenzvalene-synthesen aus Diels-Alder-addukten des benzvalens und NMR-spektroskopie. Chem. Ber. 118: 979–999.
60. Y. Sugihara, S. Wakabayashi, N. Sailo and L. Murata. 1986. 6-Cyanotetracyclo[5.5.0.0^{2,4}.0^{3,5}]dodeca-6,8,10,12-tetraene. J. Am. Chem. Soc. 108: 2773–2775.

61. M. Christi and E. Herzog. 1987. 3-(Phenylsulfonyl)tricyclo[4.1.0.0^{2,7}]hept-4-en-3-yllithium. Tetrahedron Lett. 28: 187–190.
62. M. Christi and A. Kraft. 1988. Tricyclo[3.1.0.0^{2,6}]hexanedione (the valene of *o*-benzoquinone), bicyclo[2.1.1]hexane-2,3-dione, and valenes of a quinoxaline, of phenazine, and of a benzophenazine. Angew. Chem. Int. Ed. 27: 1369–1370.
63. M. Christ, S. Krimm and A. Kraft. 1990. Some valenes of benzannelated five-membered heteroarenes - synthesis and NMR spectra. Angew. Chem. Int. Ed. 29: 675–677.
64. E.J. Taylor and A.L. Sabb. 1988. Studies on the molybdenum cofactor: Model synthetic routes directed at form B. J. Org. Chem. 53: 5839–5847.
65. A.M. Delgado. 1999. Ensenaza de las ciencias. 17: 493–502.
66. P. Venkateswarlu and S.C. Venkata. 2004. Studies on organophosphorus compounds: Reactions of benzosuberones with 2,4-bis(*p*-methoxyphenyl)-1,3,2,4-dithiadiphosphetane-2,4-disulfide (Lawesson's reagent). Tetrahedron Lett. 45: 3207–3209.
67. M.I. Hegab. 2000. Reaction of hindered α,β-unsaturated ketones with 2,4-bis(4-methoxyphenyl)-1,3,2,4-dithiadiphosphetane-2,4-disufide. Phosphorus, Sulfur, Silicon Relat. Elem. 166: 137–148.
68. S. Kametani, H. Ohmura, H. Tanaka and S. Motoki. 1982. Reaction of chalcones with 2,4-bis(4-methoxyphenyl)-1,3,2,4-dithiadiphosphetane-2,4-disulfide. Chem. Lett. 11: 793–796.
69. Ger. Qffen Patent N^0 1, 460, 783 [CA. 1976, 85, 1238991].
70. P.S. Landis. 1965. The chemistry of 1,2-dithiole-3-thiones. Chem. Rev. 65: 237–245.
71. P.S. Landis and L.A. Hamilton. 1960. Synthesis and reactions of 4-neopentyl-5-*t*-butyl-1,2-dithiole-3-thione. J. Org. Chem. 25: 1742–1744.
72. R.S. Spindt, D.R. Stevens and W.E. Baldwin. 1951. Some new alkyl 1,2-dithia-4-cyclopentene-3-thiones. J. Am. Chem. Soc. 73: 3693–3697.
73. T.J. Curphey and H.H. Joyner. 1993. Synthesis of 3*H*-1,2-dithiole-3-thiones by a novel oxidative cyclization. Tetrahedron Lett. 34: 7231–7234.
74. T.J. Curphey and A.A. Libby. 2000. Dianions of 3-oxodithioic acids: Preparation and conversion to 3*H*-1,2-dithiole-3-thiones. Tetrahedron Lett. 41: 6977–6980.
75. B.S. Pedersen and S.O. Lawesson. 1979. Studies on organophosphorus compounds - XXVIII. Tetrahedron 35: 2433–2437.
76. M.L. Aimar, J. Kreiker and R.H. Rossi. 2002. One-pot synthesis of 3*H*-1,2-dithiole-3-thione derivatives from dithiolmalonic esters. Tetrahedron Lett. 43: 1947–1949.
77. R.A. Stockman. 2003. Heterocyclic chemistry. Annu. Rep. Prog. Chem. Sect. B 99: 161–182.
78. R. Gompper, R. Knieler and K. Polborn. 1993. Thieno[2,3-*b*]thiopheno[3,2-*c*;4,5-*c*']-bis(1,2-dithiole)-3,6-dithione - an octamer of carbon sulfide. Z. Naturforsch. B: Chem. Sci. 48: 1621–1624.
79. W. Zhang and Y. Henry. 2001. A new synthetic route to 3,4-bridged 1,6,6aλ^4-trithiapentalenes. Synlett 7: 1129–1130.
80. N.M. Krstic, M.S. Bjelakovic, M.M. Dabovic and V.D. Pavlovic. 2010. Thionation of some α,β-unsaturated steroidal ketones. Molecules 15: 3462–3477.

Chapter 6
S-Heterocycle Synthesis

6.1 Introduction

Heterocycles have always been the center of attraction because of their applications in medicinal chemistry [1–3]. The heterocyclic moieties are fundamental part of widespread diversity of biologically active natural products and synthetic compounds [4–7]. The overwhelming majority of commercially accessible synthetic drugs (up to 80%) have a heterocyclic structural constituent. Because of the extensive importance of heterocycles, the formation of these compounds has always been the most important research field in synthetic chemistry. In several cases, the classical method afforded reliable access to heterocyclic compounds; however, they are now not accepted by ecological and safety standards [8, 9]. Modern developments in discovery and process chemistry emphasize novel sustainable synthetic pathways, needing fast and ecologically acceptable substitutes to the classical approaches. The development of sustainable synthetic processes to substitute the efficient but slightly outdated classical approaches started few decades ago and such approaches are in high demand till date [10, 11].

The heterocyclic compounds which contain nitrogen and sulfur have a massive importance in the area of pharmaceutical chemistry. They show many biological properties like antitubercular, antifungal, antibacterial, analgesic, and anti-inflammatory. Some of these are in the development phase because of the versatility of the skeleton, availability, and chemical simplicity [12–15].

The thiazines are six-membered heterocyclic compounds that contain a nitrogen atom and sulfur atom in their structure. The thiazines are very beneficial in the areas of medicinal and pharmaceutical chemistry and exhibit many biological activities. The 1,3-thiazines are of great importance because they form part of the scaffold of cephalosporins (3,6-dihydro-2*H*-1,3-thiazine) and are also present in some other pharmaceutically important compounds such as xylazin (agonist at the α2 group of adrenergic receptor is employed for anesthesia, sedation, analgesia, and muscle relaxation in animals), chlormezanone (used as a muscle relaxant and an anxiolytic) etc. [16–24].

LR is commercially accessible and expensive. However, it can be conveniently synthesized in large amounts from easily accessible P_2S_5 and anisole. The drawbacks of employing this technique are the need for anhydrous reaction conditions (the reagent is hygroscopic), the expensive reagent, and purification problems during the isolation of products from phosphorus containing side-products [25–29].

LR is 2,4-bis(p-methoxyphenyl)-1,3-dithia-2,4-diphosphetane-2,4-dithione. It is generally employed to transform the CO groups to thiones. Additional reagent which has been utilized effectively to achieve the thiations is P_2S_5, but this has been employed more commonly for replacing the oxygen atom of a furan ring with sulfur atom [30, 31].

6.2 Synthesis of Five-Membered S,N-Heterocycles

The reaction of chiral bis-N-acylamino alcohols with Lawesson's reagent in toluene at reflux temperature afforded chiral bis-thiazolines in good yields (Scheme 6.1). The chiral thiazoline was also prepared by the reaction of N-picolinoylamino-2-phenylethanol. The chiral bis-thiazolines thus formed, sulfur analogues of known

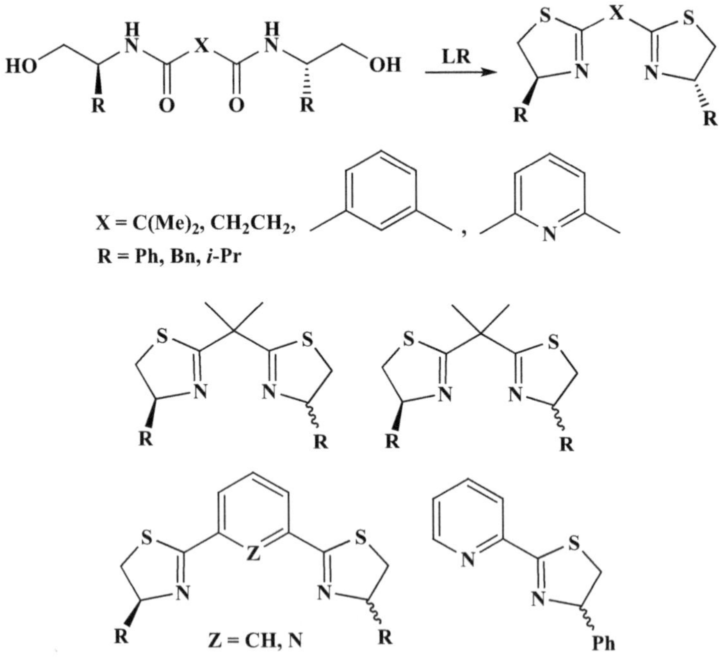

Scheme 6.1 Synthesis of bis-thiazolines

6.2 Synthesis of Five-Membered S,N-Heterocycles

oxazolines, was expected to be new chiral ligands for metals in asymmetric Diels–Alder reactions [32].

As the chiral oxazolines proved effective ligands for numerous catalytic asymmetric C–C bond formations, and especially for Diels–Alder cycloadditions, Nishio and coworkers [32] described the ability of thiazolines to achieve this conversion. The ligands were prepared by reacting bis-N-acylamino alcohols with LR (Scheme 6.2). This ligand was only evaluated with Zn(OTf)$_2$ as a catalyst in the test Diels–Alder reaction. The *endo*-product was formed with 92% enantiomeric excess and 88% diastereoselectivity. Yamakuchi et al. [33] synthesized sterically crowded "roofed" 2-thiazolines by thermal [4 + 2]-cycloadditions of 2-thiazolone and cyclic dienes, subsequent hydrolytic ring-cleavage with barium hydroxide, and final thiazoline ring synthesis following the general process for the formation of oxazoline ligands. They

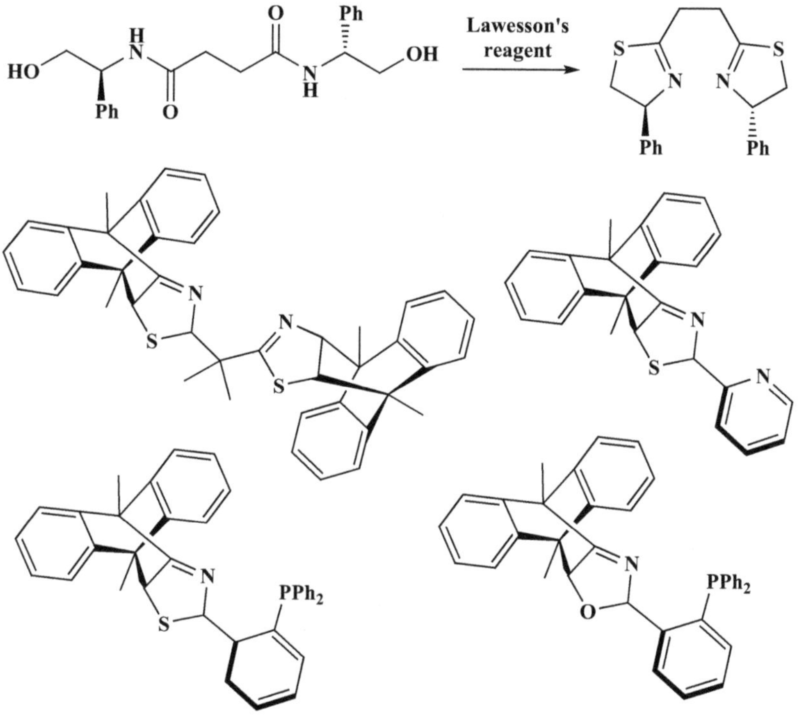

L	temp.	time	yield (%)	ee endo (%)
1	0	3	97	2 (R)
2	0	24	88	16 (S)
3	0	1	92	76 (R)
4	-60	36	81	92 (R)
5	-60	24	94	73 (R)

Scheme 6.2 Synthesis of bis-thiazoline

prepared a bis-thiazoline derivative, a pyridylthiazoline, and (2-diphenylphosphino)-phenylthiazoline (along with its oxazoline analogue for comparison) in high yield. The Diels–Alder reaction was achieved with these ligands and Cu(OTf)$_2$ in CH$_2$Cl$_2$ to provide the *endo*-product as a major isomer within a few hours. The bis-thiazoline and pyridylthiazoline afforded low enantioselectivities, whereas the phosphinothiazoline led to product with 76% enantiomeric excess by carrying out the reaction at 0 °C. Decreasing the temperature to −60 °C allowed a considerable increase in the enantioselectivity of the reaction, since the product was isolated with 92% enantiomeric excess. The similar phosphinooxazoline ligand remained less enantioselective under similar reaction conditions, justifying the favorable influence of the sulfur atom in the heterocyclic ring on the *N,P*-chelating behavior to the copper center [34].

The substituted thiazoles were synthesized in 50% yield. Many secondary aromatic propargylic alcohols participated well in the reaction, affording propargylation/sulfuration/cyclization products with complete regioselectivity. Both aliphatic and aromatic amides were efficiently introduced into thiazole scaffold. The structure of substituted thiazoles was entirely different from substituted thiazoles. The thiazoles were also synthesized by reacting starting compound with LR without 10 mol% iron(III) chloride (Scheme 6.3). This outcome evidently exhibited that iron(III) chloride was not required in the cyclization step of the sequential reactions [35, 36].

The formation of southern hemisphere model started with the production of thiazole derived from a modified glutamate residue (fragment D) that was assembled [37]. The stereocontrolled hydroxylation of glutamate under Hanessian conditions [38] afforded 4-hydroxy glutamate, immediately transformed into its *t*-butyldimethylsilyl ether, which on purification was obtained as a single diastereomer over two steps in 68% yield. The hydrogenolysis of benzyl ester was followed by transformation of acid into amide. The reaction of amide with LR afforded thioamide, which underwent Hantzsch reaction with 3-bromopyruvic acid [39] to provide the desired fragment over six steps in 34% overall yield (Scheme 6.4) [40].

A tryptamine thiazole alkaloid (+)-bacillamide B, isolated from a new bacterium *Bacillus endophyticus*, was prepared by a Hantzsch synthesis of (*S*)-2-acetoxyethylthiazole-4-carboxylate (Scheme 6.5) [41, 42].

Three differently substituted thiazoles have been prepared successfully from easily accessible propargylic alcohols, amides, and LR [36]. Various secondary or tertiary propargylic alcohols possessing not only terminal alkyne groups but also internal

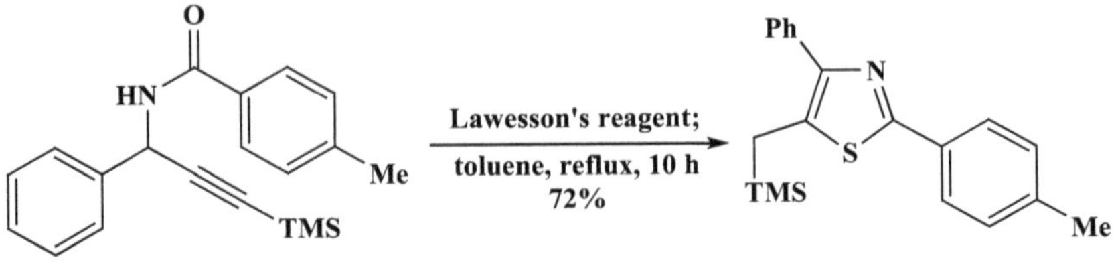

Scheme 6.3 Synthesis of thiazole

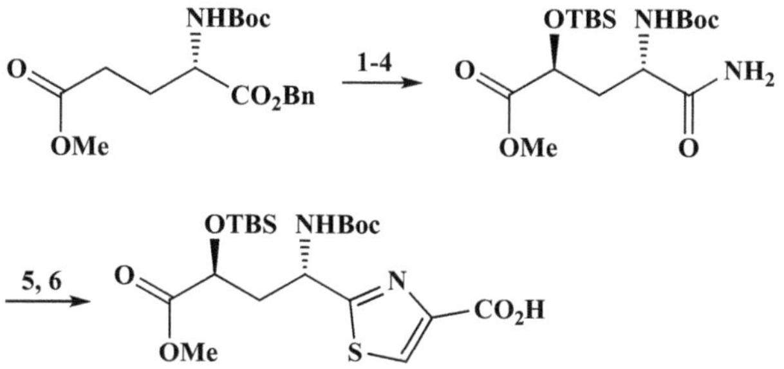

6.2 Synthesis of Five-Membered S,N-Heterocycles

Reagents and conditions: (1) LHMDS, 2-benzenesulfonyl-3-phenyloxaziridine, THF, -78 °C, (2) TBSCl, imidazole, DMF, 68% over 2 steps, (3) H$_2$, Pd-C (10%), MeOH, (4) EtO$_2$CCl, Et$_3$N, THF then 30% aq. NH$_4$OH, 88% over 2 steps, (5) Lawesson's reagent, THF, (6) 3-bromopyruvic acid, EtOH, CaCO$_3$, 57% over 2 steps.

Scheme 6.4 Synthesis of thiazole

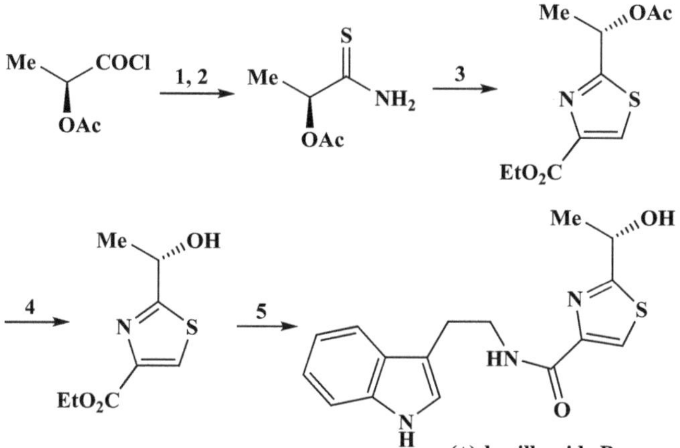

Reagents and conditions: (1) NH$_3$, CH$_2$Cl$_2$, rt, (2) Lawesson's reagent, dioxane, (3) ethyl bromopyruvate, ethyloxirane, *i*-PrOH, 60 °C, then TFAA, rt, (4) LiOH, THF, MeOH, (5) dipyridyl sulfide, PPh$_3$, CH$_2$Cl$_2$.

Scheme 6.5 Synthesis of tryptamine thiazole alkaloid (+)-bacillamide B

alkyne groups can be utilized successfully, and a number of functional groups like cyclohexenyl, cyclopropyl, chloro, bromo, methoxy, and ester were tolerated under reaction conditions (Scheme 6.6).

The substituted thiazoles were prepared directly from amides, propargylic alcohols, and LR in a one-pot process (Scheme 6.7) [36]. The Fe-catalyzed substitution reaction of propargylic alcohol with amide to provide the propargylic

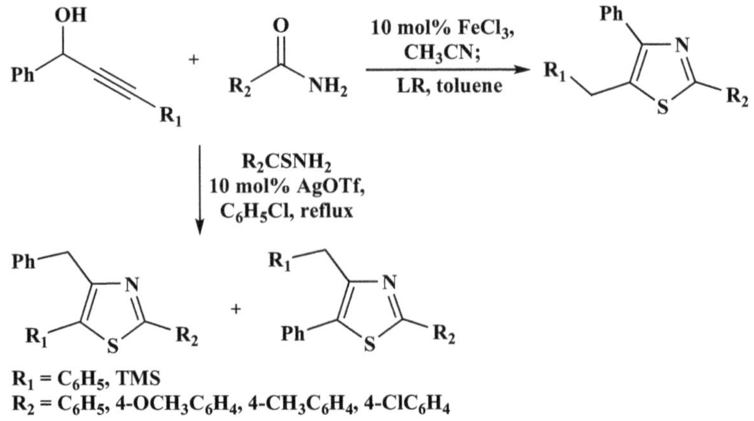

Scheme 6.6 Synthesis of thiazoles

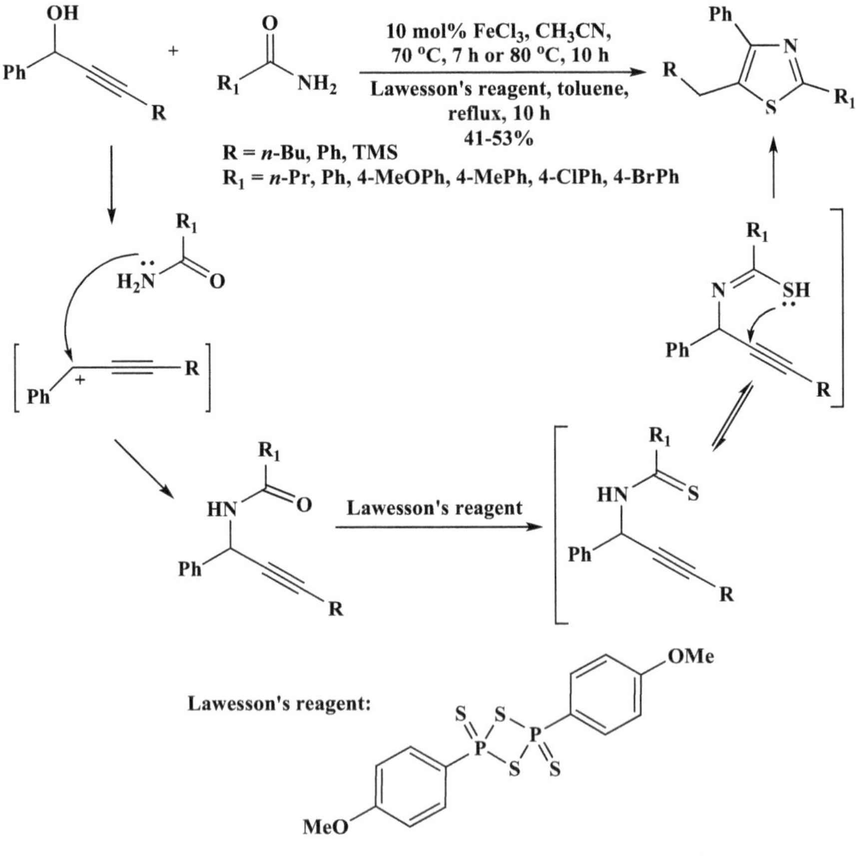

Scheme 6.7 Synthesis of thiazoles

amide was followed by sulfuration with LR to afford the thioamide. Subsequently, cycloisomerization of thioamides afforded final product with complete regioselectivity.

Hydroxyl amide was selected as a starting compound for the formation of thiazoline. First, thionation of amide with LR was carried out in hexamethylphosphoramide [26] but the conversion was very low and the desired thioamide could not be isolated from the crude reaction mixture. In a modified method, hydroxy amide was first transformed into *t*-butyldimethylsilyl ether; subsequent reaction with LR in toluene at 85 °C overnight, followed by acidic work-up and flash chromatography, was successful and afforded thiazoline in 78% yield. The thionation of *t*-butyldimethylsilyl ether was accompanied by the removal of *t*-butyldimethylsilyl protecting group and the intermediate alcohol was derivatized with LR [28], to generate a leaving group (through a phosphorus-oxygen bond formation), followed by cyclization to synthesize the thiazoline with inversion of configuration (Scheme 6.8) [43].

The reaction consisted of elaboration of the dipeptide fragment and its coupling to the acid corresponding to ester. The formation of dipeptide started with the coupling of known protected threonine with *N*-carboxy 2-amino-2-butenoic acid anhydride and after that ammonolysis to afford the amide (Scheme 6.9) [44]. Lawesson reaction afforded thioamide, which generated thiazole under modified-Hantzsch conditions. A sequence of protection/deprotection steps resulted in the formation of trifluoroacetic acid salt.

The enone was prepared from glycolonitrile (Scheme 6.10); reaction with H_2S provided thioamide, which underwent a Hantzsch reaction with ethyl bromopyruvate (EBP) in refluxing EtOH. The emerging thiazolyl ester was subjected to ammonolysis and after that protection of free alcohol provided amide in 64% yield. The amide was converted into aldehyde in three uneventful steps featuring yet another Hantzsch

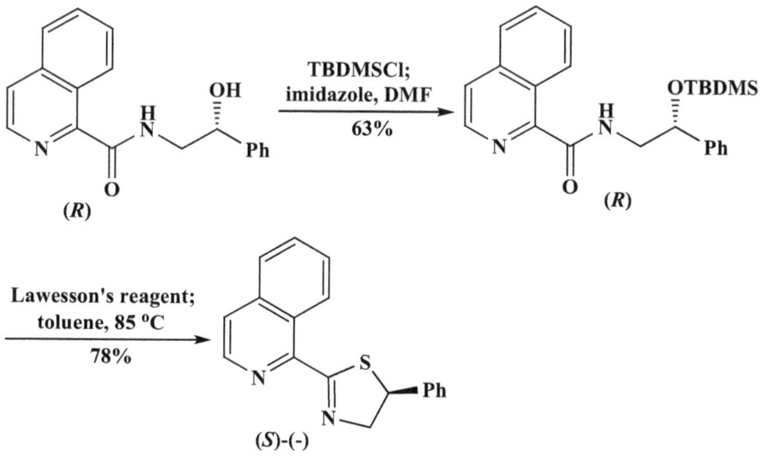

Scheme 6.8 Synthesis of thiazoline

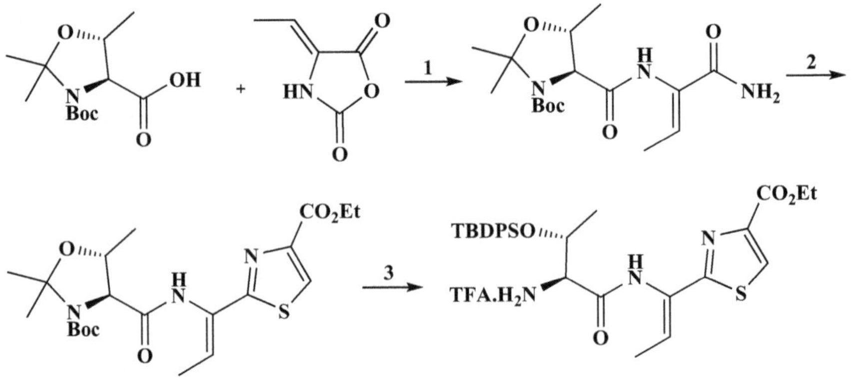

Reagents and conditions: (1) DCC, DMAP, THF then NH$_4$OH, 82% over 2 steps, (2) a) Lawesson's reagent, DME, 49%, b) EBP, KHCO$_3$, EtOH, then TFAA, Py, 70%, (3) a) TFA, DCM, 87%, b) TBDPSCl, imidazole, DCM, 0 °C to rt, 92%, c) TFA, DCM, 4 Å mol. sieves, no yield furnished.

Scheme 6.9 Synthesis of thiazole

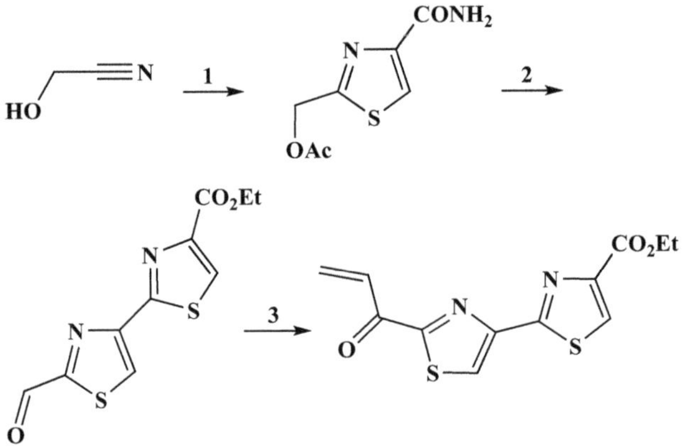

Reagents and conditions: (1) a) H$_2$S, H$_2$O, Py, Et$_3$N, b) EBP, EtOH, reflux, 95% over 2 steps, c) NH$_3$, MeOH, d) Ac$_2$O, Py, 67% over 2 steps, (2) a) Lawesson's reagent, toluene, reflux, b) EBP, EtOH, reflux, then K$_2$CO$_3$, EtOH, c) PCC, DCM, 40% from (4-carbamoylthiazol-2-yl)methyl acetate, (3) a) H$_2$C=CHMgBr, THF, -78 °C, b) activated MnO$_2$, EtOAc, 72% over 2 steps

Scheme 6.10 Synthesis of bis-thiazole

6.2 Synthesis of Five-Membered S,N-Heterocycles

thiazole synthesis. An addition of vinylmagnesium bromide and oxidation of allylic alcohol transformed aldehyde into enone [45].

The thioamide, needed for the formation of thiazole, is very polar and H_2O-soluble, complicating purification by aqueous extractions or column chromatography. The originally designed process [46] required the reaction of a commercial 55% aqueous solution of glycolonitrile with gaseous hydrogen sulfide in Py and TEA, then the evaporation of reaction mixture to a thick oil. This left a residue of crude thioamide, which was utilized directly in a Hantzsch thiazole synthesis. Because the synthesis of thiazole ring was consistently quite efficient on various thioamides, it was surmised that the yield of thiazole reflected the yield of thioamide reported in the previous step (Scheme 6.11) [45].

A de novo formation of pyridine unit circumvented many preceding problems. It needed the development of appropriate pyridine synthesizing approach. A substantial demonstration was furnished by Bagley et al. [47] in the formation of promothiocin A. A modified Bohlmann-Rahtz reaction involving the union of ynone with enamine synthesized pyridine in a single step. The procedure involved Michael-type addition of enamine to ynone and then cyclization/aromatization of formed intermediate. Pyridine was rapidly advanced to oxazole–thiazole bearing pyridine, which was further elaborated to promothiocin A (Scheme 6.12).

The formation of final compound began from L-threonine derivative [48]. The ammonolysis and then chemoselective transformation of the amide into thioamide set a stage for Hantzsch thiazole formation that afforded oxazole–thiazole. The condensation of oxazole–thiazole with anion of easily accessible methylthiazole (synthesized in three steps from commercially accessible thioacetamide) afforded desired product in 3:177 mixture of enol (major) and keto tautomers (Scheme 6.13).

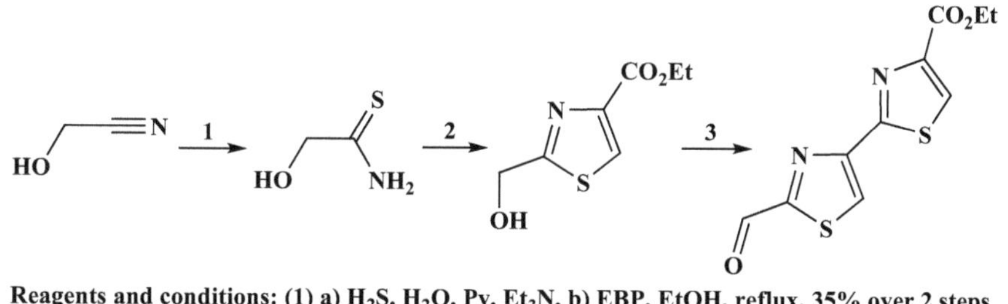

Reagents and conditions: (1) a) H_2S, H_2O, Py, Et_3N, b) EBP, EtOH, reflux, 35% over 2 steps, (2) a) NH_3, MeOH, b) Ac_2O, Py, c) Lawesson's reagent, xylene, reflux, d) EBP, EtOH, reflux, then K_2CO_3, (3) PCC, DCM, 41% over 5 steps.

Scheme 6.11 Synthesis of bis-thiazole

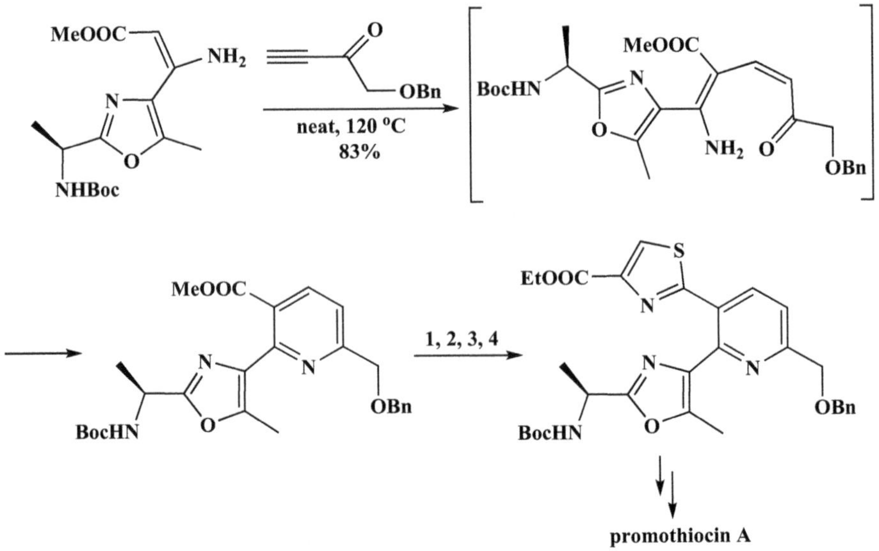

Reagents and conditions: (1) LiOH, 92%, (2) EtO$_2$CCl, Et$_3$N, then NH$_4$OH, 85%, (3) Lawesson's reagent, 59%, (4) EBP, 72%.

Scheme 6.12 Synthesis of promothiocin A

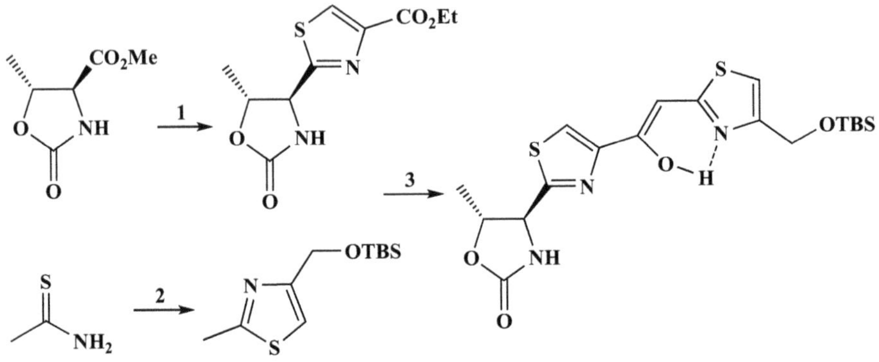

Reagents and conditions: (1) a) NH$_3$, MeOH, b) Lawesson's reagent, benzene, reflux, c) EBP, EtOH, reflux, 82% overall, (2) a) EBP, EtOH, reflux, 96%, b) LAH, Et$_2$O, 0 °C, c) TBSCl, imidazole, DMF, 94% over 2 steps, (3) 3 eq. 4-(((*t*-butyldimethylsilyl)oxy)methyl)-2-methylthiazole, 3 eq. *n*-BuLi, THF, -78 °C, then 1 eq. ethyl 2-((4*S*,5*R*)-5-methyl-2-oxooxazolidin-4-yl)thiazole-4-carboxylate, 90%.

Scheme 6.13 Synthesis of oxazole appended bis-thiazole

6.3 Synthesis of Five-Membered S,N-Polyheterocycles

The 2-iodoaniline and benzoyl chloride were reacted in the presence of LR (Scheme 6.14). The reaction was at first analyzed with 5 mol% copper(I) iodide, 10 mol% 1,10-phenanthroline, and 2 eq. 1,4-diazabicyclo[2.2.2]octane in toluene. Under these conditions, the benzothiazole was obtained in 65% yield. A control experiment showed that employment of ligand was not necessary for the reaction. The yield enhanced to 92% when CH_2Cl_2 was added as a cosolvent. Further screening of bases showed that 1,8-diazabicyclo[5.4.0]undec-7-ene or 1,4-diazabicyclo[2.2.2]octane was the best choice in the reaction. Similar results were obtained when the amount of copper(I) iodide was decreased to 1 mol%. The reaction carried out without copper(I) iodide afforded benzothiazole in satisfactory yield, which confirmed that the mechanism of ring-closure was not metal-assisted. It was rationalized that a S_NAr2 nucleophilic substitution was involved in the reaction procedure on the basis of following factors: (1) after deprotonation, the imine is EWG, (2) the sulfur anion is a very strong nucleophile, and (3) an intramolecular ring-closure to a five-membered ring is highly preferred [49].

The scope of this cascade one-pot reaction was examined under optimized reaction conditions (2 eq. 1,4-diazabicyclo[2.2.2]octane, dichloromethane/toluene). Since many 2-iodoanilines and acid chlorides are synthetically available or commercially accessible, this reaction sequence can be used to prepare a small library of benzothiazoles. The benzothiazoles were formed in good to excellent yields with a variety of 2-iodoanilines and acid chlorides. In addition to benzoyl chloride, reaction of 2-iodoaniline with furan-2-carbonyl chloride or CH_3COCl also afforded products in good yields. All reactions proceeded easily to synthesize the benzothiazoles in good yields (Scheme 6.15) [49].

Bose and Idrees [50] reported a regioselective and metal-free cascade methodology for the formation of substituted benzothiazoles in a regioselective manner from benzoyl chlorides, 2-fluroanilines, and LR under conventional thermal conditions or MWI for 5 min (Scheme 6.16) [51].

The benzothiazole was prepared from substituted anilines. The reaction of anilines with 4-nitrobenzoylchloride afforded benzanilides, which on reaction with

Scheme 6.14 Synthesis of 2-phenylbenzothiazole

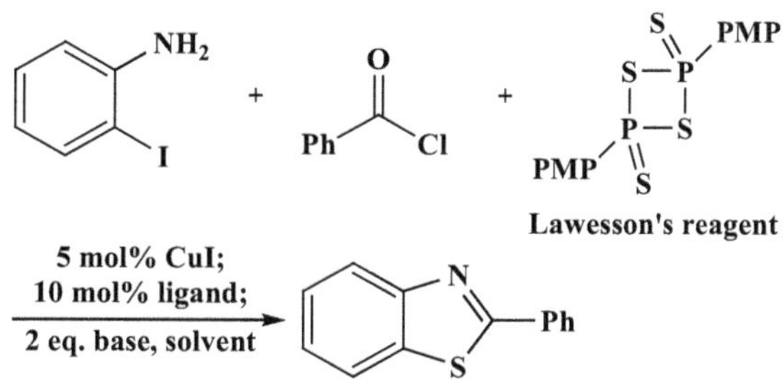

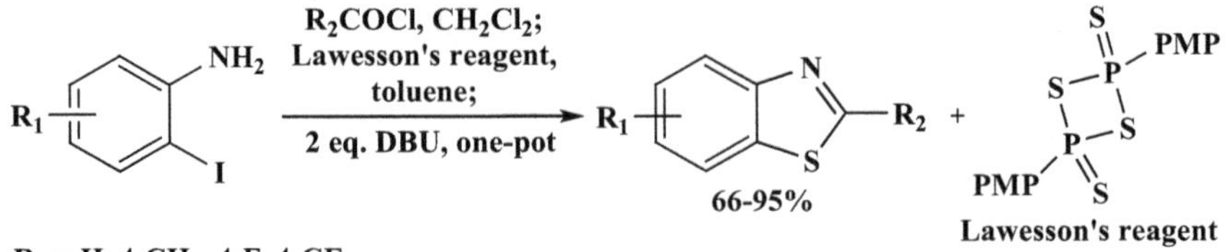

R₁ = H, 4-CH₃, 4-F, 4-CF₃
R₂ = C₆H₅, 4-MeOC₆H₄, 3-MeC₆H₄, 2-MeOC₆H₄, 3,5-(CF₃)₂C₆H₅, 2-furyl, Me

Scheme 6.15 Synthesis of benzothiazoles

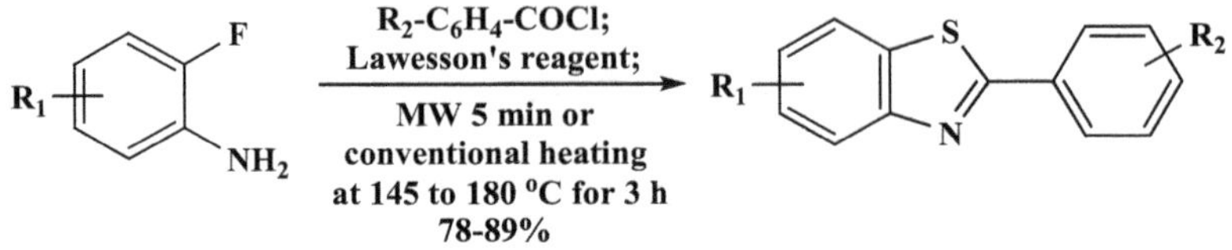

Scheme 6.16 Synthesis of benzothiazoles

LR furnished thiobenzanilides. The thiobenzanilides were transformed to 2-(4-nitrophenyl)-6-benzothiazoles via Jacobson's cyclization employing potassium ferricyanide as a reagent under reflux for 30 min. The benzothiazoles were prepared by the reduction of nitro group with tin(II) chloride (Scheme 6.17) [52–55].

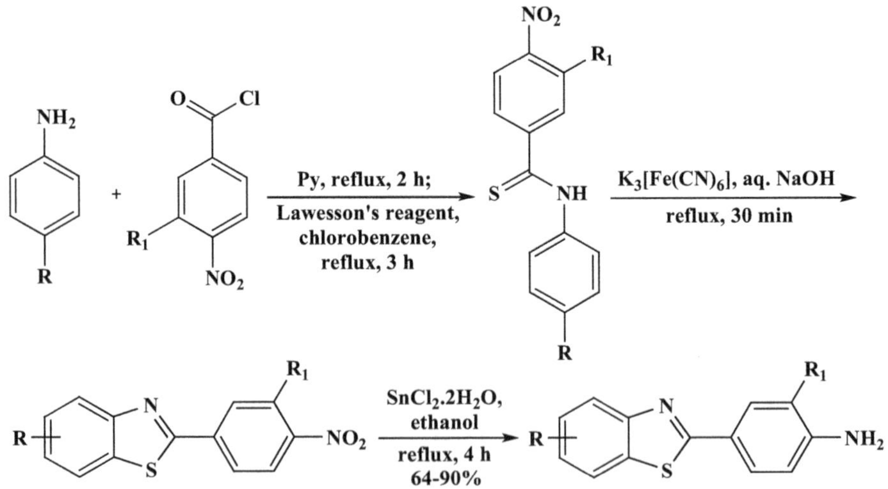

R = 4-F, 6-F, 4,5-di-F, 4,6-di-F, 5,7-di-F, 5-F, 7-F, 5,6-di-F, 6,7-di-F
R₁ = H, CH₃

Scheme 6.17 Synthesis of benzothiazoles

Hutchinson et al. [56] employed an analogous approach for the formation of benzothiazoles using fluorinated nitrobenzanilides as substrates. All the fluoro-substituted nitrobenzanilides afforded benzothiazoles in excellent yield except 4,6-difluoro- and 5,7-difluoronitrobenzanilides [57].

6.4 Synthesis of Five-Membered Fused *S,N*-Heterocycles

The thiazolo[5,4-*b*]pyridines and oxazolo[5,4-*b*]pyridines with *p*-*N,N*-dimethylaminophenyl group were synthesized (Scheme 6.18). The reduction

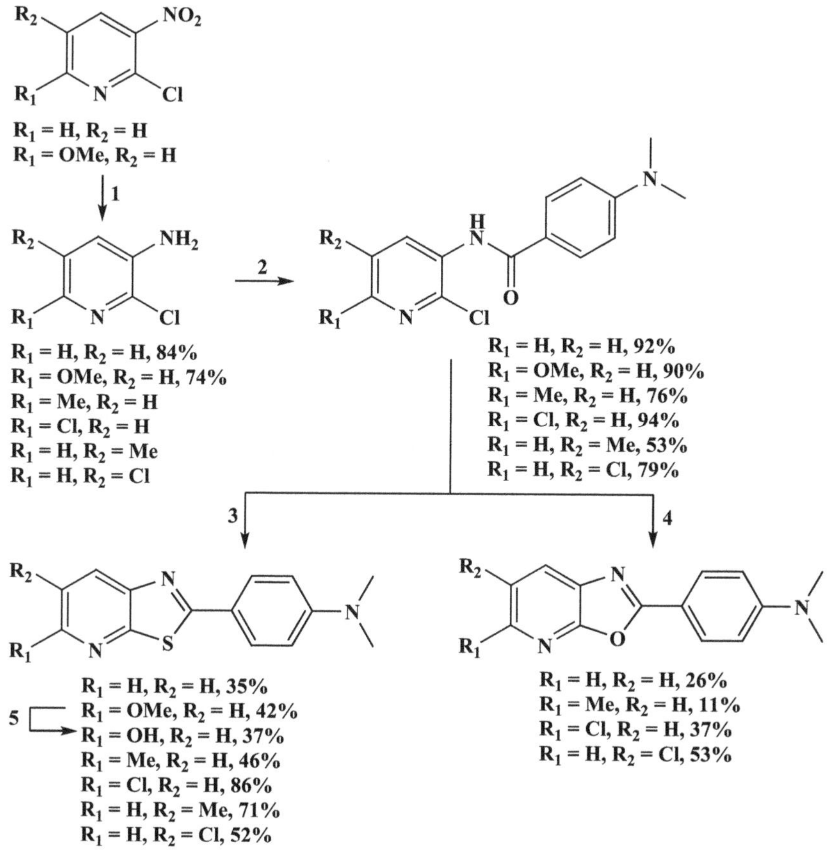

Reagents and conditions: (1) $SnCl_2$, EtOH, reflux, 30 min, 2 h, (2) 4-dimethylaminobenzoyl chloride, Py, rt, 18 h, (3) Lawesson's reagent, chlorobenzene, reflux, 3 h, (4) P_2O_5, hexamethyldisiloxane, 1,2-dichlorobenzene, 140 °C, 2 d or polyphosphoric acid, dichloromethane, 140 °C, 1 d, (5) BBr_3, dichloromethane, reflux, 12 h.

Scheme 6.18 Synthesis of thiazolo[5,4-*b*]pyridines and oxazolo[5,4-*b*]pyridines

of nitro group of 3-nitropyridines with SnCl$_2$ [58] afforded amino compounds and consequent amidation of amino compounds with 4-dimethylaminobenzoyl chloride in pyridine gave benzamides. The transformation of CO group of benzamides to thiocarbonyl group with LR and then ring-closure afforded thiazolo[5,4-b]pyridines [59]. The reaction of benzamides with P$_2$O$_5$ or H$_3$PO$_4$ afforded oxazolo[5,4-b]pyridines [60]. The OH compounds were prepared from methoxy compound by demethylation employing BBr$_3$ [61, 62].

Different from the formation of 2-arylbenzothiazoles [63, 64], there are very few synthetic pathways for the synthesis of 2-arylthiazolo[4,5-b]pyridines in the literature. The acylation of 2-amino-3-bromopyridine derivatives with benzoic acids afforded benzamides. The cyclization of benzamides was carried out with LR to afford the 2-arylthiazolo[4,5-b]pyridines (Scheme 6.19). Although the conversion was attained in one-pot reaction in the absence of metal catalyst, multisteps were needed to synthesize the 2-amino-3-bromopyridine utilizing excess of bromine or N-bromosuccinimide. This cyclization reaction also suffered from low yields and high reaction temperature [65].

Scheme 6.20 describes the formation of thiazolo[5,4-b]pyridines with p-N-methylaminophenyl or p-aminophenyl group. An amidation of amino compounds and ring-closure of amides took place. The reduction of nitro group of thiazole with SnCl$_2$ resulted in the amino compounds, which were methylated with paraformaldehyde and NaBH$_4$ in CH$_3$ONa to afford the terminal secondary amines [62, 66].

Fruit and coworkers [67] synthesized thiazolo[4,5-b]pyrazine derivatives from amidopyrazines by reacting with LR (Scheme 6.21). The thiazolo[4,5-b]pyrazines show many biological activities [68–70] and act as fluorophores. Because benzo-fused thiazolo[4,5-b]pyrazines have fluorescence character [71–79], the thiazolo[4,5-b]pyrazine derivatives may be key compounds for the formation of beneficial fluorophores. A phenyl group was incorporated at C-2 of thiazolo[4,5-b]pyrazine ring to examine the substituent effects on the spectroscopic properties. The 2-phenyl derivatives were isolated from *Cypridina oxyluciferin* analogue.

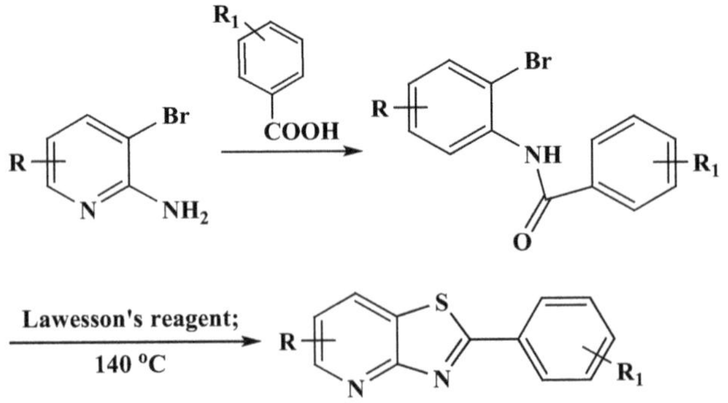

Scheme 6.19 Synthesis of 2-arylthiazolo[4,5-b]pyridines

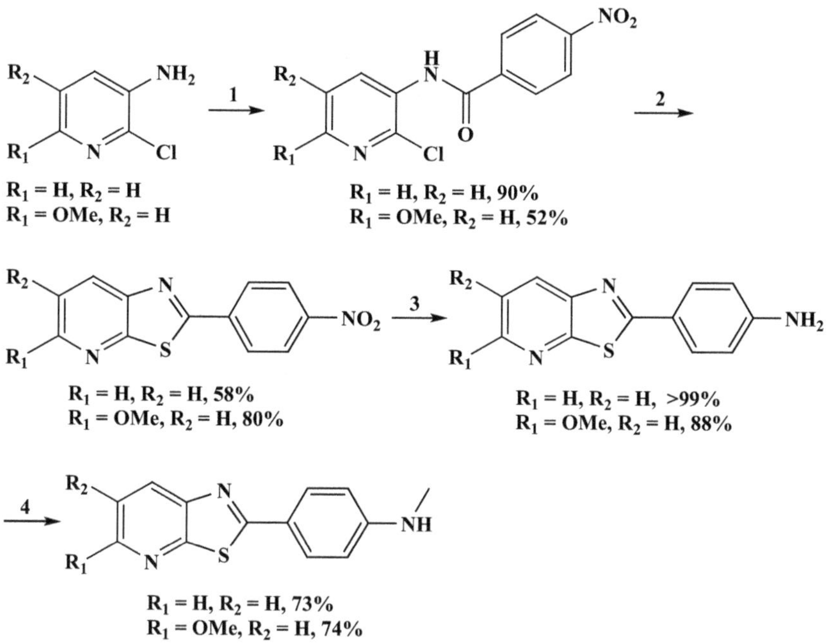

6.4 Synthesis of Five-Membered Fused S,N-Heterocycles

Reagents and conditions: (1) 4-nitrobenzoyl chloride, Py, rt, 18 h, (2) Lawesson's reagent, chlorobenzene, reflux, 3 h, (3) $SnCl_2$, EtOH, reflux, 6–18 h, (4) $(CH_2O)n$, NaOMe, MeOH, $NaBH_4$, reflux, 5 h.

Scheme 6.20 Synthesis of thiazolo[5,4-b]pyridines

The formation of disaccharide thiazoline began from peracetylated Manβ1,4GlcNAc derivative [80, 81]. The reaction of α/β mixture of peracetylated Manβ1,4GlcNAc derivative (α:β ca. 3:1) with LR [2,4-bis(4-methoxyphenyl)-1,3-dithia-2,4-diphosphetane-2,4-disulfide] in toluene at 80 °C afforded thioacetamide α-isomer as the main product along with thiazoline as the minor product. The thioacetamide α-isomer could not be cyclized to synthesize the thiazoline derivative under the reaction conditions. This result was consistent with the observations reported with monosaccharide derivatives that the α-anomers of peracetylated GlcNAc and GalNAc were resistant to next bimolecular nucleophilic substitution-type cyclization due to the unfavorable configuration of 1-α-OAc leaving group [82, 83]. Many Lewis acids have been investigated as catalysts to transform the thioacetamide α-isomer into thiazoline derivative. A combination of trimethylsilyl chloride and boron trifluoride etherate was the best catalyst that was capable to convert the thioacetamide to thiazoline in an excellent yield. Other catalysts like boron trifluoride etherate or trimethylsilyl bromide would either led to decomposition or low yield of thiazoline product. The de-O-acetylation of thiazoline afforded disaccharide thiazoline (Scheme 6.22) [28, 84].

The tetrasaccharide thiazoline was synthesized. The reaction of tetrasaccharide derivative [80] with LR and then treatment with trimethylsilyl chloride and boron

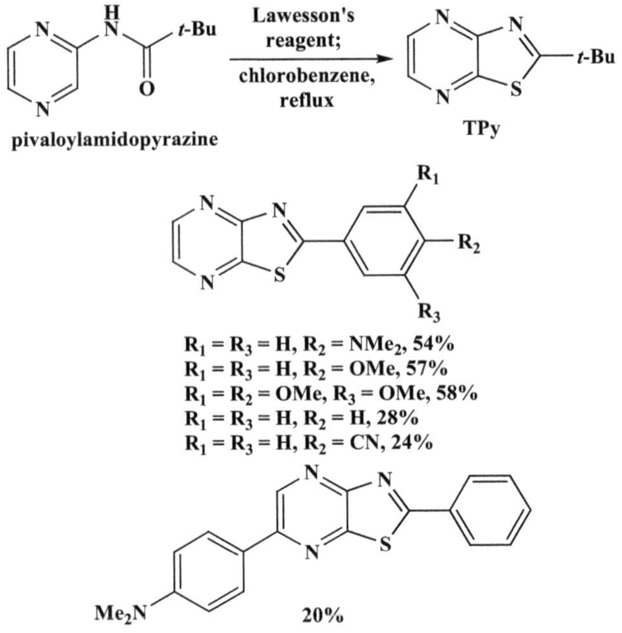

Scheme 6.21 Synthesis of thiazolo[4,5-b]pyrazine

trifluoride etherate using 2,4,6-collidine afforded thiazoline derivative in 76% yield in two steps. The thiazoline derivative was de-*O*-acetylated to afford the tetrasaccharide thiazoline (Scheme 6.23) [84].

An efficient and less time-consuming synthetic pathway for 2,3-bis[2(4-halomethyl)thiazolyl]methyloxynaphthalene was needed to produce the NTB18C6 and NTN18C6. The bromomethylthiazole was prepared from thioamide in three sequential steps: treatment with ethyl bromopyruvate in dry EtOH to synthesize the thiazole, reduction of thiazole with lithium aluminum hydride to afford the hydroxylmethylthiazole and subsequent bromination with carbon tetrabromide–triphenylphosphine to afford the bromomethylthiazole in 44% overall yield. The reaction of thioamide with 1,3-dichloroacetone was examined for the rapid synthesis of halomethylthiazole. The reaction of these two reagents in refluxing C_6H_6 afforded chloromethylthiazole in 87% yield. The cyclization of chloromethylthiazole or bromomethylthiazole with 2,3-dihydroxynapthalene or catechol and potassium carbonate in CH_3COCH_3 afforded NTN18C6 and NTB18C6 in 70% and 92% yield, respectively (Scheme 6.24) [85].

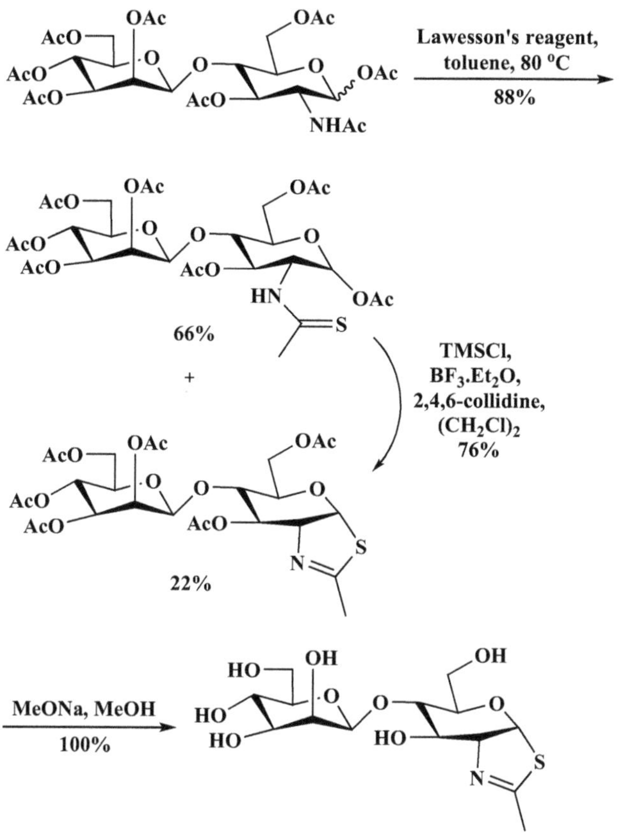

6.5 Synthesis of Six-Membered S-Heterocycles

Scheme 6.22 Synthesis of disaccharide thiazoline

6.5 Synthesis of Six-Membered S-Heterocycles

The thionation of chalcone afforded "D-dimer" of thiachalcone. This dimer underwent a retro-Diels–Alder reaction with Ph_2CN_2 to synthesize the 2,3,5,5-tetraphenyl-2,3-dihydrothiophene through 1,5-dipolar electrocyclization of intermediate thiocarbonyl ylide with an extended π-system. Li et al. [86] described that researchers were not able to reproduce the outcomes of Saito et al. [87] and were unsuccessful to isolate the "D-dimer." The single product obtained by thionation of chalcone was "T-dimer." Although "T-dimer" is less appropriate as a substrate of thiachalcone than "D-dimer," the reaction was performed with Ph_2CN_2. A reaction temperature of 50–60 °C and slow addition of diazo compound to the solution of "T-dimer" in C_6H_6 were selected as reaction conditions to make a retro-Diels–Alder reaction possible. But "T-dimer" was reacted with Ph_2CN_2 at the C=S group before the retro-Diels–Alder reaction occurred and afforded thiirane (Scheme 6.25). Unexpectedly, thiirane

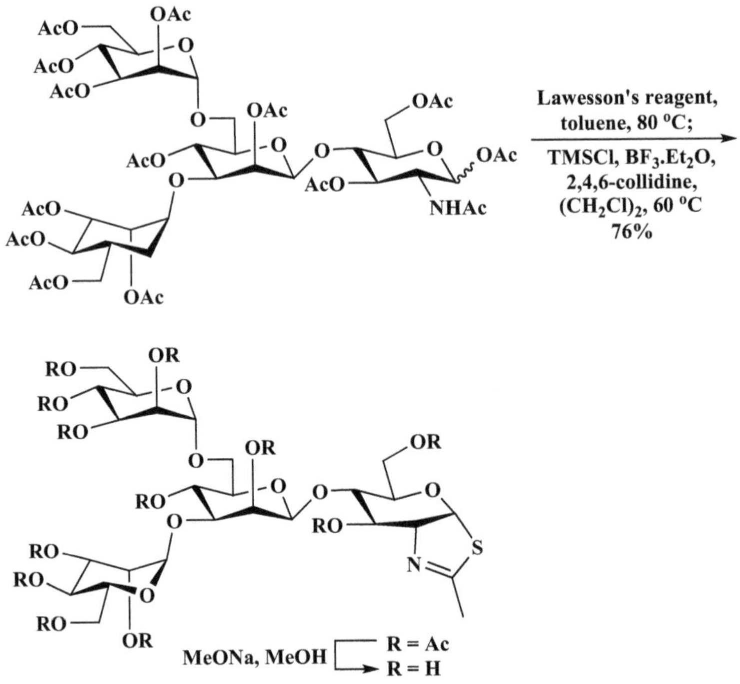

Scheme 6.23 Synthesis of tetrasaccharide thiazolines

could not be desulfurized with triphenylphosphine to afford the alkene, most likely because the sulfur atom is very much hindered for a reaction of triphenylphosphine.

The reaction of acylketene dithioacetal with sodium borohydride afforded unsaturated alcohol, and then with Lawesson's reagent led to a dimerized product in 57% yield (Scheme 6.26) [27].

The thiopyran-4-thiones with hydryl or methyl substituents were sensitive to air, and the subsequent anaerobic thionations were carried out with phosphorus pentasulfide. A one-pot protocol directly afforded thiopyran-4-thione, isolable by benchtop chromatography, from the parent pyrone in the presence of Lawesson's reagent (Scheme 6.27). It is not well understood at that time what structural properties of dithiomaltol led to its increased stability under aerobic conditions. In order to examine this unique substitution, other pyrones were reacted with Lawesson's reagent in excess amount and their products were examined [88].

The reaction of thioacylsilanes bearing ferrocene functionality attached to thiocarbonyl group is shown in Scheme 6.28 [89]. Unlike ketones and acylsilanes, which required more reaction time and high temperatures for the substitution of C=O with C=S moiety, acylsilanes bearing ferrocene were transformed easily into thioacylsilanes with LR in high yields in a few minutes at rt [90, 91].

The carbinol acetal, synthesized by NaBH$_4$ reduction of acyl ketene dithioacetal, was treated with LR (Scheme 6.29). A reddish brown viscous liquid, obtained after

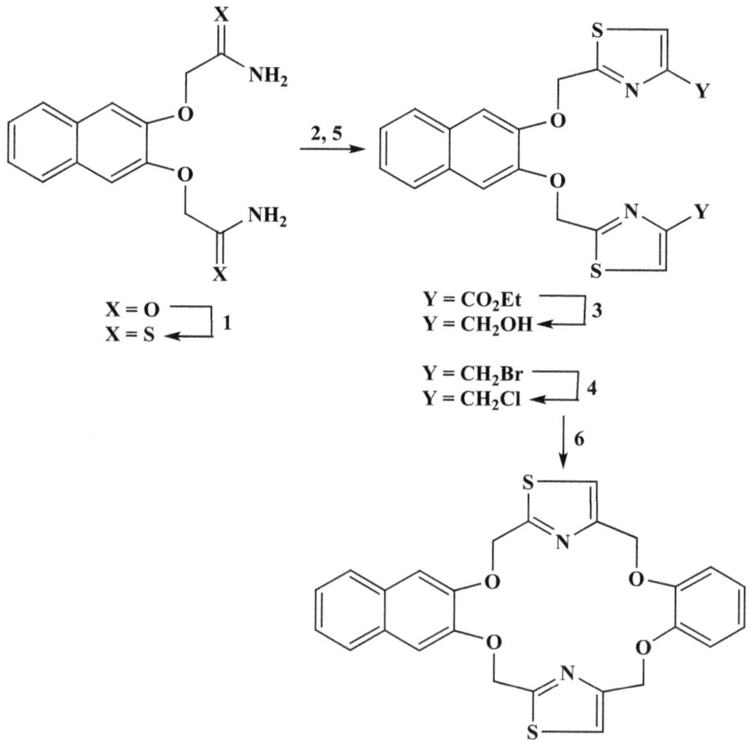

6.5 Synthesis of Six-Membered S-Heterocycles

Reagents and conditions: (1) Lawesson's reagent, THF, (2) BrCH$_2$COCO$_2$Et, EtOH, (3) LiAlH$_4$, THF, (4) CBr$_4$, PPh$_3$, CH$_2$Cl$_2$, (5) 1,3-dichloroacetone, benzene, (6) K$_2$CO$_3$, acetone, catechol.

Scheme 6.24 Synthesis of NTB18C6 and NTN18C6

column chromatography via silica gel, was identified as methyl-3,4-dihydro-2,4-dimethyl-6-methylthio-2H-thiopyran dithiocarboxylate. The methyl-3,4-dihydro-2,4-dimethyl-6-methylthio-2H-thiopyran dithiocarboxylate was prepared by a [4 + 2]-cycloaddition of α,β-unsaturated dithioester with dienophile functionality of a second molecule of the dithioester [92, 93].

The 5-ketoamides afforded 6-phenyl-3,4-dihydro-2H-thiine-2-thione in less yields (Scheme 6.30) [94, 95].

6.6 Synthesis of Six-Membered Fused S-Heterocycles

The 1,3-cyclohexadione was reacted with Lawesson's reagent at rt in toluene to synthesize its thione derivative (Scheme 6.31) [96, 97]. A dimerized product was isolated when the analogous reaction was performed in refluxing toluene [27].

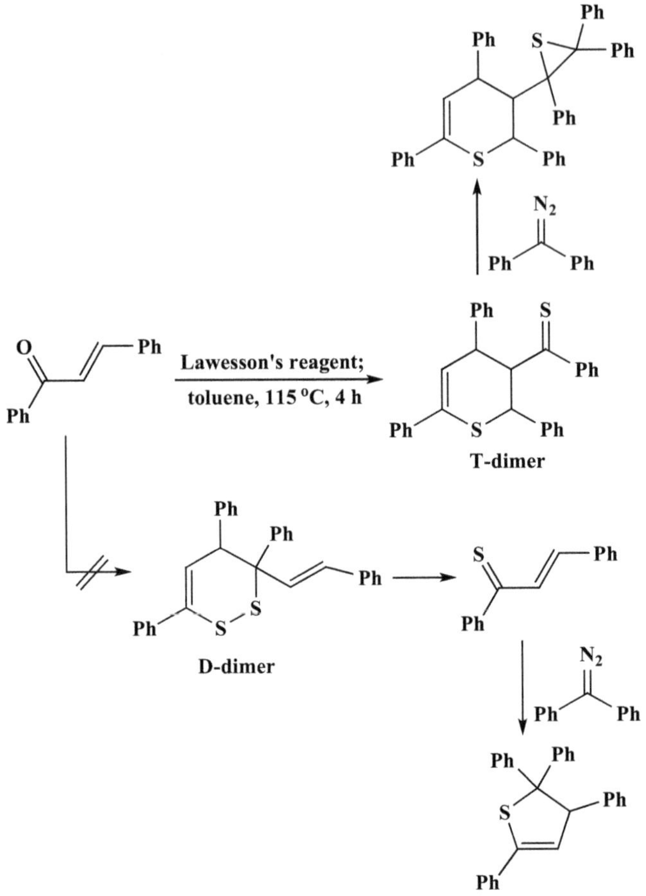

Scheme 6.25 Synthesis of 2,4,6-triphenyl-3-(2,3,3-triphenylthiiran-2-yl)-3,4-dihydro-2H-thiopyran and 2,2,3,5-tetraphenyl-2,3-dihydrothiophene

Scheme 6.26 Synthesis of thiopyran

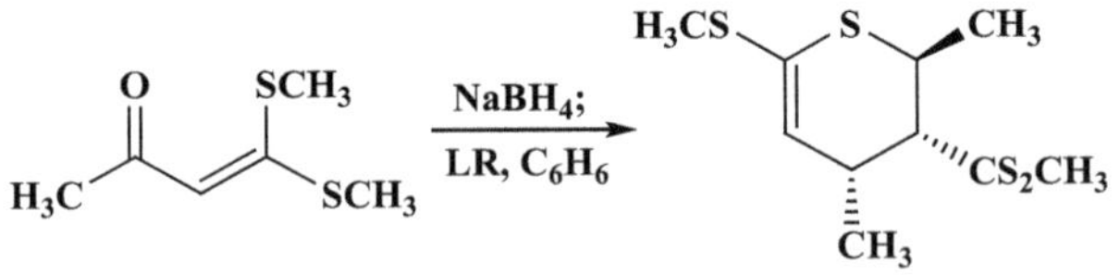

There are less examples of sulfur containing six-membered heterocyclic compounds. Unlike N-containing heterocyclic compounds, which are mostly fused to one of the rings of triterpenic frame, S-containing heterocyclic compounds were synthesized through direct incorporation of sulfur into triterpene structure (Scheme 6.32). Three thiatriterpenes were synthesized by the reaction of oxo derivatives with LR (Scheme 6.33) [98]. Two isomeric 1,3,2-oxathiaphosphinines were

6.6 Synthesis of Six-Membered Fused S-Heterocycles

Scheme 6.27 Synthesis of thiopyran-4-thiones

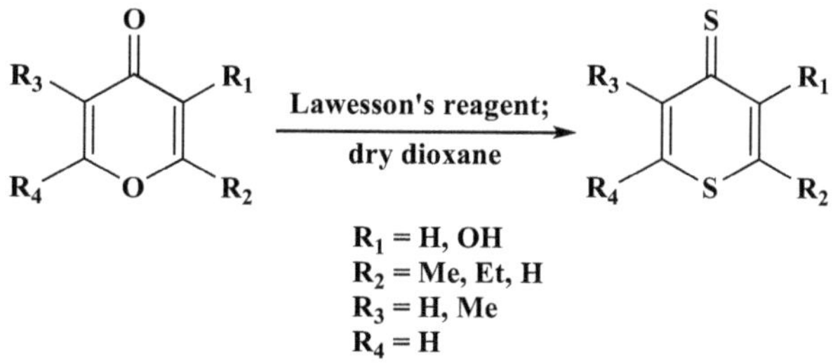

R_1 = H, OH
R_2 = Me, Et, H
R_3 = H, Me
R_4 = H

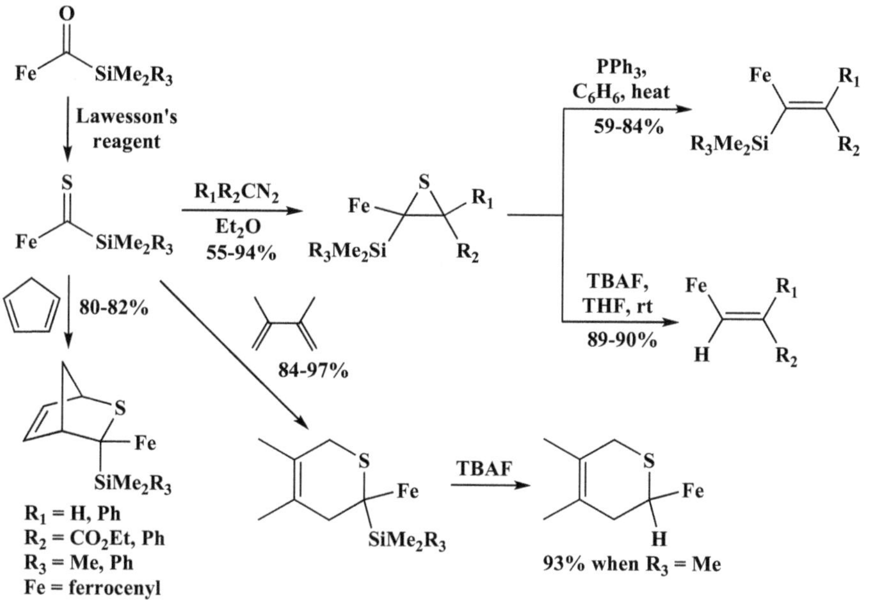

R_1 = H, Ph
R_2 = CO_2Et, Ph
R_3 = Me, Ph
Fe = ferrocenyl

Scheme 6.28 Synthesis of thiopyrans

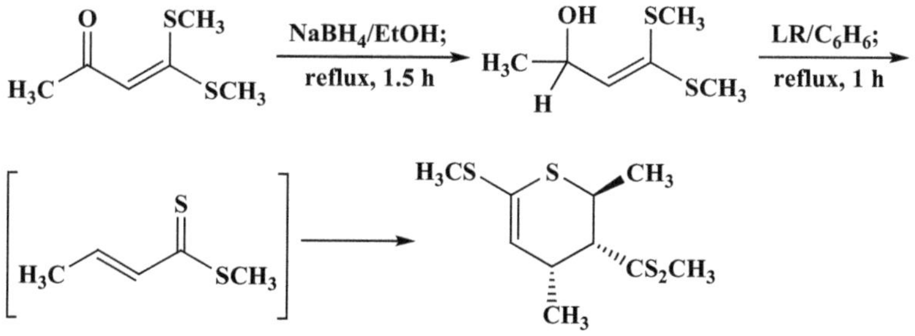

Scheme 6.29 Synthesis of methyl-3,4-dihydro-2,4-dimethyl-6-methylthio-2H-thiopyran dithiocarboxylate

Scheme 6.30 Synthesis of 6-phenyl-3,4-dihydro-2H-thiine-2-thione

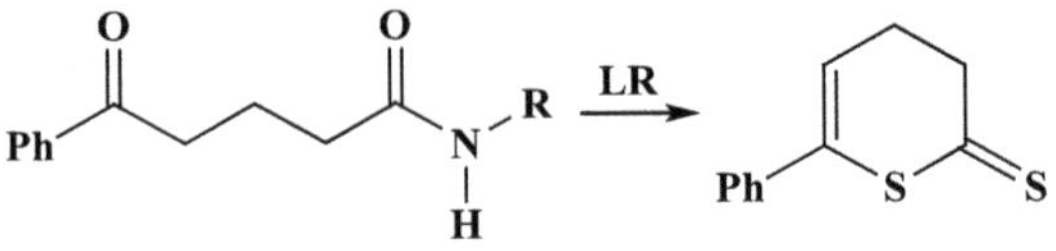

Scheme 6.31 Synthesis of thioxanthene

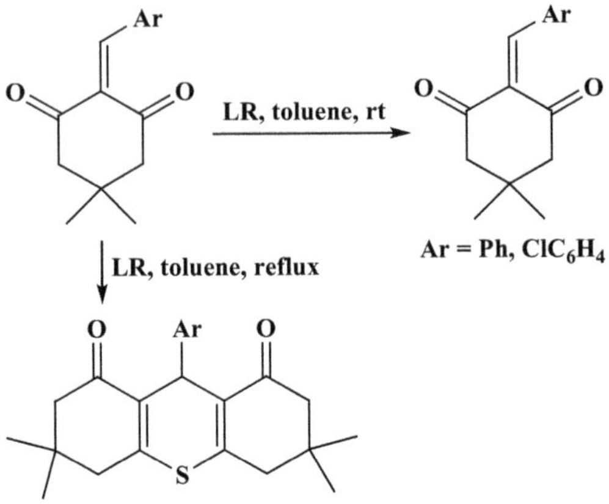

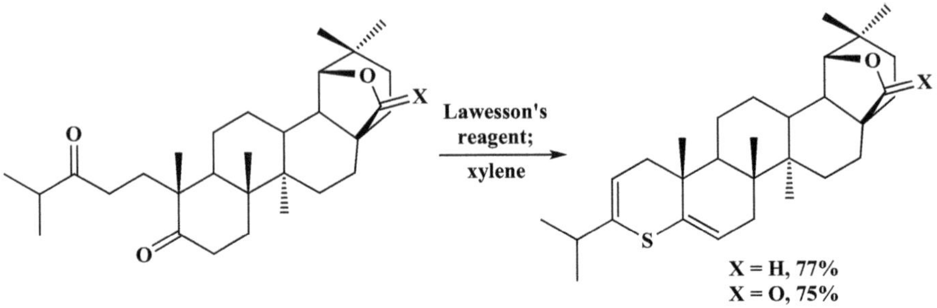

Scheme 6.32 Synthesis of thiopyrans

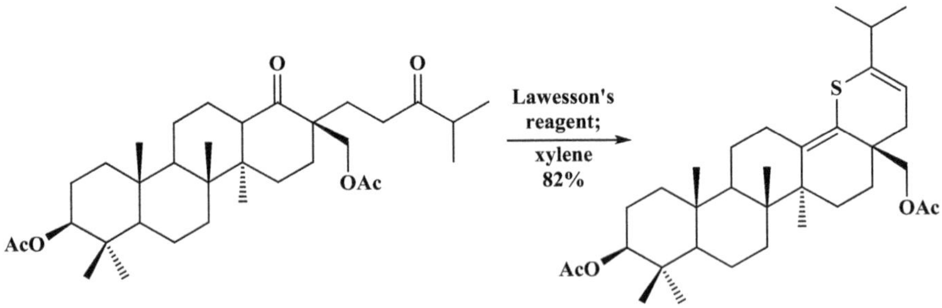

Scheme 6.33 Synthesis of thiatriterpenes

6.6 Synthesis of Six-Membered Fused S-Heterocycles

obtained from unsaturated aldehyde (Scheme 6.34) [99].

The first cyclic trithioanhydride was obtained where an acid chloride derived from 1,8-naphthylic anhydride was thionated with LR in refluxing C_6H_5Cl (Scheme 6.35) [100]. The chloride derivative of 1,8-naphthalic anhydride was reacted with LR to afford the dithioanhydride and then an isomerization of dithioanhydride occurred with a small amount of TEA in cold dimethylformamide. A second reaction with LR under refluxing conditions provided cyclic trithioanhydride with a naphthyl framework.

The thionation of starting compounds with either phosphorus pentasulfide or Lawesson's reagent in refluxing xylene for 2 h afforded spiro-type products, respectively, along with by-product (Scheme 6.36) [27, 101].

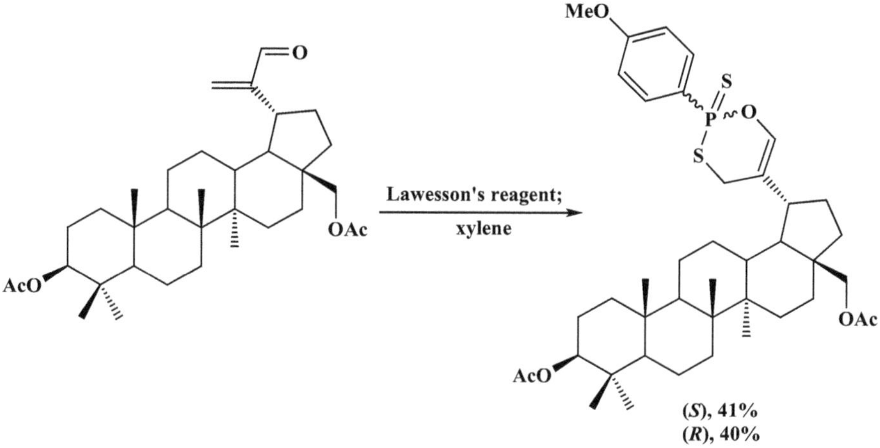

Scheme 6.34 Synthesis of 1,3,2-oxathiaphosphinines

Scheme 6.35 Synthesis of benzoisothiochromene-1,3-dithione

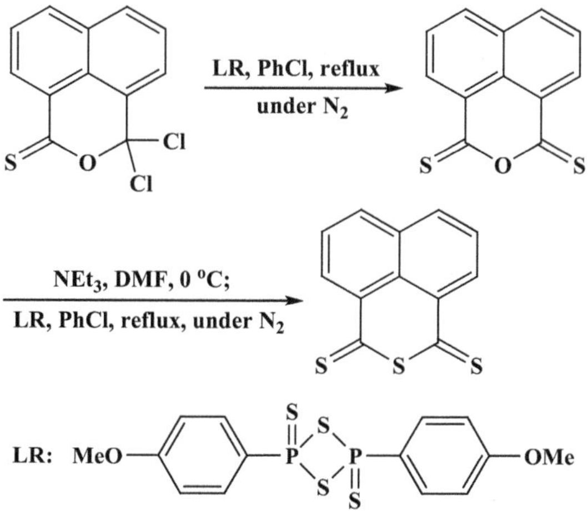

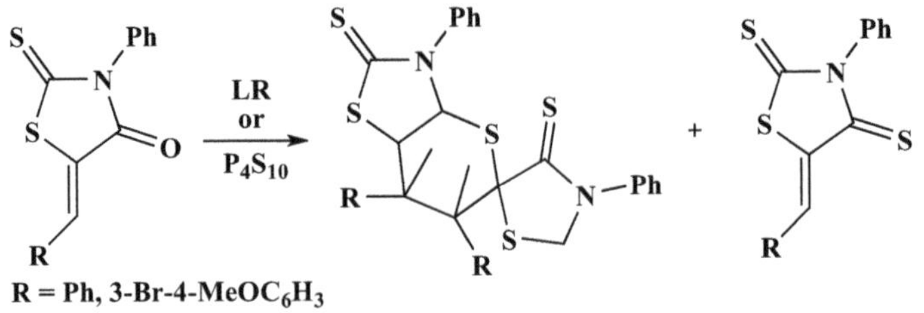

LR
or
P_4S_{10}
Ph
R = Ph, 3-Br-4-MeOC6H3

Scheme 6.36 Synthesis of thiopyrans and thiazoles

6.7 Synthesis of Six-Membered *S,S*- and *S,S,S*-Heterocycles

The reaction of α,β-unsaturated ketone with Lawesson's reagent in refluxing carbon disulfide under dinitrogen led to dimerization to synthesize the 3,4-dihydro-1,2-dithiins (Scheme 6.37) [27, 102].

The α,β-unsaturated ketone, 2-(phenylthio)methylene-1-tetralone, was reacted with LR to afford the 3,4-dihydro-1,2-dithiins analogue (Scheme 6.38) [27, 103, 104].

The reaction of indanone with Lawesson's reagent in refluxing toluene afforded final compound in 95% yield, the structure of which was determined after a single-crystal X-ray analysis (Scheme 6.39) [27, 105].

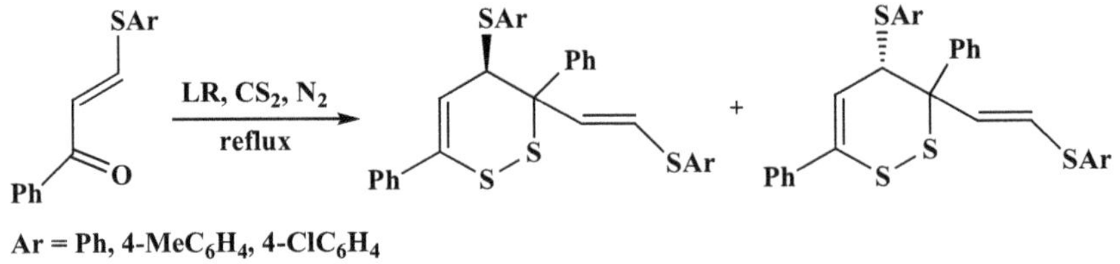

LR, CS2, N2
reflux
SAr
Ph
Ar = Ph, 4-MeC6H4, 4-ClC6H4

Scheme 6.37 Synthesis of dihydro-1,2-dithiins

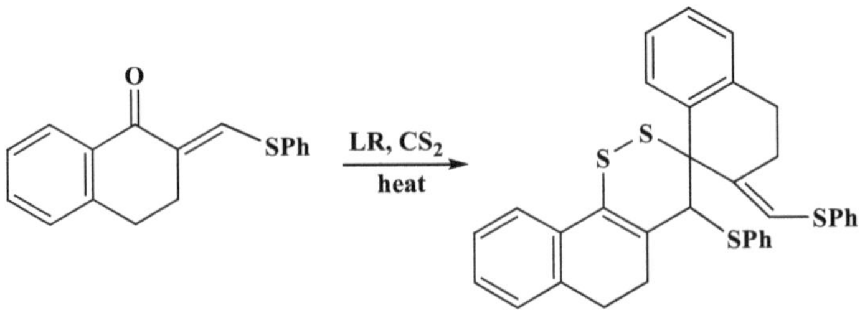

SPh
LR, CS2
heat

Scheme 6.38 Synthesis of dihydro-1,2-dithiin

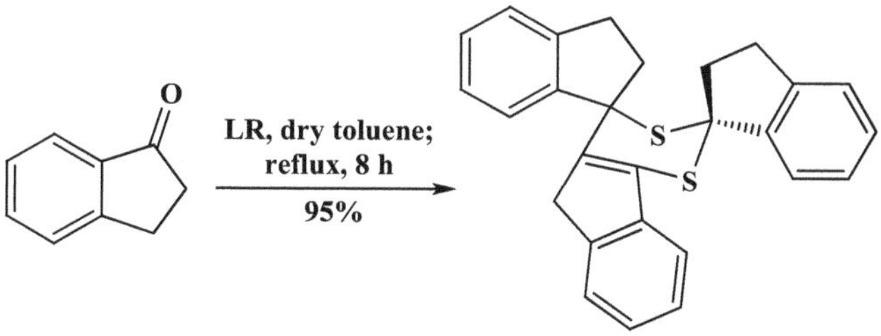
O
LR, dry toluene;
reflux, 8 h
95%
S
S

Scheme 6.39 Synthesis of dithiin

The reaction of 1,8-diketones with phosphorus pentasulfide in refluxing dioxane or toluene and in the presence or absence of *p*-toluenesulfonic acid or sodium bicarbonate afforded vinylene analogue of ethylenedioxythiphene (EDOT) along with dithienothiophene (DTT) in 51–75% yield (Scheme 6.40) [106, 107].

The precursors of 1,4-dithiin ring derivative and thiophene derivative were synthesized from BTDT [108]. The precursors were reacted with LR to afford two different types of products based on the nature of substituent. The reaction of precursors bearing Me groups with Lawesson's reagent afforded 6-membered 1,4-dithiin ring derivative, whereas the reaction of substrate with Lawesson's reagent provided an unpredictable thiophene derivative (Scheme 6.41) [109–111].

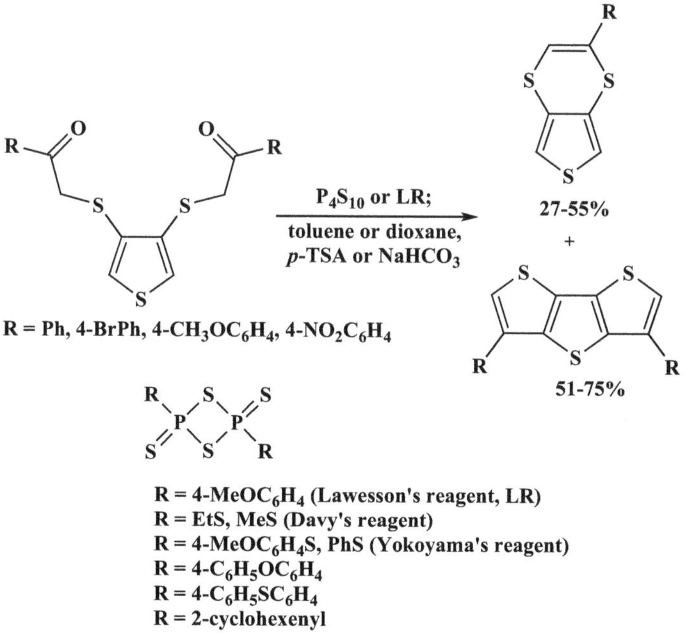
R
O
O
R
S
S
S
P_4S_{10} or LR;
toluene or dioxane,
p-TSA or $NaHCO_3$
R = Ph, 4-BrPh, 4-$CH_3OC_6H_4$, 4-$NO_2C_6H_4$
27-55%
+
51-75%
R = 4-$MeOC_6H_4$ (Lawesson's reagent, LR)
R = EtS, MeS (Davy's reagent)
R = 4-$MeOC_6H_4S$, PhS (Yokoyama's reagent)
R = 4-$C_6H_5OC_6H_4$
R = 4-$C_6H_5SC_6H_4$
R = 2-cyclohexenyl

Scheme 6.40 Synthesis of thienodithiin and dithienothiophenes

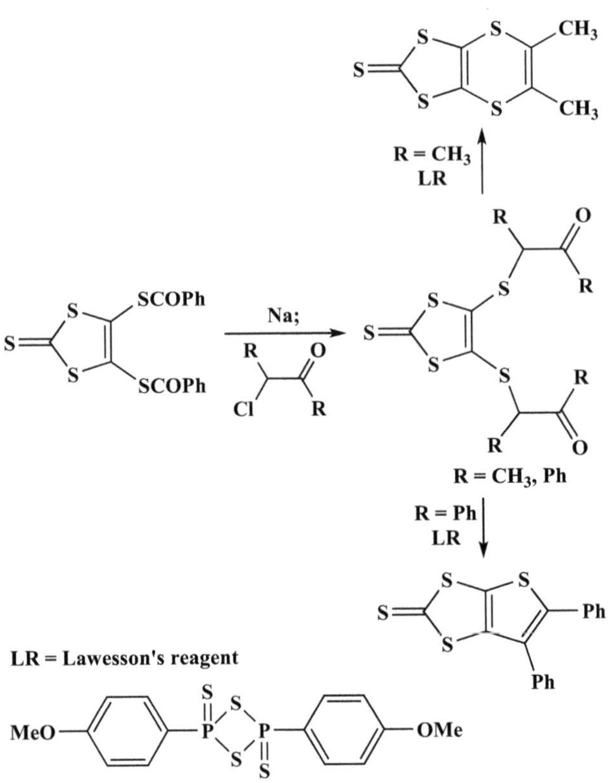

Scheme 6.41 Synthesis of 5,6-dimethyl-[1,3]dithiolo[4,5-*b*][1,4]dithiin-2-thione and 5,6-diphenylthieno[2,3-*d*][1,3]dithiole-2-thione

The reaction of 1,8-diketones with either phosphorus pentasulfide or Lawesson's reagent in refluxing toluene furnished fused dithiins as major and thiophenes as minor products (Scheme 6.42) [107, 112–119].

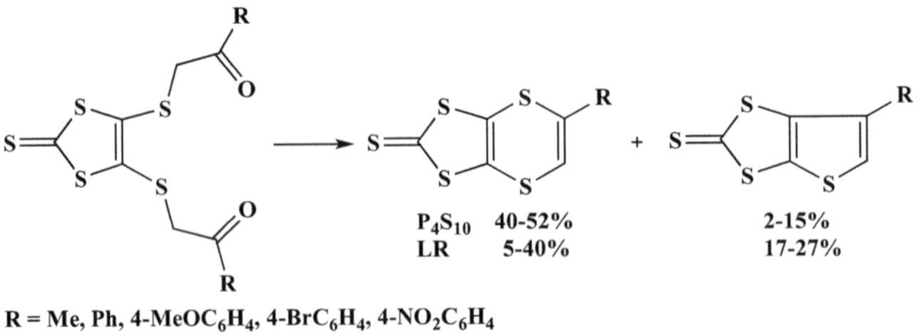

R = Me, Ph, 4-MeOC$_6$H$_4$, 4-BrC$_6$H$_4$, 4-NO$_2$C$_6$H$_4$

Scheme 6.42 Synthesis of dithiolodithiins and thienodithiole

6.7 Synthesis of Six-Membered *S,S*- and *S,S,S*-Heterocycles

The 5,6-diphenyl[1,3]dithiolo[4,5-*b*][1,4]dithiin-2-thione and its coupling product, which is a fully unsaturated analogue of BEDT-TTF, were synthesized. The 1,8-diketone was smoothly prepared in 90% yield by the reaction of 1 eq. dianion and 2 eq. desyl chloride in dry EtOH at rt for 3 h (Scheme 6.43) [108, 114, 120].

The reaction of a series of 1,8-diketones with Lawesson's reagent or phosphorus pentasulfide was further explored in 2003 [113]. With both reactants, 1,4-dithiin was obtained as a major and thiophene as a minor product along with by-products (Scheme 6.44). Depending on the ERGs or EWGs present on starting compound,

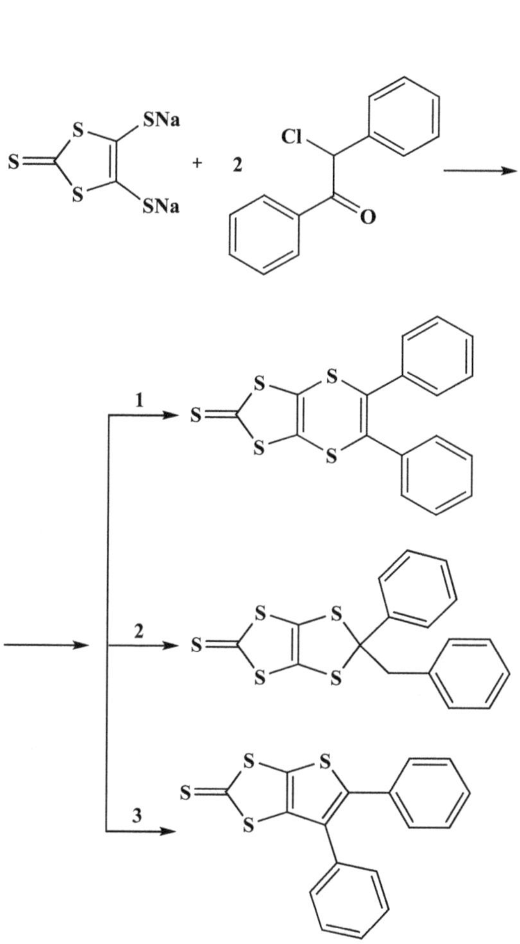

Reagents and conditions: (1) P_4S_{10}, toluene, reflux, dark, 3 h, (2) P_4S_{10}, toluene, reflux, 3 h, (3) LR, toluene, reflux, overnight.

Scheme 6.43 Synthesis of 5,6-diphenyl[1,3]dithiolo[4,5-*b*][1,4]dithiin-2-thione, 5-benzyl-5-phenyl[1,3]dithiolo[4,5-*d*][1,3]dithiole-2-thione, and 5,6-diphenylthieno[2,3-*d*][1,3]dithiole-2-thione

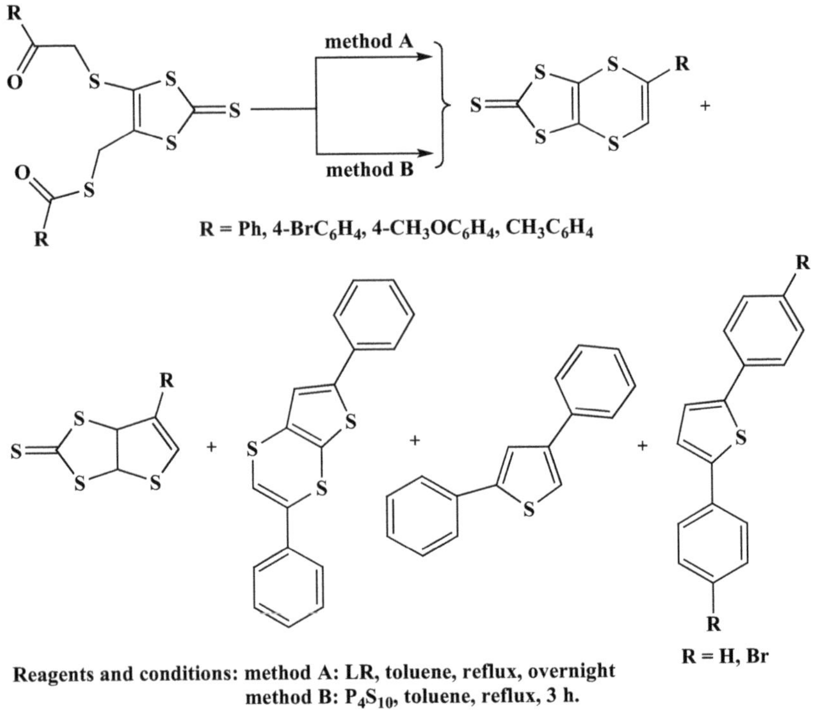
method A
method B
R = Ph, 4-BrC6H4, 4-CH3OC6H4, CH3C6H4
R = H, Br
Reagents and conditions: method A: LR, toluene, reflux, overnight
method B: P4S10, toluene, reflux, 3 h.

Scheme 6.44 Synthesis of 1,4-dithiins and thiophenes

the yields of 1,4-dithiin and thiophene, with Lawesson's reagent varied from 35 to 52% and from not detected to 18%, respectively, phosphorus pentasulfide afforded 1,4-dithiin and thiophene in yields ranging from 5 to 49% and not detected to 27%, respectively. Both of the reagents afforded dithiin as a major product [120].

The 1,8-diketones were reacted with Lawesson's reagent to synthesize the five- and six-membered *S*-heterocyclic compounds with the removal of some carbon atoms (Schemes 6.45, 6.46, and 6.47) [95, 112].

Totally different products were obtained by carrying out the reaction at rt in CH_2Cl_2 or by employing strongly electron-withdrawing aromatic residues. In the first

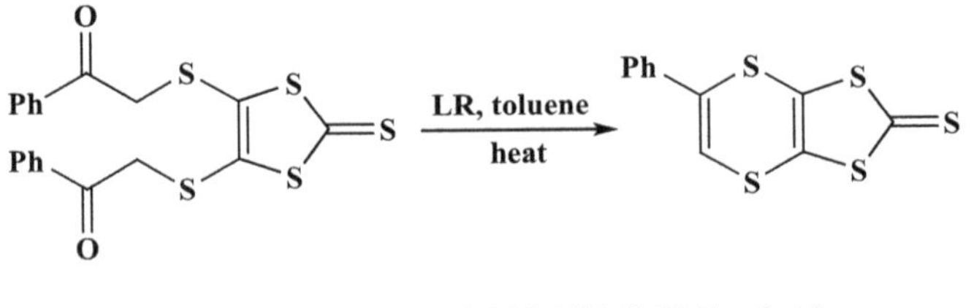
LR, toluene
heat

Scheme 6.45 Synthesis of 5-phenyl[1,3]dithiolo[4,5-*b*][1,4]dithiine-2-thione

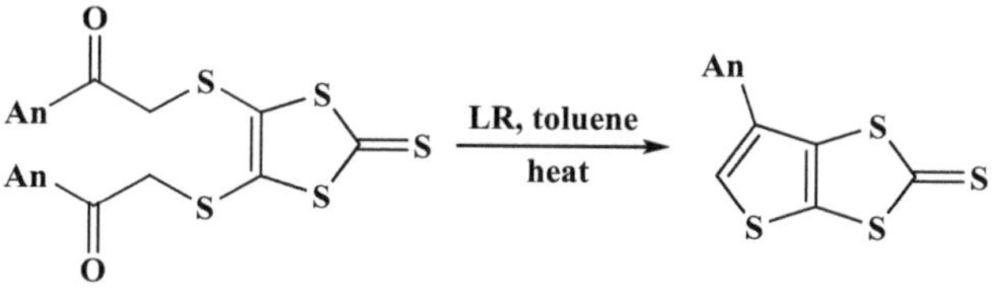

6.7 Synthesis of Six-Membered S,S- and S,S,S-Heterocycles

Scheme 6.46 Synthesis of thienodithiole

Scheme 6.47 Synthesis of 2-phenylbenzo-1,4-dithiin

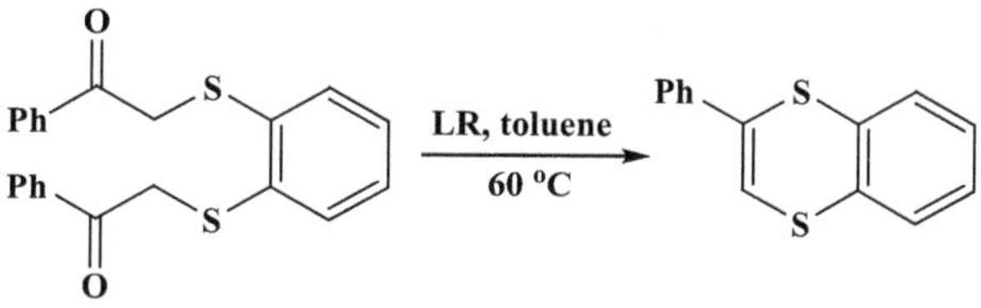

case, 3,3′-linked thienylfuran derivatives and in the other case unusual cage-like structures containing acetal–thioacetal functionalities were synthesized (Scheme 6.48). Similar Se-containing cage-like products were obtained when Woollins' reagent [121, 122] was utilized as a source of Se [123].

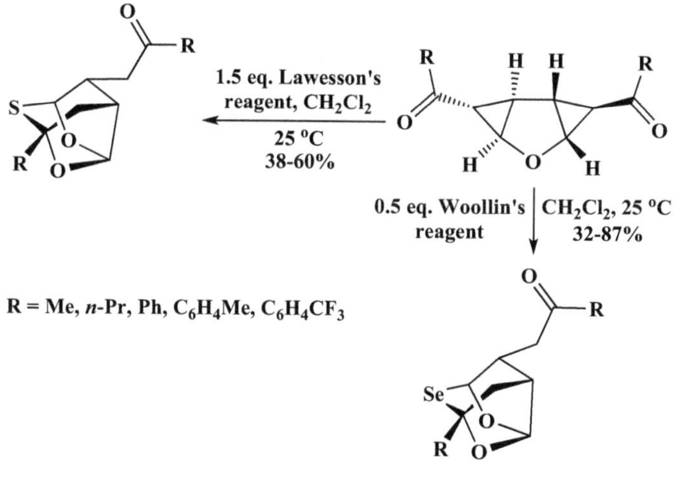

R = Me, n-Pr, Ph, C_6H_4Me, $C_6H_4CF_3$

Scheme 6.48 Synthesis of acetal-thioacetal functionalities

6.8 Synthesis of Six-Membered S,N-Heterocycles

The 6-bromoisatoic anhydride was reacted with LR to prepare the 6-bromo-1H-3,1-benzothiazine-2,4-dithione in 60% yield (Scheme 6.49) [124, 125].

The 2-iodoaniline was condensed with ethyl acrylate or acrylonitrile and alkyl or aryl aldehydes in catalytic amounts of an organopalladium reagent to provide the substituted alkene. The substituted alkene was cyclized with LR through an intramolecular S-conjugate addition to afford the 4H-3,1-benzothiazines (Scheme 6.50) [125].

The reaction of imidazole, possessing amide and ester groups, with Lawesson's reagent afforded purine analogue with a dithiolactone ring (Scheme 6.51) [27, 126].

The 3,4-dibenzoylamino-2,5-dicarbethoxythieno-2,3-thiophene was reacted with NH$_3$ to afford the 3,4-dibenzoylaminothieno[2,3-b]thiophene-2,5-dicarboxamide. These compounds were proved to be good starting compounds for the formation of numerous poly-fused thienothiophenes, where 3,4-dibenzoylamino-2,5-dicarbethoxythieno[2,3]thiophene was reacted with LR

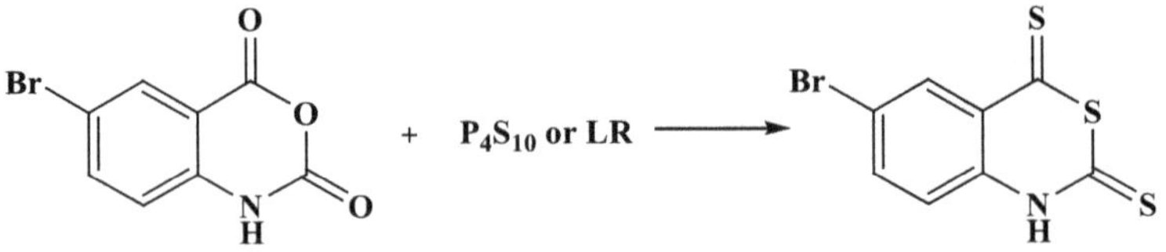

Scheme 6.49 Synthesis of 6-bromo-1H-3,1-benzothiazine-2,4-dithione

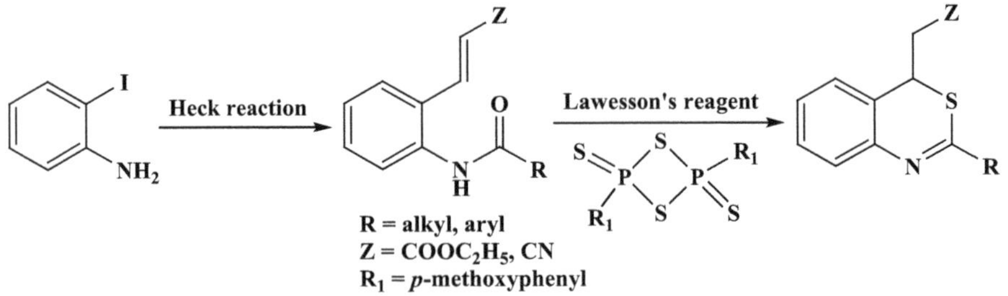

Scheme 6.50 Synthesis of 3,1-benzothiazines

Scheme 6.51 Synthesis of imidazo-1,3-thiazines

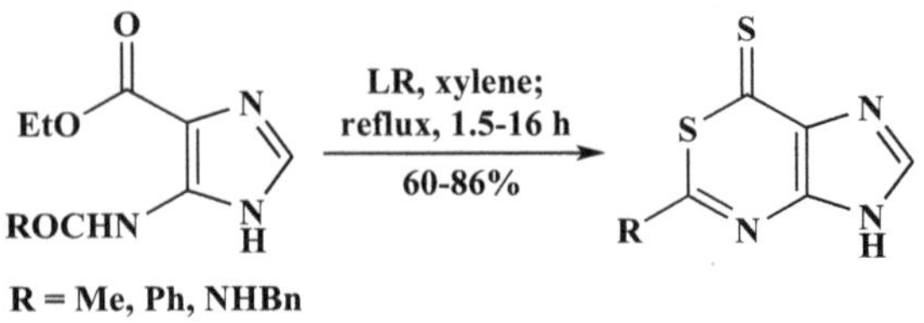

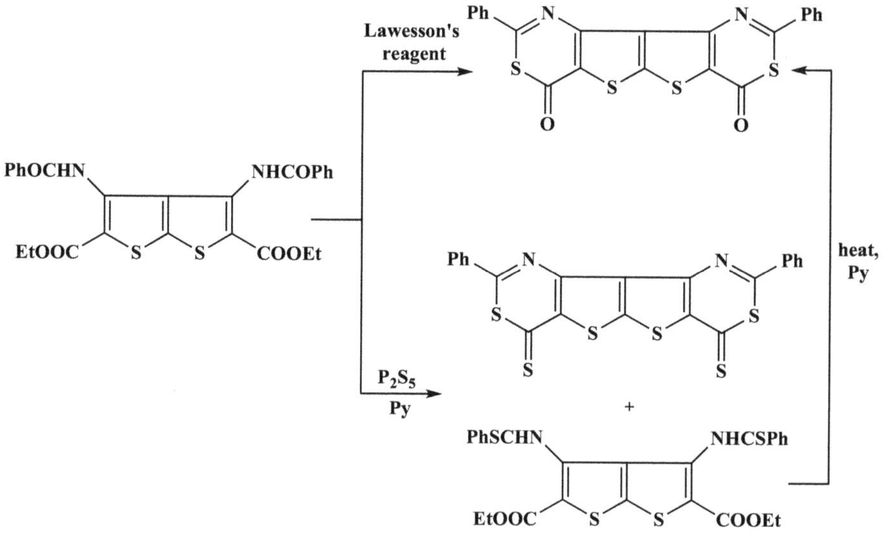

6.8 Synthesis of Six-Membered S,N-Heterocycles

Scheme 6.52 Synthesis of bis-thieno-1,3-thiazines and thienothiophene

to afford the bis-phenylthieno[2,3-d][1,3]thiazine. However, the reaction of 3,4-dibenzoylamino-2,5-dicarbethoxythieno-2,3-thiophene in refluxing pyridine afforded bis-4-thiono-2-phenylthieno[2,3-d][1,3]thiazine (48%) and 3,4-thiobenzamido-2,5-dicarbethoxythieno-2,3-thiophene (32%), respectively. The 3,4-dibenzoylaminothieno[2,3-b]thiophene-2,5-dicarboxamide was reacted with LR or phosphorus pentasulfide to provide the bis-4-thiono-2-phenylthieno[2,3-d][1,3]thiazine] or bis-phenylthieno[2,3-d][1,3]thiazine and bis-4-thiono-2-phenylthieno[2,3-d][1,3]thiazine, respectively. The 3,4-thiobenzamido-2,5-dicarbethoxythieno-2,3-thiophene was transformed into bis-phenylthieno[2,3-d][1,3]thiazine on heating in dry pyridine (Schemes 6.52 and 6.53) [127].

6.9 Synthesis of Six-Membered S,N,N-Heterocycles

The 2,2-bis(2-fluoroethyl)-4-bromobenzhydrazide was reacted with LR in p-xylene followed by treatment of filtrate with gaseous HCl to afford the 2-aryl-4-fluoroethyl-5,6-dihydro-4H-1,3,4-thiadiazine (Scheme 6.54) [128].

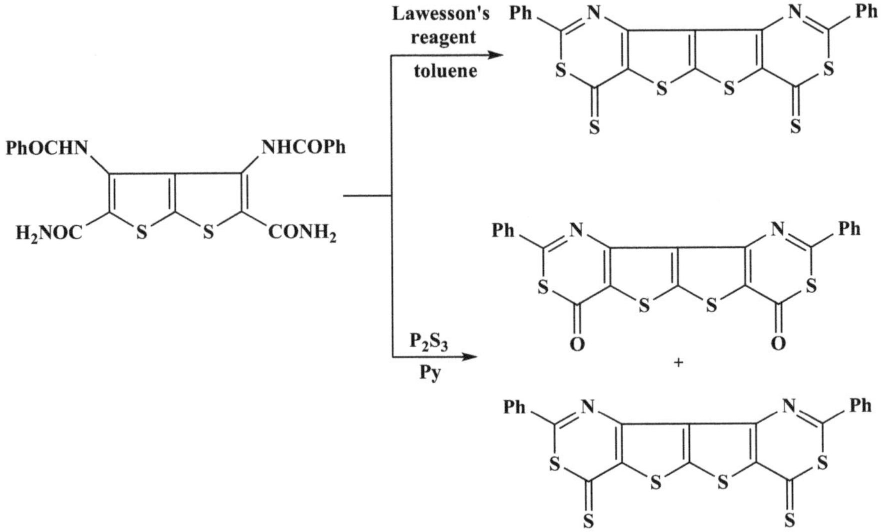

Scheme 6.53 Synthesis of bis-thieno-1,3-thiazines

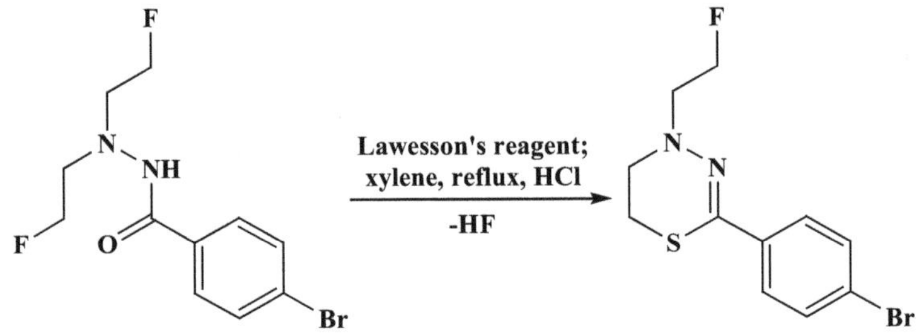

Scheme 6.54 Synthesis of 2-aryl-4-fluoroethyl-5,6-dihydro-4H-1,3,4-thiadiazine

6.10 Synthesis of Seven-Membered S-Heterocycles

The 2H-pyran-2-ones in boiling toluene was reacted with LR to yield the trifluoromethylated 2H-pyran-2-thiones in 83 and 58% yield. Their reaction with nitrosobenzene provided unexpectedly adducts, which proved to be isomeric with the expected primary Diels–Alder cycloadducts (Schemes 6.55, 6.56, and 6.57) [129, 130].

6.10 Synthesis of Seven-Membered S-Heterocycles

Scheme 6.55 Synthesis of 6-phenyl-4-(trifluoromethyl)-8-oxa-2-thia-6-azabicyclo[3.2.1]oct-3-en-7-one

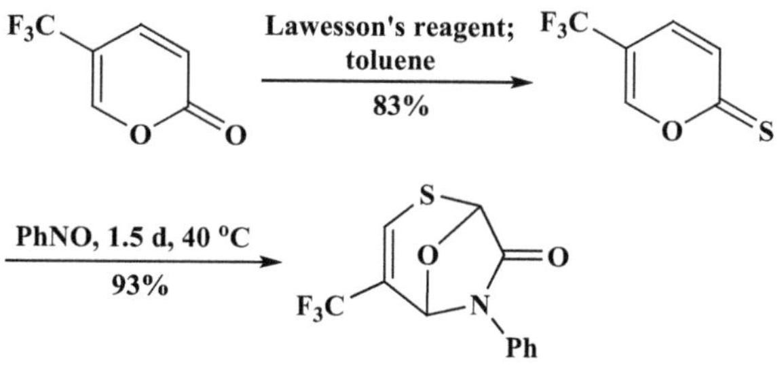

Scheme 6.56 Synthesis of methyl 7-oxo-6-phenyl-4-(trifluoromethyl)-8-oxa-2-thia-6-azabicyclo[3.2.1]oct-3-ene-3-carboxylate

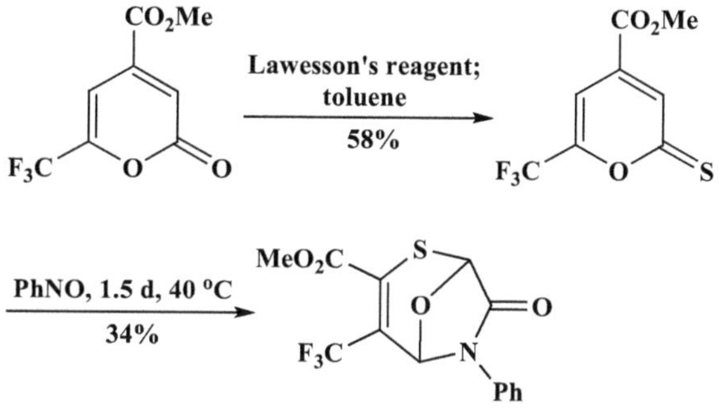

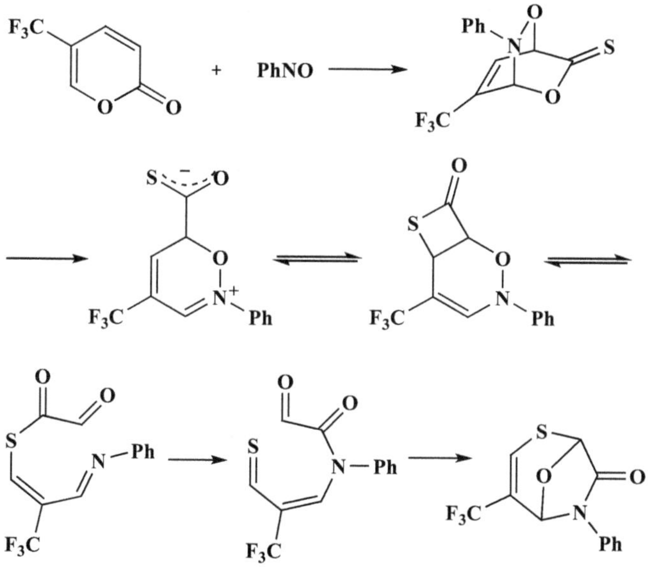

Scheme 6.57 Synthesis of 6-phenyl-4-(trifluoromethyl)-8-oxa-2-thia-6-azabicyclo[3.2.1]oct-3-en-7-one

References

1. (a) N. Kaur, M. Devi, Y. Verma, P. Grewal, P. Bhardwaj, N. Ahlawat and N.K. Jangid. 2019. Photochemical synthesis of fused five-membered O-heterocycles. Curr. Green Chem. 6: 155–183. (b) N. Kaur, M. Devi, Y. Verma, P. Grewal, P. Bhardwaj, N. Ahlawat and N.K. Jangid. 2019. Applications of metal and non-metal catalysts for the synthesis of oxygen containing five-membered polyheterocycles: A mini review. SN Appl. Sci. 1: 1–32. (c) G.W. Gribble. 2000. Recent developments in indole ring synthesis - methodology and applications. J. Chem. Soc. Perkin Trans. 1 7: 1045–1075.
2. (a) N. Kaur, N. Ahlawat, Y. Verma, P. Bhardwaj, P. Grewal and N.K. Jangid. 2020. Rhodium catalysis in the synthesis of fused five-membered N-heterocycles. Inorg. Nano Met. Chem. 50: 1260–1289. (b) N. Kaur. 2019. Multiple nitrogen-containing heterocycles: Metal and non-metal assisted synthesis. Synth. Commun. 49: 1633–1658. (c) T.A. Gilchrist. 1998. Synthesis of aromatic heterocycles. J. Chem. Soc. Perkin Trans. 1 3: 615–628.
3. (a) A. Nobuyoshi, O. Akihiko, M. Chikara, T. Tatsuya, O. Masami and S. Hiromitsu. 1999. Synthesis and antitumor activity of duocarmycin derivatives: Modification of segment-A of A-ring pyrrole compounds. J. Med. Chem. 42: 2946–2960. (b) N. Kaur, N. Ahlawat, P. Grewal, P. Bhardwaj and Y. Verma. 2019. Organo or metal complex catalyzed synthesis of five-membered oxygen heterocycles. Curr. Org. Chem. 23: 2822–2847. (c) N. Kaur, R. Tyagi and D. Kishore. 2014. Expedient protocols for the installation of 1,5-benzoazepino-based privileged templates on the 2-position of 1,4-benzodiazepine through a phenoxyl spacer. J. Heterocycl. Chem. 51: E340-E343.
4. (a) P.S. Baran, J.M. Richter and D.W. Lin. 2005. Short, enantioselective total synthesis of stephacidin A. Angew. Chem. Int. Ed. 44: 606–609. (b) N. Kaur. 2013. Recent trends in the chemistry of privileged scaffold: 1,4-Benzodiazepine. Int. J. Pharm. Biol. Sci. 4: 485–513. (c) N. Kaur, N. Ahlawat, Y. Verma, P. Grewal and P. Bhardwaj. 2019. A review of ruthenium catalyzed C-N bond formation reactions for the synthesis of five-membered N-heterocycles. Curr. Org. Chem. 23: 1901–1944.
5. (a) M. Torok, M. Abid, S.C. Mhadgut and B. Torok. 2006. Organofluorine inhibitors of amyloid fibrillogenesis. Biochemistry 45: 5377–5383. (b) N. Kaur, P. Bhardwaj, M. Devi, Y. Verma and P. Grewal. 2019. Gold-catalyzed C-O bond forming reactions for the synthesis of six-membered O-heterocycles. SN Appl. Sci. 1: 1–37. (c) N. Kaur. 2019. Ionic liquid assisted synthesis of six-membered oxygen heterocycles. SN Appl. Sci. 1: 1–20.
6. (a) P. Kohling, A.M. Schmidt and P. Eilbracht. 2003. Tandem hydroformylation/Fischer indole synthesis: A novel and convenient approach to indoles from olefins. Org. Lett. 5: 3213–3216. (b) N. Kaur, N.K. Jangid and V. Sharma. 2018. Metal- and nonmetal-catalyzed synthesis of five-membered S,N-heterocycles. J. Sulfur Chem. 39: 193–236. (c) N. Kaur, N.K. Jangid and V. Rawat. 2018. Synthesis of heterocycles through platinum-catalyzed reactions. Curr. Catal. 7: 3–25.
7. S.J. Yang and W.A. Denny. 2002. A new short synthesis of 3-substituted 5-amino-1-(chloromethyl)-1,2-dihydro-3H-benzo[e]indoles (amino-CBIs). J. Org. Chem. 67: 8958–8961.
8. K. Masatane and T.J. Yutaka. 1981. The chemistry of carbazoles. VII. Synthesis of methylcarbazoles. J. Heterocycl. Chem. 18: 709–714.
9. H.J. Knolker and K.R. Reddy. 2002. Isolation and synthesis of biologically active carbazole alkaloids. Chem. Rev. 102: 4303–4427.
10. L. Ackermann, L. Kaspar and C.J. Gschrei. 2004. Hydroamination/Heck reaction sequence for a highly regioselective one-pot synthesis of indoles using 2-chloroaniline. Chem. Commun. 24: 2824–2825.
11. J. Zhao and R.C. Larock. 2005. Synthesis of substituted carbazoles by a vinylic to aryl palladium migration involving domino C-H activation processes. Org. Lett. 7: 701–704.
12. Z. Liu and R.C. Larock. 2004. Synthesis of carbazoles and dibenzofurans via cross-coupling of o-iodoanilines and o-iodophenols with silylaryl triflates. Org. Lett. 6: 3739–3741.

13. X. Cai and V. Snieckus. 2004. Combined directed *ortho* and remote metalation-cross-coupling strategies. General method for benzo[*a*]carbazoles and the synthesis of an unnamed indolo[2,3-*a*]carbazole alkaloid. Org. Lett. 6: 2293–2295.
14. B.K. Banik, S. Samajdar and I. Banik. 2004. Simple synthesis of substituted pyrroles. J. Org. Chem. 69: 213–216.
15. A. Dastan, A. Kulkarni and B. Torok. 2012. Environmentally benign synthesis of heterocyclic compounds by combined microwave-assisted heterogeneous catalytic approaches. Green Chem. 14: 17–37.
16. Simerpreet and C.S. Damanjit. 2013. Synthesis and biological evaluation of 1,3-thiazines - a review. Pharmacophore 4: 70–88.
17. G. Vincent, B.V. Mathew, J. Joseph, M. Chandran, A.R. Bhat and K.K. Kumar. 2014. A review on biological activities of thiazine derivatives. Int. J. Pharm. Chem. Sci. 3: 341–348.
18. S. Jupudi, K. Padmini, P.J. Preethi, P.V.P.D. Bharadwaj and P.V. Rao. 2013. An overview on versatile molecule: 1,3-Thiazines. Asian J. Res. Pharm. Sci. 3: 170–177.
19. V.K. Rai, B.S. Yadav and L.S.D. Yadav. 2009. The first ionic liquid-promoted one-pot diastereoselective synthesis of 2,5-diamino-/2-amino-5-mercapto-1,3-thiazin-4-ones using masked amino/mercapto acids. Tetrahedron 65: 1306–1315.
20. L. Fu, Y. Li, D. Ye and S. Yin. 2010. Synthesis and calming activity of 6H-2-amino-4-aryl-6-(4-β-D-allopyranosyloxyphenyl)-1,3-thiazine. Chem. Nat. Compd. 46: 169–172.
21. F. Haider and Z. Haider. 2012. Synthesis and antimicrobial screening of some 1,3-thiazines. J. Chem. Pharm. Res. 4: 2263–2267.
22. Z. Hossaini, M. Nematpour and I. Yavari. 2010. Ph$_3$P-Mediated one-pot synthesis of functionalized 3,4-dihydro-2H-1,3-thiazines from N,N'-dialkylthioureas and activated acetylenes in water. Monatsh. Chem. 141: 229–232.
23. R.S. Ganorkar, R.P. Ganorkar and V.V. Parhate. 2013. Synthesis, characterization and antibacterial activities of some new bromo/nitro 1,3-thiazenes. Rasayan J. Chem. 6: 65–67.
24. S.S. Didwagh and P.B. Piste. 2013. Green synthesis of thiazine and oxazine derivatives - a short review. Int. J. Pharm. Sci. Res. 4: 2045–2061.
25. S. Scheibye, B.S. Pdersen and S.O. Lawesson. 1978. Studies on organophosphorus compounds. XXII. The dimer of *p*-methoxyphenylthionophosphine sulfide as a thiation reagent. A new route to *o*-substituted thioesters and dithioesters. Bull. Soc. Chim. Belg. 87: 293–297.
26. S. Scheibye, B.S. Pederson and S.O. Lawesson. 1978. Studies on organophosphorus compounds XXI. The dimer of *p*-methoxyphenylthionophosphine sulfide as thiation reagent. A new route to thiocarboxamides. Bull. Soc. Chim. Belg. 87: 229–238.
27. T. Ozturk, E. Ertas and O. Mert. 2007. Use of Lawesson's reagent in organic synthesis. Use of Lawesson's reagent in organic syntheses. Chem. Rev. 107: 5210–5278.
28. M.P. Cava. 1985. Thionation reactions of Lawesson's reagents. Tetrahedron 41: 5061–5087.
29. H.Z. Lecher, R.A. Greenwood, K.C. Whitehouse and T.H. Chao. 1956. The phosphonation of aromatic compounds with phosphorus pentasulfide. J. Am. Chem. Soc. 78: 518–522.
30. N. Takehiko. 1995. Photoreactions of 1,3-dihydroisobenzofuran-1-thiones and 1,3-dihydro-2-benzothiophene-1-thiones with alkenes. Chem. Soc. Perkin Trans. 1 5: 561–568.
31. M.P. Cava, N.M. Pollack, O.A. Marner and M.J. Mitchell. 1971. Synthetic route to benzo[*c*]thiophene and the naphtho[*c*]thiophenes. J. Org. Chem. 36: 3932–3937.
32. T. Nishio, Y. Kodama and Y. Tsurumi. 2005. Synthesis of chiral bis-thiazolines and asymmetric Diels-Alder reactions. Phosphorus, Sulfur, Silicon Relat. Elem. 180: 1449–1450.
33. M. Yamakuchi, H. Matsunaga, R. Tokuda, T. Ishizuka, M. Nakajima and T. Kunieda. 2005. Sterically congested 'roofed' 2-thiazolines as new chiral ligands for copper(II)-catalyzed asymmetric Diels-Alder reactions. Tetrahedron Lett. 46: 4019–4022.
34. M. Mellah, A. Voituriez and E. Schulz. 2007. Chiral sulfur ligands for asymmetric catalysis. Chem. Rev. 107: 5133–5209.
35. Z.-P. Zhan, J.-L. Yu, H.-J. Liu, Y.-Y. Cui, R.-F. Yang, W.-Z. Yang and J.-P. Li. 2006. A general and efficient FeCl$_3$-catalyzed nucleophilic substitution of propargylic alcohols. J. Org. Chem. 71: 8298–8301.

36. X. Gao, Y.-M. Pan, M. Lin, L. Chen and Z.-P. Zhan. 2010. Facile one-pot synthesis of three different substituted thiazoles from propargylic alcohols. Org. Biomol. Chem. 8: 3259–3266.
37. T. Belhadj, A. Nowicki and C.J. Moody. 2006. Synthesis of the 'Northern-hemisphere' fragments of the thiopeptide antibiotic nosiheptide. Synlett 18: 3033–3036.
38. S. Hanessian and R. Margarita. 1998. 1,3-Asymmetric induction in dianionic allylation reactions of amino acid derivatives - synthesis of functionally useful enantiopure glutamates, pipecolates and pyroglutamates. Tetrahedron Lett. 39: 5887–5890.
39. R.C. Kelly, I. Gebhard and N. Wicnienski. 1986. Synthesis of (R)- and (S)-(glu)thz and the corresponding bisthiazole dipeptide of dolastatin 3. J. Org. Chem. 51: 4590–4594.
40. M.C. Kimber and C.J. Moody. 2008. Construction of macrocyclic thiodepsipeptides: Synthesis of a nosiheptide 'Southern hemisphere' model system. Chem. Commun. 5: 591–593.
41. C.D. Bray and J. Olasoji. 2010. A total synthesis of (+)-bacillamide B. Synlett 4: 599–601.
42. M. Ishikura, T. Abe, T. Choshi and S. Hibino. 2013. Simple indole alkaloids and those with a non-rearranged monoterpenoid unit. Nat. Prod. Rep. 30: 694–752.
43. C. Lempereur, N. Ple, A. Turck, G. Queguiner, F. Corbin, C. Alayrac and P. Metzner. 1998. Selective thiophilic addition of alkyl- and aryl lithiums to dithio esters and a sulfine in the pyridine series. Heterocycles 48: 2019–2034.
44. K. Umemura, H. Noda, J. Yoshimura, A. Konn, Y. Yonezawa and C.-G. Shin. 1997. The synthesis of fragment A of an antibiotic, nosiheptide. Tetrahedron Lett. 38: 3539–3542.
45. D. Lefranc. 2008. Total synthesis of micrococcin P1. PhD Thesis, The University of British Columbia.
46. M.A. Ciufolini and Y.-C. Shen. 1997. Studies toward thiostrepton antibiotics: Assembly of the central pyridine-thiazole cluster of micrococcins. J. Org. Chem. 62: 3804–3505.
47. M.C. Bagley, K.E. Bashfod, C.L. Hesketh and C.J. Moody. 2000. Total synthesis of the thiopeptide promothiocin A. J. Am. Chem. Soc. 122: 3301–3313.
48. N. Xi and M.A. Ciufolini. 1995. A protection scheme for the preparation of acid chlorides of serine and threonine. Tetrahedron Lett. 36: 6595–6598.
49. Q. Ding, X.-G. Huang and J. Wu. 2009. Facile synthesis of benzothiazoles via cascade reactions of 2-iodoanilines, acid chlorides and Lawesson's reagent. J. Comb. Chem. 11: 1047–1049.
50. D.S. Bose and M. Idrees. 2010. Intramolecular cascade C-S bond formation: Regioselective synthesis of substituted benzothiazoles. Synthesis 12: 1983–1988.
51. K. Hemming. 2011. Heterocyclic chemistry. Annu. Rep. Prog. Chem. Sect. B 107: 118–137.
52. A. Kamal, S. Faazil, M.J. Ramaiah, M. Ashraf, M. Balakrishna, S.N.C.V.L. Pushpavalli, N. Patel and M. Pal-Bhadra. 2013. Synthesis and study of benzothiazole conjugates in the control of cell proliferation by modulating Ras/MEK/ERK-dependent pathway in MCF-7 cells. Bioorg. Med. Chem. Lett. 23: 5733–5739.
53. A. Kamal, B.A. Kumar, P. Suresh, N. Shankaraiah and M.S. Kumar. 2011. An efficient one-pot synthesis of benzothiazolo-4β-anilino-podophyllotoxin congeners: DNA topoisomerase-II inhibition and anticancer activity. Bioorg. Med. Chem. Lett. 21: 350–353.
54. A. Kamal, K.S. Reddy, M.N.A. Khan, R.V.C.R.N.C.M. Shetti, J. Ramaiah, S.N.C.V.L. Pushpavalli, C. Srinivas, M. Pal-Bhadra, M. Chourasia, G.N. Sastry, A. Juvekar, S. Zingde and M. Barkume. 2010. Synthesis, DNA-binding ability and anticancer activity of benzothiazole/benzoxazole-pyrrolo[2,1-c][1,4]benzodiazepine conjugates. Bioorg. Med. Chem. 18: 4747–4761.
55. H.A. Bhuva and S.G. Kini. 2010. Synthesis, anticancer activity and docking of some substituted benzothiazoles as tyrosine kinase inhibitors. J. Mol. Graph. Model. 29: 32–37.
56. I. Hutchinson, M.S. Chua, H.L. Browne, V. Trapani, T.D. Bradshaw, A.D. Westwell and M.F.G. Stevens. 2001. Antitumor benzothiazoles. 14.1 Synthesis and in vitro biological properties of fluorinated 2-(4-aminophenyl)benzothiazoles. J. Med. Chem. 44: 1446–1455.
57. R.K. Gill, R.K. Rawal and J. Bariwal. 2015. Recent advances in the chemistry and biology of benzothiazoles. Arch. Pharm. Chem. Life Sci. 348: 155–178.
58. K. Jouve and J. Bergman. 2003. Oxidative cyclization of N-methyl- and N-benzoylpyridylthioureas. Preparation of new thiazolo[4,5-b] and [5,4-b]pyridine derivatives. J. Heterocycl. Chem. 40: 261–268.

59. A. Couture and P. Grandclaudon. 1984. 2-Aryl-oxazolo- and thiazolopyridines. Synthesis via cyclization of N-(2-chloro-3-pyridinyl)arylamides and thioamides. Heterocycles 22: 1383–1385.
60. C. Flouzat and G. Guillaumet. 1990. A new convenient synthesis of 2-aryl- and 2-heteroaryloxazolo[5,4-b]pyridines. Synthesis 1: 64–66.
61. K. Hiroya, N. Suzuki, A. Yasuhara, Y. Egawa, A. Kasano and T. Sakamoto. 2000. Total synthesis of three natural products, vignafuran, 2-(4-hydroxy-2-methoxyphenyl)-6-methoxybenzofuran-3-carboxylic acid methyl ester, and coumestrol from a common starting material. J. Chem. Soc. Perkin Trans. 1 24: 4339–4346.
62. Y.R. Lee, D.J. Kim, I. Mook-Jung and K.H. Yoo. 2008. Synthesis of thia(oxa)zolopyridines and their inhibitory activities for β-amyloid fibrillization. Bull. Korean Chem. Soc. 29: 2331–2336.
63. H.Y. Guo, J.C. Li and Y.L. Shang. 2009. A simple and efficient synthesis of 2-substituted benzothiazoles catalyzed by H_2O_2/HCl. Chin. Chem. Lett. 20: 1408–1410.
64. L.Y. Fan, Y.H. Shang, X.X. Li and W.J. Hua. 2015. Yttrium-catalyzed heterocyclic formation via aerobic oxygenation: A green approach to benzothiazoles. Chin. Chem. Lett. 26: 77–80.
65. Y.-Q. Peng, L.-C. Luo, J. Gong, J. Huang and Q. Sun. 2015. New synthetic approach for the preparation of 2-aryl-thiazolo[4,5-b]pyridines via Liebeskind-Srogl reaction. Chin. Chem. Lett. 26: 1016–1018.
66. J. Barluenga, A.M. Bayon and G. Asensio. 1984. A new and specific method for the monomethylation of primary amines. J. Chem. Soc. Chem. Commun. 20: 1334–1335.
67. C. Fruit, A. Turck, N. Ple and G. Queguiner. 2002. Synthesis of N-diazinyl thiocarboxamides and of thiazolodiazines. Metalation studies. Diazines XXXIII. J. Heterocycl. Chem. 39: 1077–1082.
68. G.B. Barlin. 1983. Purine analogues as amplifiers of phleomycin. VIII. Some thiazolo[4,5-b]pyrazines and related compounds. Aust. J. Chem. 36: 983–992.
69. G.B. Barlin, S.J. Ireland and B.J. Rowland. 1984. Purine analogues as amplifiers of phleomycin. IX. Some 2- and 6-substituted thiazolo[4,5-b]pyrazines, 2-substituted thiazolo[4,5-c]- and thiazolo[5,4-b]pyridines and related compounds. Aust. J. Chem. 37: 1729–1737.
70. B. Koren, B. Stanovik and M. Tisler. 1987. Transformations of 1,2,4-thiadiazolo[2,3-x]azines. Heterocycles 26: 689–697.
71. R.C. Phadke and D.W. Rangnekar. 1986. Synthesis of 2-(alkyl and aryl)thiazolo[4,5-b]quinoxaline derivatives and study of their fluorescent properties. Bull. Chem. Soc. Jpn. 59: 1245–1247.
72. A.R. Katritzky and W.Q. Fan. 1988. Some novel quinone-type dyes containing naphthoquinone and related fused ring systems. J. Heterocycl. Chem. 25: 901–906.
73. R.C. Phadke and D.W. Rangnekar. 1989. Synthesis of 1,4-dioxino[2,3-b]quinoxaline derivatives and a study of their fluorescent properties. Dyes Pigm. 10: 159–164.
74. M. Matsuoka, A. Iwamoto, N. Furukawa and T. Kitao. 1992. Synthesis of polycyclic 1,4-dithiines and related heterocycles. J. Heterocycl. Chem. 29: 439–443.
75. D.W. Rangnekar, N.D. Sonawane and R.W. Sabnis. 1998. Synthesis of 2-styrylthiazolo[4,5-b]quinoxaline based fluorescent dyes. J. Heterocycl. Chem. 35: 1353–1356.
76. S.A. Khan. 2008. Synthesis, characterization and in vitro antibacterial activity of new steroidal 5-en-3-oxazolo and thiazoloquinoxaline. Eur. J. Med. Chem. 43: 2040–2044.
77. M. Shaikh, J. Mohanty, P.K. Singh, A.C. Bhasikuttan, R.N. Rajule, V.S. Satam, S.R. Bendre, V.R. Kanetkar and H. Pal. 2010. Contrasting solvent polarity effect on the photophysical properties of two newly synthesized aminostyryl dyes in the lower and in the higher solvent polarity regions. J. Phys. Chem. A 114: 4507–4519.
78. Y.A. Sonawane, R.N. Rajule and G.S. Shankarling. 2010. Synthesis of novel diphenylamine-based fluorescent styryl colorants and study of their thermal, photophysical, and electrochemical properties. Monatsh. Chem. 141: 1145–1151.
79. T. Nakagawa, M. Yamaji, S. Maki, H. Niwa and T. Hirano. 2014. Substituent effects on fluorescence properties of thiazolo[4,5-b]pyrazine derivatives. Photochem. Photobiol. Sci. 13: 1765–1772.

80. B. Li, Y. Zeng, S. Hauser, H. Song and L.X. Wang. 2005. Highly efficient endoglycosidase-catalyzed synthesis of glycopeptides using oligosaccharide oxazolines as donor substrates. J. Am. Chem. Soc. 127: 9692–9693.
81. Y. Zeng, J. Wang, B. Li, S. Hauser, H. Li and L.X. Wang. 2006. Glycopeptide synthesis through *endo*-glycosidase-catalyzed oligosaccharide transfer of sugar oxazolines: Probing substrate structural requirement. Chem. Eur. J. 12: 3355–3364.
82. S. Knapp, D. Vocadlo, Z. Gao, B. Kirk, J. Lou and S.G. Withers. 1996. NAG-thiazoline, an *N*-acetyl-β-hexosaminidase inhibitor that implicates acetamido participation. J. Am. Chem. Soc. 118: 6804–6805.
83. T.K. Ritter and C.H. Wong. 2001. Synthesis of *N*-acetylglucosamine thiazoline/lipid II hybrids. Tetrahedron Lett. 42: 615–618.
84. B. Li, K. Takegawa, T. Suzuki, K. Yamamoto and L.-X. Wang. 2008. Synthesis and inhibitory activity of oligosaccharide thiazolines as a class of mechanism-based inhibitors for *endo*-β-*N*-acetylglucosaminidases. Bioorg. Med. Chem. 16: 4670–4675.
85. H.-S. Kim, K.S. Do, K.S. Kim, J.H. Shim, G.S. Cha and H. Nam. 2004. Ammonium ion binding property of naphtho-crown ethers containing thiazole as subcyclic unit. Bull. Korean Chem. Soc. 25: 1465–1470.
86. G.M. Li, S. Niu, M. Segi, K. Tanaka, T. Nakajima, R.A. Zingaro, J.H. Reibenspies and M.B. Hall. 2000. On the behavior of α,β-unsaturated thioaldehydes and thioketones in the Diels-Alder reaction. J. Org. Chem. 65: 6601–6612.
87. T. Saito, H. Fujii, S. Hayashibe, T. Matsushita, H. Kato and K. Kobayashi. 1996. Uncatalysed (thermal) and Lewis acid-promoted asymmetric hetero-Diels-Alder reaction of 1-thiabuta-1,3-dienes (thiochalcones) with di-(-)-menthyl fumarate. Configuration determination by X-ray crystallographic analysis of (2S,3R,4R)-(+)-2,3-bis[(-)-menthoxycarbonyl]-4,6-diphenyl-3,4-dihydro-2H-thiopyran and conversion of cycloadducts into optically pure diols. J. Chem. Soc. Perkin Trans. 1 15: 1897–1903.
88. D. Brayton, F.E. Jacobsen, S.M. Cohen and P.J. Farmer. 2006. A novel heterocyclic atom exchange reaction with Lawesson's reagent: A one-pot synthesis of dithiomaltol, a novel heterocyclic atom exchange reaction with Lawesson's reagent: A one-pot synthesis of dithiomaltol. Chem. Commun. 2: 206–208.
89. B.F. Bonini, M.C. Franchini, M. Fochi, G. Mazzanti, A. Ricci and G. Varchi. 1999. Extremely facile formation and high reactivity of new thioacylsilanes containing the ferrocene moiety. Tetrahedron Lett. 40: 6473–6476.
90. D. Brillon. 1992. Recent developments in the area of thionation methods and related synthetic applications. Sulfur Rep. 12: 297–332.
91. A.F. Patrocinio and P.J.S. Moran. 2001. Acylsilanes and their applications in organic chemistry. J. Braz. Chem. Soc. 12: 7–31.
92. P. Gosselin, S. Masson and A. Thuillier. 1978. Synthesis of saturated and unsaturated dithioesters. Tetrahedron Lett. 30: 2715–2716.
93. K.R. Lawson, A. Singleton and G.H. Whitham. 1984. Preparation and cycloaddition reactions of some α,β-unsaturated dithioesters. J. Chem. Soc. Perkin Trans. 1 0: 859–864.
94. T. Nishio. 1998. Sulfur-containing heterocycles derived by reaction of ω-keto amides with Lawesson's reagent. Helv. Chim. Acta 81: 1207–1214.
95. M.S.J. Foreman and J.D. Woollins. 2000. Organo-P-S and P-Se heterocycles, organo-P-S and P-Se heterocycles. J. Chem. Soc. Dalton Trans. 10: 1533–1543.
96. N.R. Mohamed, M.M.T. El-Saidi, T.A. Abdallah and A.A. Nada. 2004. Studies on organophosphorus compounds VI1-5: Utility of Lawesson's reagent for synthesis of thiaphosphorine, thioxanthene, and thiaphosphole derivatives. Phosphorus, Sulfur, Silicon Relat. Elem. 179: 2387–2394.
97. N.R. Mohamed. 2000. Studies on organophosphorus compounds V 1-4. Reaction of 2,4-bis(4-methoxyphenyl)-1,3,2,4-dithia diphosphetane-2,4-disulfide with cyclic and heterocyclic ketones. Phosphorus, Sulfur, Silicon Relat. Elem. 161: 123–134.
98. M. Kvasnica, I. Rudovska, I. Cisarova and J. Sarek. 2008. Reaction of lupane and oleanane triterpenoids with Lawesson's reagent, reaction of lupane and oleanane triterpenoids with Lawesson's reagent. Tetrahedron 64: 3736–3743.

99. M. Kvasnica, M. Urban, N.J. Dickinson and J. Sarek. 2015. Pentacyclic triterpenoids with nitrogen- and sulfur containing heterocycles: Synthesis and medicinal significance, pentacyclic triterpenoids with nitrogen- and sulfur-containing heterocycles: Synthesis and medicinal significance. Nat. Prod. Rep. 32: 1303–1330.
100. M.V. Lakshmikantham, P. Carroll, G. Furst, M.I. Levinson and M.P. Cava. 1984. Thione analogs of 1,8-naphthalic anhydride. The first cyclic trithioanhydride. J. Am. Chem. Soc. 106: 6084–6085.
101. M.T. Omar, A. El-Khamry, A.M. Youssef and S. Ramadan. 2003. Synthesis and stereochemistry of thiapyranothiazoles as Diels-Alder adducts obtained from spirodimers of 1,3-thiazolidines with cinnamic acid and its ester. Phosphorus, Sulfur, Silicon Relat. Elem. 178: 721–735.
102. T. Karakasa, S. Satsumabayashi and S. Motoki. 1986. Preparation and thermal isomerization of β-arylthio α,β-unsaturated thioketone dimer. Bull. Chem. Soc. Jpn. 59: 335–337.
103. S. Moriyama, T. Karakasa and S. Motoki. 1990. Syntheses and reactions of conjugated dienic thioketones. Bull. Chem. Soc. Jpn. 63: 2540–2548.
104. T. Karakasa, S. Moriyama and S. Motoki. 1988. Synthesis of new type stable heterotriene. Conjugated dienic thioketone. Chem. Lett. 17: 1029–1032.
105. K.A. Rufanov, A.S. Stepanov, D.A. Lemenovskii and A.V. Churakov. 1999. Synthesis and structure of anhydroindanone-1 disulfide. Heteroat. Chem. 10: 369–371.
106. I. Osken, O. Sahin, A.S. Gundogan, H. Bildirir, A. Capan, E. Ertas, M.S. Eroglu, J.D. Wallis, K. Topal and T. Ozturk. 2012. Selective syntheses of vinylenedithiathiophenes (VDTTs) and dithieno[2,3-*b*;2',3'-*d*]thiophenes (DTTs); building blocks for π-conjugated systems. Tetrahedron 68: 1216–1222.
107. T. Ozturk, E. Ertas and O. Mert. 2010. A Berzelius reagent, phosphorus decasulfide (P_4S_{10}), in organic syntheses. Chem. Rev. 110: 3419–3478.
108. H.-J. Lee, Y.-Y. Kim and D.-Y. Noh. 1998. Thiophene-fused tetrathiafulvalenes as a new molecular donor. Bull. Korean Chem. Soc. 19: 1011–1013.
109. D.-Y. Noh, H.-J. Lee, J. Hong and A.E. Underhill. 1996. New synthesis of phenyl-substituted 2,3-dihydro-1,3-dithiolo[4,5-*e*][1,4]dithiin-6-thione. Tetrahedron Lett. 37: 7603–7606.
110. H.-J. Lee and D.-Y. Noh. 1998. Synthesis, X-ray crystal structure and coupling reactions of 4,5-(1',2'-diphenylethylenedithio)-1,3-dithiole-2-thione (dPhEDT-DTT). Bull. Korean Chem. Soc. 19: 340–344.
111. Y.-Y. Kim, H.-J. Lee, S.K. Namgoong, J. Hong and D.-Y. Noh. 2000. Synthesis and characterization of two different 1,3-dithiole-2-thiones as the precursors of TTF donor molecule. J. Korean Chem. Soc. 44: 513–517.
112. T. Ozturk. 1996. An unusual reaction of Lawesson's reagent with 1,8-diketones: A synthesis of fused 1,4-dithiins and thiophenes. Tetrahedron Lett. 37: 2821–2824.
113. F. Turksoy, J.D. Wallis, U. Tunca and T. Ozturk. 2003. An in depth study of the formation of new tetrathiafulvalene derivatives from 1,8-diketones. Tetrahedron 59: 8107–8116.
114. E. Ertas and T. Ozturk. 2000. An important application of the 1,8-diketone ring formation reaction: A concise synthesis of 5,6-diphenyl[1,3]dithiolo[4,5-*b*][1,4]dithiine-2-thione and its coupling product. Chem. Commun. 20: 2039–2040.
115. T. Ozturk, F. Turksoy and E. Ertas. 1999. A highly functionalized BEDT-TTF derivative. Phosphorus, Sulfur, Silicon Relat. Elem. 153: 417–418.
116. F.B. Kaynak, S. Ozbey, T. Ozturk and E. Ertas. 2001. 5,6-Diphenyl[1,3]dithiolo[4,5-*b*]dithiine-2-thione. Acta Crystallogr. C57: 926–992.
117. F.B. Kaynak, S. Ozbey, T. Ozturk and E. Ertas. 2001. 5-Benzyl-5-phenyl[1,3]dithiolo[4,5-*d*][1,3]dithiole-2-thione. Acta Crystallogr. C57: 319–320.
118. F.B. Kaynak, S. Ozbey, T. Ozturk and E. Ertas. 2001. 5,6-Diphenylthieno[2,3-*d*][1,3]dithiole-2-thione. Acta Crystallogr. C57: 1125–1126.
119. E. Ertas, F.B. Kaynak, S. Ozbey, I. Osken and T. Ozturk. 2008. Synthesis of highly functionalized BEDT-TTF analogues incorporating 1,4-dithiin rings from 1,8-diketones. Tetrahedron 64: 10581–10589.

120. E. Ertas, I. Demirtas and T. Ozturk. 2015. Bis(vinylenedithio)tetrathiafulvalene analogues of BEDT-TTF, bis(vinylenedithio)tetrathiafulvalene analogues of BEDT-TTF. Beilstein J. Org. Chem. 11: 403–415.
121. P. Bhattacharyya and J.D. Woollins. 2001. Selenocarbonyl synthesis using Woollins reagent. Tetrahedron Lett. 42: 5949–5951.
122. I.P. Gray, P. Bhattacharyya, A.M.Z. Slawin and J.D. Woollins. 2005. A new synthesis of $(PhPSe_2)_2$ (Woollins reagent) and its use in the synthesis of novel P-Se heterocycles. Chem. Eur. J. 11: 6221–6227.
123. J. Kaschel, C.D. Schmidt, M. Mumby, D. Kratzert, D. Stalke and D.B. Werz. 2013. Donor-acceptor cyclopropanes with Lawesson's and Woollins' reagents: Formation of bisthiophenes and unprecedented cage-like molecules, donor-acceptor cyclopropanes with Lawesson's and Woollins' reagents: Formation of bisthiophenes and unprecedented cage-like molecules. Chem. Commun. 49: 4403–4405.
124. S. Leistner, K. Hentschel and G. Wagner. 1983. Ring transformations reaktionen von 6-brom-3,1-benzothiazin-2,4"$1H$"-dithion zu chinazolin-2,4"$1H,3H$"-dithionen und tricyclischen chinazolinen. Z. Chem. 23: 215–216.
125. M.-G.A. Shvekhgeimer. 2001. Synthesis of heterocyclic compounds based on isatoic anhydrides ($2H$-3,1-benzoxazine-2,4-diones) (review). Chem. Heterocycl. Compd. 37: 385–443.
126. C. Kaneko, S. Hara, H. Matsumoto, T. Takeuchi, T. Mori, K. Ikeda and Y. Mizuno. 1991. A series of novel acyclic nucleosides. IV. Synthesis of N1-sulfur analogues of acyclovir, directed toward improved antiviral activities. Chem. Pharm. Bull. 39: 871–875.
127. H. Abdel-Ghany and A. Khodairy. 2000. Synthesis of polyfused thieno(2,3-b)thiophenes. Part 2: Synthesis of thienopyrimidine, thienothiazine, thienopyrrolopiperazine, and thienothiazaphospholine derivatives. Phosphorus, Sulfur, Silicon Relat. Elem. 166: 45–56.
128. (a) M.A. Dekeyser and P.T. McDonald. 1998. Pesticidal thiadiazines. PCT Int. Appl. WO1998038181A1. (b) M.A. Dekeyser, P.T. McDonald and G.W. Angle. 1996. Synthesis and insecticidal activity of heterocyclic carbothioamides against *Sogatodes orizicola*. J. Agric. Food Chem. 44: 1177–1179.
129. A. Defoin, G. Augelmann, H. Fritz, G. Geffroy, C. Schmidlin and J. Streith. 1985. Cycloaddition of $2H$-pyran-2-thiones with nitroso derivatives. An unexpected cycloaddition-rearrangement reaction. Helv. Chim. Acta 68: 1998–2014.
130. V. Sosnovskikh. 2013. Sulfur analogs of fluorinated pyrones, chromones and coumarins. J. Sulfur Chem. 34: 432–443.

Chapter 7
O- and *N*-Heterocycles Synthesis

7.1 Introduction

The heterocyclic compounds are established as the largest and most diverse class of organic chemistry. Nowadays, a lot of heterocyclic compounds are known; day by day, number is increasing rapidly because of the massive synthetic research and also their synthetic applications. The heterocyclic compounds have utility in most of the areas of sciences like pharmaceutical chemistry, medicinal chemistry, biochemistry, and also in other areas of science. Most active biological heterocyclic compounds prepared or isolated from the plants exhibit anti-inflammatory, antioxidant, antifungal, antibacterial, anticonvulsant, anti-allergic, herbicidal, and anticancer activities [1–5].

The dimer of phosphorus pentasulfide remained as the key reagent from the 2nd half of the nineteenth century until the beginning of systematic study of the utilization of Lawesson's reagent by Lawesson et al. [6–9]. Although many reagents including the analogues of Lawesson's reagent and H_2S have been employed, in general with limited success [10–12]; Lawesson's reagent has remained as the most important reagent in thionation chemistry and was followed by P_4S_{10}. Generally, it is claimed that Lawesson's reagent has benefits over dimer of phosphorus pentasulfide with respect to the needs for excess P_4S_{10} and longer reaction time. It could also be true when the number of publications published every year on both reagents is taken into consideration. Both reagents have their own advantages and disadvantages over specific reactions, and both of them deserve to be utilized [13, 14].

7.2 Synthesis of Five-Membered *O*-Heterocycles

The 1,4-diketones were reacted with Lawesson's reagent to afford the furans and thiophenes [15–17]. The reaction afforded improved yields and took place under milder conditions as compared to the reaction when P_4S_{10} was utilized (Scheme 7.1) [4].

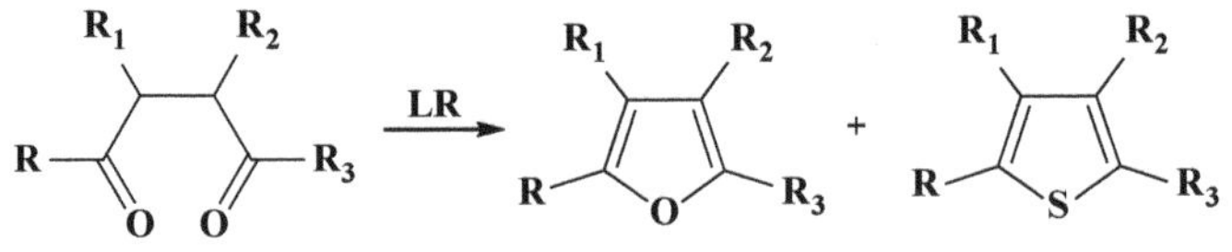

Scheme 7.1 Synthesis of furans and thiophenes

7.3 Synthesis of Five-Membered *O,N*-Heterocycles

The L-serine methyl ester was transformed into thioamide in four steps with 54% yield (Scheme 7.2) [18]. This thioamide acted as a nucleophile in a subsequent Hantzsch thiazole synthesis, which is a particularly practical strategy for the formation of thiazoles first reported in 1887 [19].

The thiazole-5-tertiary-amide was not provided by thionation–cyclization of tertiary amide (derived from ring-opening of 2-phenyl-4-[(4-methoxyarylidene)(methylthio)]oxazolone with piperidine). The obtained compound was identified as 2-phenyl-4-[(4-methoxyarylidene)(methylthio)]oxazolone, which was synthesized by thermal-eliminative cyclization probably due to steric issues (Scheme 7.3) [14].

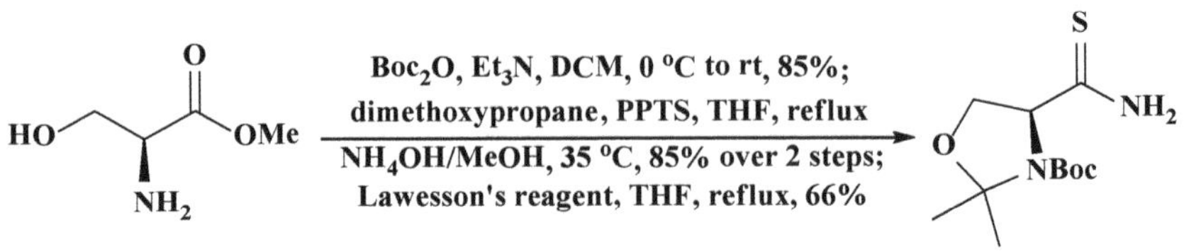

Scheme 7.2 Synthesis of oxazole

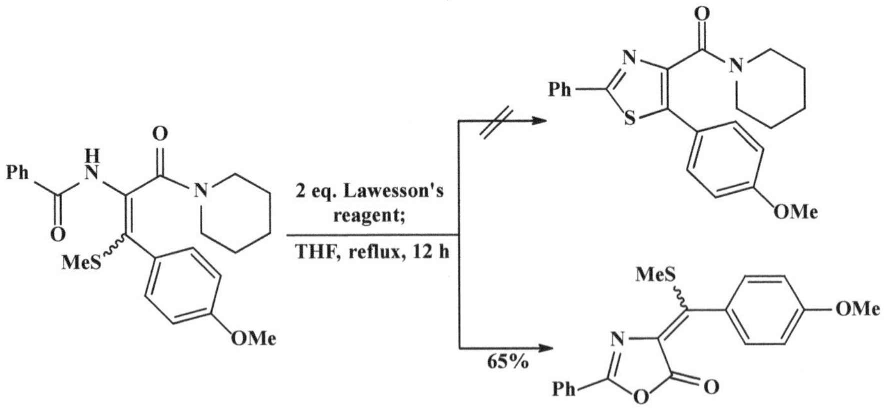

Scheme 7.3 Synthesis of 2-phenyl-4-[(4-methoxyarylidene)(methylthio)]oxazolone

7.3 Synthesis of Five-Membered O,N-Heterocycles

The cyclization methods were used in the formation of thiazole, imidazole, or oxazole (Scheme 7.4). The ester group of starting compound was hydrolyzed with lithium hydroxide to afford an acid, which underwent amidation with a substituted 2-oxo-ethylamine to generate the key intermediate. Following the cyclization steps, the benzyl group was removed with trimethylsilyl iodide in CH_3CN to accomplish the synthetic sequence [5].

Seijas and coworkers [20] described that LR is an efficient promoter in the solvent-free MW-assisted formation of 2-substituted benzoxazoles and benzothiazoles from carboxylic acids and 2-aminophenol or 2-aminothiophenol, respectively. Numerous heteroaromatic, aromatic, and aliphatic carboxylic acids reacted under reaction conditions with good yields (Scheme 7.5).

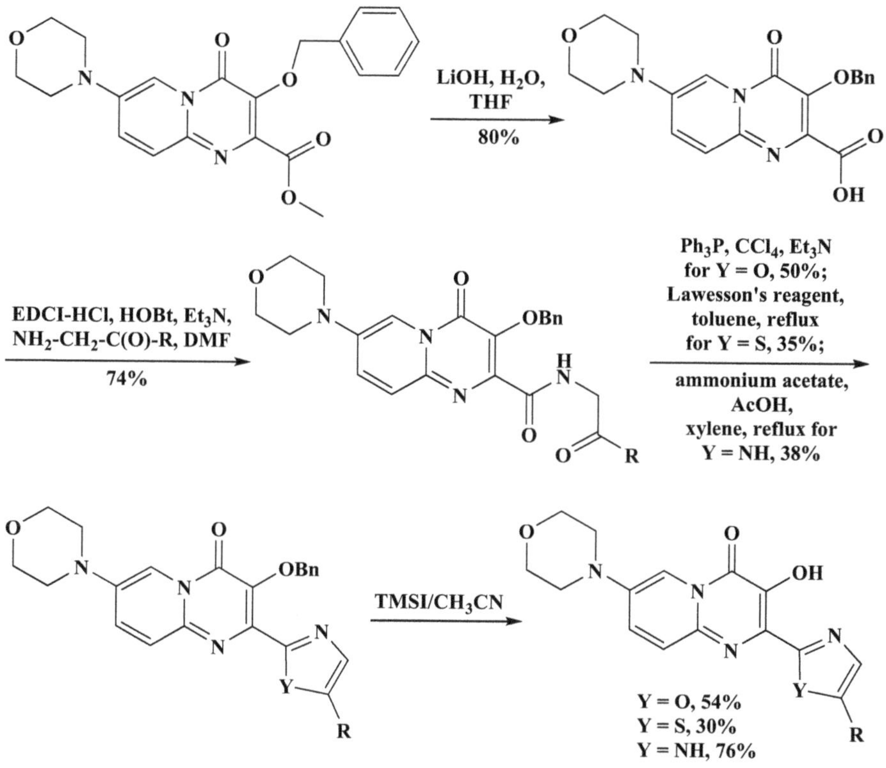

Scheme 7.4 Synthesis of oxazoles, thiazoles, and imidazoles

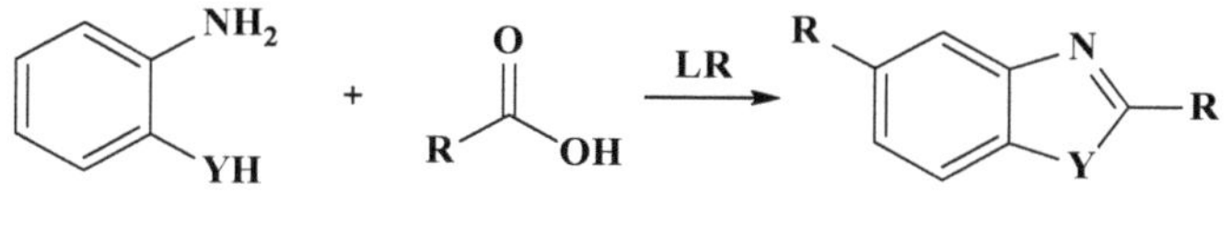

Scheme 7.5 Synthesis of 2-substituted benzoxazoles and benzothiazoles

Due to the problems in the removal of benzyl protecting group in model analysis, the phenol and alcohol were reprotected as acetates (Scheme 7.6). The succeeding important step was the formation of oxazolidine ring which was completed according to the approach reported by Pelletier et al. [21], in which oxazolidine was synthesized by heating imine with oxirane. This reaction was achieved by a two-step sequence. The synthesis of thiolactam with LR followed by desulfurization in the presence of deactivated Raney-Ni provided imine. With oxazolidine ring in hand, the final four steps afforded (±)-cyanocycline A. This sequence included the hydrolysis of acetals and amine deprotection, followed by N-methylation to afford an advanced

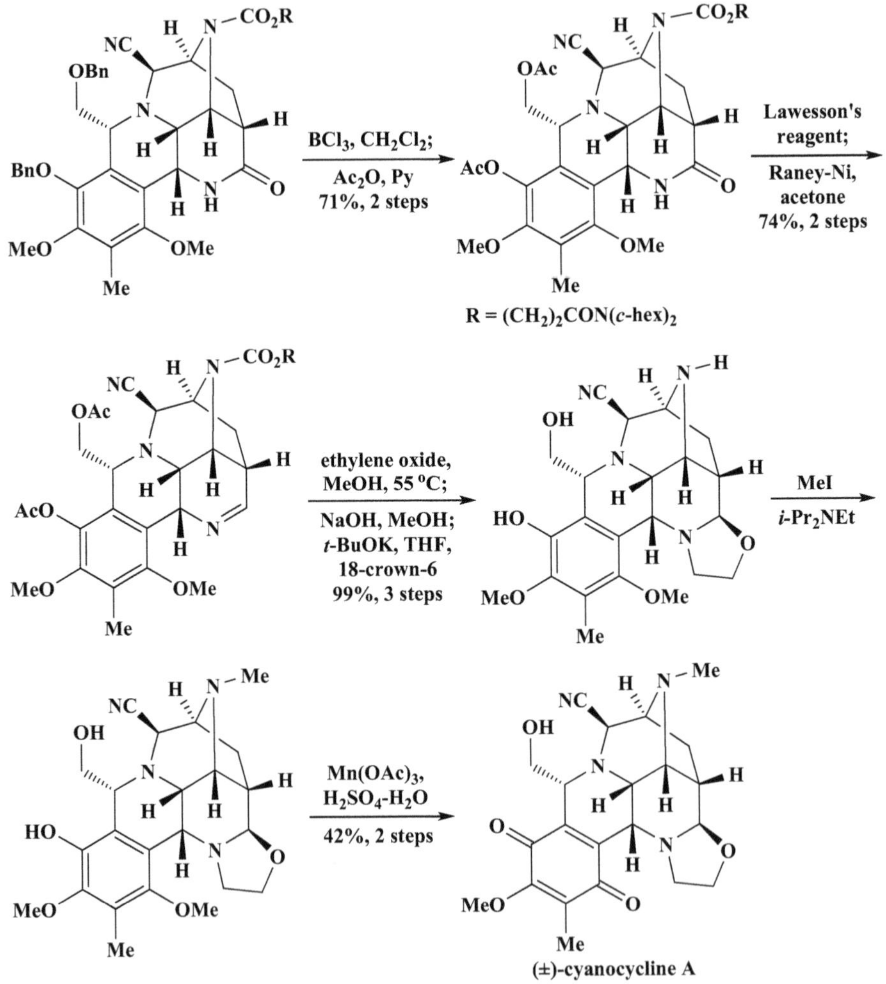

Scheme 7.6 Synthesis of (±)-cyanocycline A

7.3 Synthesis of Five-Membered O,N-Heterocycles

intermediate which on oxidation with Mn(OAc)$_3$ afforded (±)-cyanocycline A. This racemic formation was accomplished in 32 steps in 0.85% overall yield.

After the synthesis of the C ring of bioxalomycin, the next tasks were the synthesis of two oxazolidine rings F and G. Since pentacyclic lactam intermediate was very identical to Fukuyama's advanced intermediate, this approach was used for the formation of F ring. The pentacyclic lactam was transformed to thiolactam with LR (Scheme 7.7). The sensitive thiolactam was immediately transformed to imine with deactivated Raney nickel in (CH$_3$)$_2$CO in 63% total yield. This stable imine was transformed to oxazolidine by following Pelletier's approach [21] in which the oxazolidine was prepared by heating imine with ethylene oxide. Although the deprotection of benzyl groups through hydrogenolysis without reducing the CN functionality was

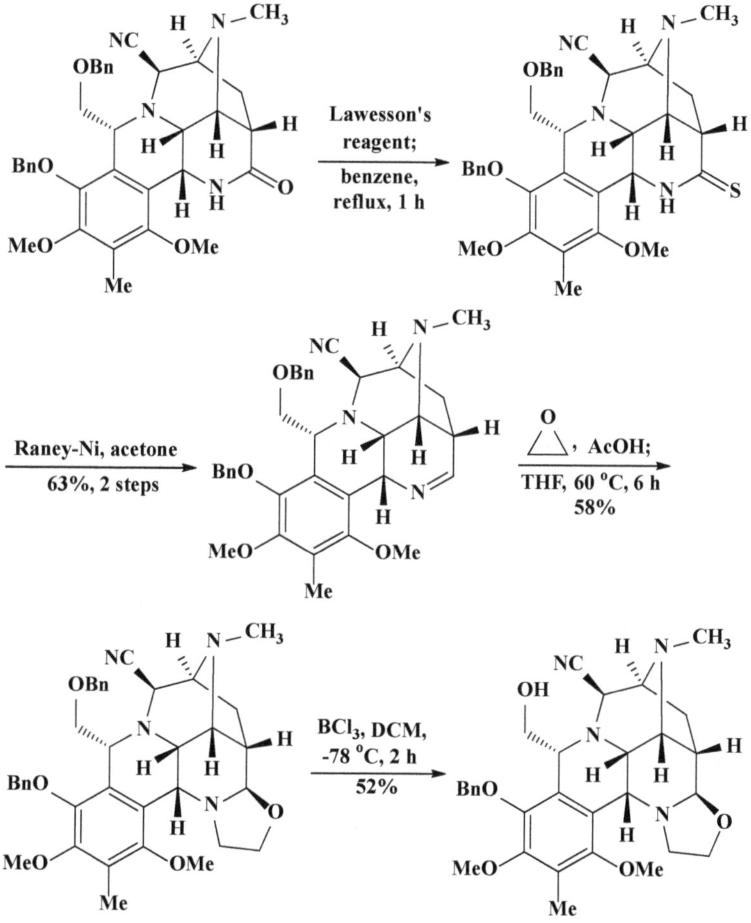

Scheme 7.7 Synthesis of oxazolidine ring bearing hexacycle

shown in similar structures [22], this approach was unsuccessful. The hydrogenolysis of oxazolidine in methanol in the presence of palladium/carbon catalyst at rt and 1 atm not proceeded and starting oxazolidine was recovered. However, deprotection of benzyl groups was carried out with excess of boron trichloride in dichloromethane at -78 °C. This reaction smoothly afforded hexacycle in 52% yield.

7.4 Synthesis of Five-Membered *O,N,N*-Heterocycles

As described in Scheme 7.8, the intermediate, generated from bromoacetophenone via azide and amine intermediates followed by coupling with orthogonally protected α-methyl serine or from 3-trifluoromethyl-4-fluorobenzoic acid through hydramide synthesis and coupling reaction with α-methyl serine, was cyclized with LR either in toluene or DCM to provide the thiazole and thiadiazole analogues after deprotection with trifluoroacetic acid. This synthetic pathway was developed into a fully telescoped procedure for the formation of thiadiazole at multikilogram scale [23]. The intermediates were transformed to oxazole or 1,3,4-oxadiazole with PPh$_3$ and hexachloroethane or PPh$_3$ and CCl$_3$CN after deprotection with trifluoroacetic acid [24].

The key starting compound for azoles was methyl ester [25]. The ester group reacted easily with hydrazine hydrate to afford the monohydrazide, which was then acylated with an acid chloride to provide an intermediate (Scheme 7.9). The cyclization of hydrazide group and after that benzyl deprotection with trimethylsilyl iodide as Lewis acid afforded 1,3,4-oxadiazole. On the other hand, the use of LR as cyclizing agents resulted in the formation of 1,3,4-thiadiazole. The triazole was also synthesized from ester, which easily underwent amidation with aqueous NH$_4$OH. The amide was treated with LR, and subsequently with methyl iodide to form the S-Me thioamide intermediate that cyclized with 4-F-phenyl acetohydrazide to provide the triazole. The synthesis of asymmetrical 1,2,4-oxadiazole needed an another approach. In the first step, ester was transformed into nitrile through amidation in the presence of NH$_3$ and dehydration with trichlorotriazine. Further, the nitrile group was treated with NH$_2$OH·HCl to afford the *N*-hydroxyamidine, which was acylated with a suitable phenylacetyl chloride to provide the precursor. The 1,2,4-oxadiazole was synthesized by refluxing intermediate in toluene [5].

The formation of cathepsin K inhibitors (cathepsin K) was described that provokes human breast cancer (Scheme 7.10) [26]. The hydrazides, prepared regioselectively from acids, were precursors of heterocyclization. The reaction was carried out with 1-methoxy-*N*-trimethylammoniosulfonylmethanimidate (Burgess reagent) under MWI or in tetrahydrofuran solution at 40 °C with Lawesson's reagent. The 1,3,4-oxadiazoles were then converted into amides [27].

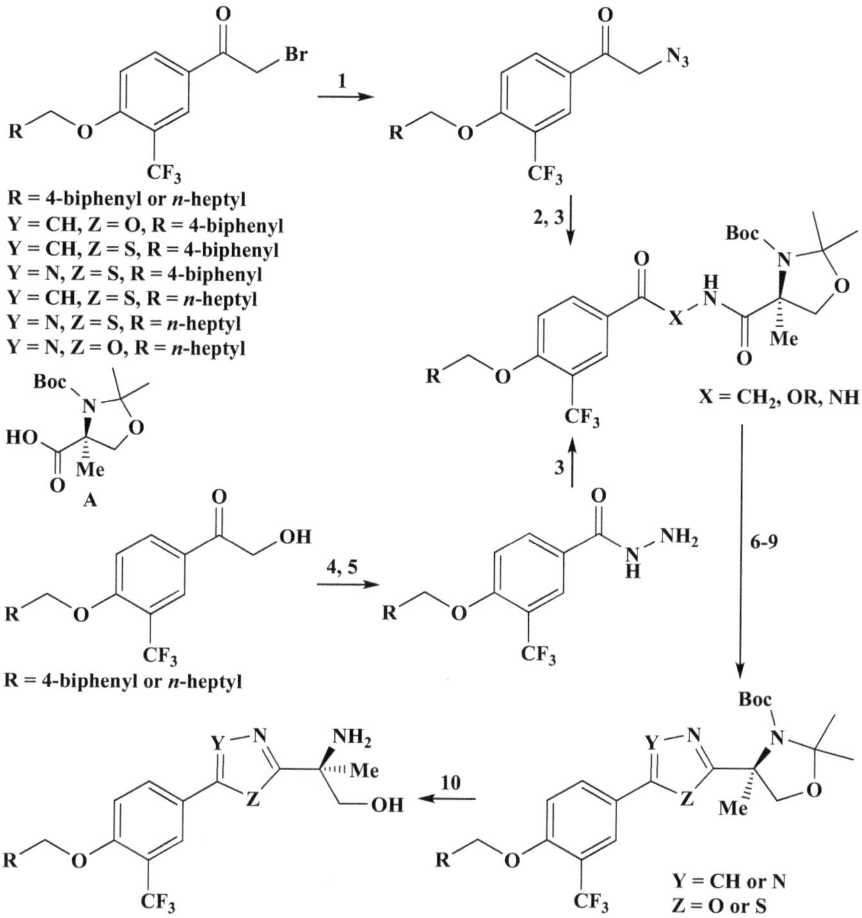

7.5 Synthesis of Six-Membered *N*-Heterocycles

R = 4-biphenyl or *n*-heptyl
Y = CH, Z = O, R = 4-biphenyl
Y = CH, Z = S, R = 4-biphenyl
Y = N, Z = S, R = 4-biphenyl
Y = CH, Z = S, R = *n*-heptyl
Y = N, Z = S, R = *n*-heptyl
Y = N, Z = O, R = *n*-heptyl

X = CH$_2$, OR, NH

R = 4-biphenyl or *n*-heptyl

Y = CH or N
Z = O or S

Reagents and conditions: (1) a) CuBr$_2$, EtOAc/CHCl$_3$ (1:1), reflux, b) NaN$_3$, DMF, (2) H$_2$ (gas), 5% Pd/C, MeOH, HCl, (3) A, HATU, DIPEA, DCM/DMF, (4) R$_2$CH$_2$OH, *t*-BuOK, THF, 70 °C, (5) hydrazine, HATU, DIPEA, DCM/DMF, (6) Lawesson's reagent, toluene, 120 °C, (7) Lawesson's reagent, DCM, (8) 5 eq. PPh$_3$, 2.5 eq. hexachloroethane (C$_2$Cl$_6$), 10 eq. TEA, (9) PPh$_3$, CCl$_3$CN, CH$_3$CN, MW, 120 °C, 20 min, (10) 6 N HCl, dioxane or TFA, DCM, or TsOH, MeOH, reflux.

Scheme 7.8 Synthesis of oxazoles, oxadiazoles, thiazoles, and thiadiazoles

7.5 Synthesis of Six-Membered *N*-Heterocycles

The reaction of amide with LR resulted in ring-closure to afford the imide (Scheme 7.11) [13, 28].

Duhamel et al. [29] in 1991 afforded an extended explanation of their earlier stated approach, describing the synthesis of optically pure (+)-enamine, constituting

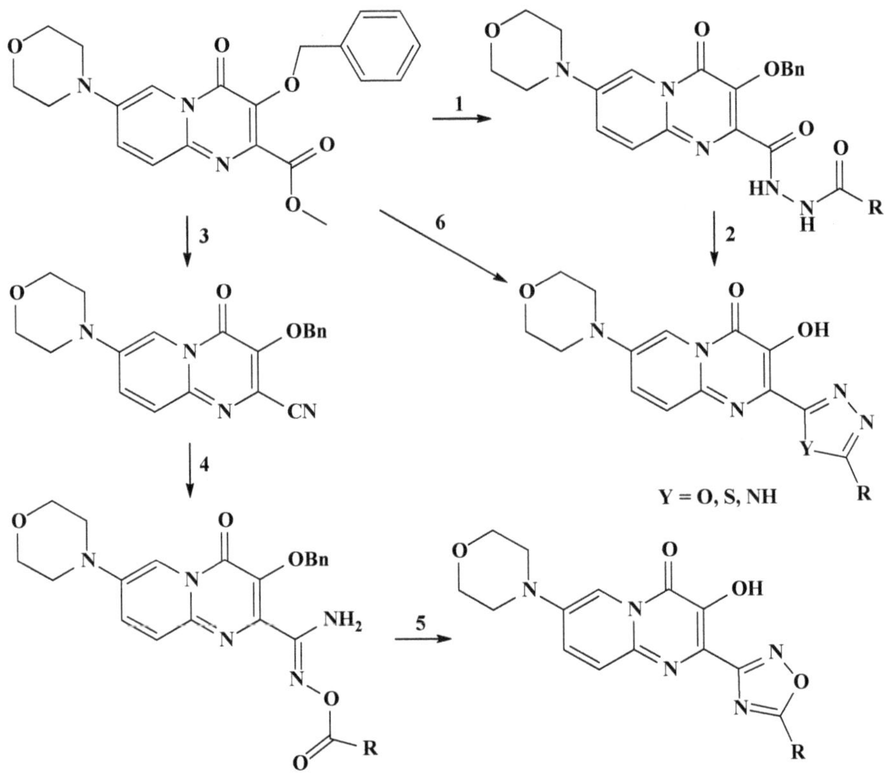

Reagents and conditions: (1) a) hydrazine hydrate, MeOH, 81%, b) RCOCl, 94%, (2) Y = O, S, a) Ph$_3$P, CCl$_4$, Et$_3$N for Y = O, 43%, Lawesson's reagent, toluene, reflux for Y = S, 35%, b) TMSI/CH$_3$CN, 41% for Y = O, 35% for Y = S, (3) a) NH$_3$, H$_2$O, 50 °C, 59%, b) 2,4,6-trichloro-1,3,5-triazine, 88%, (4) a) NH$_2$OH.HCl, NaHCO$_3$, 92% EtOH, reflux, b) RCOCl, 97%, (5) a) toluene, reflux, 75%, b) FeCl$_3$, CH$_2$Cl$_2$, 30%, (6) Y = N, a) NH$_3$, H$_2$O, 50 °C, 59%, b) Lawesson's reagent, toluene, reflux, 45%, c) MeI, 50% 4-F-phenyl acetohydrazide, AcOH, reflux, 55%, d) TMSI/CH$_3$CN, 50%.

Scheme 7.9 Synthesis of 1,2,4-oxadiazoles

a formal formation of (−)-pHTX (Scheme 7.12). The condensation of cyclohexanone with (2R)-amino-3-phenylpropan-1-ol and after that methylation of latent oxygen function led to auxiliary-bound imine. The deprotonation followed by facially selective alkylation with C$_5$H$_{11}$I led to chiral ketone with 78% enantiomeric excess after cleavage of the chiral auxiliary. This was advanced utilizing their earlier approach to produce the enantiomerically pure (+)-enamine [30].

Duhamel et al. [31] extended this approach for the total synthesis of (−)-pHTX in 1989 (Scheme 7.13). Starting from cyclic ketone, which already possess the future C-2 pentyl side-chain, they synthesized benzyl-protected lactam by an acid-catalyzed Beckmann-type ring-expansion. The benzyl-protected lactam was subjected to key

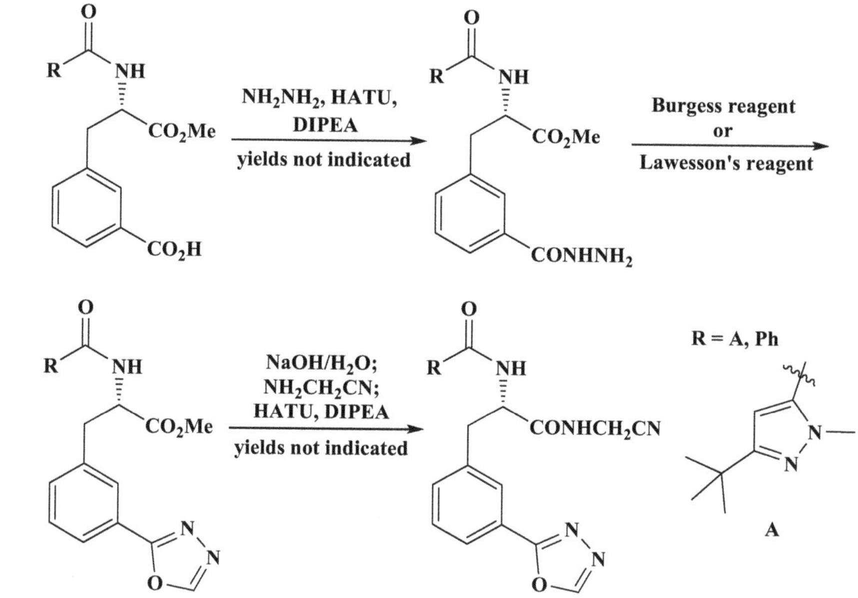

7.5 Synthesis of Six-Membered N-Heterocycles

Scheme 7.10 Synthesis of 1,3,4-oxadiazoles

Scheme 7.11 Synthesis of 3-phenyl-6-thioxopiperidin-2-one

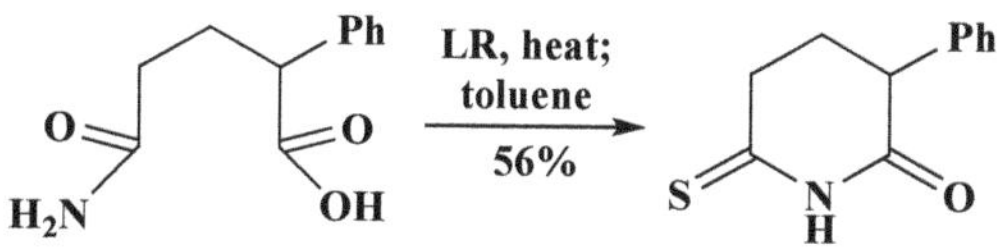

ring-contraction step; however, the main product was observed to possess an undesired *trans* relationship between the aldehyde and pentyl side chains. Further examination showed that when the methyl ester was utilized, the ring-contraction occurred to afford the *cis*-isomer exclusively in quantitative yield. A Wittig reaction then advanced this intermediate to α,β-unsaturated ketone, and hydrogenation afforded saturated ester. The diisobutylaluminum hydride reduction of ester afforded aldehyde, which underwent a base-assisted internal condensation to provide the spirocycle. An approach similar to that employed by Pearson and Ham [32] was further used to complete the synthesis, but the 1,4-organocuprate addition afforded 1,2-addition by-product in unsatisfactory amount. The desired conversion was ultimately completed by addition of TEA and TMSCl to the organocuprate before addition of the enone. The formed intermediate was directly made to react with benzeneselenyl chloride, followed by H_2O_2 in CH_3COOH, to afford the butyl enone in 60% yield. The borohydride reduction then eliminated the redundant ketone function. The Godleski's hydroboration–oxidation was performed, providing a mixture of alcohols that was debenzylated and separated to afford the 1:2 mixture of pHTX and its 7,8-epimer.

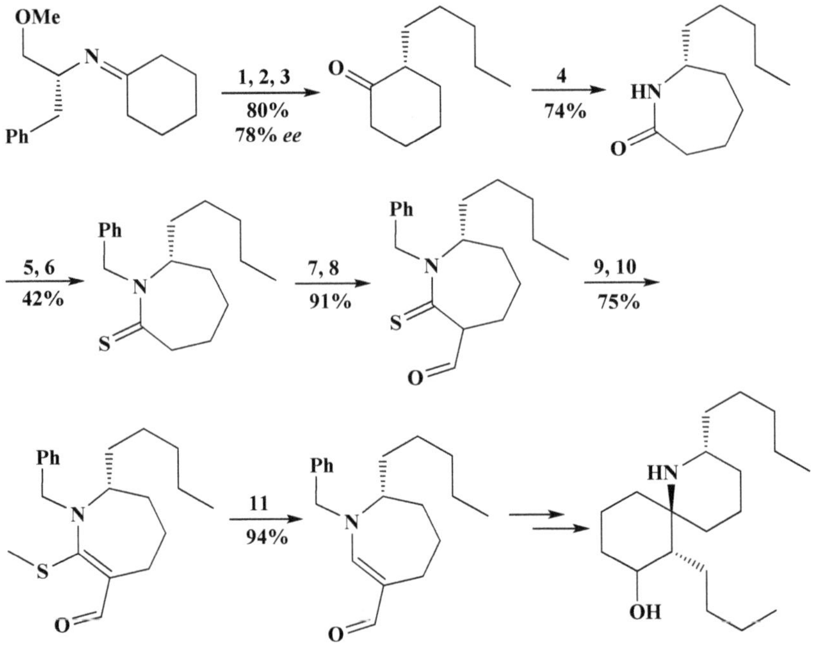

Duhamel's 1991 (-)-pHTX formal synthesis. Reagents and conditions: (1) LDA, (2) $C_5H_{11}I$, (3) AcOH, H_2O, (4) H_2NOSO_3H, HCO_2H, (5) t-BuOK, $PhCH_2Br$, (6) Lawesson's reagent, (7) t-BuOCH(NMe$_2$)$_2$, (8) 2 M HCl, (9) MeOSO$_2$F, (10) Et$_3$N, (11) Raney-Ni.

Scheme 7.12 Synthesis of optically pure (+)-enamine

The examinations into the application of acylated enamines ("enaminones" in general) and related compounds as intermediates in alkaloid formation [33–35] took benefit of their ability to function either as ambident nucleophiles or as ambident electrophiles based on the synthetic approach envisioned. These easily synthesized compounds were easily introduced into structures that possess the gross skeletal features observed in various alkaloidal systems, and they also afforded ample opportunity for controlling the enantioselectivity and diastereoselectivity. The chiral amine, prepared from lithium N-benzyl-N-[(1R)-1-phenylethyl]amide and t-butyl (2E)-oct-2-enoate by Davies protocol [36, 37], was transformed into pyrrolidine-2-thione in three steps, from which the enaminone (a vinylogous urethane) was easily synthesized by Eschenmoser sulfide contraction [38]. The chemoselective reduction of saturated ester to alcohol followed by cyclization through the iodide afforded hexahydroindolizine, which was itself an enaminone. The diastereoselective reduction of hexahydroindolizine produced the final compound (Scheme 7.14) [39].

The substrates, dimethylaminopropenoyl cyclopropanes, were synthesized in high yields (up to 91%) [40, 41]. The reaction of 1-[3-(dimethylamino)acryloyl]-N-phenyl cyclopropanecarboxamide with 0.5 eq. LR was first attempted in C_6H_6 at rt; however, no reaction was observed by thin layer chromatography of the reaction

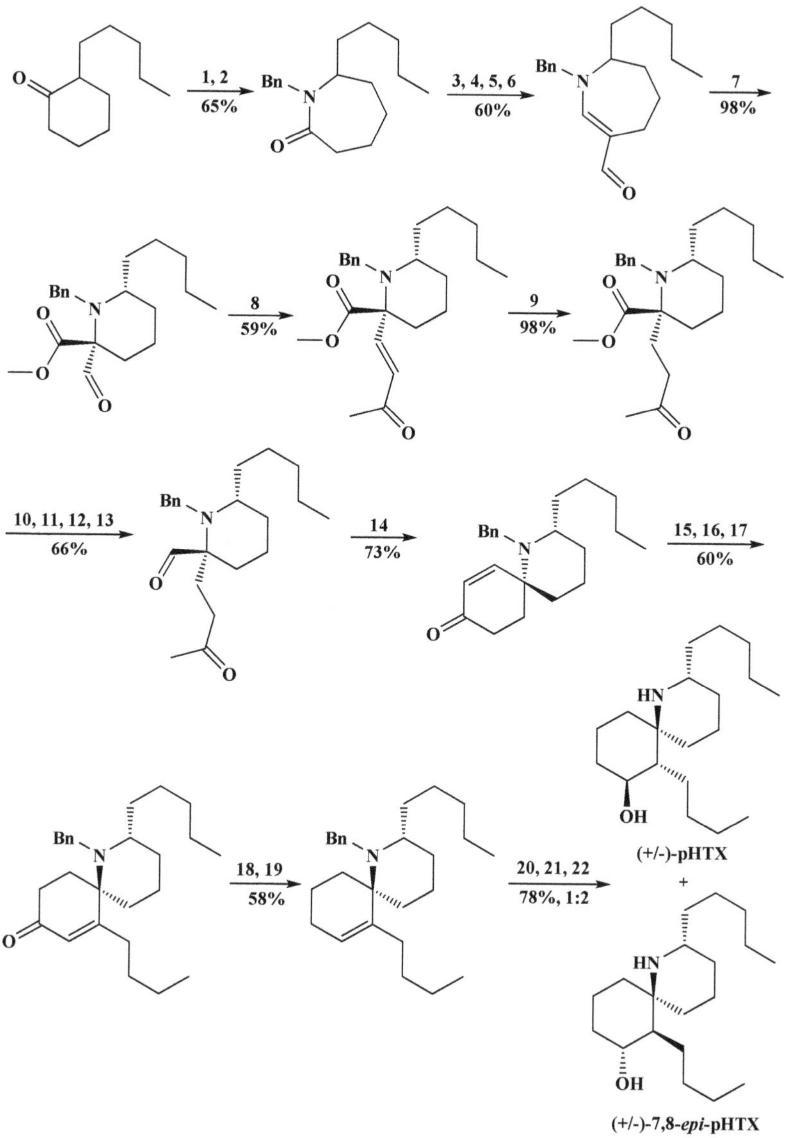

7.5 Synthesis of Six-Membered N-Heterocycles

Duhamel's 1989 (±)-pHTX total synthesis. Reagents and conditions: (1) H$_2$NOSO$_3$H, HCO$_2$H, (2) t-BuOK, PhCH$_2$Br, (3) Lawesson's reagent, (4) t-BuOCH(NMe$_2$)$_2$, (5) MeOSO$_2$F, (6) Raney Ni, (7) Br$_2$, Et$_2$O, -70 °C; MeOH, Et$_3$N, -70 °C, (8) Ph$_3$P=CHCOMe, t-BuOK, THF, -5 °C, (9) H$_2$, Pd/C, (10) (CH$_2$OH)$_2$, p-TsOH, toluene, (11) DIBAL, Et$_2$O, (12) (COCl)$_2$, DMSO, Et$_3$N, (13) 3 M HCl, reflux, (14) t-BuOK, THF, (15) Bu$_2$CuLi, Et$_3$N, TMSCl, Et$_2$O, (16) PhSeCl, THF, (17) H$_2$O$_2$, AcOH, (18) NaBH$_4$, MeOH, (19) LiAlH$_4$, AlCl$_3$, Et$_2$O, (20) BH$_3$, Me$_2$S, THF, (21) H$_2$O$_2$, NaOH, (22) H$_2$, Pd/C.

Scheme 7.13 Synthesis of pHTX

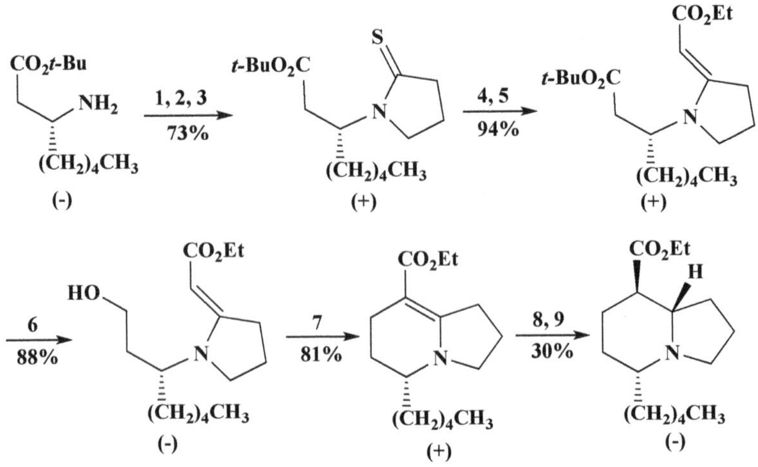

Reagents and conditions: (1) Cl(CH$_2$)$_3$COCl, NaHCO$_3$, CHCl$_3$, reflux, (2) *t*-BuOK, *t*-BuOH, rt, (3) Lawesson's reagent, PhMe, reflux, 73% over 3 steps, (4) BrCH$_2$CO$_2$Et, MeCN, rt, (5) Ph$_3$P, Et$_3$N, MeCN, rt, 94% over 2 steps, (6) LiAlH$_4$, THF, rt, 88%, (7) I$_2$, imidazole, Ph$_3$P, PhMe, 110 °C, 81%, (8) 1 atm H$_2$, PtO$_2$, AcOH, rt, (9) NaOEt (catalyst), EtOH, reflux, 30% over 2 steps.

Scheme 7.14 Synthesis of hexahydroindolizine

mixture. When the mixture was heated to reflux for 24 h, the reaction proceeded and provided a white solid along with intact 1-[3-(dimethylamino)acryloyl]-*N*-phenyl cyclopropanecarboxamide in small amounts after work-up and purification by column chromatography. The product was characterized as 5-phenyl-2,3-dihydrothieno[3,2-*c*]pyridin-4(5*H*)-one (54%) by its analytical and spectral data (Scheme 7.15). These results inspired to optimize the reaction conditions, including reaction temperature, solvent, and the ratio of Lawesson's reagent to 1-[3-(dimethylamino)acryloyl]-*N*-phenyl cyclopropanecarboxamide. A series of experiments showed that the reaction proceeded in other solvents, like toluene and xylene, and 0.5 eq. Lawesson's reagent was sufficient for the dihydrothieno[3,2-*c*]pyridin-4(5*H*)-one formation. However, much more Lawesson's reagent, for instance more than 0.6 eq., would afford a complex mixture. The optimum results

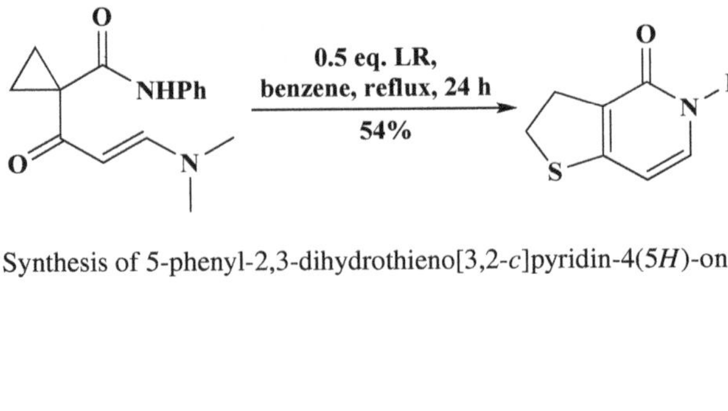

Scheme 7.15 Synthesis of 5-phenyl-2,3-dihydrothieno[3,2-*c*]pyridin-4(5*H*)-one

7.5 Synthesis of Six-Membered N-Heterocycles

were observed when 1-[3-(dimethylamino)acryloyl]-*N*-phenyl cyclopropanecarboxamide was reacted with 0.5 eq. Lawesson's reagent in dry toluene under reflux for 10 h, whereby the reaction produced 5-phenyl-2,3-dihydrothieno[3,2-*c*]pyridin-4(5*H*)-one exclusively in 82% yield (Scheme 7.16). Having established the optimum conditions for the dihydrothieno[3,2-*c*]pyridin-4(5*H*)-one formation, its scope was determined in terms of R_1, R_2, and R_3 groups. A series of reactions of aminopropenoyl cyclopropanes having different aryl and alkyl amide groups was carried out with Lawesson's reagent under the same conditions as for 5-phenyl-2,3-dihydrothieno[3,2-*c*]pyridin-4(5*H*)-one. It was found that all the reactions occurred easily to produce the dihydrothieno[3,2-*c*]pyridin-4(5*H*)-ones in moderate-to-good yields. The versatility of this dihydrothieno[3,2-*c*]pyridin-4(5*H*)-one formation was then examined by reacting aminopropenoyl cyclopropane and Lawesson's reagent under same reaction conditions. The efficiency of cyclization proved to be appropriate for aminopropenoyl cyclopropane to give the substituted dihydrothieno[3,2-*c*]pyridin-4(5*H*)-one in good yield. In this case, a single regioisomer was exclusively formed, which indicated that the ring-opening reaction or ring-expansion proceeded in a regioselective fashion [42–44].

A one-pot formation of thieno[3,2-*c*]pyridin-4(5*H*)-ones was attempted from aminopropenoyl cyclopropanes. The reaction of aminopropenoyl cyclopropanes and Lawesson's reagent was carried out in toluene under reflux for 10 h followed by an addition of 2,3-dichloro-5,6-dicyano-1,4-benzoquinone to the formed mixture, which was left under reflux for one more hour. The product was characterized as 5-phenylthieno[3,2-*c*]pyridin-4(5*H*)-one by spectral and analytical data, and the overall yield reached to 75%. Similarly, some chosen aminopropenoyl cyclopropanes were reacted with Lawesson's reagent and 2,3-dichloro-5,6-dicyano-1,4-benzoquinone in toluene under reflux to give the thieno[3,2-*c*]pyridin-4(5*H*)-ones in moderate-to-good total yields (Scheme 7.17) [44–48].

An efficient method to indole alkaloids involved a key step, i.e., a cyclocondensation reaction of racemic aldehyde diester, which was envisioned as a synthetic equivalent of secologanin, with (*S*)-tryptophanol provided enantiopure lactam in 62% yield [49]. Three stereogenic centers with a well-defined configuration have been produced in a single synthetic step. The consequent closure of C ring from

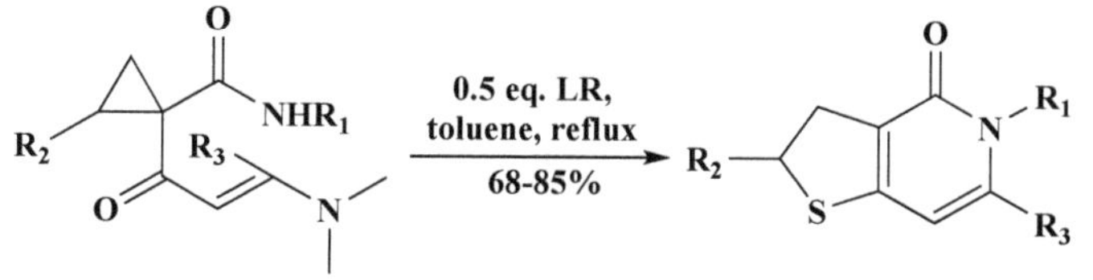

R_1 = Ph, 2-MeC$_6$H$_4$, 4-MeC$_6$H$_4$, 2,4-Me$_2$C$_6$H$_3$, 2-MeOC$_6$H$_4$, 4-MeOC$_6$H$_4$, 2-ClC$_6$H$_4$, 4-ClC$_6$H$_4$, 4-CF$_3$C$_6$H$_4$, Bn
R_2 = H, Ph
R_3 = H, Me

Scheme 7.16 Synthesis of dihydrothienopyridinones

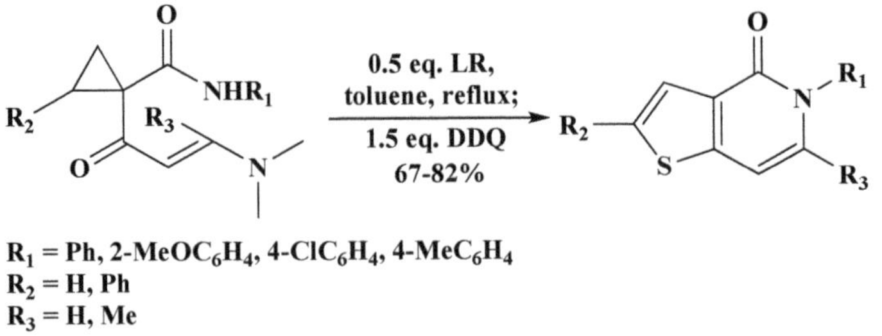

Scheme 7.17 Synthesis of thienopyridinones

lactam via thioamide generated pentacyclic compound, which contains tetracyclic scaffold of *Corynanthe* alkaloids (Scheme 7.18). The dihydrocorynantheine was synthesized enantioselectively from pentacyclic compound.

An electrocyclization of azatrienes afforded dihydropyridines, which were smoothly oxidized to pyridines. An overall procedure is especially utilized in the formation of pyridine bearing natural products. Meketa and Weinreb [50] employed aza-6π electrocyclization for the formation of ageladine A, which exhibit high inhibitory activity against zinc matrix metalloproteinase (MMPs). The iodide was reacted with amide using copper(I) iodide to produce the amide in excellent yield,

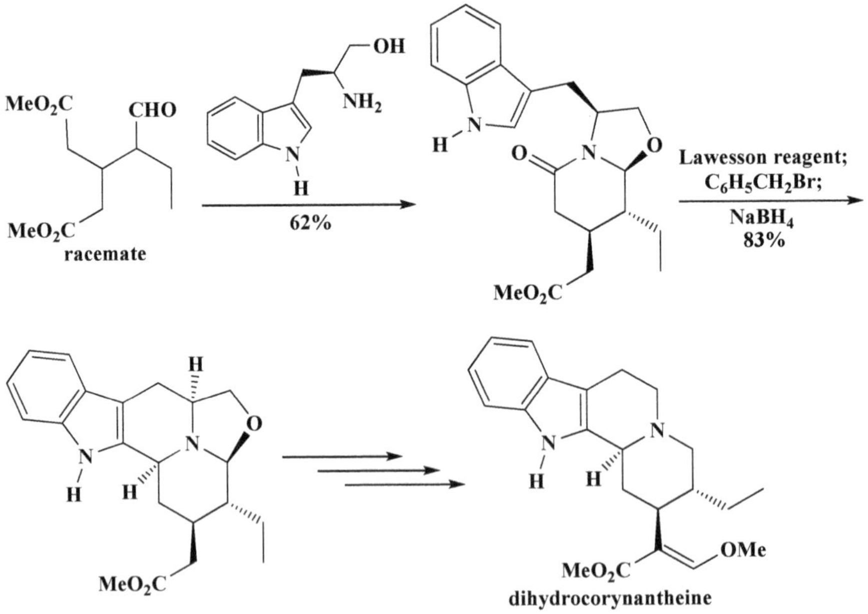

Scheme 7.18 Synthesis of dihydrocorynantheine

7.5 Synthesis of Six-Membered N-Heterocycles

then amide was transformed to key intermediate in two steps. The thermal electrocyclization of azatriene generated a key intermediate in moderate yield, which provided ageladine A (Scheme 7.19).

The formation of newly designed cycloaddition substrate started with the reaction of commercially accessible (−)-2,3-*O-i*-propylidine-D-erythronolactone and (trimethylsilyl)methyl amine in 97% yield. The Parikh–Doering oxidation of primary alcohol provided a hemiaminal, which was acetylated to give the aminal in 69% yield within two steps. The Lewis acid-promoted Mannich reaction between aminal and trimethylsilylketene acetal of $CH_3COOC_2H_5$ (72%) was followed by hydrolysis of ester to afford the carboxylic acid in 96% yield. The thionation of lactam functionality within carboxylic acid was carried out by heating with LR (95%), and the C-8 carboxyl group in the formed product was further transformed to *N*-methoxy-*N*-methyl amide (62%). The Eschenmoser sulfide contraction (81%), followed by reaction with ethynylmagnesium bromide and ethane thiolate (58% yield in two steps), installed the needed functionality for the intramolecular [3 + 2]-azomethine ylide cycloaddition. In this case, sequential reaction with triflic anhydride and TBAT afforded polycyclic alkaloid in 71% yield (Scheme 7.20). This result includes the

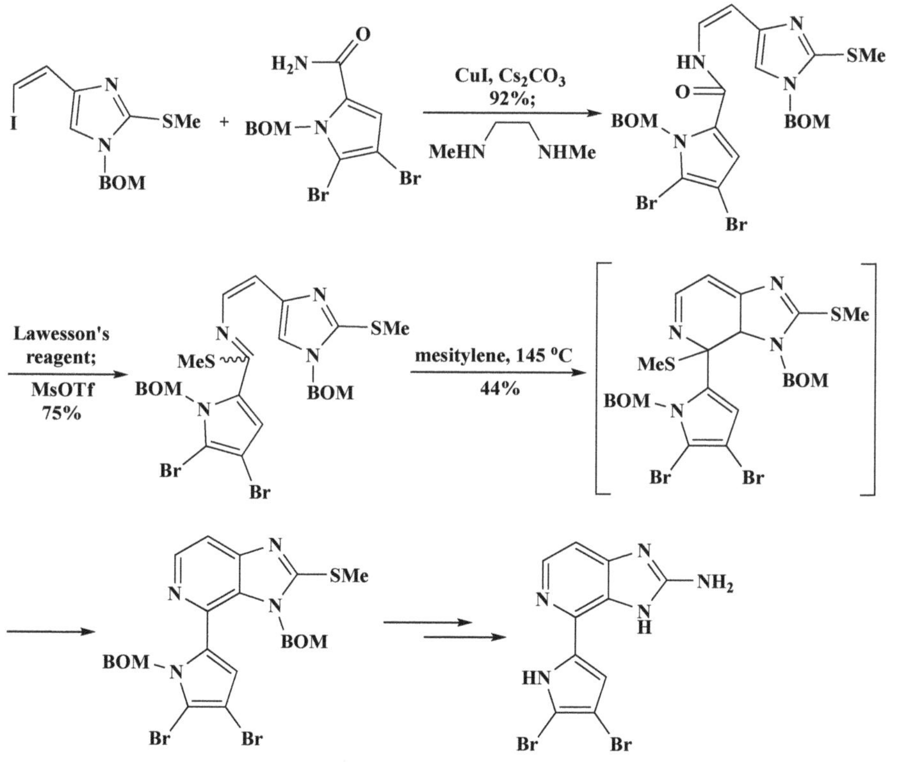

Scheme 7.19 Synthesis of ageladine A

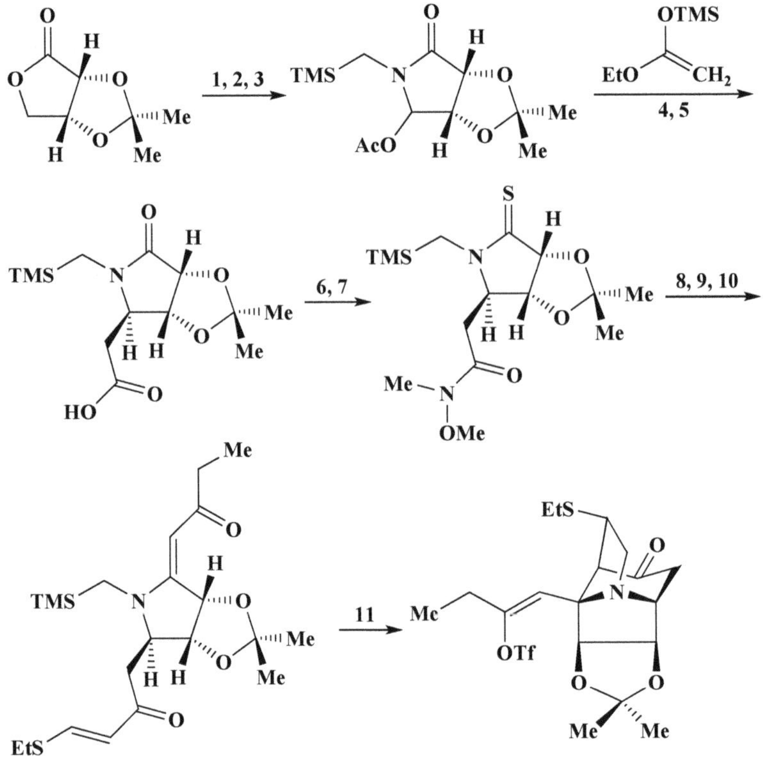

Reagents and conditions: (1) TMSCH$_2$NH$_3$Cl, Et$_3$N, THF, 70 °C, 97%, (2) SO$_3$·Py, Et$_3$N, DMSO, 23 °C, (3) Ac$_2$O, Py, 23 °C, 69%, 2 steps, (4) TMSOTf, CH$_2$Cl$_2$, 23 °C, 72%, (5) LiOH, THF, H$_2$O, 23 °C, 96%, (6) Lawesson's reagent, PhMe, 65 °C, 95%, (7) MeONHMe·HCl, EDC, Et$_3$N, CH$_2$Cl$_2$, 23 °C, 62%, (8) 1-bromo-2-butanone, Ph$_3$P, Et$_3$N, CH$_3$CN, 23 °C, 81%, (9) HCCMgCl, THF, 23 °C, (10) EtSH, Et$_3$N, CH$_2$Cl$_2$, 23 °C, 58%, 2 steps, (11) Tf$_2$O, TBAT, CHCl$_3$, -45 to 23 °C, 71%.

Scheme 7.20 Synthesis of polycyclic alkaloid

first nonracemic formation of fully oxygenated bridged pyrrolizidine core of the stemofoline alkaloids [51].

7.6 Synthesis of Six-Membered *N,N*-Heterocycles

The reductive amination of aldehyde and amine hydrobromide salt afforded γ-lactam in 71% yield, as a 1:1 ratio of diastereomers. The reaction of aldehyde and amine to afford the γ-lactam needed an excess of amine because of the competing reductive amination with the benzaldehyde side-product. It should be reported, however,

7.6 Synthesis of Six-Membered *N,N*-Heterocycles

that in this case this secondary reductive amination procedure transformed excess amine into its *N*-benzyl derivative, which was the immediate precursor of amine. The use of 1.5 eq. amine in the reductive amination of aldehyde afforded *N*-benzylamine in 24% yield (72% recovery of excess amine, which was recycled back to amine thereby improving the total efficiency of the procedure), in addition to lactam in 81% yield. The thiolactam was synthesized by the reaction of γ-lactam with 0.55 eq. LR in refluxing toluene. The thiolactam was isolated in 56% yield, and the dithio compound in 10% yield. The yield of thiolactam was found to be higher because the methoxycarbonyl group sterically shielded the phthaloyl group, thus decelerating the synthesis of dithio compound. Initial efforts to remove the phthaloyl group of thiolactam by hydrazinolysis with hydrazine hydrate in refluxing C_2H_5OH failed, and the starting compound was recovered. After experimentation with many approaches, it was observed that the reaction of thiolactam with anhydrous hydrazine in methanol afforded free amine in satisfactory yield. With the functionalized aminopropyl-thiolactam in hand, formation of bicyclic amidine was examined. The reaction of thiolactam with $HgCl_2$ in refluxing tetrahydrofuran proceeded slowly, with some starting compound still remaining after 2 d. Therefore, dioxane was replaced with tetrahydrofuran, and the reaction utilizing dioxane as solvent completed in 24 h. The purification of amidine proved problematic with the solvent system (chloroform/methanol/trifluoroacetic acid). The confusion was solved using reverse-phase column chromatography or normal-phase silica chromatography eluting with dichloromethane/methanol/*i*-propylamine, with the amidine isolated in 62% yield as a 1:1 mixture of diastereomers (Scheme 7.21) [52].

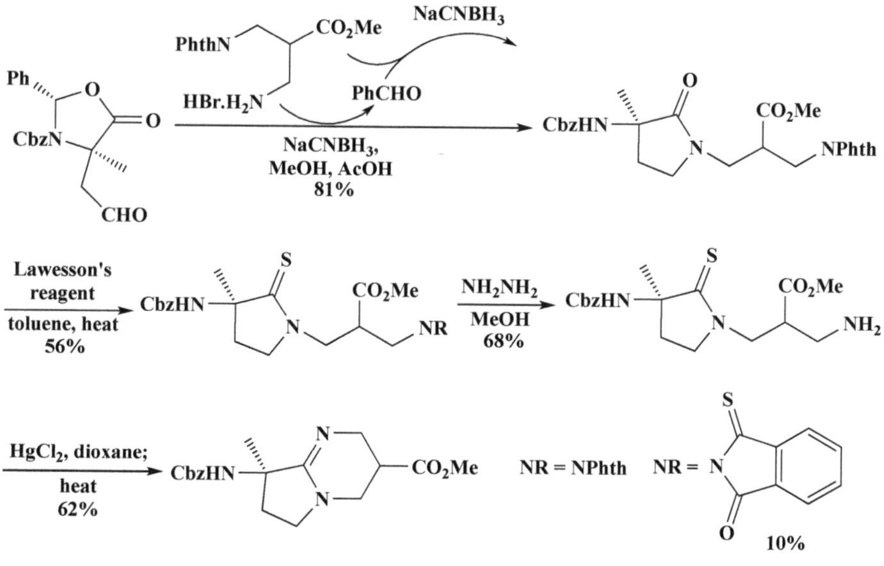

Scheme 7.21 Synthesis of amidine

An elimination of phthaloyl group from γ-lactam occurred through the reaction with hydrazine hydrate in C_2H_5OH to produce the amine. The subsequent efforts to cyclodehydrate the aminopropyl-γ-lactam to bicyclic amidine, including reflux in high boiling point solvents with catalytic acid, and through the O-silyl- and O-alkyl-imidates, were not successful [53]. Accordingly, γ-lactam was transformed to thiolactam by reaction with LR [6]. The phthalimidopropyl thiolactam was isolated in moderate yield, together with different amounts of side-product, assumed to be the dithio compound. Different conditions were applied in an effort to increase the yield of thiolactam, however, variation of amount of LR and the reaction time not resulted in significant improvements in the yield of thiolactam (optimum 40%, with 20% unreacted starting compound). The reaction of thiolactam with hydrazine hydrate afforded free amine in 77% yield. The amine was then reacted with $HgCl_2$ in tetrahydrofuran [54, 55] to afford the bicyclic amidine in 67% yield. The amidine was unstable as its free base and was thus purified and stored as its trifluoroacetate salt (Scheme 7.22) [52].

The enantiopure DBN-analogues were synthesized from easily accessible starting compounds. The (S)-malic acid was transformed into N-(2-cyanoethyl)imide by a one-pot procedure [56]. The regioselective reduction followed by acetylation afforded diacetylpyrrolidinone as a 5:1 (cis/trans) mixture. The extra stereocenter was then removed by triethylsilane reduction of N-acyliminium ion produced from diacetylpyrrolidinone utilizing boron trifluoride etherate. The protective group interconversion took place uneventfully to afford the thexyldimethylsilyl ether in 63% yield from diacetylpyrrolidinone. The extensive experimentation ultimately showed that the best approach for the reduction of the nitrile function was one described by Echavarren and coworkers [57]. This approach included the reaction with 2 eq. $COCl_2$ and a large excess of sodium borohydride, added in portions, to CH_3OH as

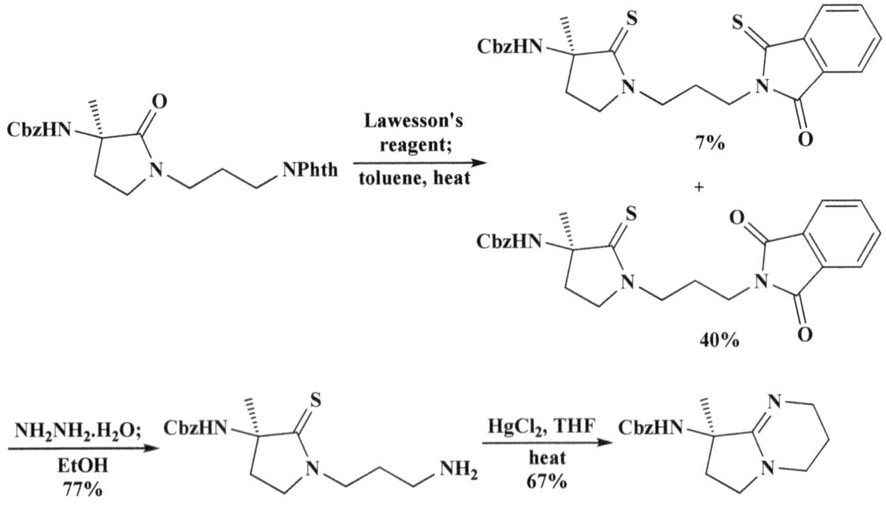

Scheme 7.22 Synthesis of bicyclic amidine

7.6 Synthesis of Six-Membered N,N-Heterocycles

the solvent containing 3% NH_3 to avoid the synthesis of dimers [58]. The amine was protected with t-Boc function to isolate the reduction product from xyldimethylsilyl ether. The amidine ring was closed under mild conditions, and for this purpose, the CO function was activated by treatment with LR to thiolactam and neat MeI to afford the methylthioininium salt. Further, the amine was deprotected with TFA to NH_4^+ salt. An immediate cyclization took place to afford the amidine when the solution of NH_4^+ salt was made alkaline with sodium hydroxide (Scheme 7.23) [59, 60].

The (S)-pyroglutamic acid was transformed to silyl-protected 5-(hydroxymethyl)pyrrolidinone in three steps [61]. Michael addition to acrylonitrile provided nitrile in high yield [62]. The reduction of nitrile function, protection, and thiolactam synthesis generated thione, which at this stage was desilylated utilizing fluoride to alcohol. The cyclization of alcohol proceeded to synthesize the amidine (Scheme 7.24) [60, 63].

Both alcohols produced amidines (Scheme 7.25). The OH groups were protected as t-butyldimethylsilyl ethers with TBDSMOTf, followed by same series of steps to give the hydroxyamidines in comparable yields [64].

The formation of target molecule, hydroxyamidine, is depicted in Scheme 7.26. The reaction of sulfone with allyltrimethylsilane and boron trifluoride etherate afforded allylated lactam in quantitative yield. This reaction proceeded through N-acyliminium ion [65]. The alcohol group was introduced by ozonolysis of the double bond and subsequent in situ reduction with sodium borohydride to provide the alcohol

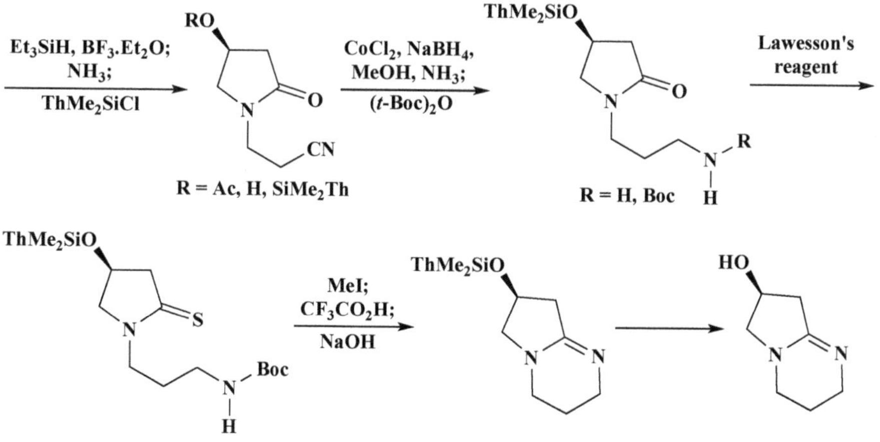

Scheme 7.23 Synthesis of amidine

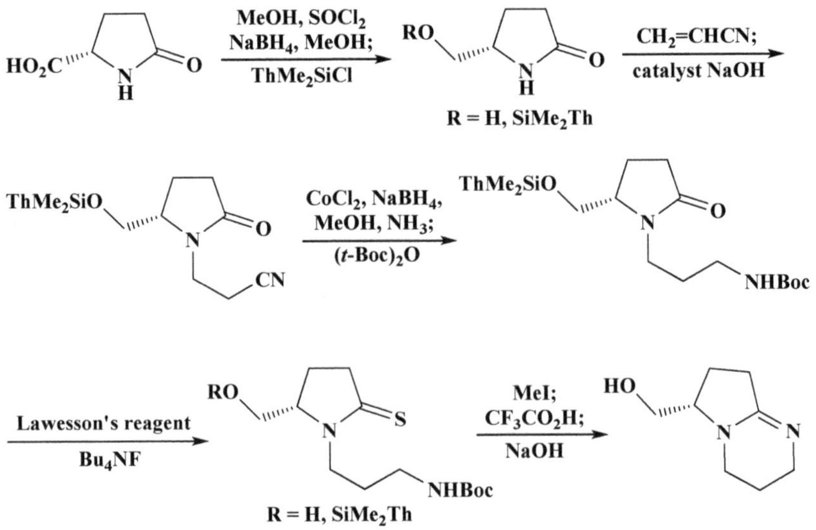

Scheme 7.24 Synthesis of amidine

in 81% yield. The primary alcohol was protected with *t*-butyldimethylsilyl group to afford the *t*-butyldimethylsilyl-substituted alcohol in 95% yield, and then the amidine functionality was incorporated [60]. This involved a series of reactions, the first of which was the reduction of nitrile function. The reaction of *t*-butyldimethylsilyl-substituted alcohol with 2 eq. cobalt(II) chloride and an excess of sodium borohydride in methanol/ammonia (to avoid the synthesis of dimers) [66] provided amine which was not purified, but immediately protected with a *t*-Boc group to give the *t*-Boc-protected amine in 58% yield after purification by column chromatography. A direct cyclization to lactam proved impossible at this stage. Therefore, a mild cyclization pathway was followed. The reaction of *t*-butyldimethylsilyl-substituted alcohol with LR [57] gave thiolactam in 81% yield. Later when the alcohol group was deprotected with tetrabutylammonium fluoride, the *t*-Boc-protected amine alcohol was treated with neat methyl iodide to afford the methylthioininium salt. The *t*-Boc group was then removed using TFA; cyclization of formed amine onto the activated lactam was facilitated by an excess of triethylamine to provide the hydroxyamidine in 81% yield [59, 64, 67, 68].

Many naturally occurring alkaloids contain a benzodiazepine ring fused with a quinazoline ring. The benzodiazepine–quinazoline framework has been synthesized following diverse approaches. In 1987, the formation of asperlicins C and E was first reported by Bock and coworkers [69]. In their methodology, the quinazoline ring was constructed from 1,4-benzodiazepine-2,5-dione and methyl anthranilate (Scheme 7.27). The benzodiazepinedione was transformed into methyl imino thioether to activate the C2-position.

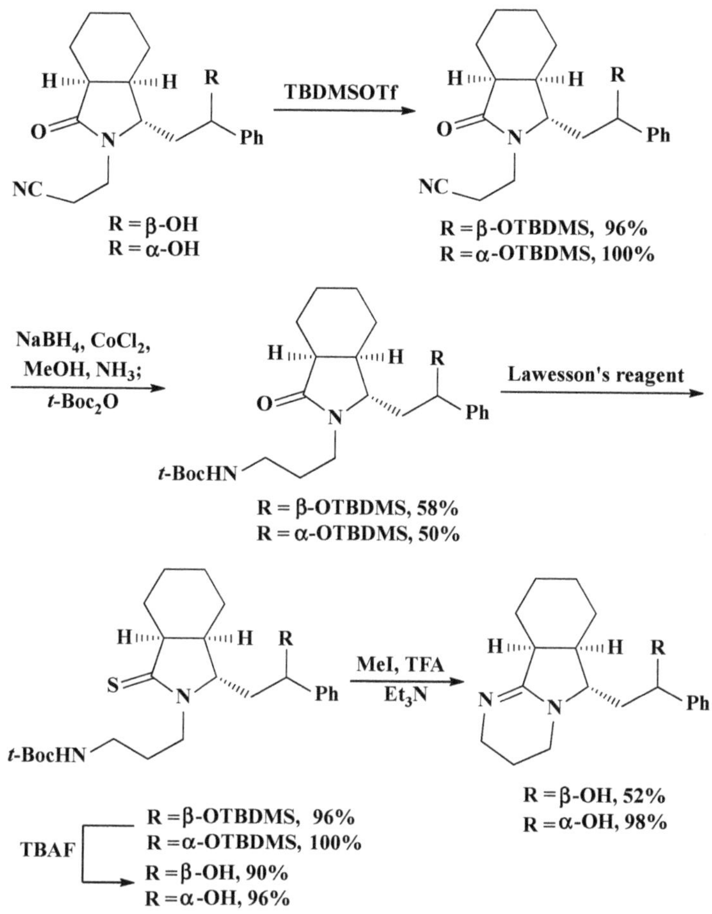

Scheme 7.25 Synthesis of amidines

7.7 Synthesis of Seven-Membered *N*-Heterocycles

The reaction of amide with Lawesson's reagent afforded ring-closed imide (Scheme 7.28) [13, 28].

Kamal et al. [70] prepared DC-81 analogue, and the thione derivative, as depicted in the Scheme 7.29. The synthesis began by the reaction of proline with activated 3,4,5-substituted 2-nitrobenzoic acid to afford the coupled product. The esterification process followed by reduction with diisobutylaluminum hydride afforded (2*S*)-*N*-(2-nitrobenzoyl)-pyrrolidine-2-carboxaldehydes. An intramolecular cyclization with Fe and a mixture of AcOH/tetrahydrofuran as solvent afforded DC-81 analogue in 65–75% yield. The thione derivative was synthesized from (2*S*)-*N*-(2-nitrobenzoyl)-pyrrolidine-2-carboxaldehydes by reaction with LR. The thione

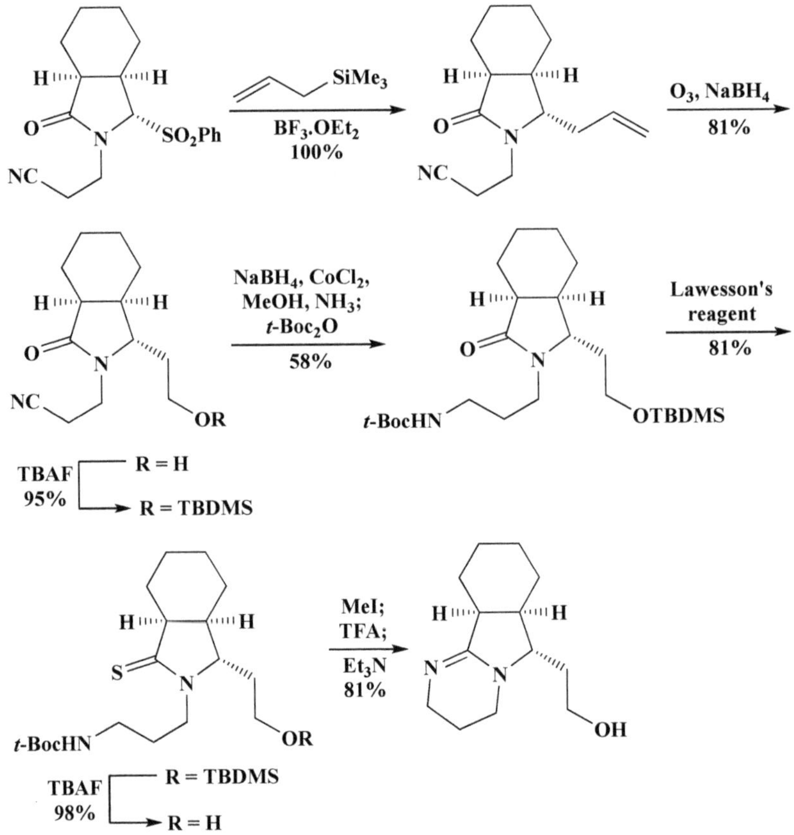

Scheme 7.26 Synthesis of hydroxyamidine

derivative underwent intramolecular cyclization with Fe in AcOH/tetrahydrofuran to yield the benzodiazepines.

7.8 Synthesis of Seven-Membered *O*-Heterocycles

Nicolaou et al. [71] reported a photochemical reaction of thiocarbonyl compounds to afford the oxepine intermediates as part of the attempt to prepare brevitoxin B. Scheme 7.30 describes a new approach for the synthesis of oxepane systems from acyclic substrates. The diester was transformed to its dithiono counterpart under standard Lawesson's conditions [6, 72, 73], and the dithiono compound was irradiated, probably producing the radical species, and then the 1,2-dithietane system [74–76]. Under irradiation conditions, the 1,2-dithietane system lost sulfur to give the oxepene which was then regioselectively hydrolyzed to oxepanone [77]. Despite

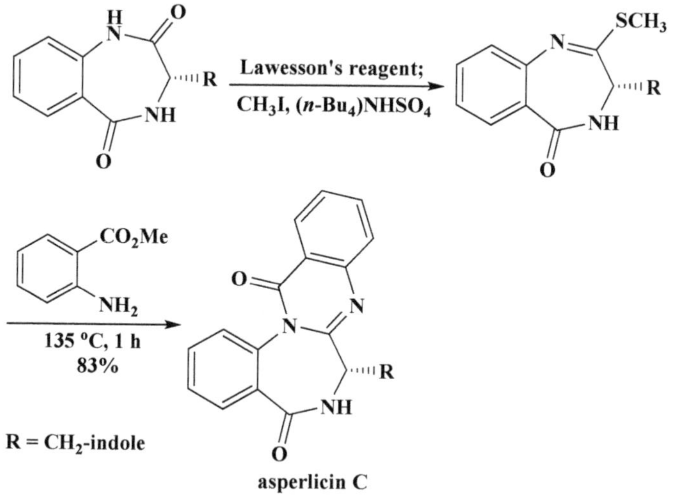

7.8 Synthesis of Seven-Membered O-Heterocycles

Scheme 7.27 Synthesis of benzodiazepinediones

Scheme 7.28 Synthesis of ring closed imide

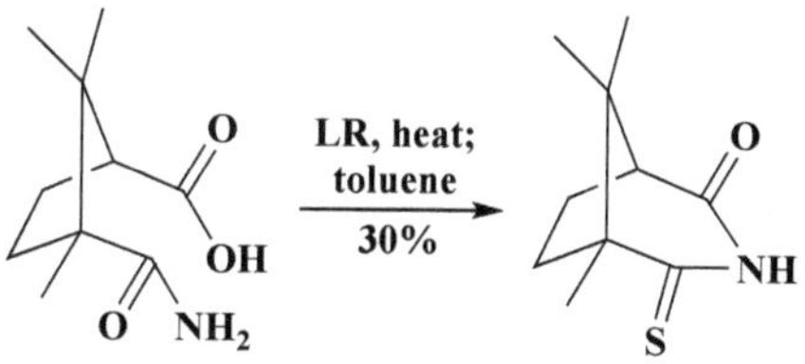

the mixture of isomers in oxepene, the final product oxepanone was formed, through equilibration under applied conditions, as single stereoisomer with two stereocenters flanking the carbonyl group firmly established on pseudo-equatorial positions. Many dithionoesters acted as precursors to a series of oxepenes and oxepanones. The thionations were performed in 50–55% yield utilizing excess of LR and 1,1,3,3-tetramethylthiourea at 150 to 160 °C. The final hydrolytic step was completed in 75–95% yield either under acidic conditions (hydrochloric acid–water) or with fluoride (tetra-n-butylammonium fluoride-tetrahydrofuran). This approach was also possibly utilized in a late stage connection of two complex fragments in the synthesis of a natural product.

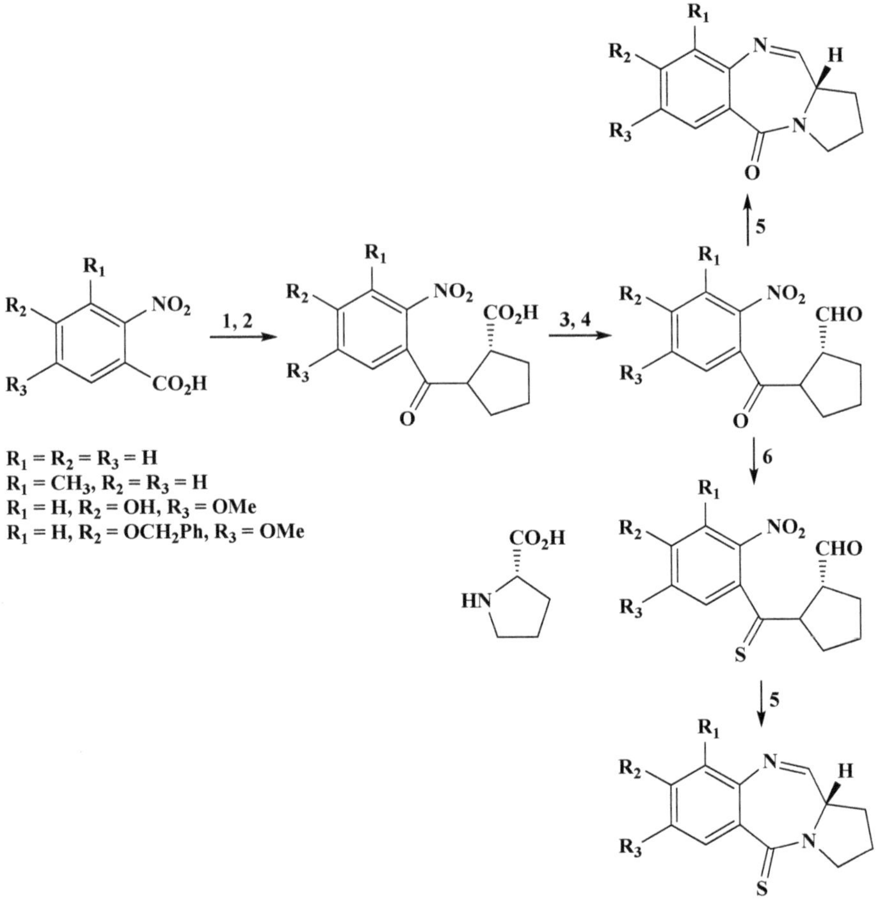

$R_1 = R_2 = R_3 = H$
$R_1 = CH_3, R_2 = R_3 = H$
$R_1 = H, R_2 = OH, R_3 = OMe$
$R_1 = H, R_2 = OCH_2Ph, R_3 = OMe$

Reagents and conditions: (1) $SOCl_2$, benzene, rt, 3-4 h, (2) L-proline, Et_3N, THF, 0 °C, 1 h, (3) H^+, MeOH, reflux, 2-3 h, (4) DIBAL-H, DCM, -78 °C, 45 min, (5) Fe, AcOH, THF, rt, 3-6 h, (6) Lawesson's reagent, toluene, 80 °C, 2-3 h.

Scheme 7.29 Synthesis of benzodiazepines

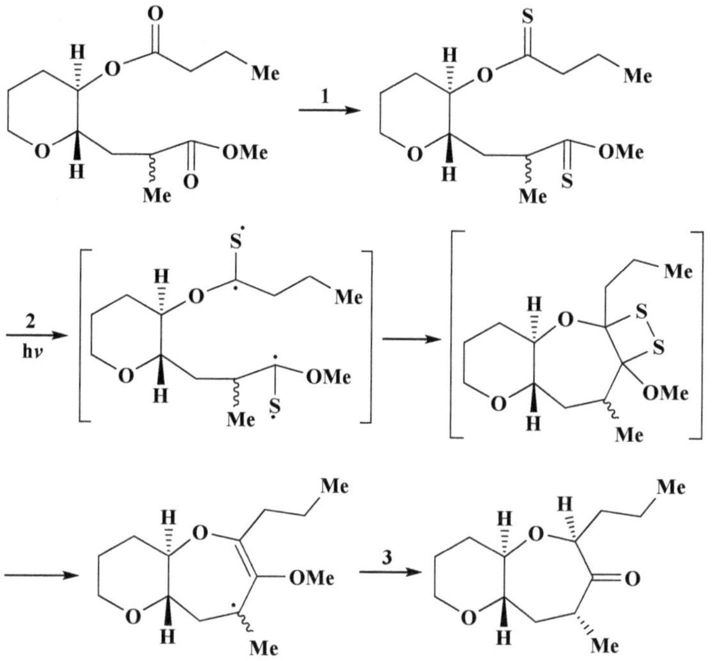

7.8 Synthesis of Seven-Membered *O*-Heterocycles

Reagents and conditions: (1) 3 eq. Lawesson's reagent, 3 eq. 1,1,3,3-tetramethylthiourea, xylene, 160 °C, 2 h, 47%, (2) hν, Hanovia 450 W UV lamp, pyrex filter, toluene, 70 °C, 2 h, 63%, (3) 2 M HCl, 25 °C, 2 h, 80%.

Scheme 7.30 Synthesis of pyranooxepinone

References

1. (a) N. Kaur. 2018. Mercury-catalyzed synthesis of heterocycles. Synth. Commun. 48: 2715–2749. (b) N. Kaur. 2015. Greener and expeditious synthesis of fused six-membered *N,N*-heterocycles using microwave irradiation. Synth. Commun. 45: 1493–1519. (c) N. Kaur. 2019. Seven-membered *N*-heterocycles: Metal and non-metal assisted synthesis. Synth. Commun. 49: 987–1030. (d) N. Kaur, P. Bhardwaj, M. Devi, Y. Verma and P. Grewal. 2019. Synthesis of five-membered *O,N*-heterocycles using metal and non-metal. Synth. Commun. 49: 1345–1384. (e) N. Kaur. 2013. Expedient protocol for the installation of thiadiazole on 2-position of 1,4-benzodiazepin-5-carboxamide through a phenoxyl spacer. Int. J. Pharm. Biol. Sci. 4: 366–373. (f) N. Kaur and D. Kishore. 2013. Application of chalcones in heterocycles synthesis: Synthesis of 2-(isoxazolo, pyrazolo and pyrimido) substituted analogues of 1,4-benzodiazepin-5-carboxamides linked through an oxyphenyl bridge. J. Chem. Sci. 125: 555–560.
2. (a) N. Kaur. 2018. Photochemical-mediated reactions in five-membered *O*-heterocycles synthesis. Synth. Commun. 48: 2119–2149. (b) N. Kaur. 2019. Gold and silver assisted synthesis of five-membered oxygen and nitrogen containing heterocycles. Synth. Commun. 49: 1459–1485. (c) N. Kaur. 2019. Synthesis of six-membered *N*-heterocycles using ruthenium catalysts. Catal. Lett. 14: 1513–1539. (d) N. Kaur. 2018. Green synthesis of three to five-membered *O*-heterocycles using ionic liquids. Synth. Commun. 48: 1588–1613. (e) N. Kaur. 2018. Synthesis of seven- and higher-membered nitrogen-containing heterocycles using

photochemical irradiation. Synth. Commun. 48: 2815–2849. (f) N. Kaur. 2018. Ruthenium-catalyzed synthesis of five-membered *O*-heterocycles. Inorg. Chem. Commun. 99: 82–107. (g) N. Kaur, P. Sharma, R. Sirohi and D. Kishore. 2012. Microwave assisted synthesis of 2-hetryl amino substituted novel analogues of 1,4-benzodiazepine-5-piperidinyl carboxamides. Arch. Appl. Sci. Res. 4: 2256–2260.
3. (a) N. Kaur. 2018. Solid-phase synthesis of sulfur-containing heterocycles. J. Sulfur Chem. 39: 544–577. (b) N. Kaur. 2018. Photochemical reactions as key steps in five-membered *N*-heterocycles synthesis. Synth. Commun. 48: 1259–1284. (c) N. Kaur. 2018. Recent developments in the synthesis of nitrogen-containing five-membered polyheterocycles using rhodium catalysts. Synth. Commun. 48: 2457–2474. (d) N. Kaur, Y. Verma, P. Grewal, P. Bhardwaj and M. Devi. 2019. Application of titanium catalysts for the syntheses of heterocycles. Synth. Commun. 49: 1847–1894. (e) N. Kaur. 2019. Cobalt-catalyzed C-N, C-O, C-S bond formation: Synthesis of heterocycles. J. Iran. Chem. Soc. 16: 2525–2553. (f) N. Kaur. 2019. Nickel catalysis: Six membered heterocycle syntheses. Synth. Commun. 49: 1103–1133. (g) R. Tyagi, N. Kaur, B. Singh and D. Kishore. 2013. A noteworthy mechanistic precedence in the exclusive formation of one regioisomer in the Beckmann rearrangement of ketoximes of 4-piperidones annulated to pyrazolo-indole nucleus by organocatalyst derived from TCT and DMF. Synth. Commun. 43: 16–25.
4. M.S.J. Foreman and J.D. Woollins. 2000. Organo-P-S and P-Se heterocycles. J. Chem. Soc. Dalton Trans. 10: 1533–1543.
5. G. Le, N. Vandegraaff, D.I. Rhodes, E.D. Jones, J.A.V. Coates, N. Thienthong, L.J. Winfield, L. Lu, X. Li, C. Yu, X. Feng and J.J. Deadman. 2010. Design of a series of bicyclic HIV-1 integrase inhibitors. Part 2: Azoles: Effective metal chelators. Bioorg. Med. Chem. Lett. 20: 5909–5912.
6. B.S. Pedersen and S.O. Lawesson. 1979. Studies on organophosphorus compounds - XXVIII. Tetrahedron 35: 2433–2437.
7. B.S. Pedersen, S. Scheibye, N.H. Nilson and S.-O. Lawesson. 1978. Studies on organophosphorus compounds XX. Syntheses of thioketones. Bull. Soc. Chim. Belg. 87: 223–228.
8. S. Scheibye, B.S. Pedersen and S.-O. Lawesson. 1978. Studies on organophosphorus compounds XXIII. Synthesis of salicylthioamides and 4*H*-1,3,2-benzoxaphoshorine derivatives from the dimer of *p*-methoxyphenylthionophosphine sulfide and salicylamides. Bull. Soc. Chim. Belg. 87: 299–306.
9. B.S. Pedersen, S. Scheibye, N.H. Nilson, K. Clausen and S.-O. Lawesson. 1978. Studies on organophosphorus compounds XXII. The dimer of *p*-methoxyphenylthionophosphine sulfide as thiation reagent. A new route to *O*-substituted thioesters and dithioesters. Bull. Soc. Chim. Belg. 87: 293–297.
10. L. Henry. 1869. About a new form of formation and presentation of nitriles. Ann. Chem. Pharm. 152: 148–152.
11. J. Wislicenus. 1869. Ueber die beta-oxybuttersäure. Ann. Chem. Pharm. 149: 205–215.
12. V. Polshettiwar. 2004. Phosphorus pentasulfide (P_4S_{10}). Synlett 12: 2245–2246.
13. T. Ozturk, E. Ertas and O. Mert. 2007. Use of Lawesson's reagent in organic syntheses. Chem. Rev. 107: 5210–5278.
14. S.V. Kumar, G. Parameshwarappa and H. Ila. 2013. Synthesis of 2,4,5-trisubstituted thiazoles via Lawesson's reagent-mediated chemoselective thionation-cyclization of functionalized enamides. J. Org. Chem. 78: 7362–7369.
15. D.R. Shridhar, M. Jogibhukta, P.S. Rao and V.K. Handa. 1982. An improved method for the preparation of 2,5-disubstituted thiophenes. Synthesis 12: 1061–1062.
16. F. Freeman, D.S.H.L. Kim and E. Rodriguez. 1992. Preparation of 1,4-diketones and their reactions with bis(trialkyltin) or bis(triphenyltin) sulfide-boron trichloride. J. Org. Chem. 57: 1722–1727.
17. A. Merz and F. Ellinger. 1991. Convenient synthesis of α-terthienyl and α-quinquethienyl via a Friedel-Crafts route. Synthesis 6: 462–464.
18. C.-C. Lin, W. Tantisantisom and S.R. McAlpine. 2013. Total synthesis and biological activity of natural product urukthapelstatin A. Org. Lett. 15: 3574–3577.

19. A. Hantzsch and J.H. Weber. 1887. Ueber verbindungen des thiazols (pyridins der thiophenreihe). Berichte der Deutschen Chemischen Gesellschaft 20: 3118–3132.
20. J.A. Seijas, M.P. Vázquez-Tato, M.R. Carballido-Reboredo, C.J. Crecente and L. Romar. 2007. Lawesson's reagent and microwaves: A new efficient access to benzoxazoles and benzothiazoles from carboxylic acids under solvent-free conditions. Synlett 2: 313–316.
21. S.W. Pelletier, J. Nowacki and N.V. Mody. 1979. A convenient method for constructing oxazolidine and thiazolidine rings in C20-diterpenoid alkaloid derivatives. Synth. Commun. 9: 201–206.
22. H. Saito, S. Kobayashi, Y. Uosaki, A. Sato, K. Fujimoto, K. Miyoshi, T. Ashizawa, M. Morimoto and T. Hirata. 1990. Synthesis and biological evaluation of quinocarcin derivatives. Chem. Pharm. Bull. 38: 1278–1285.
23. M.S. Anson, J.P. Graham and A.J. Roberts. 2011. Development of a fully telescoped synthesis of the S1P1 agonist GSK1842799. Org. Process Res. Dev. 15: 649–659.
24. H. Deng, S.G. Bernier, E. Doyle, J. Lorusso, B.A. Morgan, W.F. Westlin and G. Evindar. 2013. Discovery of clinical candidate GSK1842799 as a selective S1P1 receptor agonist (prodrug) for multiple sclerosis. ACS Med. Chem. Lett. 4: 942–947.
25. E. Jones, N. Vandegraff, G. Le, N. Choi, W. Issa, K. Macfarlane, N. Thienthong, L. Winfiled, J. Coates, L. Lu, X. Li, X. Feng, C. Yu, D. Rhodes, J. Deadman. 2010. Design of a series of bicyclic HIV-1 integrase inhibitors. Part 1: Selection of the scaffold. Bioorg. Med. Chem. Lett. 20: 5913–5917.
26. P.A. Bethel, S. Gerhardt, E.V. Jones, P.W. Kenny, G.I. Karoutchi, A.D. Morley, K. Oldham, N. Rankine, M. Augustin, S. Krapp, H. Simader and S. Steinbacher. 2009. Design of selective cathepsin inhibitors. Bioorg. Med. Chem. Lett. 19: 4622–4625.
27. Z. Ribkovskaia and F. Macaev. 2012. Synthesis of mono-substituted and symmetrically 2,5-disubstituted 1,3,4-oxadiazoles. Chem. J. Moldova Gen. Ind. Ecol. Chem. 7: 136–142.
28. M.J. Milewska, T. Bytner and T. Polonski. 1996. Synthesis of lactams by regioselective reduction of cyclic dicarboximides. Synthesis 12: 1485–1488.
29. P. Duhamel, M. Kotera and B. Marabout. 1991. Stereoselective synthesis of (R)-7-pentylhexahydroazepine-2-thione: Formal precursor of (−)-perhydrohistrionicotoxin. Tetrahedron: Asymmetry 2: 203–206.
30. A. Sinclair and R.A. Stockman. 2007. Thirty-five years of synthetic studies directed towards the histrionicotoxin family of alkaloids. Nat. Prod. Rep. 24: 298–326.
31. P. Duhamel, M. Kotera, T. Monteil and B. Marabout. 1989. Highly stereoselective ring contraction of heterocyclic enamines: Total synthesis of perhydrohistrionicotoxin and its 2,6-epimer. J. Org. Chem. 54: 4419–4425.
32. A.J. Pearson and P. Ham. 1983. Flexible synthetic approach to histrionicotoxin congeners. Formal total synthesis of (±)-perhydrohistrionicotoxin. J. Chem. Soc. Perkin Trans. 0: 1421–1425.
33. J.P. Michael, C.B. de Koning, D. Gravestock, G.D. Hosken, A.S. Howard, C.M. Jungmann, R.W.M. Krause, A.S. Parsons, S.C. Pelly and T.V. Stanbury. 1999. Enaminones: Versatile intermediates for natural product synthesis. Pure Appl. Chem. 71: 979–988.
34. J.P. Michael, C.B. de Koning, T.J. Malefetse and I. Yillah. 2004. Preparation and reductive reactions of vinylogous sulfonamides (β-sulfonyl enamines), and application to the synthesis of indolizidines. Org. Biomol. Chem. 2: 3510–3517.
35. J.P. Michael and D. Gravestock. 2000. Vinylogous urethanes in alkaloid synthesis. Applications to the synthesis of racemic indolizidine 209B and its (5R*,8S *,8aS *)-(±) diastereomer, and to (-)-indolizidine 209B. J. Chem. Soc. Perkin Trans. 1 12: 1919–1928.
36. S.G. Davies and O. Ichihara. 1991. Asymmetric synthesis of R-β-amino butanoic acid and S-β-tyrosine: Homochiral lithium amide equivalents for Michael additions to α,β-unsaturated esters. Tetrahedron: Asymmetry 2: 183–186.
37. J.F. Costello, S.G. Davies and O. Ichihara. 1994. Origins of the high stereoselectivity in the conjugate addition of lithium(α-methylbenzyl)benzylamide to t-butyl cinnamate. Tetrahedron: Asymmetry 5: 1999–2008.

38. M. Roth, P. Dubs, E. Gotschi and A. Eschenmoser. 1971. Sulfidkontraktion via alkylative kupplung: Eine methode zur darstellung von β-dicarbonylderivaten. Über synthetische methoden, 1. Mitteilung. Helv. Chim. Acta 54: 710–734.
39. J.P. Michael, C.B. de Koning and C.W. van der Westhuyzen. 2005. Studies towards the enantioselective synthesis of 5,6,8-trisubstituted amphibian indolizidine alkaloids via enaminone intermediates. Org. Biomol. Chem. 3: 836–847.
40. R. Zhang, Y. Liang, G. Zhou, K. Wang and D. Dong. 2008. Ring-enlargement of dimethylaminopropenoyl cyclopropanes: An efficient route to substituted 2,3-dihydrofurans. J. Org. Chem. 73: 8089–8092.
41. R. Zhang, Y. Zhou, Y. Liang, Z. Jiang and D. Dong. 2009. Ring-opening/recyclization reactions of (dimethylamino)propenoyl-substituted cyclopropanes: Facile synthesis of halogenated pyridin-2(1H)-ones. Synthesis 15: 2497–2500.
42. P. Gopinath and S. Chandrasekaran. 2011. Synthesis of functionalized dihydrothiophenes from doubly activated cyclopropanes using tetrathiomolybdate as the sulfur transfer reagent. J. Org. Chem. 76: 700–703.
43. Z. Zhang, Q. Zhang, S. Sun, T. Xiong and Q. Liu. 2007. Domino ring-opening/recyclization reactions of doubly activated cyclopropanes as a strategy for the synthesis of furoquinoline derivatives. Angew. Chem. Int. Ed. 46: 1726–1729.
44. P. Huang, R. Zhang, Y. Liang and D. Dong. 2012. Lawesson's reagent-initiated domino reaction of aminopropenoyl cyclopropanes: Synthesis of thieno[3,2-c]pyridinones. Org. Biomol. Chem. 10: 1639–1644.
45. K. Yamagata, M. Takaki, K. Ohkubo and M. Yamazaki. 1993. Studies on heterocyclic enaminonitriles, XIV. Preparation of dihydrothiophenium-1-bis(ethoxycarbonyl)methylides and their recyclization with a base. Liebigs Ann. Chem. 12: 1263–1267.
46. K. Yamagata, Y. Tomioka, M. Yamazaki, T. Matsuda and K. Noda. 1982. Studies on heterocyclic enaminonitriles. II. Synthesis and aromatization of 2-amino-3-cyano-4,5-dihydrothiophenes. Chem. Pharm. Bull. 30: 4396-4401.
47. E. Nagashima, K. Suzuki and M. Sekiya. 1982. Thermal rearrangements of allyl 2,2-dichloro-, 1,2-dichloro- and 1,2,2-trichloro-substituted vinyl sulfides. Chem. Pharm. Bull. 30: 4384–4395.
48. P.A. Rosy, W. Hoffmann and N. Müller. 1980. Aromatization of dihydrothiophenes. Thiophenesaccharin: A sweet surprise. J. Org. Chem. 45: 617–620.
49. O. Bassas, N. Llor, M.M.M. Santos, R. Griera, E. Molins, M. Amat and J. Bosch. 2005. Biogenetically inspired enantioselective approach to indolo[2,3-a]- and benzo[a]quinolizidine alkaloids from a synthetic equivalent of secologanin. Org. Lett. 7: 2817–2820.
50. M.L. Meketa and S.M. Weinreb. 2007. A new total synthesis of the zinc matrix metalloproteinase inhibitor ageladine A featuring a biogenetically patterned 6pi-2-azatriene electrocyclization. Org. Lett. 9: 853–855.
51. R.J. Carra, M.T. Epperson and D.Y. Gin. 2008. Application of an intramolecular dipolar cycloaddition to an asymmetric synthesis of the fully oxygenated tricyclic core of the stemofoline alkaloids. Tetrahedron 64: 3629–3641.
52. C.A. Hutton and P.A. Bartlet. 2007. Preparation of diazabicyclo[4.3.0]nonene-based peptidomimetics. J. Org. Chem. 72: 6865–6872.
53. H. Oediger, F. Moller and K. Eiter. 1972. Bicyclic amidines as reagents in organic syntheses. Synthesis 11: 591–598.
54. M.G. Bock, R.M. DiPardo, B.E. Evans, K.E. Rittle, R.M. Freidinger, R.S.L. Chang and V.J. Lottit. 1988. Cholecystokinin antagonists. Synthesis and biological evaluation of 3-substituted 1,4-benzodiazepin-2-amines. J. Med. Chem. 31: 264–268.
55. M.B. Foloppe, S. Rault and M. Robba. 1992. Pyrrolo[2,1-c][1,4]benzodiazepines: A mild conversion of thiolactam into amidine. Tetrahedron Lett. 33: 2803–2804.
56. A.R. Chamberlin and J.Y.L. Chung. 1983. Synthesis of optically active pyrrolizidinediols: (+)-Heliotridine. J. Am. Chem. Soc. 105: 3653–3656.
57. A. Echavarren, A. Galan, J. de Mendoza, A. Salmeron and J.-M. Lehn. 1988. Anion-receptor molecules: Synthesis of a chiral and functionalized binding subunit, a bicyclic guanidinium group derived from L- or D-asparagines. Helv. Chim. Acta 71: 685–693.

58. W.J. Klaver, H. Hiemstra and W.N. Speckamp. 1989. Synthesis and absolute configuration of the *Aristotelia* alkaloid *Peduncularine*. J. Am. Chem. Soc. 111: 2588–2595.
59. B. Yde, N.M. Yousif, U. Pedersen, I. Thomsen and S.-O. Lawesson. 1984. Studies on organophosphorus compounds XLVII. Preparation of thiated synthons of amides, lactams and imides by use of some new *P,S*-containing reagents. Tetrahedron 40: 2047–2052.
60. J. Dijkink, K. Eriksen, K. Goubitz, M.N.A. van Zanden and H. Hiemstra. 1996. Synthesis and X-ray crystal structure of (*S*)-9-hydroxymethyl-1,5-diazabicyclo[4.3.0]non-5-ene, an enantiopure DBN-analogue. Tetrahedron: Asymmetry 7: 515–524.
61. S. Saijo, M. Wada, J. Himizu and A. Ishida. 1980. Heterocyclic prostaglandins. V. Synthesis of (12R,15S)-(-)-11-deoxy-8-azaprostaglandin E_1 and related compounds. Chem. Pharm. Bull. 28: 1449–1458.
62. F.G. Fang, M. Prato, G. Kim and S.J. Danishefsky. 1989. The aza-Robinson annulation: An application to the synthesis of iso-A58365A. Tetrahedron Lett. 30: 3625–3628.
63. G. Bertrand, R. Reed, R. Réau and F. Dahan. 1993. DBU and DBN are strong nucleophiles: X-Ray crystal structures of onio- and dionio-substituted phosphanes. Angew. Chem. Int. Ed. 32: 399–401.
64. M. Ostendorf, S. van der Neut, F.P.J.T. Rutjes and H. Hiemstra. 2000. Enantioselective synthesis of hydroxy-substituted DBN-type amidines as potential chiral catalysts. Eur. J. Org. Chem. 1: 105–113.
65. H. Hiemstra and W.N. Speckamp. 1991. Comprehensive organic synthesis. B.M. Trost and I. Fleming (Eds.). Pergamon Press, Oxford, 2: 1047.
66. A.G.M. Barrett. 1991. Comprehensive organic synthesis. B.M. Trost and I. Fleming (Eds.). Pergamon: Oxford, 8: 251.
67. R.J. Steffan, E. Matelan, M.A. Ashwell, W.J. Moore, W.R. Solvibile, E. Trybulski, C.C. Chadwick, S. Chippari, T. Kenney, A. Eckert, L. Borges-Marcucci, J.C. Keith, Z. Xu, L. Mosyaz and D.C. Harnish. 2004. Synthesis and activity of substituted 4-(indazol-3-yl)phenols as pathway-selective estrogen receptor ligands useful in the treatment of rheumatoid arthritis. J. Med. Chem. 47: 6435–6438.
68. J.O. Osby, S.W. Heinzman and B. Ganem. 1986. Studies on the mechanism of transition-metal-assisted sodium borohydride and lithium aluminum hydride reductions. J. Am. Chem. Soc. 108: 67–72.
69. M.G. Bock, R.M. DiPardo, S.M. Pitzenberger, C.F. Homnick, S.M. Springer and R.M. Freidinger. 1987. Total synthesis of nonpeptidal cholecystokinin antagonists from *Aspergillus alliaceus*. J. Org. Chem. 52: 1644–1646.
70. A. Kamal, B.S.P. Reddy and B.S.N. Reddy. 1996. A new facile procedure for the preparation of pyrrolo[2,1-*c*][1,4]benzodiazepines: Synthesis of the antibiotic DC-81 and its thio analogue. Tetrahedron Lett. 37: 2281–2284.
71. K.C. Nicolaou, C.-K. Hwang, M.E. Duggan, D.A. Nugiel, Y. Abe, K.B. Reddy, S.A. Defrees, D.R. Reddy, R.A. Awartani, S.R. Conley, F.P.J.T. Rutjes and E.A. Theodorakis. 1995. Total synthesis of brevetoxin B. 1. First generation strategies and new approaches to oxepane systems. J. Am. Chem. Soc. 117: 10227–10238.
72. K.C. Nicolaou, D.G. McGany, P.K. Somers, B.H. Kim, W.W. Ogilvie, G. Yiannikouros, C.V.C. Prasad, C.A. Veaie and R.R. Hark. 1990. Synthesis of medium-sized ring ethers from thionolactones. Applications to polyether synthesis. J. Am. Chem. Soc. 112: 6263–6276.
73. G. Lajorie, F. Lipine, L. Maziaki and B. Belleau. 1983. Facile regioselective formation of thiopeptide linkages from oligopeptides with new thionation reagents. Tetrahedron Lett. 24: 3815–3818.
74. K.C. Nicolaou, C.-K. Hwang, M.E. Duggan and P.J. Carroll. 1987. Dithiatopazine. The first stable 1,2-dithietane. J. Am. Chem. Soc. 109: 3801–3802.

75. K.C. Nicolaou, C.-K. Hwang, S. Defrees and N.A. Stylianides. 1988. Novel chemistry of dithiatopazine. J. Am. Chem. Soc. 110: 4868–4869.
76. K.C. Nicolaou, S.A. DeFreei, C.-K. Hwang, N. Stylianides, P.J. Cairoll and J.P. Snyder. 1990. Dithiatopazine and related systems. Synthesis, chemistry, X-ray crystallographic analysis, and calculations. J. Am. Chem. Soc. 112: 3029–3039.
77. K.C. Nicolaou, C.-K. Hwang and D.A. Nugiel. 1988. A photolytic entry into oxepane systems. Angew. Chem. Int. Ed. 27: 1362–1364.

Chapter 8
Phosphorus Pentasulfide in Heterocycle Synthesis

8.1 Introduction

The presence of heterocyclic compounds in all types of organic molecules of interest in pharmacology, biology, electronics, optics, material sciences, and so on is very well known. The heterocyclic chemistry is an integral part of chemical sciences and forms a considerable part of the modern researches that are ongoing throughout the world [1–3].

The P_4S_{10} is among the oldest thionating agents for organic compounds. Even it is utilized for the synthesis of the most utilized thionating agent, i.e., Lawesson's reagent. For sulfur chemistry, the P_4S_{10} has now been an indispensable reagent, specifically for transforming almost all types of oxo groups to thio groups, which are important functional groups to conduct different organic reactions or for utilization as end products in medicinal, material, chemistry, etc. Almost all types of heterocycles containing sulfur atom(s) are synthesized using P_4S_{10}. Its range varies from thiophene to thiazole, dithiazole, imidazoline, thiazoline, thiadiazole, thiazine, and pyrimidine. It finds broad applications in the thionation reactions of pyrimidines, purines, and nucleosides. Another important reaction of P_4S_{10} is the reduction of sulfoxides to sulfides. The P_4S_{10}, like Lawesson's reagent, surprises by offering unexpected reactions, the results of which led chemists to explore novel approaches and reactions. It could be a benefit to the synthetic researchers to use reagents, phosphorus pentasulfide and Lawesson's reagent, in the synthetic pathways to provide the surprising products and best results [4–10].

A sulfur heteroatom has been in the interest of many groups involved in the formation of organic compounds. This has been carried out mainly through the reactions of thionating agents, among the oldest and the most important ones of which is P_4S_{10}. In organic syntheses, phosphorus pentasulfide has been utilized broadly for a wide range of purposes, primarily as thionating agent of organic (also inorganic) compounds and for the formation of different heterocyclic compounds including dithiins, thiophenes, thiazoles, dithiazoles, thiazolines, thiadiazoles, imidazolines, thiazines, pyrimidines, and imides. The thionations of peptides, nucleosides, purines,

© The Author(s), under exclusive license to Springer Nature Singapore Pte Ltd. 2022
N. Kaur, *Lawesson's Reagent in Heterocycle Synthesis*,
https://doi.org/10.1007/978-981-16-4655-3_8

and pyrimidines, and reductions of sulfoxides to sulfides are performed with P_4S_{10}. The P_4S_{10} is a commercially accessible compound, and not only utilized for industrial applications like formation of additives for insecticides, flotation agents, oil, and lubricants, etc. It is also utilized for the thionation reactions and synthesis of heterocyclic compounds [11–18].

8.2 P_2S_5 in Heterocycle Synthesis

The thioanilides are reacted with $NaNO_2$ either in glacial CH_3COOH or in 70% CH_3COOH to synthesize the nitrobenzotriazoles in good yield (72–83%). The stability of nonbenzenoid thiocarbonylbenzotriazoles is generally poor. The aliphatic nitrated thiocarbonylbenzotriazoles were synthesized by Shalaby et al. [19, 20]. Possibly, the stability improved with the presence of EWG, i.e., nitro in the benzotriazole ring and resulted in the isolation of aliphatic thiocarbonylbenzotriazoles. Many aromatic and aliphatic thiocarbonyl-1H-6-nitrobenzotriazoles have been synthesized by following the approach of Katritzky et al. [21] (Scheme 8.1). The amides (83–99%) were afforded regioselectively by the reaction of 4-nitro-1,2-phenylendiamines with acid chlorides. The nucleophilicity of amino group in the *para*-position was lowered by resonance and inductive effect of the nitro group, leaving the *m*-amino group to attack the carbonyl of acid chloride. The intermediate amides were stirred with phosphorus pentasulfide at rt to afford the thiocarbonyl-1H-6-nitrobenzotriazoles.

The triazinone derivative was transformed into its thio analogue utilizing phosphorus pentasulfide, which was *S*-methylated and subsequently hydrazinated followed by cyclization through its reflux with HCO_2H to synthesize the 4-methyl-1,2,4-triazolo[4′,3′:4,5][1,2,4]triazino[2,3-*a*]benzimidazole (Scheme 8.2) [22–24].

The generality of this reaction was evaluated under optimized reaction conditions. The furan products were afforded with high efficiency by the treatment of a series of β-ketone allenic sulfides (R = methyl) with phosphorus pentasulfide. This reaction was not affected by the substituents on Ar. However, β-aldehyde allenic sulfide under

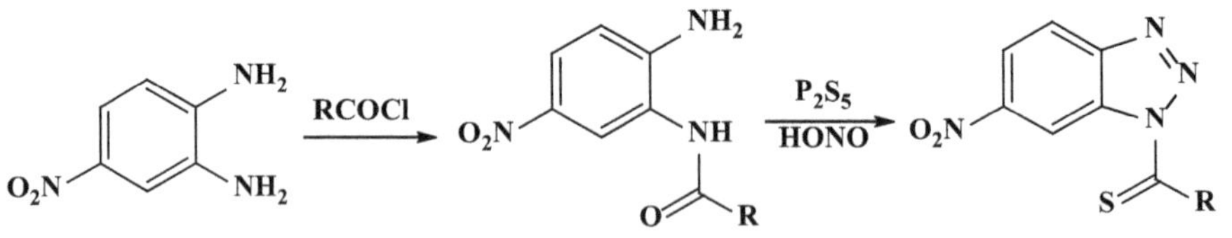

R	yield (%)	R	yield (%)
ethyl	84	4-methoxyphenyl	86
4-methylphenyl	98	4-bromophenyl	99
2-furanyl	95	pentyl	81
4-nitrophenyl	83	2-thienyl	91

Scheme 8.1 Synthesis of thiocarbonyl-1H-6-nitrobenzotriazoles

8.2 P$_2$S$_5$ in Heterocycle Synthesis

Scheme 8.2 Synthesis of 4-methyl-1,2,4-triazolo[4',3':4,5][1,2,4]triazino[2,3-a]benzimidazole

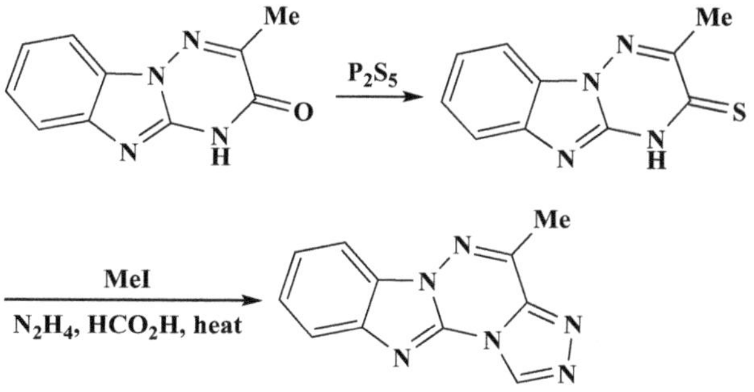

the same conditions could not be converted into furan product. The furan products were obtained with high efficiency when β-carbonyl allenic sulfides were cyclized with base. Besides, furan derivatives were obtained when β-ketone allenic sulfides were cyclized in the presence of phosphorus pentaoxide (Scheme 8.3) [25].

It was assumed that Paal thiophene synthesis took place via initial synthesis of furan through dehydration of 1,4-diketone, and after that furan was converted into thiophene because furans were generally isolated as side-products in the Paal thiophene synthesis. Campaigne and Foye [26] in 1952 were capable to prove that Paal thiophene synthesis could not take place through furan intermediate. It occurred through the formation of thione. The parallel experiments were conducted to verify this. The reactions of 2,5-hexanedione and 1,2-dibenzoylethane with phosphorus pentasulfide and the reactions of 2,5-dimethylfuran and 2,5-diphenylfuran under Paal thiophene synthesis conditions were compared. A greater yield of thiophene was obtained by the reactions using diketones suggesting that the furan was not an important intermediate in the reaction route, but rather a side-product (Scheme 8.4) [27].

Many synthetic approaches have been reported for the synthesis of various substituted thiophenes. An approach reported by Paal [28] was used for the formation of 1-phenyl-5-methylthiophene from 1-phenyl-1,4-pentadione with P$_2$S$_5$ as the sulfur source (Scheme 8.5).

Tanaka and coworkers [29] in 1982 have described the chemoselective reaction of ethoxalylaminoacetophenone with P$_2$S$_5$ to provide the ethyl 5-phenyl-1,3-thiazole-2-carboxylate (Scheme 8.6) [30, 31].

The modified conditions were reported by Erlenmeyer [32], which avoided the requirement to preform the thioamide (Scheme 8.7). This approach suffered from moderate yields.

Scheme 8.3 Synthesis of furans

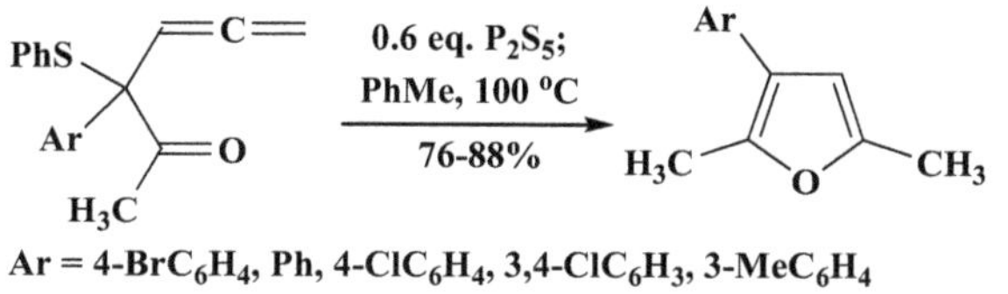

Ar = 4-BrC$_6$H$_4$, Ph, 4-ClC$_6$H$_4$, 3,4-ClC_6H$_3$, 3-MeC$_6$H$_4$

Scheme 8.4 Synthesis of thiophenes

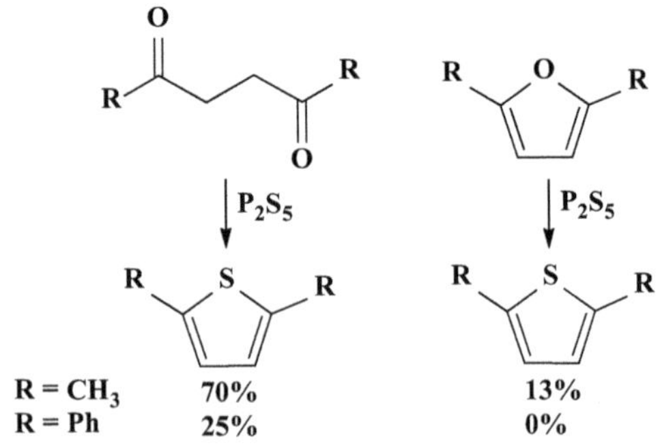

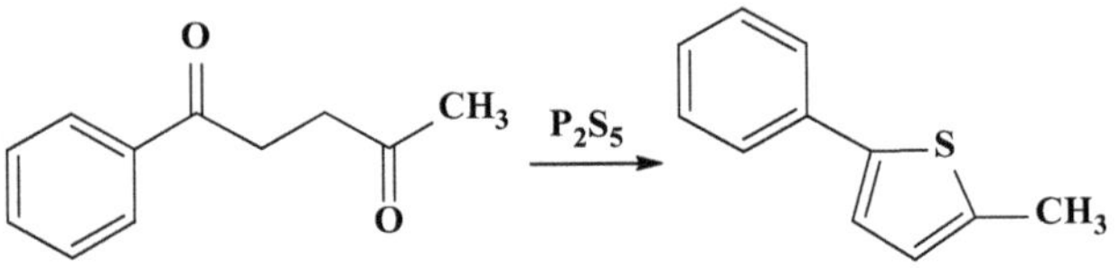

R = CH₃ 70%
R = Ph 25%

13%
0%

Scheme 8.5 Synthesis of 1-phenyl-5-methylthiophene

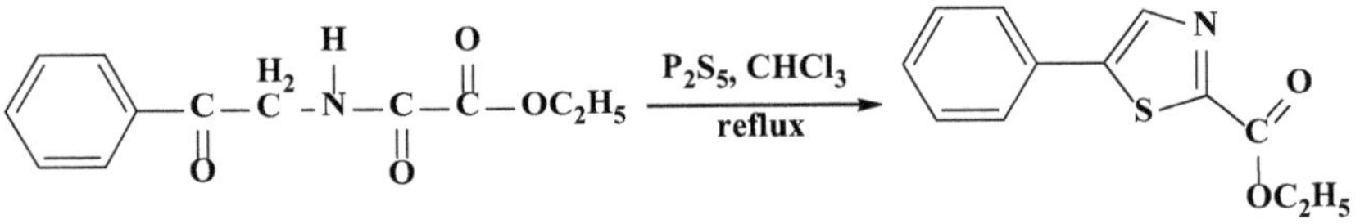

Scheme 8.6 Synthesis of ethyl 5-phenyl-1,3-thiazole-2-carboxylate

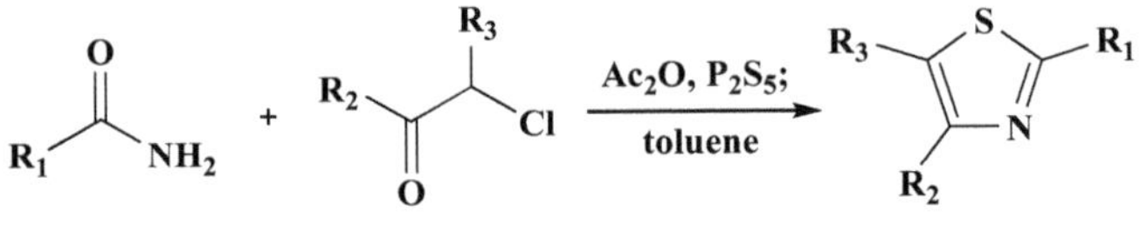

Scheme 8.7 Synthesis of thiazoles

A modification of Robinson Gabriel oxazole synthesis included the heating of reagents in the presence of P_2S_5 or sulfur to provide the thiazole product (Scheme 8.8)

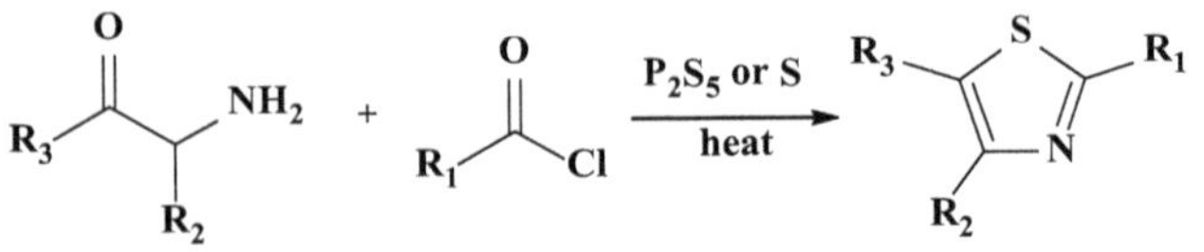

Scheme 8.8 Synthesis of thiazoles

8.2 P$_2$S$_5$ in Heterocycle Synthesis

[33].

In 1961 [34], the chemical structure of D-luciferin, isolated from firefly tails, was proposed and further confirmed by synthesis (Scheme 8.9) [35]. The D-luciferin was obtained in 9% yield from *p*-anisidine in nine steps. The 2-cyano-6-hydroxybenzothiazole is the key intermediate for the formation of D-luciferin. The starting compound *p*-anisidine, via intermediates, is converted into thio acid,

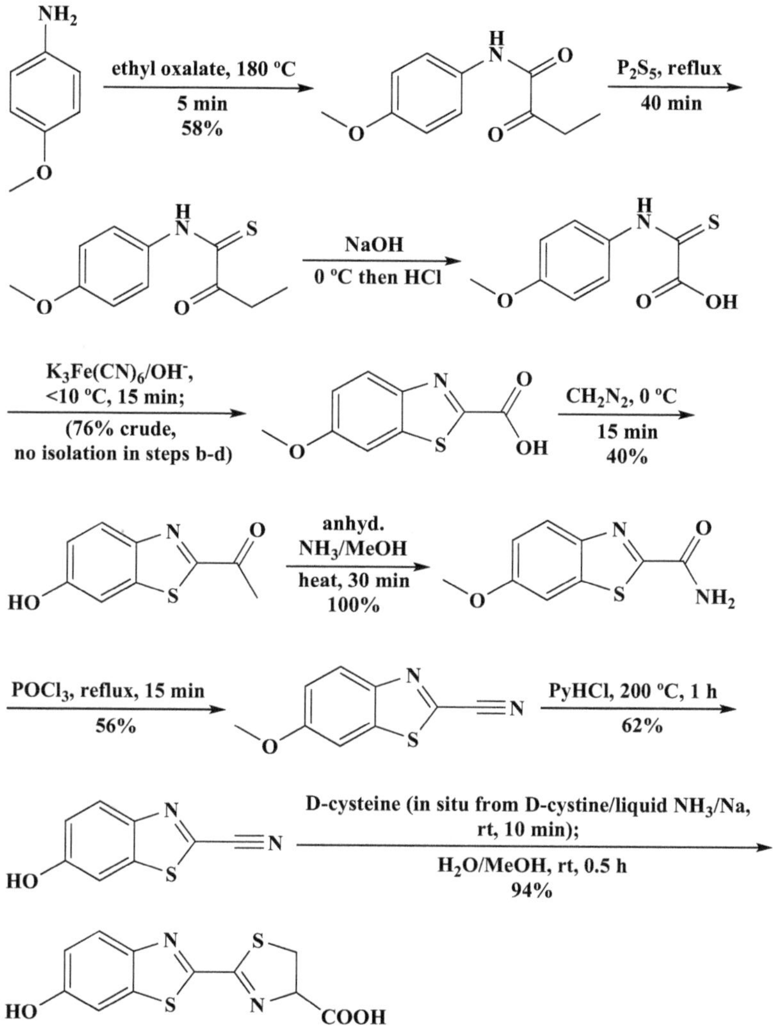

Scheme 8.9 Synthesis of 2-(6-hydroxybenzo[*d*]thiazol-2-yl)-4,5-dihydrothiazole-4-carboxylic acid

in turn cyclized to 6-methoxybenzothiazole-2-carboxylic acid. The 2-cyano-6-hydroxybenzothiazole was synthesized in four steps from this benzothiazole derivative. The key intermediate 2-cyano-6-hydroxybenzothiazole, for the formation of D-luciferin, was synthesized almost quantitatively by the reaction of D-cysteine, in situ synthesized by the reduction of D-cysteine [36].

The 2-methoxy-3,4,5,6-tetrafluorobenzoic acid was synthesized [37] and reacted (without purification) with $SOCl_2$ to provide the acid chloride. The amide was provided by the reaction of this chloride with pentafluoroaniline, which was transformed into 4,5,6,7-tetrafluoro-2-(2-methoxyphenyl)benzothiazole through cyclization with P_2S_5, and after that the ligand 4,5,6,7-tetrafluoro-2-(2-hydroxyphenyl)benzothiazole was provided by demethylation with boron tribromide (Scheme 8.10).

The formation of half fluorinated ligand 2-(2-hydroxyphenyl)-4,5,6,7-tetrafluorobenzothiazole was similar to the fully fluorinated ligand, except that

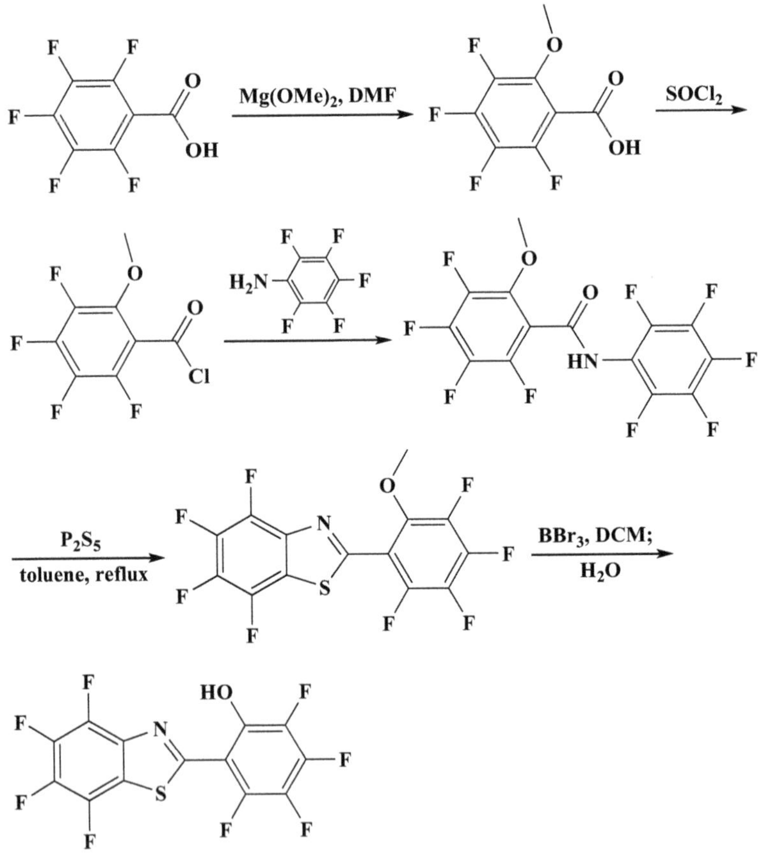

Scheme 8.10 Synthesis of 4,5,6,7-tetrafluoro-2-(2-hydroxy-3,4,5,6-tetrafluorophenyl)benzothiazole

8.2 P$_2$S$_5$ in Heterocycle Synthesis

the acylation of pentafluoroaniline took place utilizing 2-methoxybenzoyl chloride, which was synthesized from commercially accessible 2-methoxybenzoic acid and SOCl$_2$ (Scheme 8.11) [38].

Kuo and coworkers [39] utilized an in situ acyl chloride synthesis and late-stage cyclization methodology toward symmetric 1,3,4-oxadiazole and 1,3,4-thiadiazole-based calamitic liquid crystals. The 6-hydroxy-2-naphthoic acid was subjected to Fisher esterification conditions (alcohol, acid catalyst), Williamson etherification (bromoalkanes, K$_2$CO$_3$, catalytic iodide salt in CH$_3$COCH$_3$), and hydrolysis with base to provide the 6-alkoxy-2-naphthoic acid. The 6-alkoxy-2-naphthoic acid was further treated with SOCl$_2$, and after that hydrazine monohydrate in tetrahydrofuran to provide the symmetrical 2-naphthobis-hydrazides in good yield (80%). The bis-hydrazides were further cyclized to 1,3,4-thiadiazoles in good yield (70–74%) utilizing phosphorus pentasulfide. This methodology shows that good yield in these cyclization reactions can still be obtained (albeit with a 24 h reaction time) utilizing P$_2$S$_5$; however, in most of the cases, yields can be smoothly matched and reaction time reduced significantly utilizing LR (Scheme 8.12).

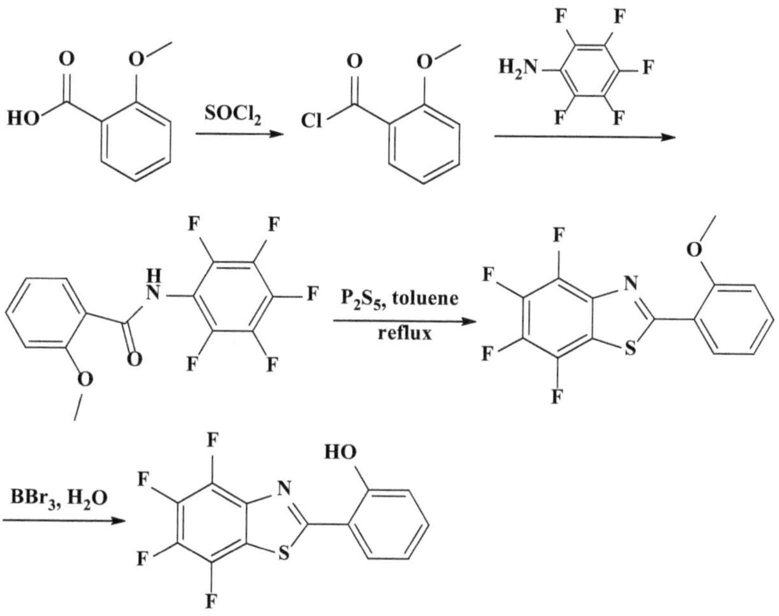

Scheme 8.11 Synthesis of 2-(2-hydroxyphenyl)-4,5,6,7-tetrafluorobenzothiazole

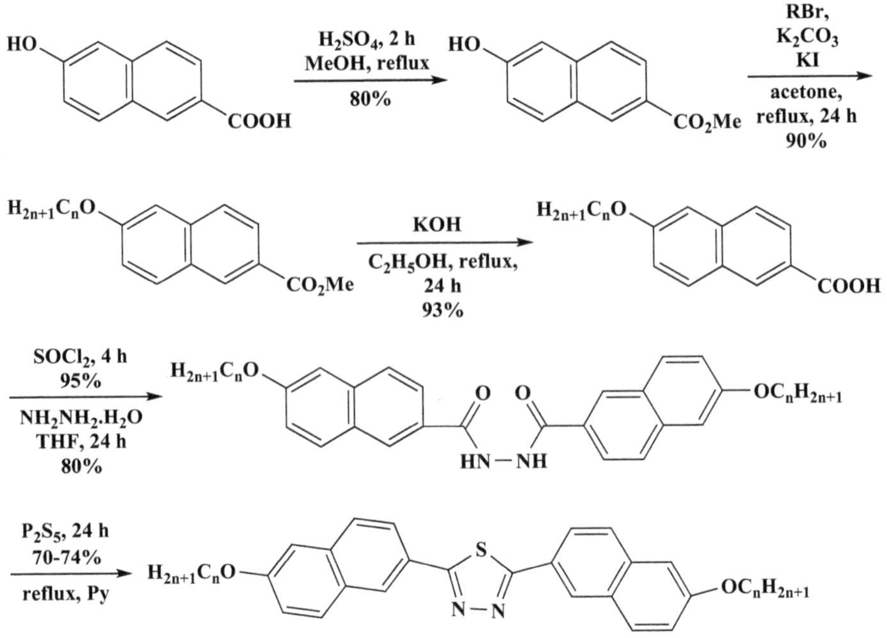

Scheme 8.12 Synthesis of 1,3,4-thiadiazoles

8.3 P_4S_{10} in Heterocycle Synthesis

8.3.1 Synthesis of Five-Membered Heterocycles

8.3.1.1 Synthesis of Five-Membered N-Heterocycles

The indoles were afforded via intermediates by the reaction of benzyl monoarylimines with phosphorus pentasulfide in refluxing toluene (Scheme 8.13) [17, 40].

The diimidazolines were obtained when alkanedinitriles were reacted with ethylenediamine and small quantity of phosphorus pentasulfide in toluene (dry) at 90 °C for 10 h (Scheme 8.14) [17, 41].

The imidazolines were obtained when a mixture of nitriles, ethylenediamine, and phosphorus pentasulfide was irradiated under MWs (Schemes 8.15 and 8.16) [17, 42]. An irradiation (720 W) for 1.25–20 min afforded imidazolines in high yield (86–98%).

The imidazolines were obtained when diamines were reacted with nitriles and phosphorus pentasulfide. The imidazolines were obtained in 72–88% yield using a small amount of phosphorus pentasulfide in the reaction of ethylenediamine with arylaminoacetonitriles at 80 to 120 °C (Scheme 8.17) [17, 43].

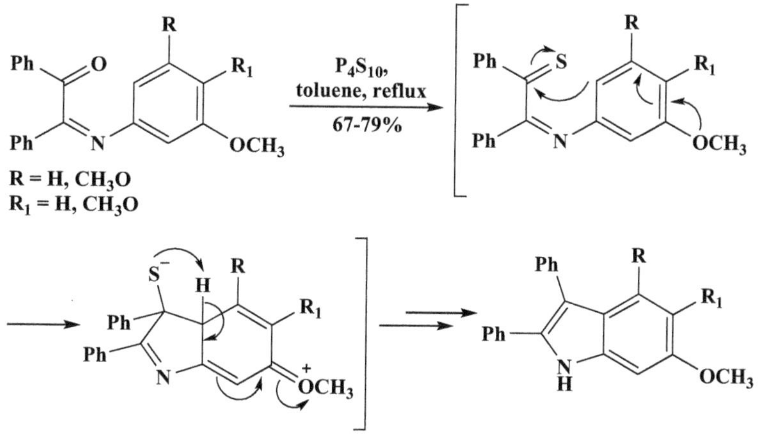

8.3 P$_4$S$_{10}$ in Heterocycle Synthesis

Scheme 8.13 Synthesis of indoles

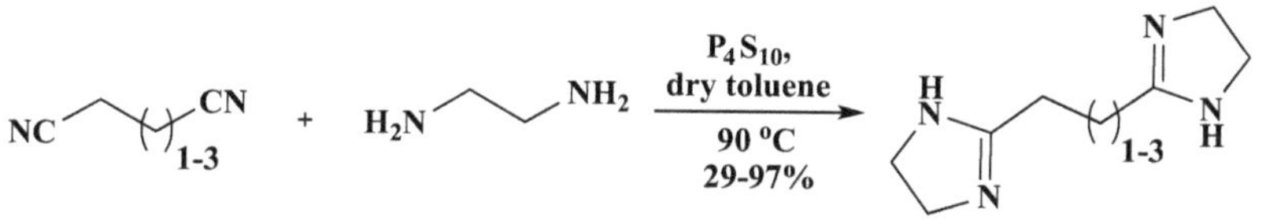

Scheme 8.14 Synthesis of diimidazolines

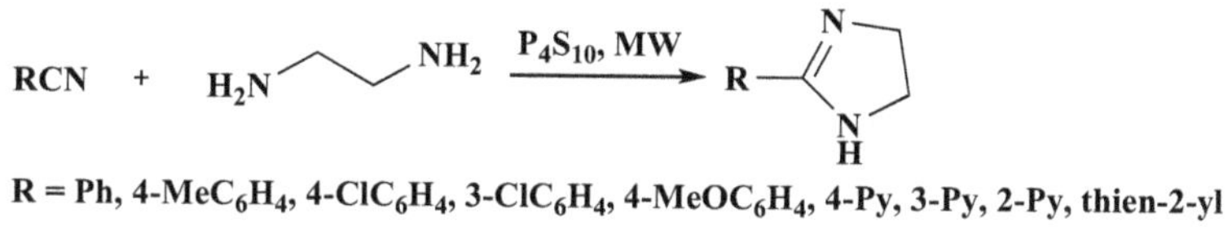

Scheme 8.15 Synthesis of imidazolines

R = Ph, 4-MeC$_6$H$_4$, 4-ClC$_6$H$_4$, 3-ClC$_6$H$_4$, 4-MeOC$_6$H$_4$, 4-Py, 3-Py, 2-Py, thien-2-yl

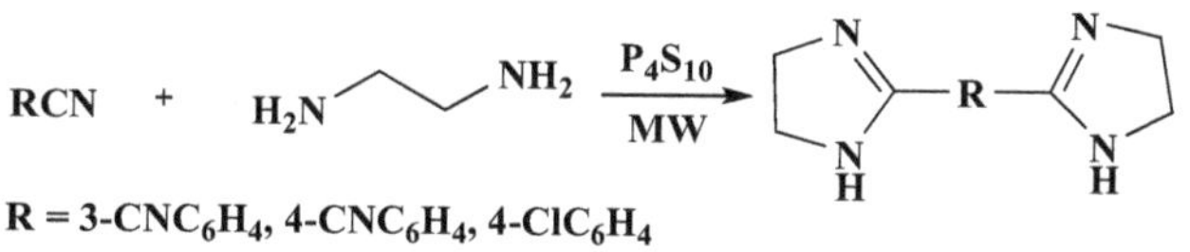

Scheme 8.16 Synthesis of imidazolines

R = 3-CNC$_6$H$_4$, 4-CNC$_6$H$_4$, 4-ClC$_6$H$_4$

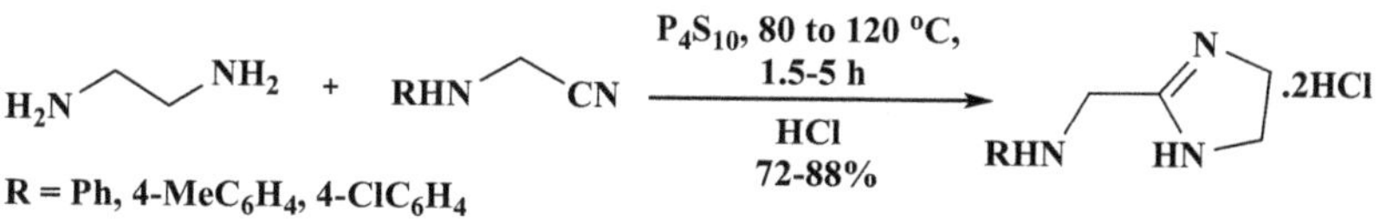

Scheme 8.17 Synthesis of imidazolines

R = Ph, 4-MeC$_6$H$_4$, 4-ClC$_6$H$_4$

A rearrangement product was synthesized by the thionation of starting material with phosphorus pentasulfide in tetrahydrofuran at 50 °C for 3 d via a ring-opening intermediate (Scheme 8.18) [17, 44].

The cyclic ureas were synthesized by reacting diamines with CO_2 (Scheme 8.19) [17].

The amide systems containing reactive o-amine groups reacted upon addition of phosphorus pentasulfide. The addition products were formed via intermediate which has an o-amino system by the reaction of amides with phosphorus pentasulfide (Scheme 8.20) [17, 45].

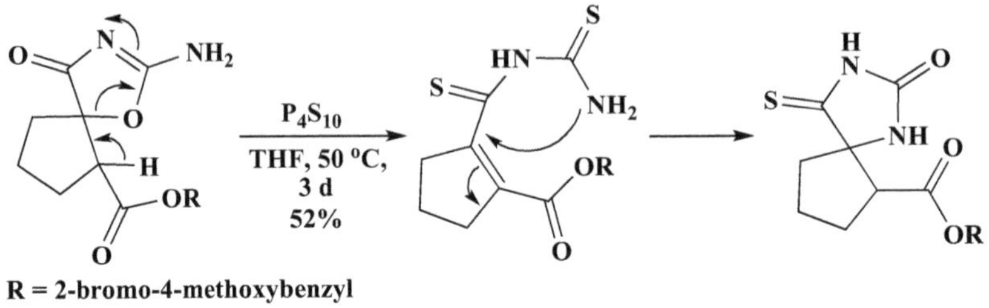

R = 2-bromo-4-methoxybenzyl

Scheme 8.18 Synthesis of imidazolidines

$R_1 = H, Me, Ph, HOCH_2CH_2, HOCH(CH_3)CH_2$
$R_2 = H, Me$

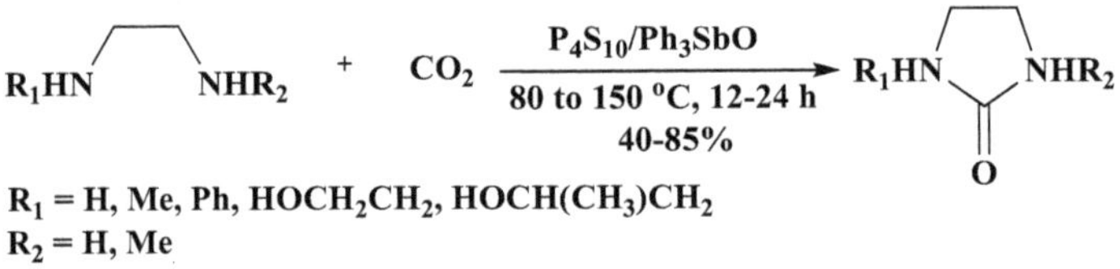

Scheme 8.19 Synthesis of imidazolidines

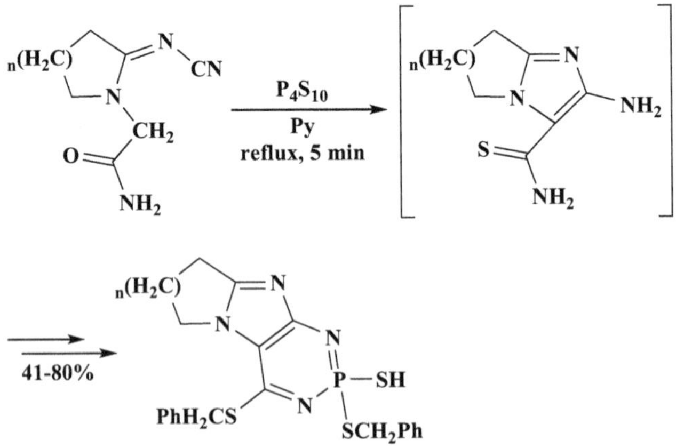

Scheme 8.20 Synthesis of imidazodiazaphosphinines

8.3 P$_4$S$_{10}$ in Heterocycle Synthesis

The triazolam, 8-chloro-6-(2'-chlorophenyl)-1-methyl-4H-s-triazolo[4,3-a][1,4]benzodiazepine, was prepared by a method that contains a key step of benzodiazepine formation by the reaction of *o*-aminobenzophenones with α-amino acid derivatives. The 7-chloro-5-(2-chlorophenyl)-2,3-dihydro-1H-1,4-benzodiazepin-2-one was provided by the reaction of 2-amino-2',5-dichlorobenzophenone with glycine ethyl ester. The 7-chloro-5-(2-chlorophenyl)-2,3-dihydro-1H-1,4-benzodiazepin-2-one was treated with P$_4$S$_{10}$ to convert the CO group into a thiocarbonyl group to provide the 7-chloro-5-(2-chlorophenyl)-2,3-dihydro-1H-1,4-benzodiazepin-2-thione. The acetylhydrazone, obtained by reacting cyclic thioamide with acetylhydrazine, was cyclized into triazolam on heating (Scheme 8.21) [46–51].

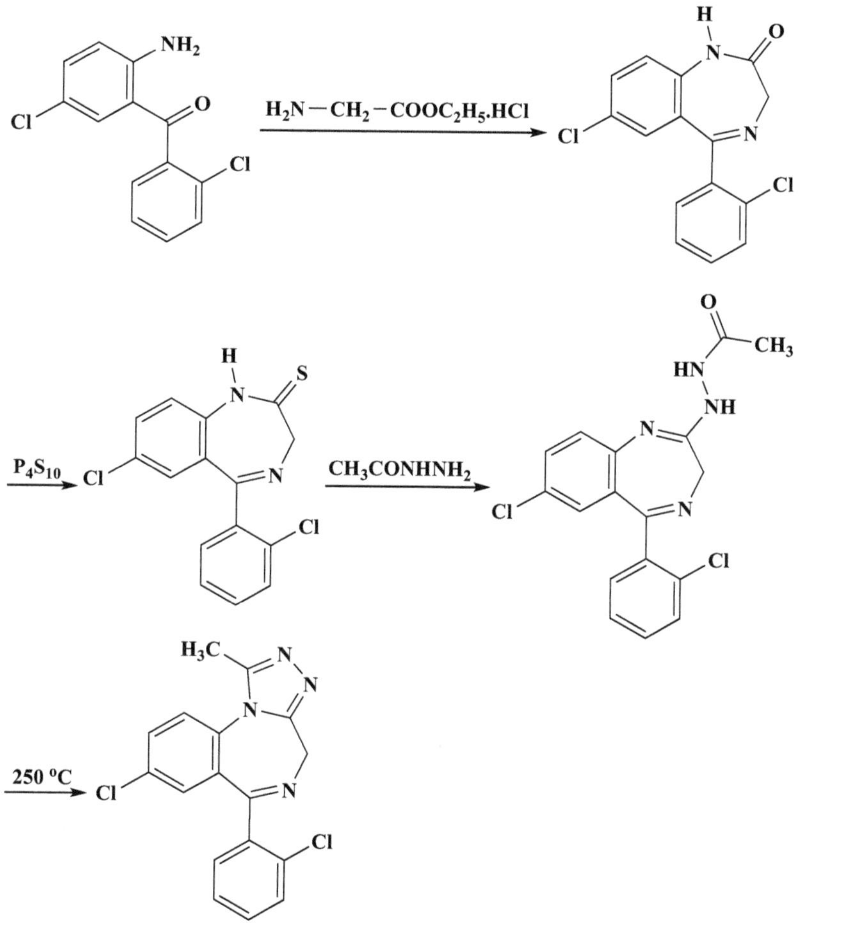

Scheme 8.21 Synthesis of triazolam

8.3.1.2 Synthesis of Five-Membered *O*-Heterocycles

The 3-furancarbothioates were obtained in 38–55% yield when ester 1,4-diketone was reacted with phosphorus pentasulfide in xylene (Scheme 8.22) [52].

The thiophene was synthesized by the reaction of 2-benzoylbenzoic acid with phosphorus pentasulfide in refluxing xylene for 4 h, whereas dimer was obtained by either refluxing the same compound in benzene (dry) for 25 h or heating the mixture neat at 115 °C for 15 min (Scheme 8.23) [17, 53].

The phenanthrene-9,10-quinone was reacted with different arylalcohols like catechol (Scheme 8.24), resorcinol (Scheme 8.25), phloroglucinol (Scheme 8.26), pyrogallol (Scheme 8.27), 1-naphthol, and 2-naphthol (Scheme 8.28) in the presence of phosphorus pentasulfide [54]. The dioxin, furans, disulfides, sulfides, ether, and alcohol were obtained when the reactions were performed at 160 and 220 °C for 2 h [17].

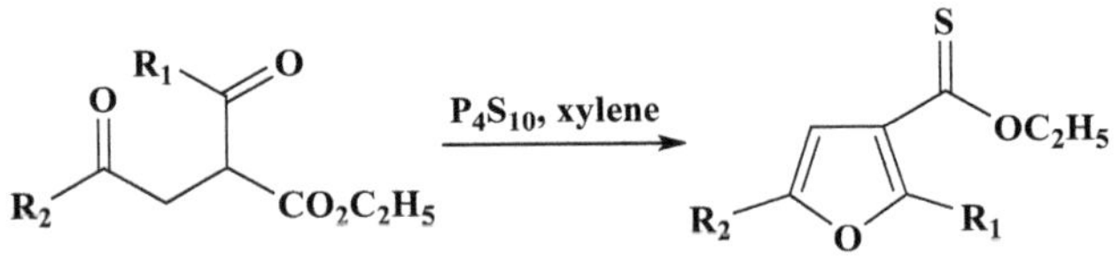

Scheme 8.22 Synthesis of 3-furancarbothioates

Scheme 8.23 Synthesis of 1-phenylbenzothiophene and isobenzofuran

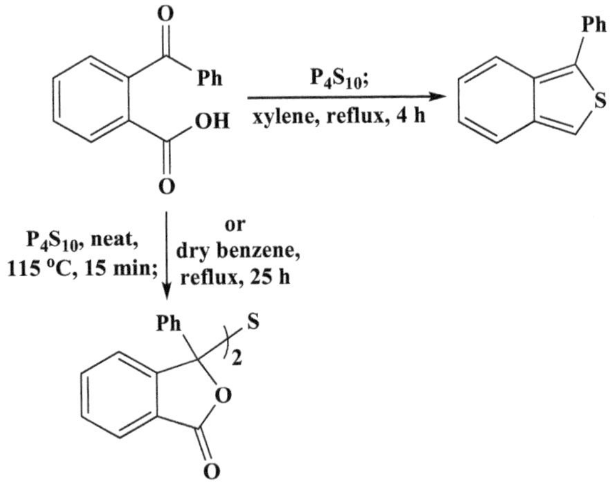

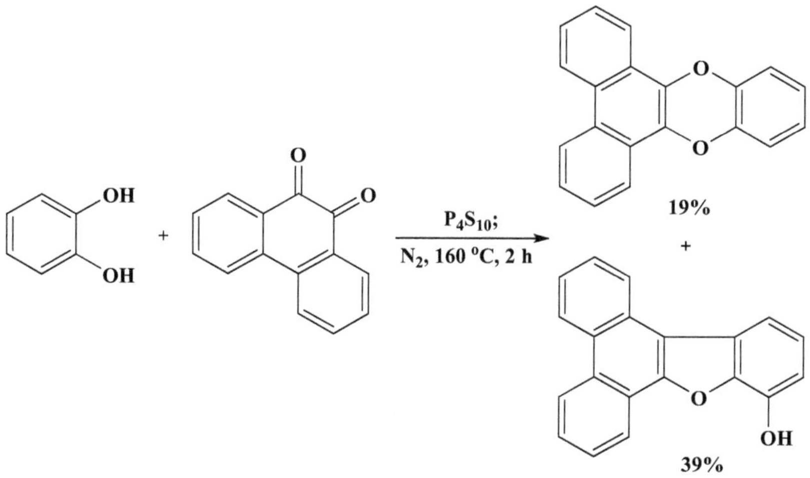

8.3 P$_4$S$_{10}$ in Heterocycle Synthesis

Scheme 8.24 Synthesis of dioxin and furan

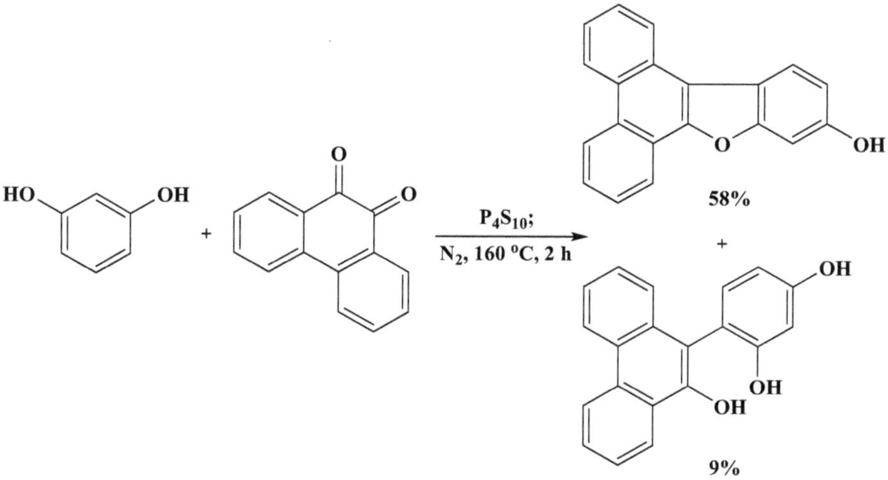

Scheme 8.25 Synthesis of furan

8.3.1.3 Synthesis of Five-Membered S-Heterocycles

Thiophene was obtained when 1,4-dicarbonyl compounds were reacted with a source of sulfur [55]. Paal and Knorr individually reported the starting examples of condensation reactions between 1,4-diketones and primary amines which is known as Paal–Knorr pyrrole synthesis (Scheme 8.29) [28]. The basic mechanism of this synthetic

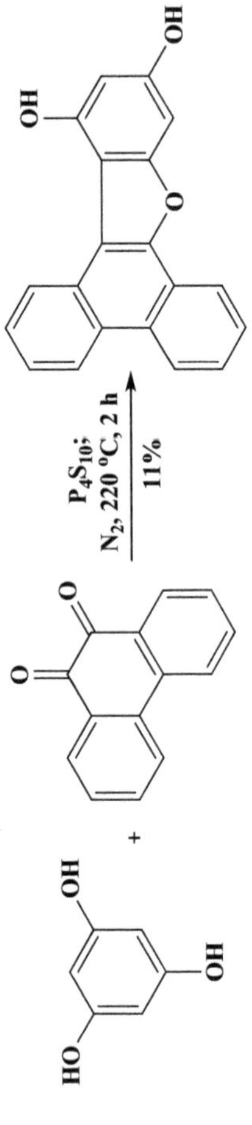

Scheme 8.26 Synthesis of furan

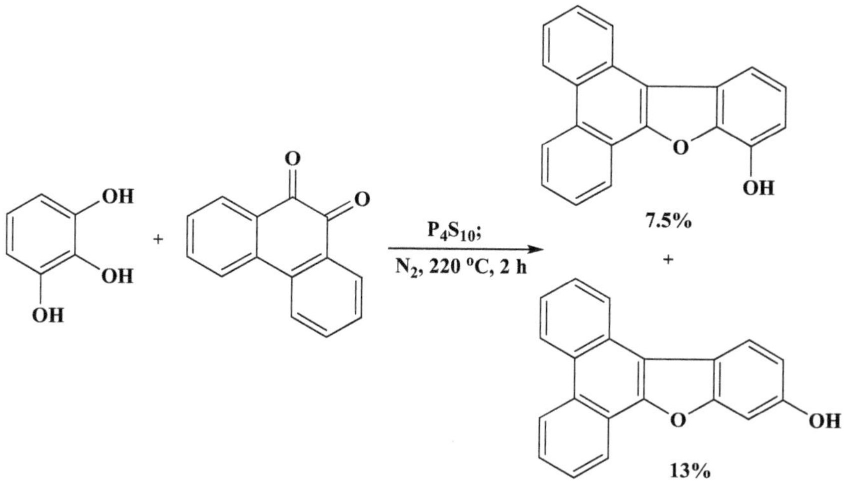

8.3 P$_4$S$_{10}$ in Heterocycle Synthesis

Scheme 8.27 Synthesis of furans

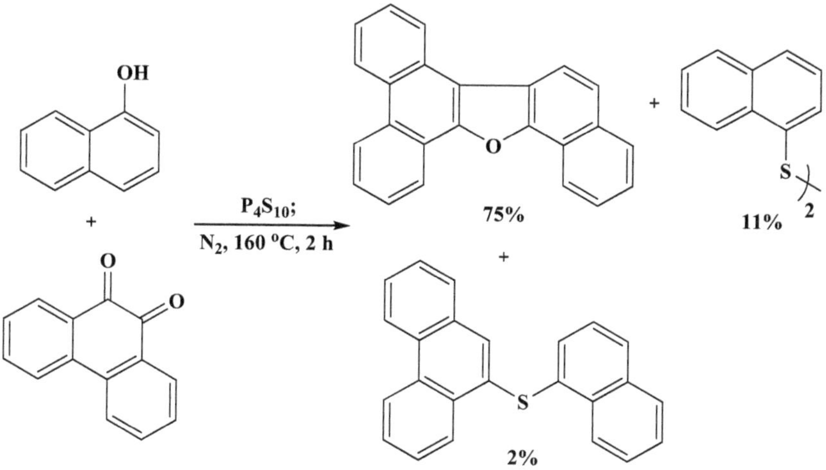

Scheme 8.28 Synthesis of furan

Scheme 8.29 Synthesis of 3-methylthiophene

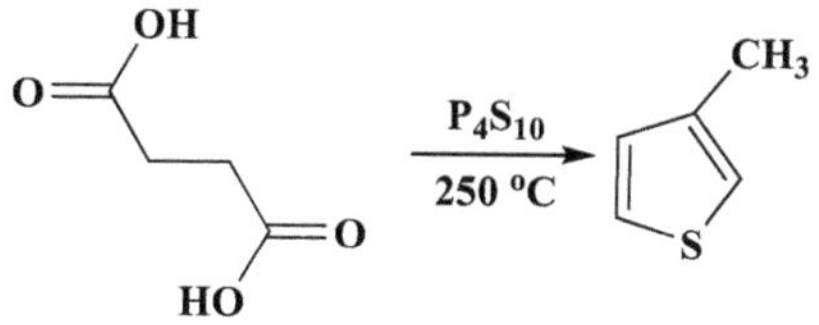

process included the cyclization of 1,4-diketones, either with a primary amine (Paal–Knorr pyrrole synthesis), with a sulfur source (Paal thiophene synthesis) or through the dehydration of diketone itself (Paal furan synthesis) [27, 56].

The 1-keto, 4-aldehydes were synthesized after the deprotection of acetal. The cyclization reactions took place without problems for both the thiophene and the pyrrole molecules (Scheme 8.30). The phosphorus pentasulfide in toluene and dry ammonium carbonate was heated with 1-keto, 4-aldehyde compounds to synthesize the heterocyclic compounds. A protected pyrrole was formed when organic amine was utilized. The unprotected pyrroles were protected with allyl bromide (not shown). The same result was observed if ammonium carbonate was not utilized as nitrogen donor, but allylamine. This is an electrophilic substitution reaction that works well with electron-rich compounds like substituted pyrrole and thiophene. In practice, however, the Stetter reaction not proceeded on any 3-alkyl-substituted heterocyclic carboxaldehyde that has been exposed to Stetter conditions before. This showed that the electron-donating effect of Me group hindered the addition of cyanide to aldehyde. No benzoin condensation was found either, and the starting materials were recovered without any sign of reaction or side reaction (Scheme 8.31) [57].

The tetralin was utilized as a solvent for the modification of classical Paal–Knorr synthesis (Scheme 8.32) [52]. In some cases, furans were formed as side-products.

This reaction has been employed for the formation of 2,5-heterocyclophanes (Scheme 8.33) and 2,4-heterophanes (Scheme 8.34) [52].

The 2,5-thiophenedi-P-propionic acid was synthesized from starting material by Paal–Knorr reaction, which was of interest for studying polymers (Scheme 8.35) [52]. The utilization of LR enhanced the yield of 2,5-disubstituted thiophenes from 1,4-diketones.

The reaction of starting material with sulfur and morpholine at 145 °C provided thiophene and morpholine-substituted thiophene in 11 and 8% yield, respectively. The reaction of morpholine containing starting material with hydrogen sulfide/hydrogen chloride provided morpholine-substituted thiophene in 26% yield;

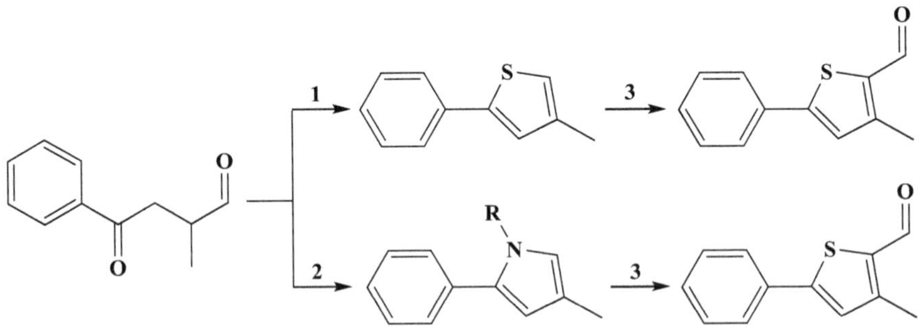

Reagents and conditions: (1) P_4S_{10}, toluene, (2) NH_4CO_3 (R = H) or allylamine (R = allyl), (3) $POCl_3$, DMF, dichloroethane.

Scheme 8.30 Synthesis of thiophenes

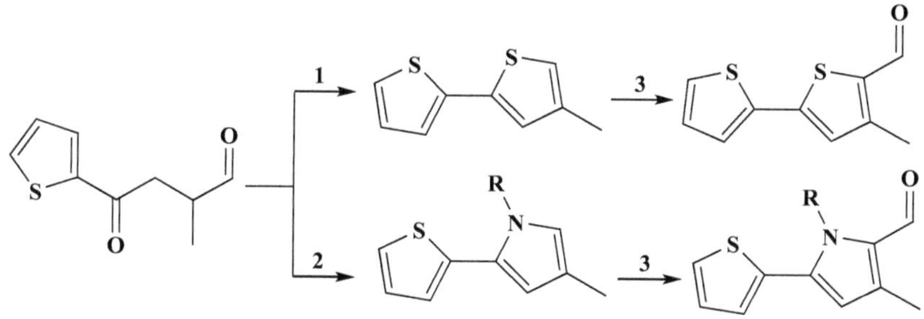

8.3 P$_4$S$_{10}$ in Heterocycle Synthesis

Reagents and conditions: (1) P$_4$S$_{10}$, toluene, (2) NH$_4$CO$_3$ (R = H) or allylamine (R = allyl), (3) POCl$_3$, DMF, dichloroethane.

Scheme 8.31 Synthesis of thiophenes

Scheme 8.32 Synthesis of thiophenes

$$RCOCH_2CH_2COR_1 \xrightarrow{P_4S_{10}} \text{thiophene}$$

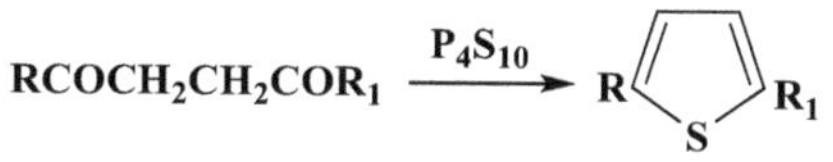

Scheme 8.33 Synthesis of thiophenes

n = 8, 51%
n = 11, 37%

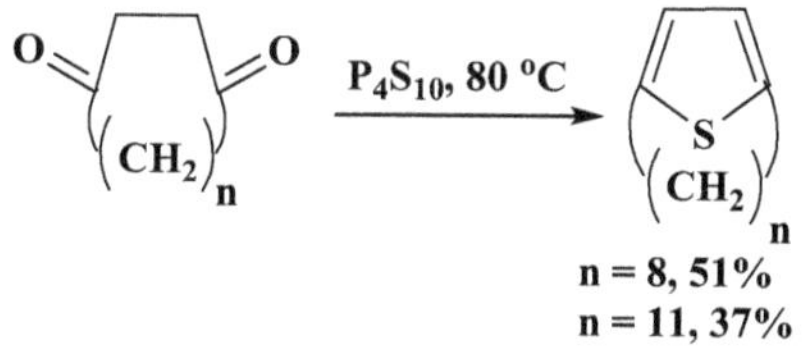

Scheme 8.34 Synthesis of thiophenes

n = 6, 61%
n = 7, 51%
n = 8, 70%

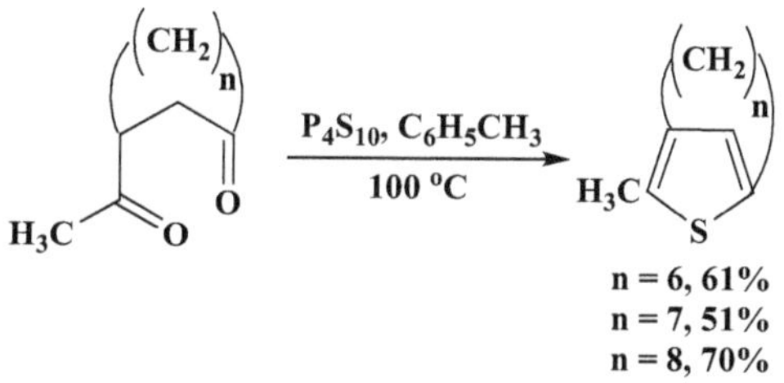

its reaction with phosphorus pentasulfide provided thiophene in 31% yield (Scheme 8.36) [52].

The electrophilic addition of certain acid chlorides to allyl or methallyl chlorides provided β,γ-dichloroketones, which afforded 2-alky- and 2,4-dialkylthiophenes upon reaction with phosphorus pentasulfide in dimethylformamide or dioxane or with potassium sulfide/hydrogen sulfide. The nucleophilic attack of sulfur atom on the carbon was the key step after thionation (Scheme 8.37) [52].

The 3-acetyl-2,5-dimethylthiophene was prepared by cyclization of 1,4-dicarbonyl compounds with P$_4$S$_{10}$ or hydrogen sulfide [58], followed by acetylation of 2,5-dimethylthiophene intermediate (Scheme 8.38).

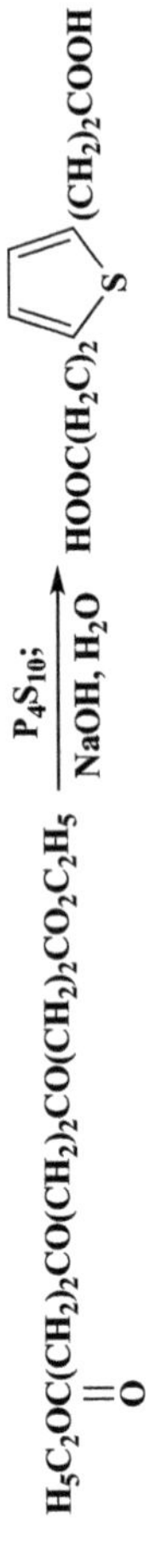

$H_5C_2OC(CH_2)_2CO(CH_2)_2CO(CH_2)_2CO_2C_2H_5 \xrightarrow[NaOH, H_2O]{P_4S_{10};} HOOC(H_2C)_2$—[thiophene]—$(CH_2)_2COOH$

Scheme 8.35 Synthesis of 2,5-disubstituted thiophenes

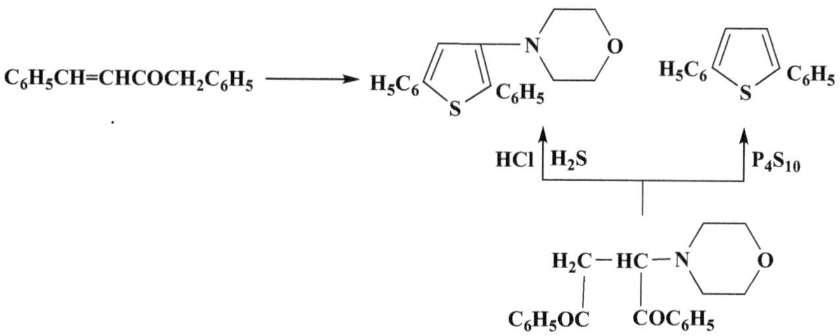

C6H5CH=CHCOCH2C6H5
H5C6
S
C6H5
N
O
H5C6
S
C6H5
HCl
H2S
P4S10
H2C−HC−N
O
C6H5OC
COC6H5

Scheme 8.36 Synthesis of thiophenes

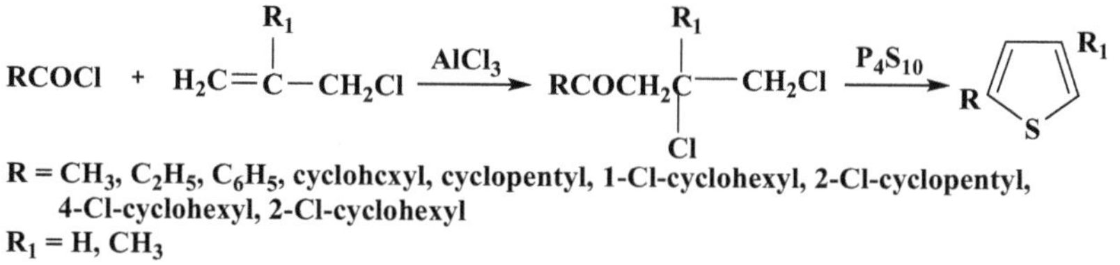

RCOCl + H2C=C−CH2Cl
R1
AlCl3
RCOCH2C—CH2Cl
R1
Cl
P4S10
R
S
R1
R = CH3, C2H5, C6H5, cyclohcxyl, cyclopentyl, 1-Cl-cyclohexyl, 2-Cl-cyclopentyl,
4-Cl-cyclohexyl, 2-Cl-cyclohexyl
R1 = H, CH3

Scheme 8.37 Synthesis of thiophenes

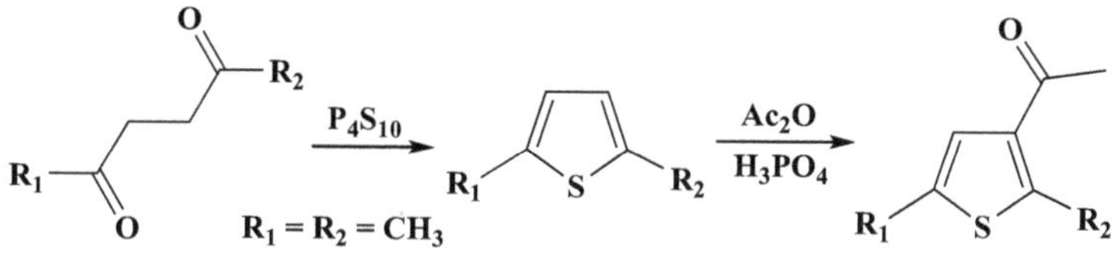

O
R2
R1
O
P4S10
R1 = R2 = CH3
R1
S
R2
Ac2O
H3PO4
O
R1
S
R2

Scheme 8.38 Synthesis of thiophenes

The thiophenes bearing trifluoromethyl and fluorine were synthesized by the reaction of unsaturated ketones, synthesized from hexafluoroacetone, with phosphorus pentasulfide (Scheme 8.39) [59]. The thiophenes were synthesized by initial replacement of oxygen with sulfur at 120 to 140 °C followed by an intramolecular 1,5-cyclization [17].

The thiophene was provided by thionation of *N*-phenylacetylthiobenzamides with phosphorus pentasulfide in boiling carbon disulfide (Scheme 8.40) [60]. A possible mechanism was described in which, in starting, the oxo group was transformed to thione, a tautomer was attached to the thione carbon of thione, and then the hydrogen sulfide removal afforded intermediate. The thiophene was obtained in 37–74% yield by intramolecular cyclization of obtained thione [17].

The reaction of γ-chloroketones with phosphorus pentasulfide in DMF or dioxane at 90 °C afforded substituted thiophenes in 65–77% yield (Scheme 8.41) [17, 61].

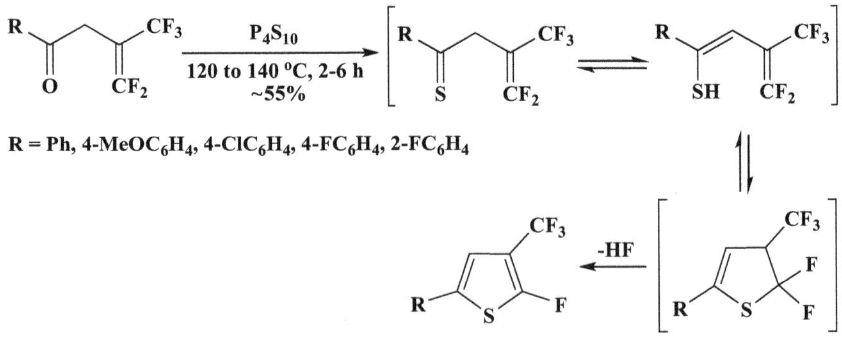

264 8 Phosphorus Pentasulfide in Heterocycle Synthesis

Scheme 8.39 Synthesis of thiophenes

R = Ph, 4-MeOC$_6$H$_4$, 4-ClC$_6$H$_4$, 4-FC$_6$H$_4$, 2-FC$_6$H$_4$

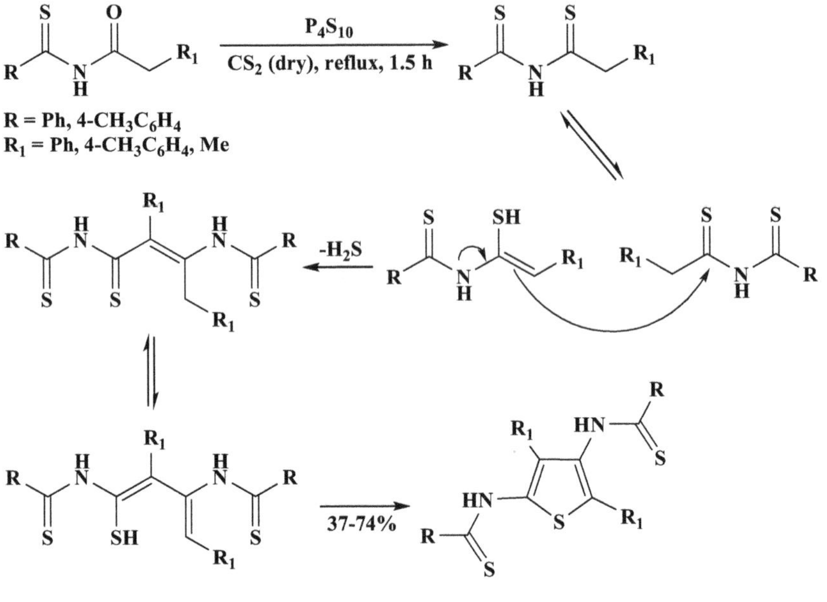

Scheme 8.40 Synthesis of thiophenes

R = Ph, 4-CH$_3$C$_6$H$_4$
R$_1$ = Ph, 4-CH$_3$C$_6$H$_4$, Me

Scheme 8.41 Synthesis of thiophenes

R = cyclohexyl, cyclopentyl, 1-chlorocyclohexyl, 1-chlorocyclopentyl, 4-chlorocyclohexyl
R$_1$ = H, Me

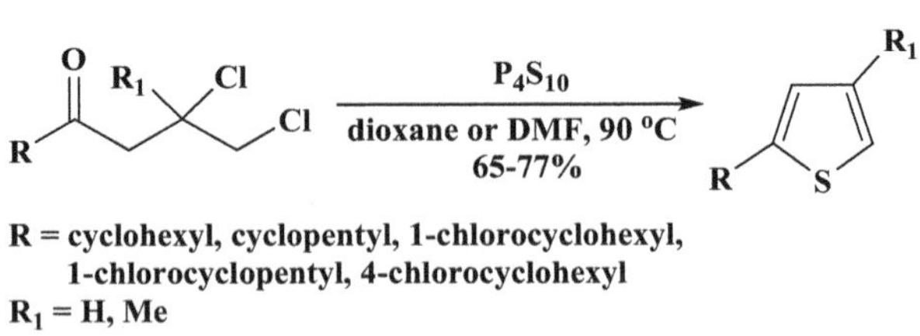

8.3 P$_4$S$_{10}$ in Heterocycle Synthesis

The 3,4- and 2,4-thiophenedicarboxylic acids were synthesized efficiently using 1,4-dialdehydes or their acetals. However, 2,3-thiophenedicarboxylic acid was obtained in only small amounts when 2-ethoxalyl-4,4-diethoxybutyronitrile was utilized in this reaction, and, instead, the isothiazole-fused compound was the major product. However, 5-methyl-2,3-thiophenedicarboxylic acid was formed in 24% yield from methyl homologue. The 2,3,4- and 2,3,5-thiophenetricarboxylic acids were synthesized by this pathway (Schemes 8.42, 8.43, 8.44, 8.45, 8.46, and 8.47) [52].

The oxazolinone was refluxed in xylene, which resulted in subsequent ring-opening and further ring formation to provide the benzothiophene in 30% yield (Scheme 8.48) [17, 62].

The PITN (poly(isothianaphthene)) was synthesized from phthalide and phthalic anhydrate utilizing phosphorus pentasulfide (Scheme 8.49) [63–67]. The same

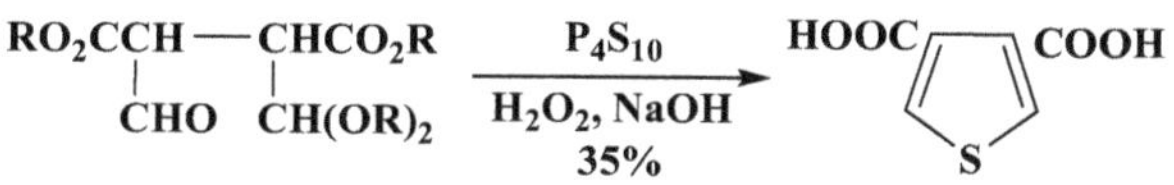

Scheme 8.42 Synthesis of thiophene dicarboxylic acid

RO$_2$CCH—CHCO$_2$R
 | |
CHO   CH(OR)$_2$

$\xrightarrow[\text{H}_2\text{O}_2, \text{NaOH}]{\text{P}_4\text{S}_{10}}$ 35%

HOOC—[thiophene]—COOH

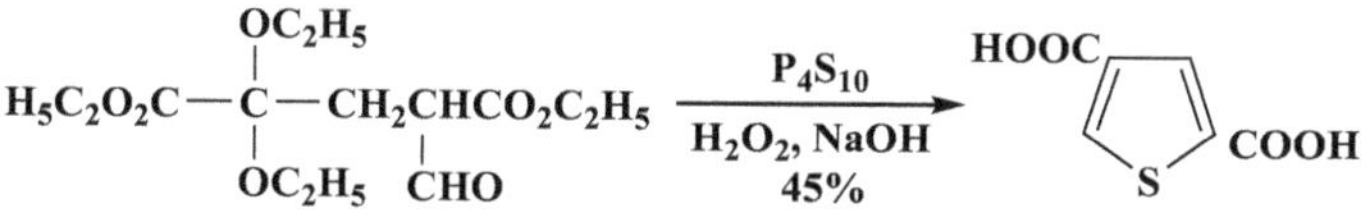

Scheme 8.43 Synthesis of thiophene dicarboxylic acid

         OC$_2$H$_5$
          |
H$_5$C$_2$O$_2$C—C—CH$_2$CHCO$_2$C$_2$H$_5$
          | |
         OC$_2$H$_5$ CHO

$\xrightarrow[\text{H}_2\text{O}_2, \text{NaOH}]{\text{P}_4\text{S}_{10}}$ 45%

HOOC—[thiophene]—COOH

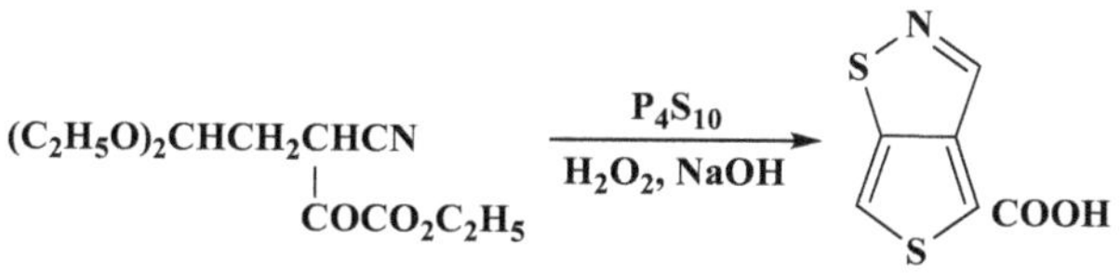

Scheme 8.44 Synthesis of thienoisothiazole carboxylic acid

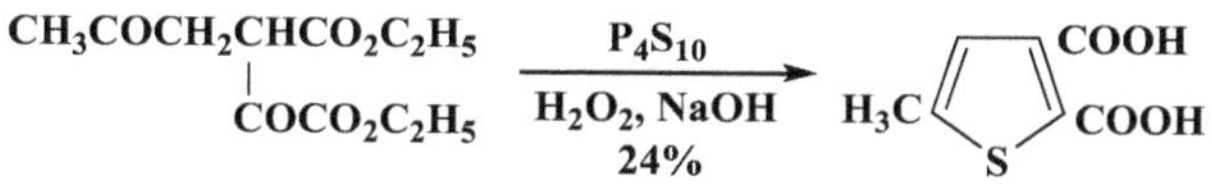

(C$_2$H$_5$O)$_2$CHCH$_2$CHCN
                   |
                   COCO$_2$C$_2$H$_5$

$\xrightarrow[\text{H}_2\text{O}_2, \text{NaOH}]{\text{P}_4\text{S}_{10}}$

[thienoisothiazole]—COOH

Scheme 8.45 Synthesis of methylthiophene dicarboxylic acid

CH$_3$COCH$_2$CHCO$_2$C$_2$H$_5$
              |
              COCO$_2$C$_2$H$_5$

$\xrightarrow[\text{H}_2\text{O}_2, \text{NaOH}]{\text{P}_4\text{S}_{10}}$ 24%

H$_3$C—[thiophene]—COOH, COOH

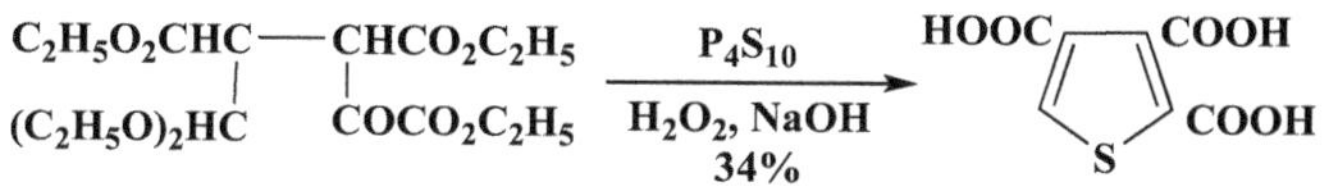

Scheme 8.46 Synthesis of thiophene tricarboxylic acid

C$_2$H$_5$O$_2$CHC—CHCO$_2$C$_2$H$_5$
       | |
(C$_2$H$_5$O)$_2$HC   COCO$_2$C$_2$H$_5$

$\xrightarrow[\text{H}_2\text{O}_2, \text{NaOH}]{\text{P}_4\text{S}_{10}}$ 34%

HOOC—[thiophene]—COOH, COOH

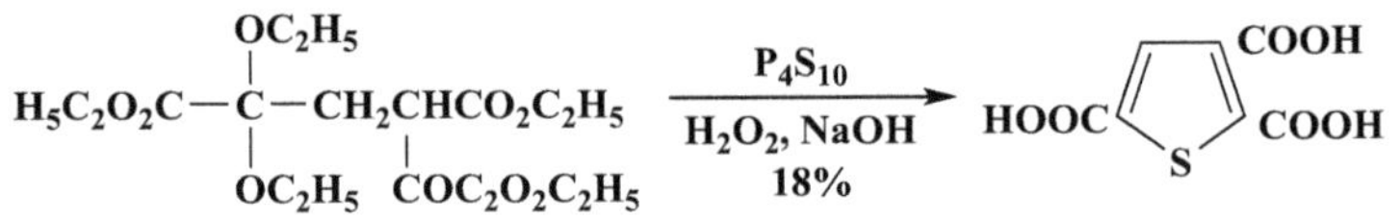

Scheme 8.47 Synthesis of thiophene tricarboxylic acid

Scheme 8.48 Synthesis of benzothiophene

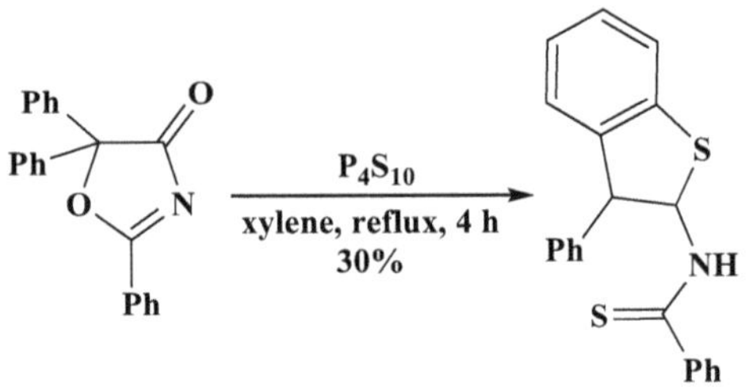

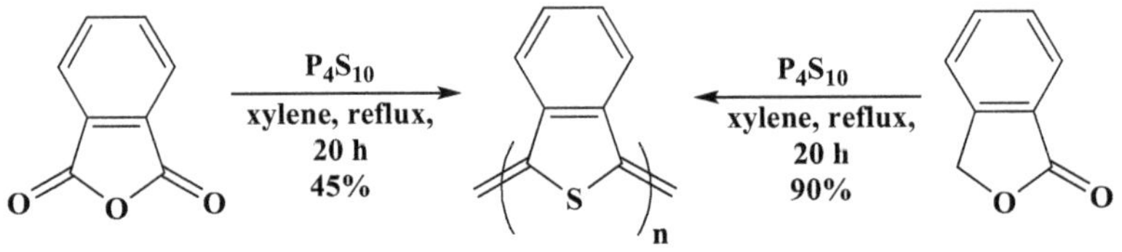

Scheme 8.49 Synthesis of poly(isothianaphthene)

product, poly(isothianaphthene), was provided by the reactions of both phthalide and phthalide anhydride with phosphorus pentasulfide in refluxing xylene for 20 h [17].

The reaction time for the polymerization of phthalide was kept shorter as in the case of phthalic anhydride. The poly(isothianaphthene) was formed only in 9% yield, and the major product was found to be thiophthalide along with dithiophthalide in small amount (Scheme 8.50) [17, 64].

Polymerization not occurred when the reaction time was kept shorter, like 3 h; instead thiophthalic anhydrides and dimers were formed along with poly(isothianaphthene) in small amounts (Scheme 8.51) [64]. The polymerization of

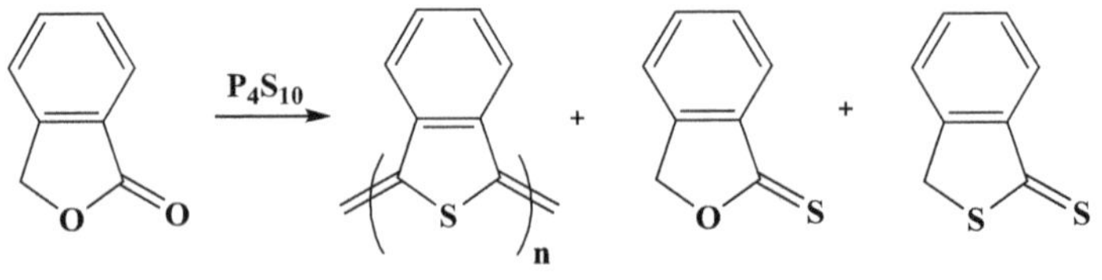

Scheme 8.50 Synthesis of poly(isothianaphthene), thiophthalide, and dithiophthalide

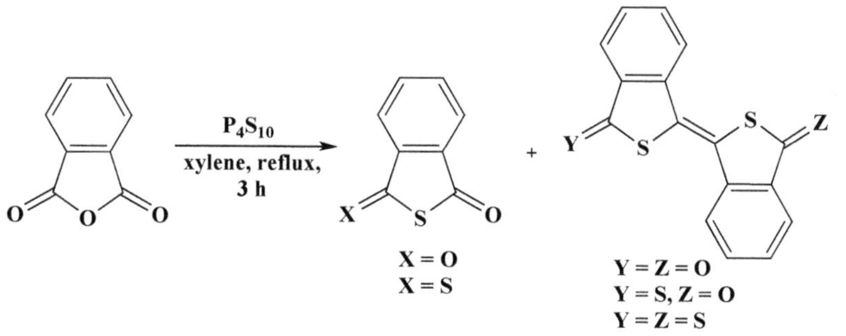

8.3 P$_4$S$_{10}$ in Heterocycle Synthesis

Scheme 8.51 Synthesis of isothianaphthene

thiophthalic anhydride provided poly(isothianaphthene) in higher yield under same reaction conditions [17].

The 1,2-diketones like benzyl (Scheme 8.52), 2,2′-thienyl (Scheme 8.53), acenaphthenequinone (Scheme 8.54), and *o*-chloranil (Scheme 8.55) were reacted with phosphorus pentasulfide in the presence of phenol at high temperatures (160 to 170 °C) to provide either fused or substituted thiophenes, although, in low yields (1.8–25%) [17, 54].

The diphenylene sulfide was synthesized by refluxing a mixture of 2-hydroxydiphenyl and phosphorus pentasulfide for 10 h (Scheme 8.56) [68]. Following a similar approach, the dibenzothiophene *S*,*S*-dioxide was obtained when

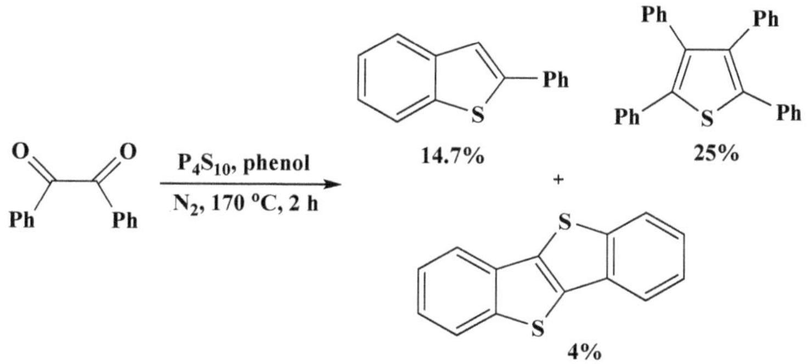

Scheme 8.52 Synthesis of fused or substituted thiophenes

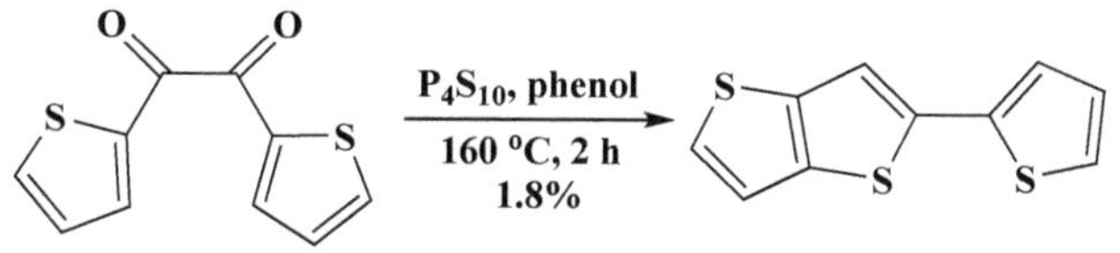

Scheme 8.53 Synthesis of fused thiophene

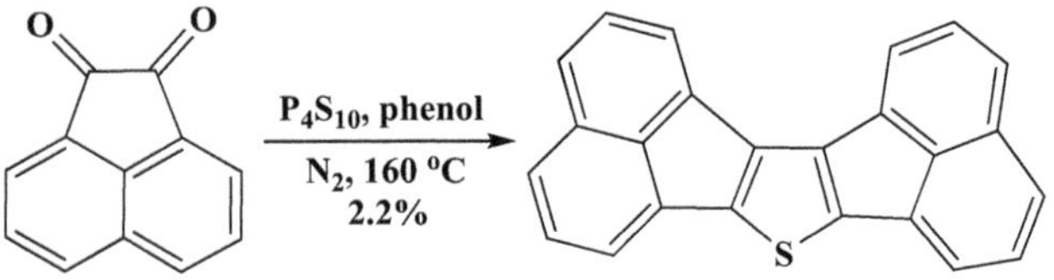

Scheme 8.54 Synthesis of fused thiophene

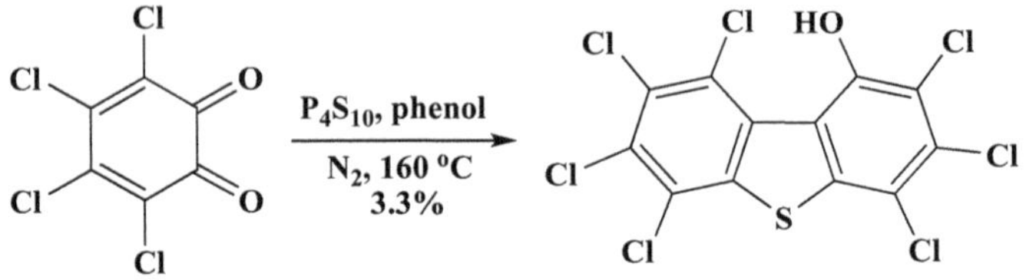

Scheme 8.55 Synthesis of fused thiophene

Scheme 8.56 Synthesis of dibenzo[b,d]thiophene

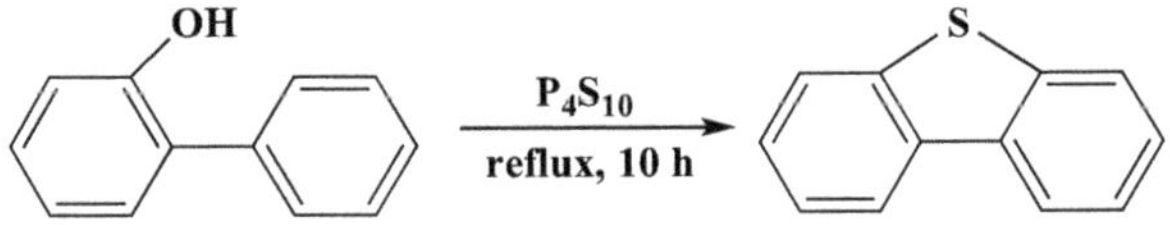

2,2′-dihydroxydiphenyl was heated with phosphorus pentasulfide at 350 °C for 15 min in carbon dioxide in an autoclave (Scheme 8.57) [17].

Strained systems have been synthesized, albeit in low yields (Schemes 8.58 and 8.59) [52].

More complex and functionalized 1,4-dicarbonyl compounds have been utilized in Paal–Knorr reactions for the formation of cyclopenta[b]thiophenes (Schemes 8.60, 8.61, and 8.62) [52].

Scheme 8.57 Synthesis of dibenzothiophene S,S-dioxide

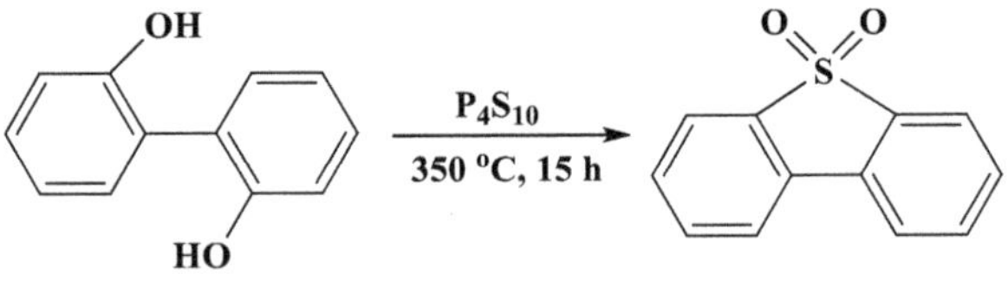

Scheme 8.58 Synthesis of thiophene

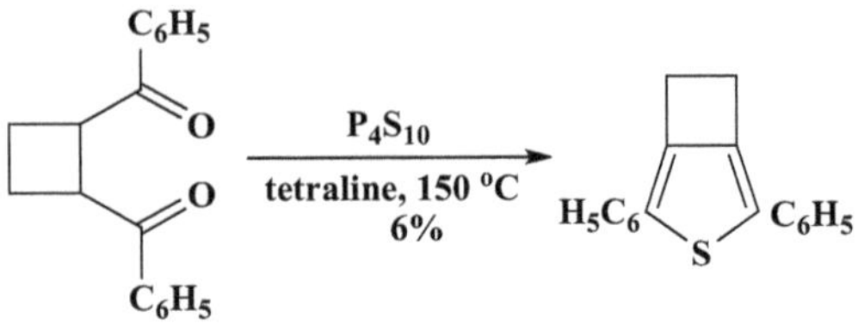

8.3 P$_4$S$_{10}$ in Heterocycle Synthesis

Scheme 8.59 Synthesis of thiophene

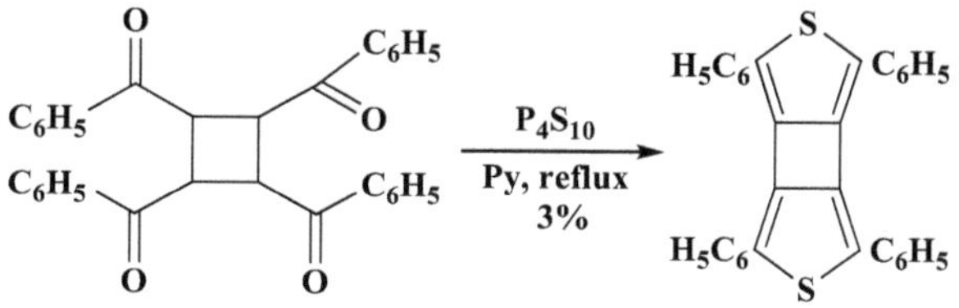

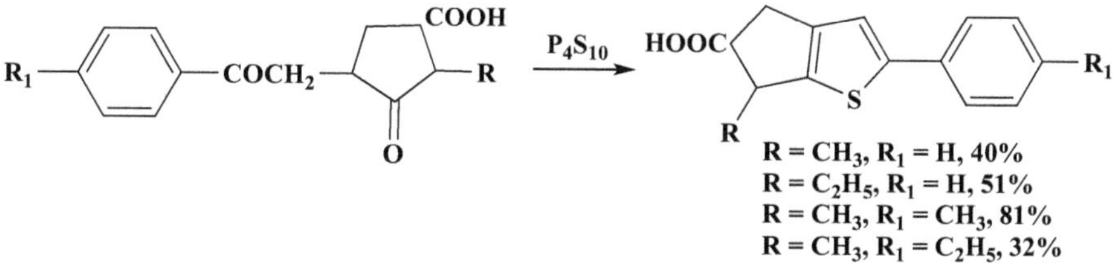

Scheme 8.60 Synthesis of cyclopenta[b]thiophenes

R = CH$_3$, R$_1$ = H, 40%
R = C$_2$H$_5$, R$_1$ = H, 51%
R = CH$_3$, R$_1$ = CH$_3$, 81%
R = CH$_3$, R$_1$ = C$_2$H$_5$, 32%

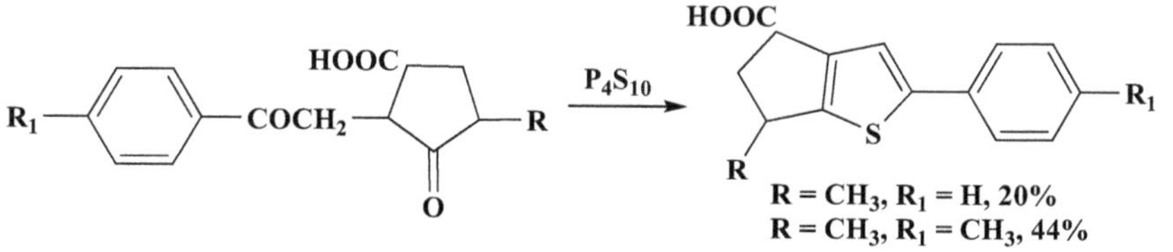

Scheme 8.61 Synthesis of cyclopenta[b]thiophenes

R = CH$_3$, R$_1$ = H, 20%
R = CH$_3$, R$_1$ = CH$_3$, 44%

Scheme 8.62 Synthesis of cyclopenta[b]thiophenes

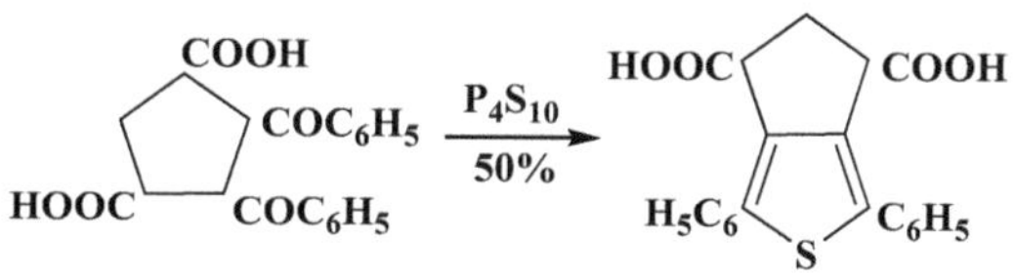

Capan et al. [69] prepared thienothiophenes, having *p*-substituted phenyl groups at C-3, in moderate to good yields by a ring-closure reaction of monoketones with phosphorus pentasulfide (Scheme 8.63) [70, 71].

Scheme 8.63 Synthesis of thienothiophenes

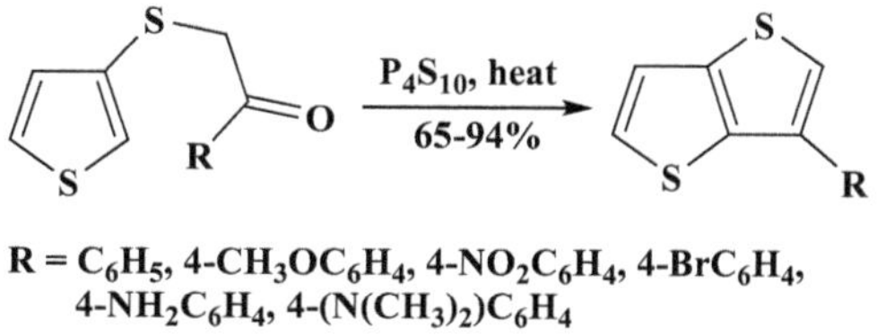

R = C$_6$H$_5$, 4-CH$_3$OC$_6$H$_4$, 4-NO$_2$C$_6$H$_4$, 4-BrC$_6$H$_4$,
4-NH$_2$C$_6$H$_4$, 4-(N(CH$_3$)$_2$)C$_6$H$_4$

The tetrabenzoylethane was reacted in three-step pathway to provide the tetraphenylthieno[3,4-c]thiophene in 38% overall yield. A double ring-closure using phosphorus pentasulfide afforded 4,6-dihydro-1,3,4,6-tetraphenylthieno[3,4-c]thiophene. The reaction of 4,6-dihydro-1,3,4,6-tetraphenylthieno[3,4-c]thiophene with sodium periodate afforded sulfoxide in 96% yield. The dehydration with Ac$_2$O afforded isolable "nonclassical" thienothiophene in 87% yield (Scheme 8.64) [71–73].

Ozturk et al. [74–76] reacted 1,8-diketones with phosphorus pentasulfide to prepare the DDTs, having aryl units (Ph, 4-BrC$_6$H$_4$, 4-MeOC$_6$H$_4$, and 4-O$_2$NC$_6$H$_4$) in 53–95% yield (Scheme 8.65). The dithienothiophene was reacted with Ag salt for the synthesis of a silver-epoxy nanocomposite [77]. The electron transfer reaction was successful between photo-excited dithienothiophene and the Ag salt under visible light irradiation [78]. Osken et al. [79, 80] also prepared dithienothiophene having 2-thienyl, and conducted its electro-copolymerization with ethylenedioxythiophene (EDOT) on a glassy carbon electrode (GCE) and platinum electrodes. The mechanism of the synthesis of dithienothiophene using phosphorus pentasulfide showed that the attack of CO oxygen on the phosphorus atom of phosphorus pentasulfide occurred resulting in an electrophilic carbon, of which, in turn, intramolecular attack from α-position of the thiophene generated an intermediate and then the dithienothiophene [71, 81].

Paal–Knorr approach has been beneficial for the formation of nonclassical thienothiophenes (Scheme 8.66). It was interesting to note that tetrabenzoylethane in xylene afforded cis- and trans-dihydro compounds, on the other hand, in pyridine, the nonclassical compound was formed [52].

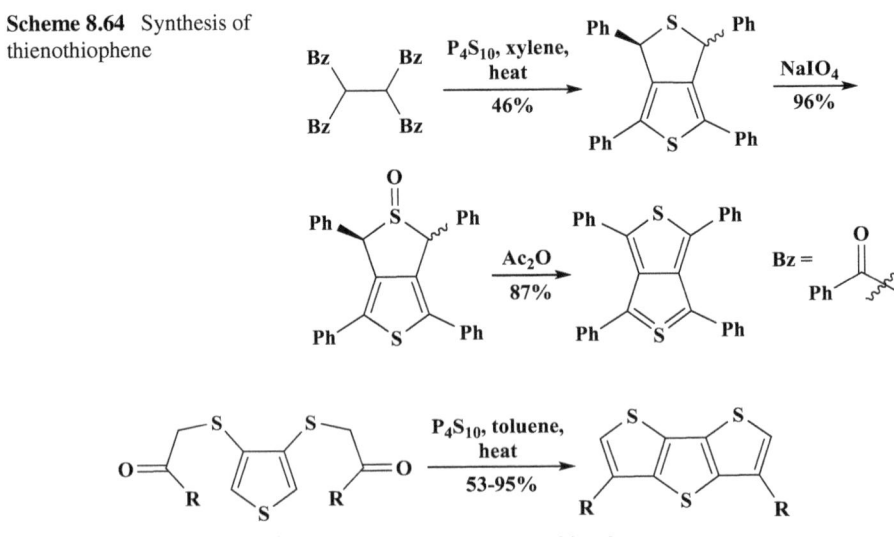

Scheme 8.64 Synthesis of thienothiophene

R = Ph, 4-MeOC$_6$H$_4$, 4-BrC$_6$H$_4$, 4-NO$_2$C$_6$H$_4$, 2-thienyl

Scheme 8.65 Synthesis of dithienothiophenes

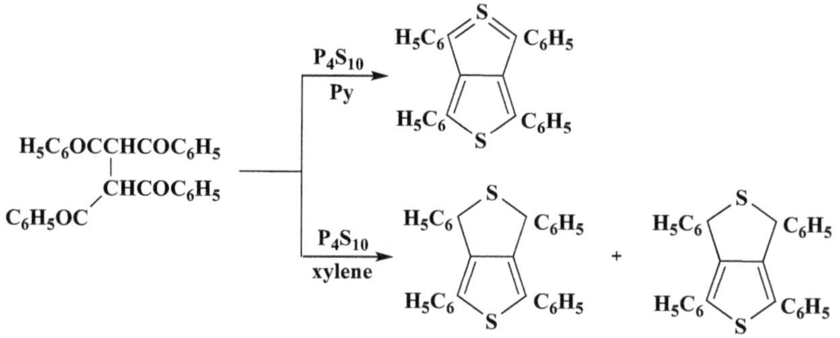

Scheme 8.66 Synthesis of thienothiophenes

One more example is the formation of thienodiazole from 3,4-dibenzoyl-1,2,5-thiadiazole in 53% yield (Scheme 8.67) [52].

The dicarbonyl compound was treated with phosphorus pentasulfide in refluxing dioxane for 4 h to provide the dithiolactone (Scheme 8.68) [17, 82].

A short, convenient, and robust method was described to prepare the 8-alkyl-2-(het)arylthieno[2,3-b]indoles from 1-alkylisatins and the acetylated (hetero)arenes which are readily available reagents, including commercially accessible ones. It is well known that the reaction of isatins with methyl ketones afforded aldol-type adducts under catalysis of mild bases, like secondary or tertiary amines. These adducts were dehydrated smoothly with acidic agents to provide the crotonic condensation products, 3-(2-oxo-2-(hetero)arylethylidene)indolin-2-ones, which underwent reduction of carbon–carbon double bond with $Na_2S_2O_4$ [83], hydrogen/palladium over carbon [84] or trimethylphosphine-water [85] into indolin-2-ones. The indolin-2-ones having 4-oxobutyramide (1,4-dicarbonyl derivative) fragment were cyclized

Scheme 8.67 Synthesis of thienothiadiazole

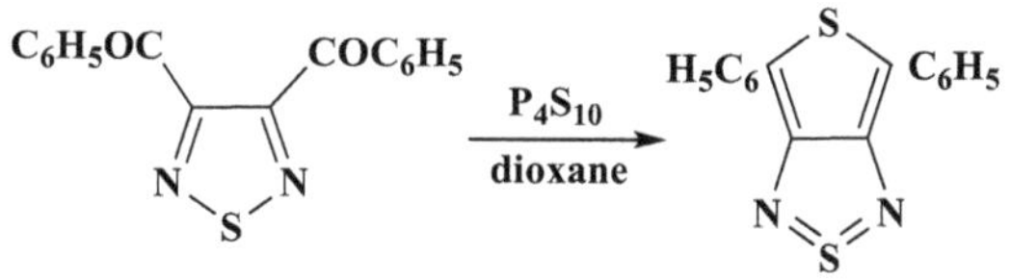

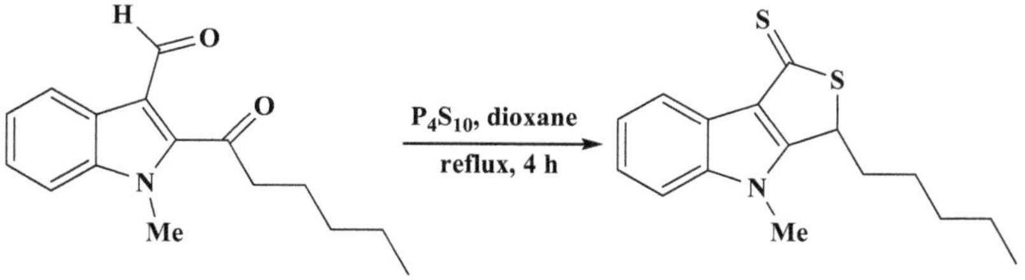

Scheme 8.68 Synthesis of 4-methyl-3-pentyl-3,4-dihydro-1H-thieno[3,4-b]indole-1-thione

into thieno[2,3-*b*]indole by Paal–Knorr reaction with thionation agents such as phosphorus pentasulfide or LR. This four-step pathway to thieno[2,3-*b*]indoles through the synthesis of indoline-2-ones from isatins and methyl ketones has been realized earlier [86, 87]. In particular, the reaction of unsubstituted isatin with CH_3COCH_3 afforded 2-methyl-8*H*-thieno[2,3-*b*]indole in 15% yield (Scheme 8.69). Although it appeared to be a very harmonious methodology, it has hardly a noteworthy synthetic interest, since the desired compounds were obtained in low yields from isatins and ketones in four steps [88].

Although thionation of indolin-2-one took place with phosphorus pentasulfide [87], different by-products were formed (Scheme 8.70) [89, 90]. The indoline-2-thione was obtained in 65% yield along with the side-products as major and minor products, respectively, when indolin-2-one was reacted with phosphorus pentasulfide in refluxing benzene for 2 h. The amount of by-products enhanced to 45% when the reaction was performed in xylene. The by-products were formed through the formation of an indoline-2-thione intermediate [17].

The fused thiophene was directly obtained during the formation of thioketones from (2-haloaryl)ketones utilizing phosphorus pentasulfide as a thionation reagent, as the 2-halophenyl group was replaced by 2-chloropyridyl group (Scheme 8.71) [17, 91].

The reaction of 3,4-dibenzoylpyrazole, which is a 1,4-diketone, with phosphorus pentasulfide in refluxing dry pyridine for 5 h provided thienopyrazole in 85% yield, and the resonance structures are depicted in Scheme 8.72 [17, 92].

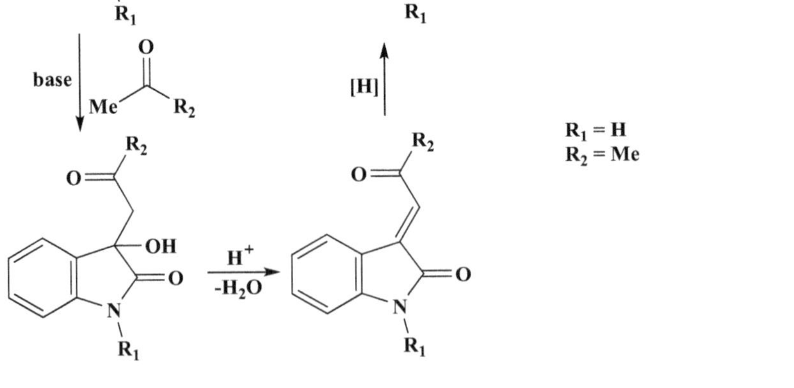

Scheme 8.69 Synthesis of 2-methyl-8*H*-thieno[2,3-*b*]indole

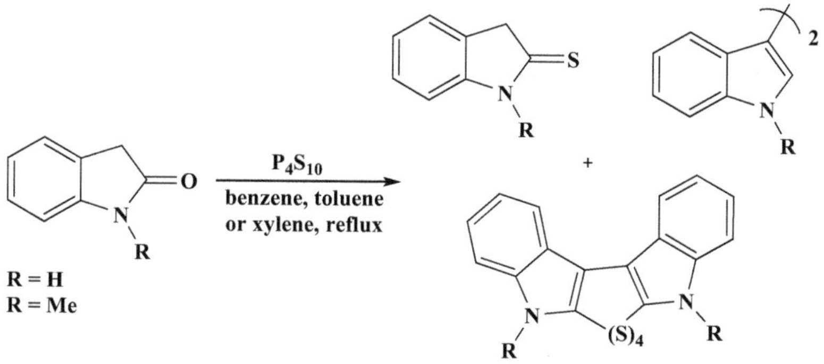

8.3 P$_4$S$_{10}$ in Heterocycle Synthesis

Scheme 8.70 Synthesis of indoline-2-thiones

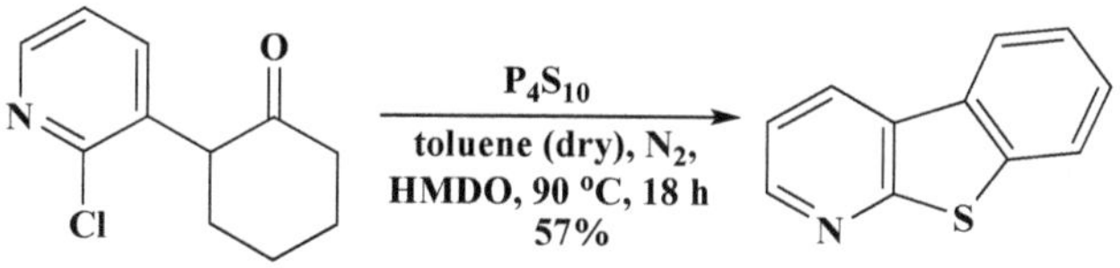

Scheme 8.71 Synthesis of fused thiophene

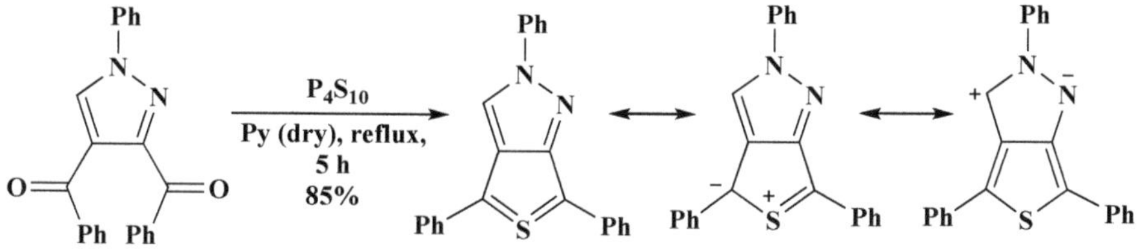

Scheme 8.72 Synthesis of thienopyrazole

The fused thiophenes were obtained in 38–64% yield by reacting γ-hydroxycarbonyls with P$_4$S$_{10}$ in refluxing pyridine (Scheme 8.73) [17, 93].

The fused thiophene was afforded by the treatment of a more complex γ-hydroxycarbonyl with phosphorus pentasulfide in thioacetamide (Scheme 8.74) [17, 94, 95].

The reaction of both unsaturated carbonyls [96] and Mannich bases provided thiophenes fused to 1,2,4-triazines (Schemes 8.75 and 8.76) [17].

The reaction of 1,2-diketone bearing two indole functionalities with phosphorus pentasulfide in refluxing pyridine provided thiophene derivative, albeit in low yield (9.5%). The X-ray crystallography was performed to determine the structure of thiophene (Scheme 8.77) [17, 97].

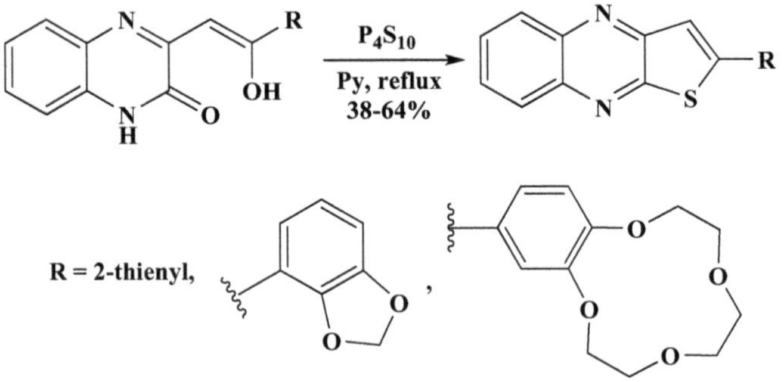

Scheme 8.73 Synthesis of thienoquinoxalines

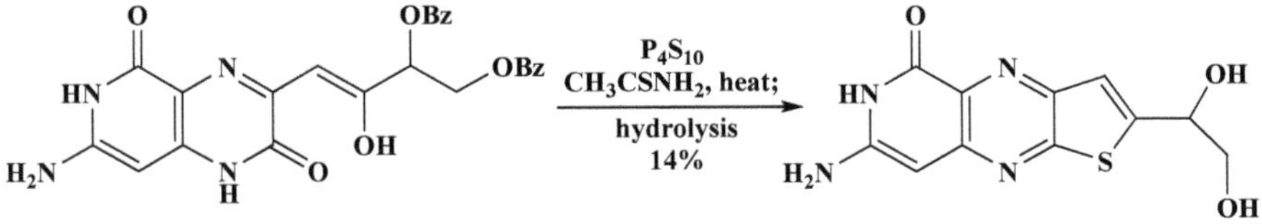

Scheme 8.74 Synthesis of pyridothienopyrazine

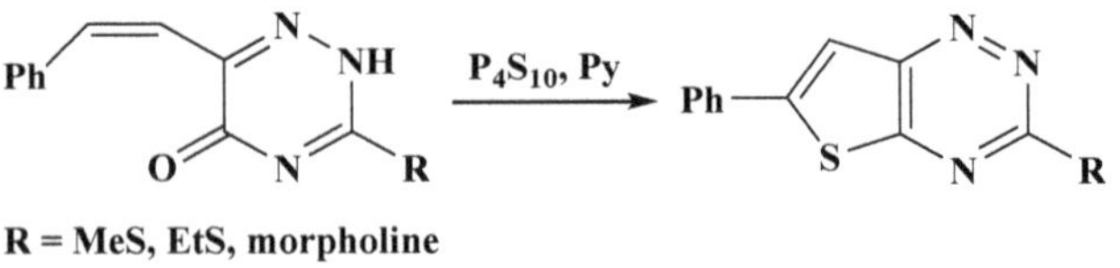

Scheme 8.75 Synthesis of thieno-1,2,4-triazines

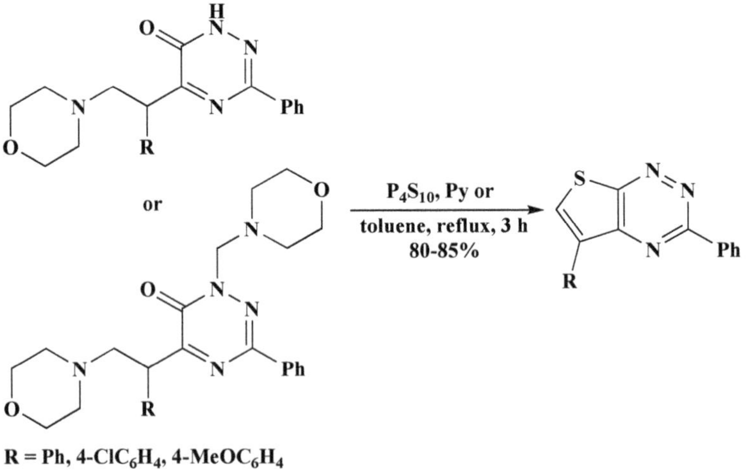

Scheme 8.76 Synthesis of thieno-1,2,4-triazines

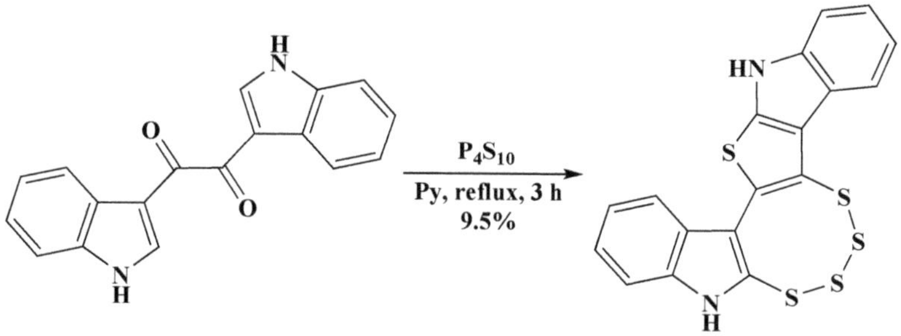

Scheme 8.77 Synthesis of fused thiophene

8.3.1.4 Synthesis of Five-Membered *S,S*-Heterocycles

The thione derivative 2,3-diphenylcyclopropenethione was obtained in 68% yield by the reaction of 2,3-diphenylcyclopropenone with phosphorus pentasulfide in benzene (dry) at 50 to 60 °C (Scheme 8.78) [98]. On the other hand, dithiolethione rather than 2,3-diphenylcyclopropenethione was obtained when 2,3-diphenylcyclopropenone was treated with phosphorus pentasulfide [99]. Further, a rather extensive study showed that both 2,3-diphenylcyclopropenethione and dithiolethione were formed in equal ratios on the treatment of 2,3-diphenylcyclopropenone with phosphorus pentasulfide in benzene at 45 °C [100]. Moreover, dithiolethione was obtained as only product in 10% yield when the mixture was refluxed for 30 min. The thione in 15% yield and dithiolethione in trace amounts were obtained when the same reaction was performed at rt [17].

The reaction of 2-thioxo-1,3-thiazine-4-one with phosphorus pentasulfide in hot pyridine provided 1,2-dithiol-3-thione in 31% yield (Scheme 8.79) [17, 101].

The thionation of diacylacetamide, prepared by the reaction of diacylacetamide with phosphorus pentaoxide, with phosphorus pentasulfide in refluxing xylene for 1 h provided 1,2-dithiole quantitatively (Scheme 8.80) [17, 102].

Scheme 8.78 Synthesis of 2,3-diphenylcycloprop-2-ene-1-thione and 4,5-diphenyl-3*H*-1,2-dithiole-3-thione

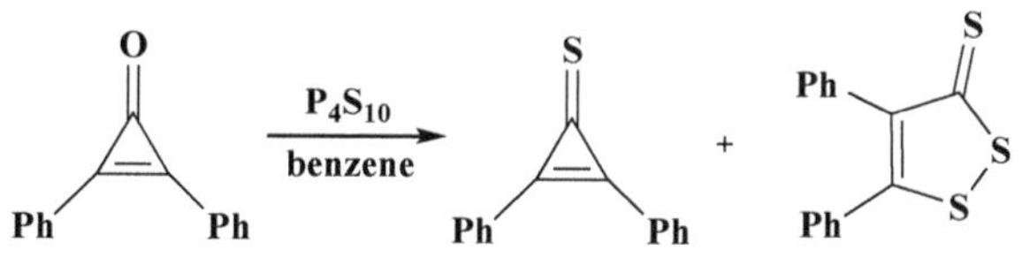

Scheme 8.79 Synthesis of 1,2-dithiol-3-thione

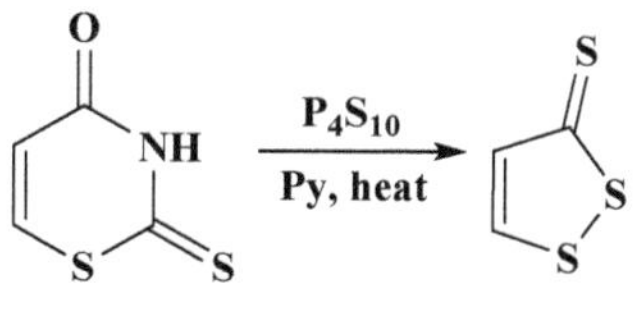

Scheme 8.80 Synthesis of 1,2-dithiole

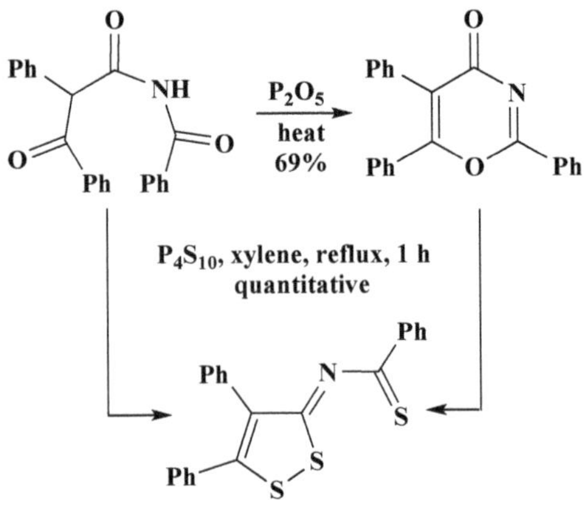

The 3H-1,2-dithiole-3-thiones are pseudoaromatic heterocyclic compounds. Many of them are pharmaceutically valuable products [103–106]. The 1,2-dithiole-3-thiones were synthesized by reacting ketoesters with phosphorous pentasulfide (Scheme 8.81) [107].

The 5-alkylthio-3H-1,2-dithiol-3-thiones were prepared by reacting dialkyl malonate ester with P_4S_{10} and sulfur in boiling xylene in the presence of 2-mercaptobenothiazole/ZnO as a catalyst. The reaction involved the initial transformation of ester moiety to thionoester groups, which underwent rearrangement to produce the thioesters. The reaction of thioesters with phosphorus pentasulfide resulted in cyclization to provide the substituted dithiole-3-thione through dithioester intermediates (Scheme 8.82) [108].

The displacement of one or both of the alkylthio groups of ketene dithioacetal provided α-oxoketene N,S-acetals or aminals, respectively. There are some reactions which convert acyl ketene dithioacetals directly into functionalized heterocyclic

Scheme 8.81 Synthesis of 1,2-dithiole-3-thiones

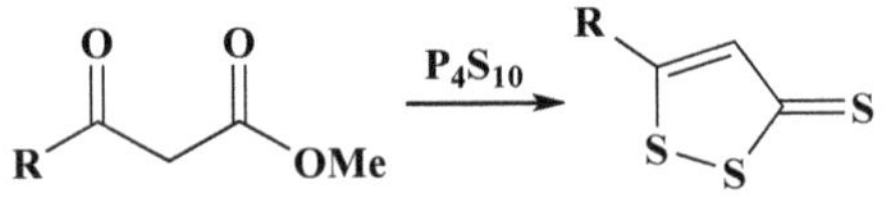

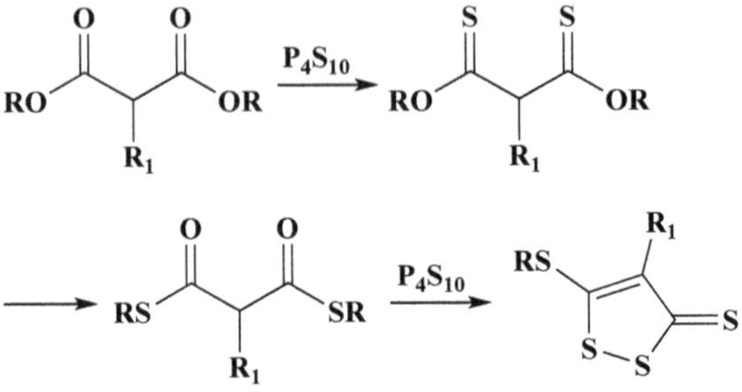

Scheme 8.82 Synthesis of 5-alkylthio-3H-1,2-dithiol-3-thiones

compounds. For example, the reaction of α-oxo ketene dithioacetals with phosphorus pentasulfide provided a convenient process for the formation of 3-thione-1,2-dithiols (Scheme 8.83) [52].

The bis(2-carboxy-3-chlorophenyl)disulfur was reacted with phosphorus pentasulfide in refluxing xylene to provide the dithiolethione in 45% yield (Scheme 8.84) [17, 109].

The mixture of TFA/CH$_3$COOH has been utilized for the cyclization of butyltrithiocarbonates with P$_4$S$_{10}$ to provide the dithiole [110]. Alternatively, the keto-alkyldithiocarbonates were reacted with P$_4$S$_{10}$ in boiling decaline to synthesize the dithiole (Scheme 8.85) [111–114].

The dithiole was afforded by an effort to thionate the ketone, containing an epoxide moiety, the mechanism of which was suggested to involve the intermediates (Scheme 8.86) [115]. The reaction of epoxyketone with phosphorus pentasulfide provided a similar result where dithiole was obtained in 50% yield (Scheme 8.87) [17, 116].

Scheme 8.83 Synthesis of 1,2-dithiol-3-thiones

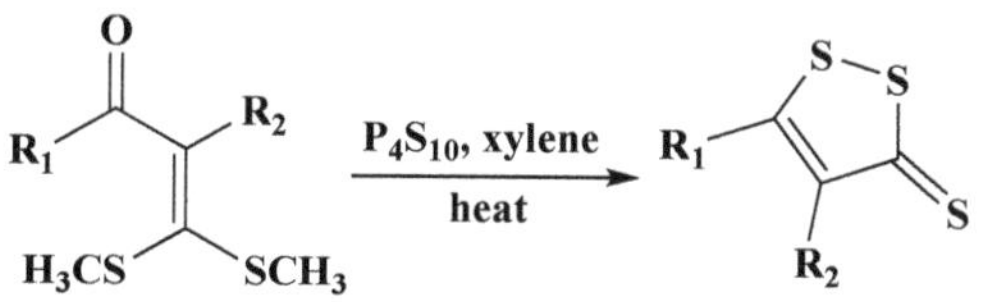

Scheme 8.84 Synthesis of dithiolethione

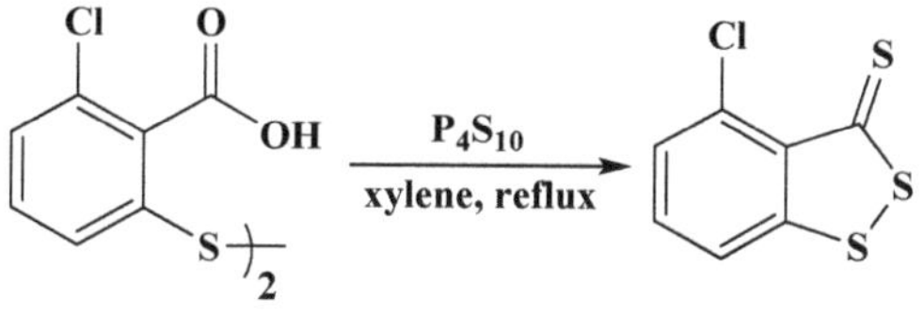

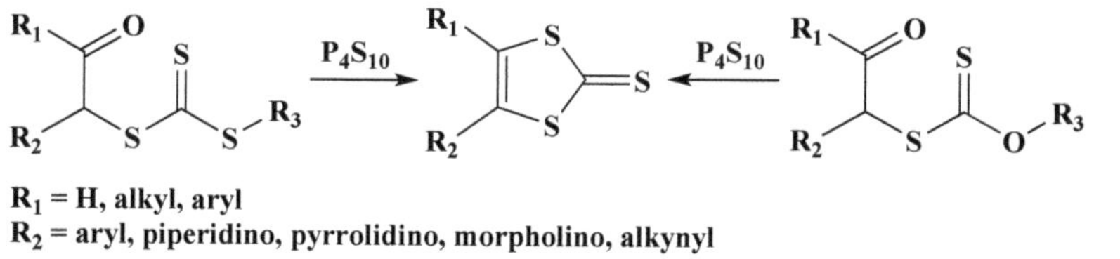

R_1 = H, alkyl, aryl
R_2 = aryl, piperidino, pyrrolidino, morpholino, alkynyl

Scheme 8.85 Synthesis of dithioles

Scheme 8.86 Synthesis of dithioles

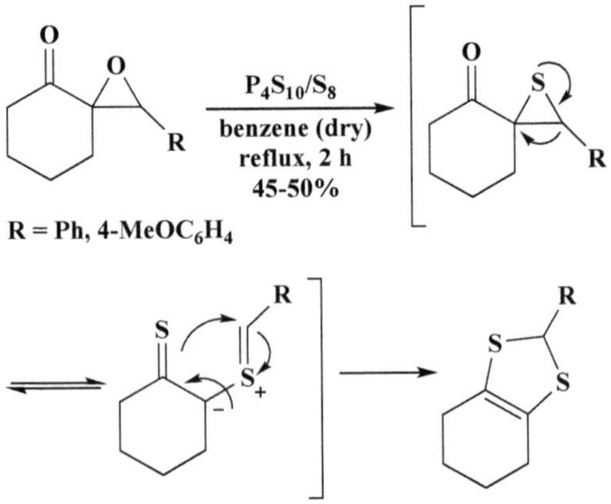

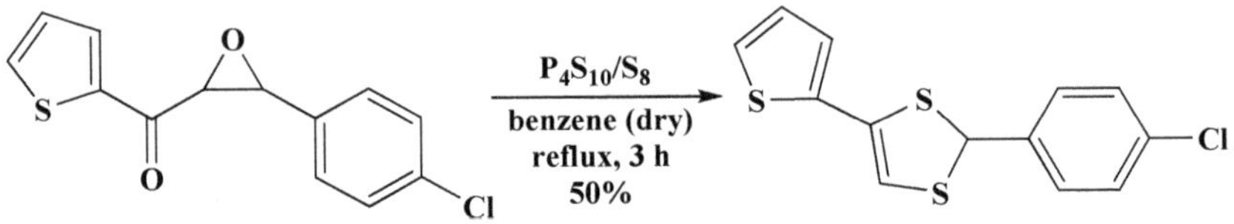

Scheme 8.87 Synthesis of dithioles

8.3.1.5 Synthesis of Five-Membered *S,S,S*-Heterocycles

The thioketene having trifluoromethylsulfanyl groups was synthesized [117]. The dimers were prepared through the formation of thioketene intermediate by the reaction of acetyl chloride, carboxylic acid, and ketene with phosphorus pentasulfide in refluxing toluene (Scheme 8.88) [17].

The 1,2,4-trithiolane is an unexpected product of phosphorus pentasulfide obtained by the reaction of some ketones with phosphorus pentasulfide. Two *cis*- and *trans*-trithiolanes were isolated (Scheme 8.89) along with the required thioketone in an attempt to transform the oxo group of ketone to thio group [17, 118].

8.3.1.6 Synthesis of Five-Membered *S,N*-Heterocycles

The reaction of *N,N*-diformylaminomethyl aryl ketones with phosphorus pentasulfide in chloroform at 60 °C for 45–60 min synthesized arylthiazoles (Scheme 8.90) [17, 119].

A mixture of thiazoles was obtained by the reaction of benzamidodiethoxypropionate with a mixture of phosphorus pentasulfide/sulfur in carbon disulfide or xylene (Scheme 8.91) [17, 120].

The reaction of tris(hydroxamide)s with phosphorus pentasulfide and triethylamine in refluxing toluene provided tris(thiazoline)s (Scheme 8.92) [17, 121].

8.3 P₄S₁₀ in Heterocycle Synthesis

Scheme 8.88 Synthesis of 1,3-dithietane and 1,2,4-trithiolane

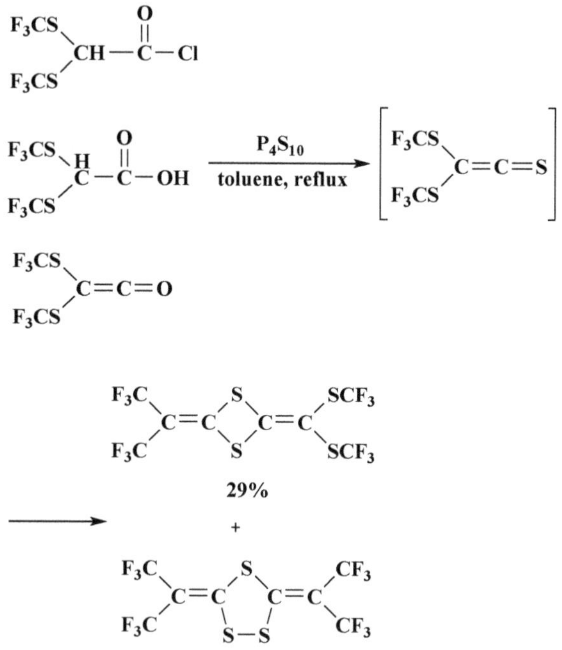

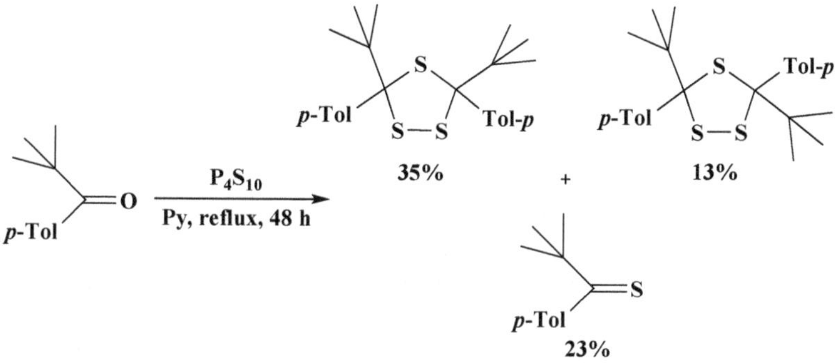

Scheme 8.89 Synthesis of trithiolanes

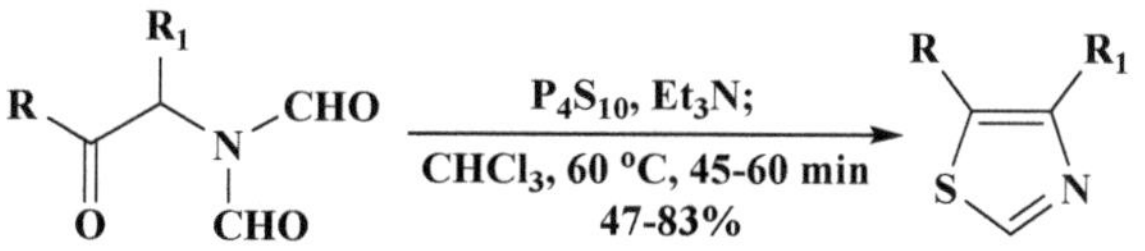

Scheme 8.90 Synthesis of thiazoles

R = 4-FC$_6$H$_4$, 2-MeOC$_6$H$_4$, 4-MeOC$_6$H$_4$, 2-MeC$_6$H$_4$, 4-MeC$_6$H$_4$, 3-NO$_2$C$_6$H$_4$, 4-ClC$_6$H$_4$, 1-naphthyl, 2-naphthyl, 2-furyl, 2-thienyl
R$_1$ = H, Me

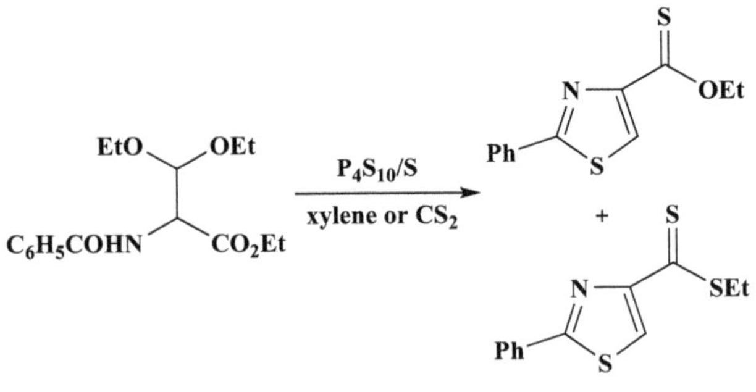

Scheme 8.91 Synthesis of thiazoles

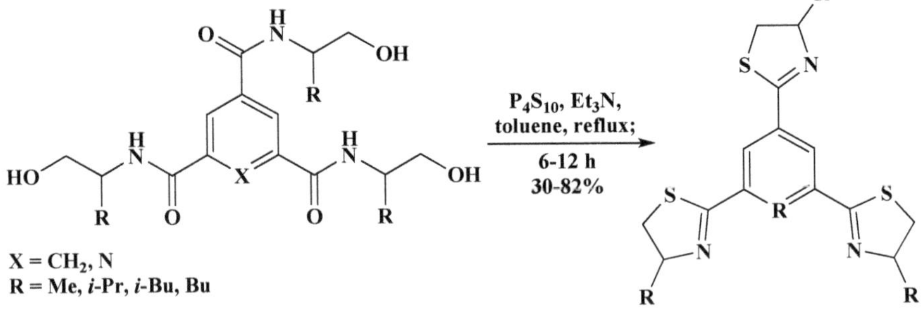

Scheme 8.92 Synthesis of tris-thiazolines

The 1,3-dioxolane analogues of thiazole nucleosides have been synthesized starting from thiazole derivative which in turn was easily accessible from 2,2-diethoxyacetamide in a one-pot reaction (Scheme 8.93) [122].

Ayer and coworkers [123] have used Hantzsch reaction as a key step in the formation of natural product camalexin. The indole-3-carboxamide was treated with P_4S_{10} in benzene for 3 h to provide the indole-3-thiocarboxamide in situ, which was further reacted with chloroacetaldehyde diethyl acetal in refluxing EtOH for 15 h to produce the camalexin in 35% yield (Scheme 8.94).

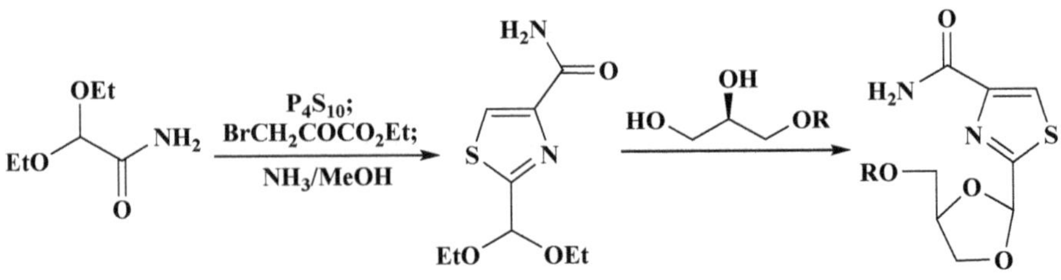

Scheme 8.93 Synthesis of dioxolanothiazoles

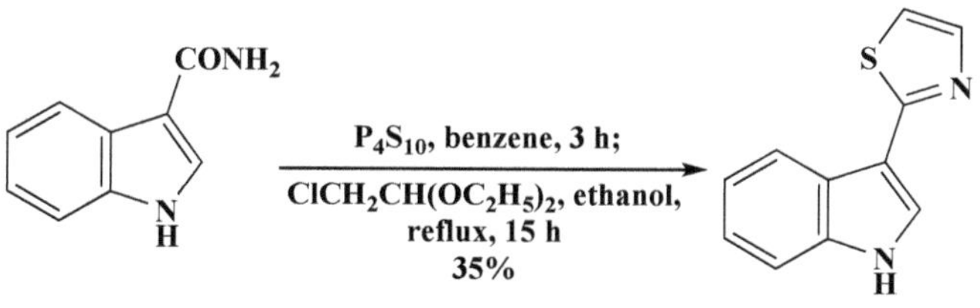

8.3 P$_4$S$_{10}$ in Heterocycle Synthesis

Scheme 8.94 Synthesis of camalexin

The thiazolines were obtained in 32–49% yield by the reaction of oxazolines with phosphorus pentasulfide in refluxing DCM for a long time (140 h) (Scheme 8.95) [17, 124].

Many 2,4-disubstituted 5-acetoxythiazoles were synthesized by reacting methyl thiobenzoate derivatives, prepared from methyl benzoate, with racemic phenylglycine. The reaction was completed utilizing a two-phase reaction mixture containing 3 N sodium hydroxide and ether. The coupled product was reacted with Ac$_2$O to synthesize the required thiazole derivatives (Scheme 8.96) [125].

The thiazoline ring was obtained in 70–77% yield by an interesting reaction of *N*-aroylaziridines with phosphorus pentasulfide in refluxing toluene for 3 h (Scheme 8.97 and 8.98) [126]. The proposed mechanism suggested the involvement of aziridine-1-thione intermediate, which was rearranged to thiazolines [17].

The isatins were utilized for the formation of fused indole derivatives. The reduction of 1-methylisatin-3-oximes with zinc in acidic media provided an

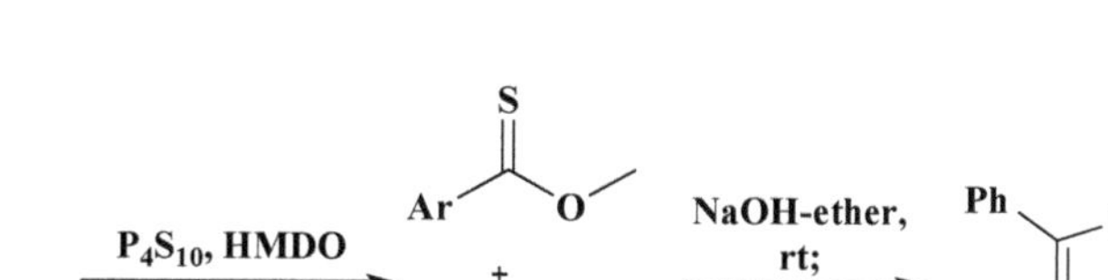

R_1 = H, Et
R_2 = H, Bn, *i*-Pr

Scheme 8.95 Synthesis of thiazolines

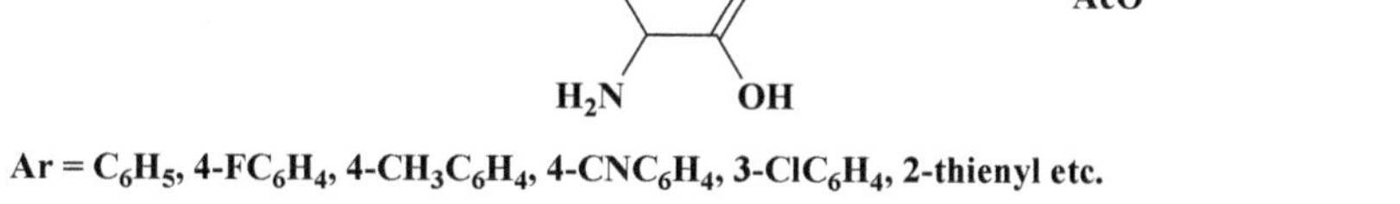

Ar = C$_6$H$_5$, 4-FC$_6$H$_4$, 4-CH$_3$C$_6$H$_4$, 4-CNC$_6$H$_4$, 3-ClC$_6$H$_4$, 2-thienyl etc.

Scheme 8.96 Synthesis of 2,4-disubstituted 5-acetoxythiazoles

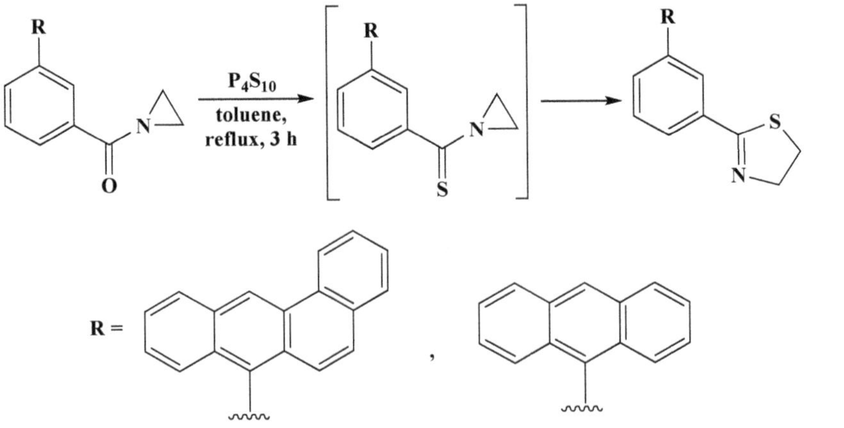

Scheme 8.97 Synthesis of thiazolines

Scheme 8.98 Synthesis of thiazolines

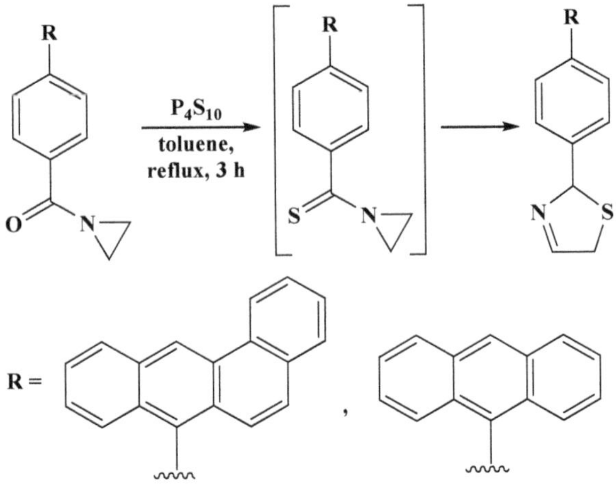

acetamidooxindole, which was treated with phosphorus pentasulfide to provide the indolothiazoles in yields ranging from moderate to good (Scheme 8.99) [127, 128].

The imidazothiazole was afforded by the reaction of hydantoin with phosphorus pentasulfide in boiling dioxane, possibly via thionoester which provided thiazole upon removal of EtOH (Scheme 8.100) [17, 129].

The thiazoline and thiazole heterocyclic compounds were provided by the reaction of 1-amide-4-hydroxyl and 1-amide-4-carbonyl systems. The reaction of 1-amide-4-halogen system with phosphorus pentasulfide provided similar results. The reaction of γ-chloro- (Scheme 8.101) and γ-bromo amides (Scheme 8.102) with phosphorus pentasulfide in pyridine at 100 °C for 2 h and neat at 145 to 150 °C (1 h) and 120 °C (2 h), respectively, provided thiazoles [17, 130, 131].

8.3 P$_4$S$_{10}$ in Heterocycle Synthesis

Scheme 8.99 Synthesis of indolothiazoles

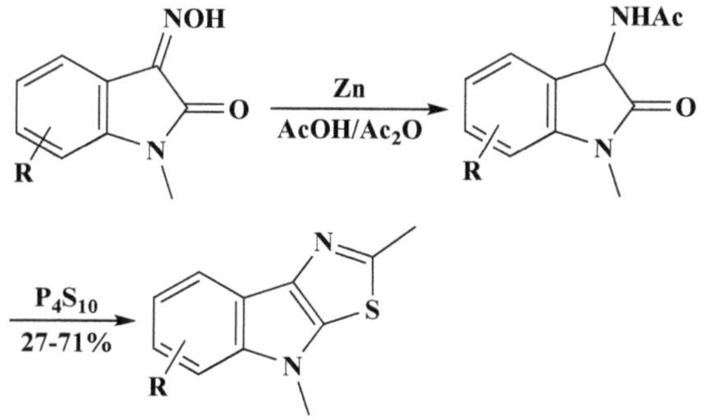

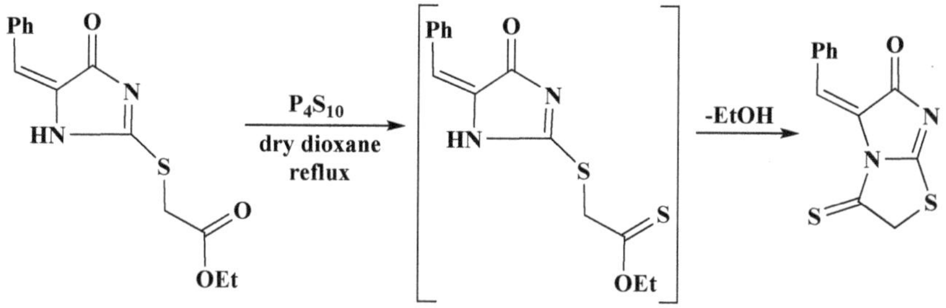

Scheme 8.100 Synthesis of imidazothiazole

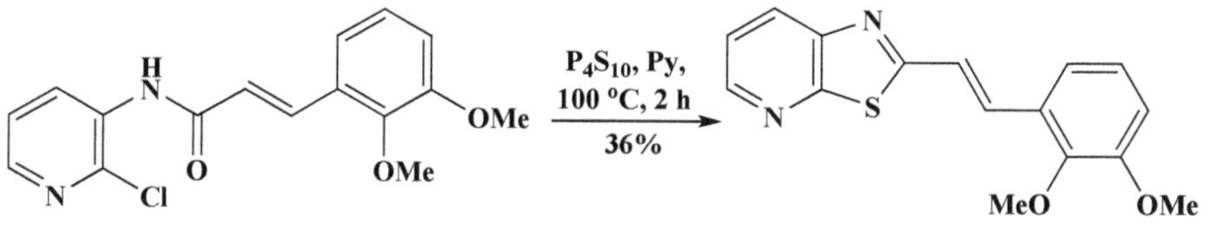

Scheme 8.101 Synthesis of thiazolopyridine

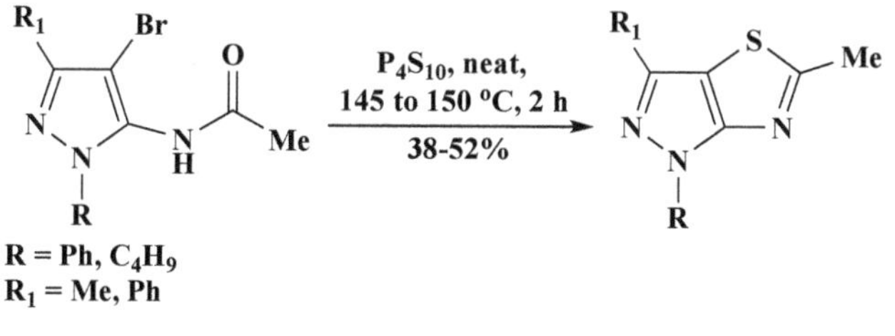

R = Ph, C$_4$H$_9$
R$_1$ = Me, Ph

Scheme 8.102 Synthesis of pyrazolothiazoles

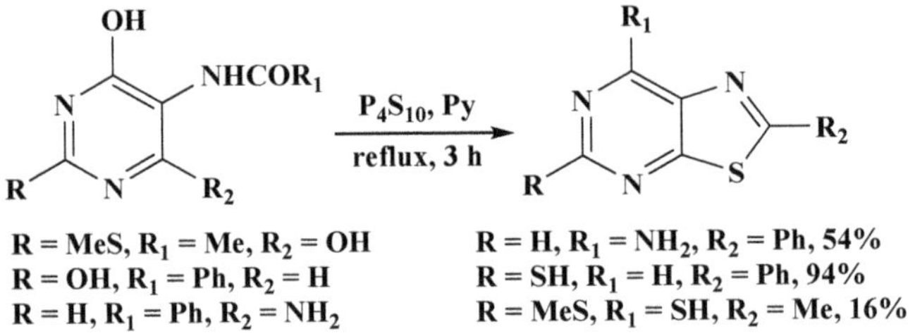
OH
NHCOR1
N
R
N
R2
P4S10, Py
reflux, 3 h
R1
N
N
R2
R
N
S
R = MeS, R1 = Me, R2 = OH
R = OH, R1 = Ph, R2 = H
R = H, R1 = Ph, R2 = NH2
R = H, R1 = NH2, R2 = Ph, 54%
R = SH, R1 = H, R2 = Ph, 94%
R = MeS, R1 = SH, R2 = Me, 16%

Scheme 8.103 Synthesis of thiazolopyrimidines

Scheme 8.104 Synthesis of thiazoline

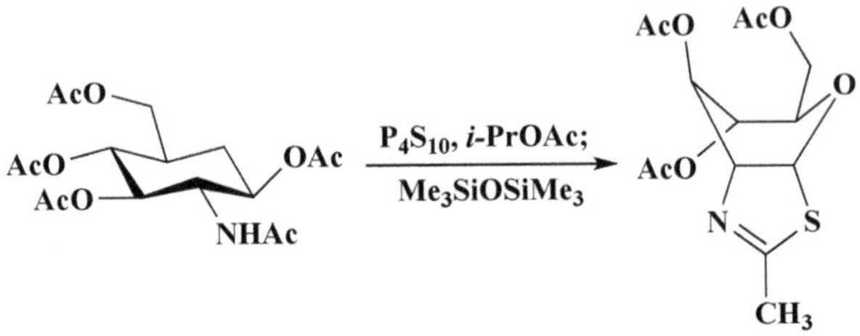
AcO
AcO
AcO
OAc
NHAc
P4S10, i-PrOAc;
Me3SiOSiMe3
AcO
AcO
O
AcO
N
S
CH3

The thiazolopyrimidines were synthesized by the reaction of pyrimidines with phosphorus pentasulfide in refluxing pyridine for 3 h (Scheme 8.103) [17, 132].

The GlcNAc-thiazoline triacetate was prepared in quantitative yield from 2-acetamido-2-deoxy-tetra-*O*-acetyl-D-glucopyranose utilizing freshly prepared LR, 2,4-bis(4-methoxyphenyl)-1,3-dithia-2,4-diphosphetane-2,4-disulfide (Scheme 8.104) [133]. Similarly, GalNAc-thiazoline triacetate was prepared [134]. This thionation/cyclization reaction was based on the phosphorus pentasulfide/HMDS process of Curphey [135], which minimized the amount of phosphorus/sulfur side-products that must be removed chromatographically. The amount of phosphorus pentasulfide relative to starting compound was enhanced to 0.4 eq. in this process to offer the thionation of the side-product, CH_3COOH, and to ensure that the reaction proceeded to completion [136].

8.3.1.7 Synthesis of Five-Membered *S,N,N*-Heterocycles

Dankova et al. [137, 138] synthesized cyanothioacetanilide by boiling amide with phosphorus pentasulfide in dioxane. The 5-anilino-1,2,3-thiadiazole-4-carbonitrile was obtained by the reaction of cyanothioacetanilide with azidobenzenesulfite. The 5-anilino-1,2,3-thiadiazole-4-carbothioamide was afforded by thiation of 5-anilino-1,2,3-thiadiazole-4-carbonitrile with H_2S (Scheme 8.105) [139].

The oxadiazolethiones were obtained in 45–64% yield by the reaction of oxadiazoleones with phosphorus pentasulfide in refluxing xylene (Scheme 8.106) [140]. The oxadiazolethiones underwent rearrangement to produce the thiadiazoleones, which

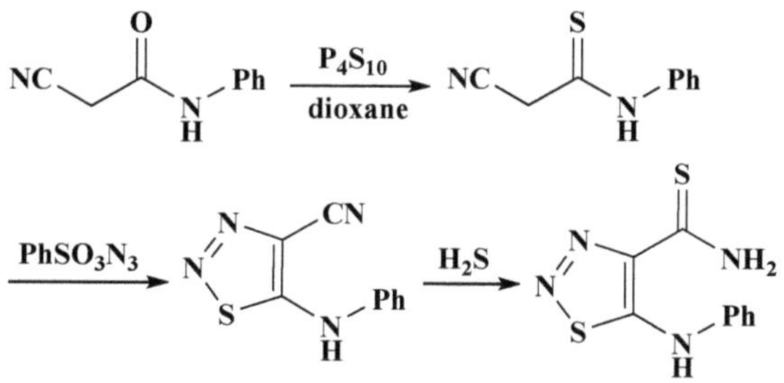

Scheme 8.105 Synthesis of thiadiazole

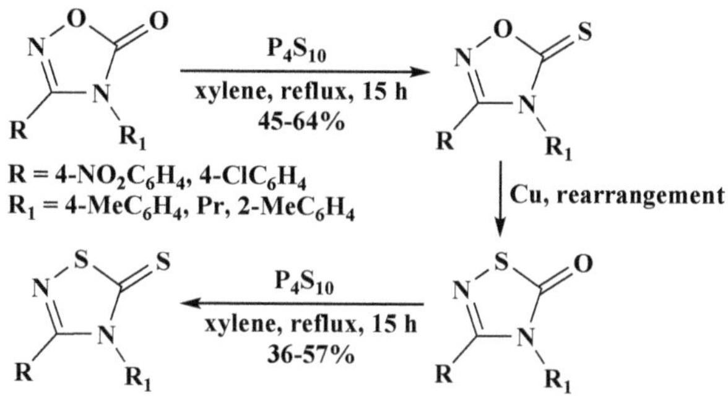

Scheme 8.106 Synthesis of thiadiazolethiones

R = 4-NO$_2$C$_6$H$_4$, 4-ClC$_6$H$_4$
R$_1$ = 4-MeC$_6$H$_4$, Pr, 2-MeC$_6$H$_4$

provided thiadiazolethiones [17] upon reaction with phosphorus pentasulfide under the same conditions.

The 5-*t*-butyl-3-[2,4-dichloro-5-(2-propynyloxy)phenyl]-1,3,4-thiadiazol-2(3*H*)-one, an arylthiadiazolone herbicide structurally related to oxadiargyl and oxadiazon, was synthesized in high yield in two steps starting from *N*-2,4-dichloro-5-(2-propynyloxy)phenyl]-*N'*-pivaloylhydrazine. The *N*-2,4-dichloro-5-(2-propynyloxy)phenyl]-*N'*-pivaloylhydrazine was converted into *N*-thiopivaloylhydrazine by treatment with P$_4$S$_{10}$ and later transformed into 5-*t*-butyl-3-[2,4-dichloro-5-(2-propynyloxy)phenyl]-1,3,4-thiadiazol-2(3*H*)-one upon reaction with trichloromethyl chloroformate in dioxane at rt for 3 h (Scheme 8.107) [141].

The thiadiazoles were synthesized by the reaction of hydrazides and triethylorthoformates with phosphorus pentasulfide in alumina (phosphorus pentasulfide/aluminum oxide) under MWI (Scheme 8.108) [17, 142].

The reaction of 1,3,4-oxadiazolinones with phosphorus pentasulfide in refluxing xylene provided 1,3,4-thiadiazolo[3,2-*a*]benzimidazoles (Scheme 8.109) [143]. The phosphorus pentasulfide-refluxing xylene system after an induction period of approximately 10 h synthesized hydrogen sulfide which, in turn, had potential of reducing the nitro group [144].

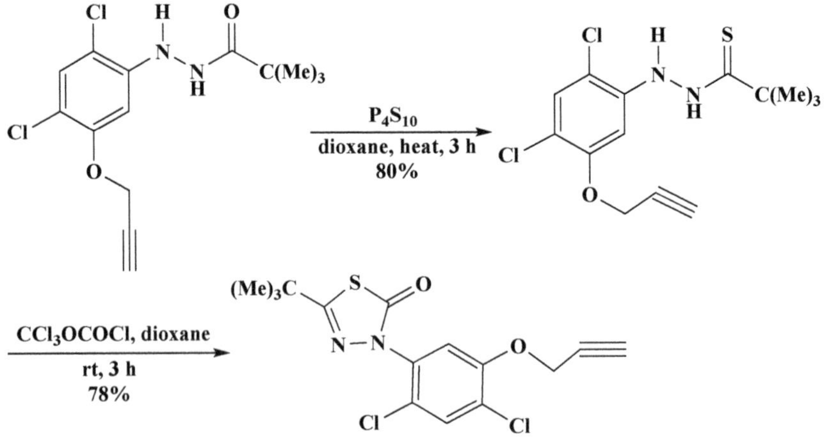

Scheme 8.107 Synthesis of 5-*t*-butyl-3-[2,4-dichloro-5-(2-propynyloxy)phenyl]-1,3,4-thiadiazol-2(3*H*)-one

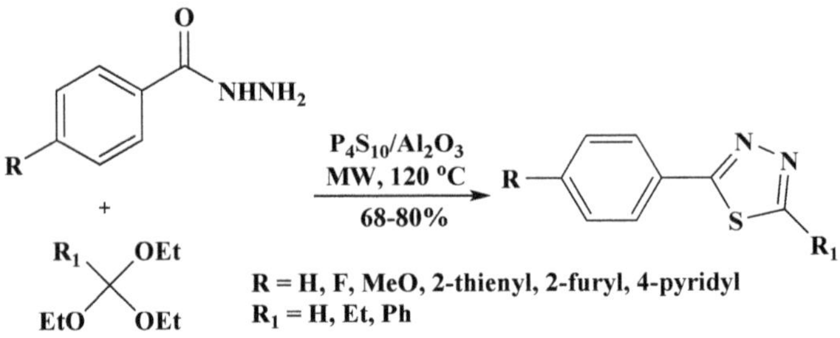

Scheme 8.108 Synthesis of thiadiazoles

8.3.1.8 Synthesis of Five-Membered *S,S,N*-Heterocycles

The benzothiazathiolium chloride was synthesized in good yield by the addition of chlorine to a mixture of phosphorus pentasulfide and 4-chloroaniline in Ac$_2$O at 75 °C (Scheme 8.110) [17, 145].

The reaction of oxathiazine-*S*-oxides with phosphorus pentasulfide in refluxing toluene for 1 h provided dithiazole, although it was evident that a somewhat higher yield was obtained utilizing Lawesson's reagent (Scheme 8.111) [17, 146].

The sodium *t*-amylate-mediated condensation of 2 eq. CS$_2$ with caprolactam, when quenched with excess MeI, provided adduct. The replacement of caprolactam with pyrrolidinone provided homolog. The tricyclic products were obtained in 60% and 41% yield, respectively, when adduct was reacted with P$_4$S$_{10}$ followed by perchloric acid (Scheme 8.112) [147].

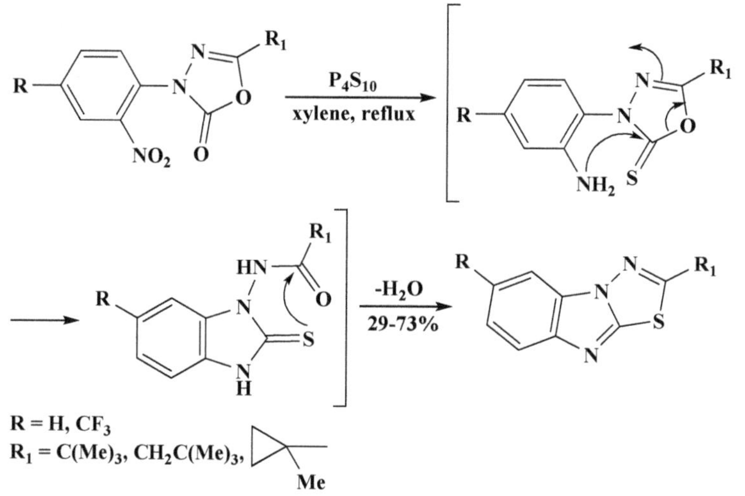

8.3 P$_4$S$_{10}$ in Heterocycle Synthesis

Scheme 8.109 Synthesis of 1,3,4-thiadiazolo[3,2-a]benzimidazoles

R = H, CF$_3$
R$_1$ = C(Me)$_3$, CH$_2$C(Me)$_3$, ▷—Me

Scheme 8.110 Synthesis of benzothiazathiolium chloride

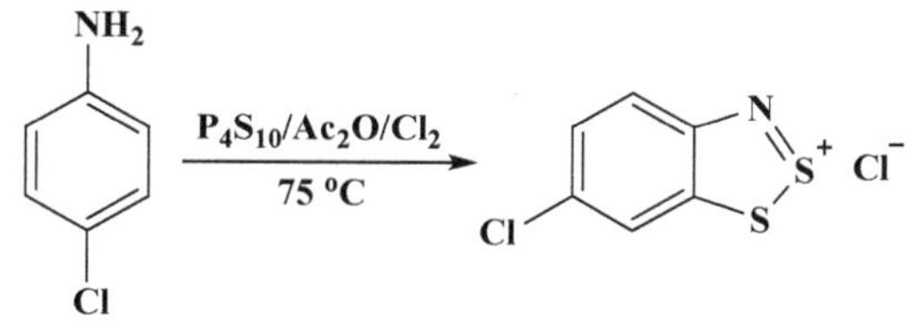

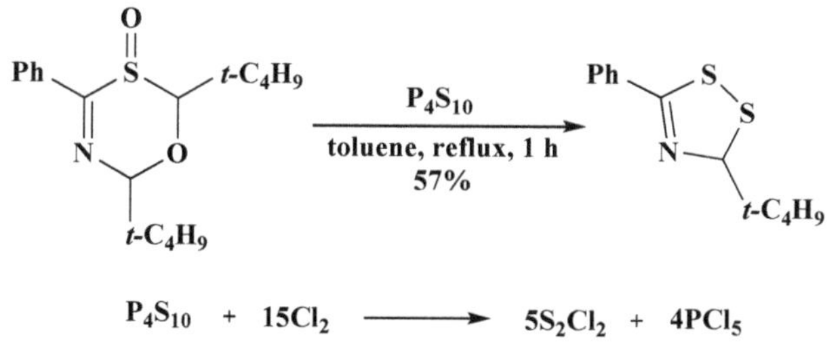

Scheme 8.111 Synthesis of dithiazole

P$_4$S$_{10}$ + 15Cl$_2$ ⟶ 5S$_2$Cl$_2$ + 4PCl$_5$

8.3.2 Synthesis of Six-Membered Heterocycles

During the synthesis of *Peripentadenia* alkaloids, Michael et al. [148] was interested to see if he could cyclize the pyrrolidine intermediate to synthesize the indolizidine skeleton (Scheme 8.113). The intermediate was treated with 2 eq. *t*-BuOK in THF at rt to provide the indolizidine in 75% yield as a possible mixture of diastereomers. The indolizidine was isolated as a mixture of diastereomers in 66% yield when the same

Scheme 8.112 Synthesis of dithiolo-1,2,4-dithiazolium ions

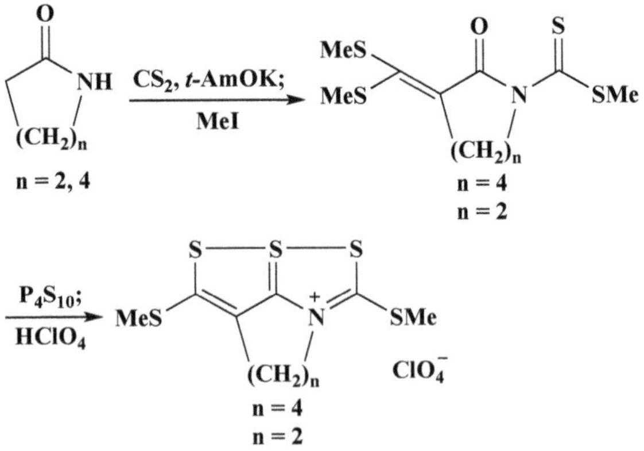

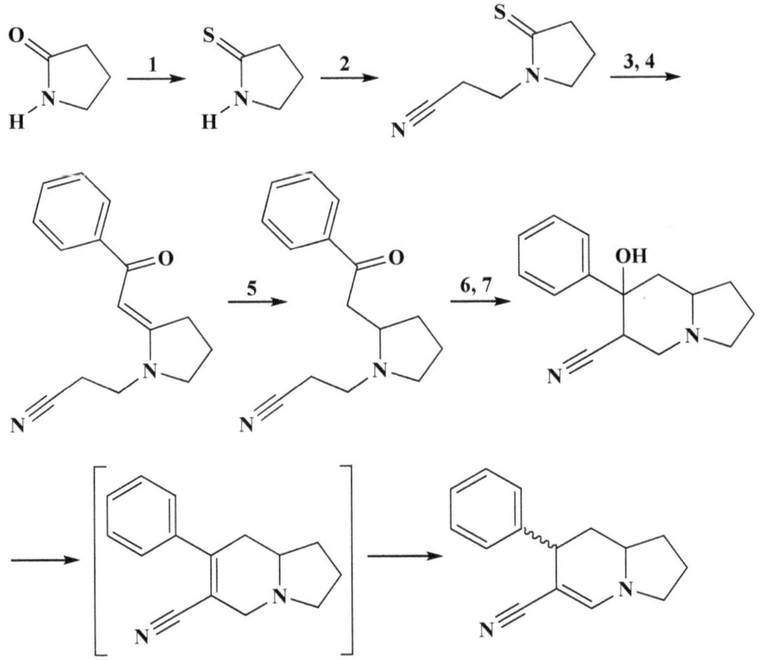

Reagents and conditions: (1) P_4S_{10}, Na_2CO_3, THF, (2) acrylonitrile, NaOH (catalyst), THF, (3) phenacyl bromide, acetone, (4) PPh_3, Et_3N, MeCN, (5) $LiAlH_4$, THF, 0 °C, (6) t-BuOK, THF, rt, (7) t-BuOK, THF, heat.

Scheme 8.113 Synthesis of indolizidine

reaction was performed at rt followed by heating under reflux. Clearly, base-induced elimination occurred, but instead a vinylogous cyanamide was obtained [149].

Starting from the racemic lactam, the carbonyl was thionated to provide the thiolactam in 81% yield. The alkylation of thiolactam with ethyl crotonate took place to provide the ester as a mixture of diastereomers and in an unsatisfactory yield (23%). Following the sulfide contraction and generation of the key intermediate, vinylogous urethane, the cyclization reaction occurred and separable diastereomers were isolated in a 1:1 ratio. Disappointingly the following two steps, reduction of the enaminone and decarboxylation, were performed on minimal material and enough desired products were not recovered to provide the conclusive characterization. However, there was a spectroscopic proof that ester indolizidinone and indolizidinone were formed. The final step in the synthesis, defunctionalization of the keto group, was never performed (Scheme 8.114) [150].

The compounds containing diimidazoline and dipyrimidine moieties were prepared (Scheme 8.115) [41]. The dipyrimidines were obtained when alkanedinitriles were treated with propylenediamine in toluene (dry) at 90 °C for 10 h in the

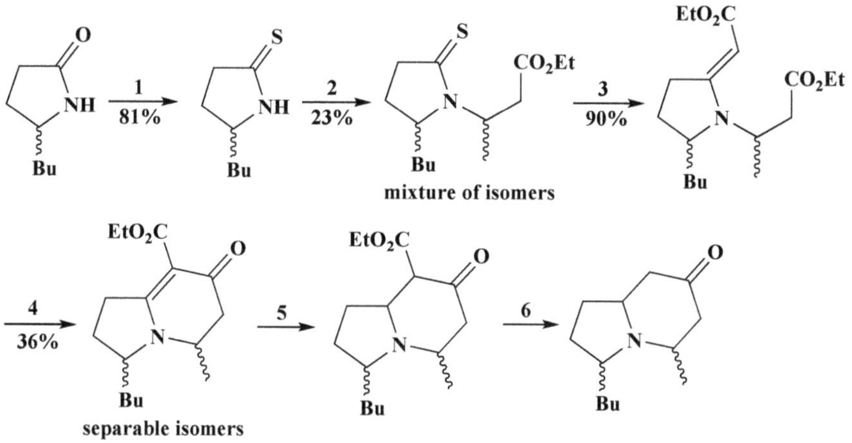

Reagents and conditions: (1) P_4S_{10}, THF, Na_2CO_3, (2) NaH, THF, ethyl crotonate, 12 h, then reflux for 5 h, (3) a) ethyl bromoacetate, CH_3CN, 0 °C, 12 h, b) PPh_3, Et_3N, 2 h, (4) a) NaOH, H_2O, reflux, b) Ac_2O, MeCN, 60 °C, (5) $LiAlH_4$, THF, 5 h, (6) a) KOH, reflux 2 h, b) HCl, reflux, 1 h.

Scheme 8.114 Synthesis of indolizidinone

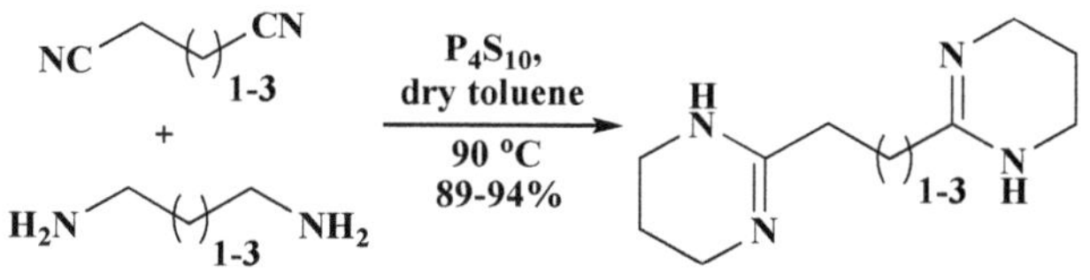

Scheme 8.115 Synthesis of dipyrimidines

presence of a small amount of phosphorus pentasulfide. Similar results were obtained using Lawesson's reagent, S_8 or $Na_2S \cdot 9H_2O$ in place of phosphorus pentasulfide [17].

The 1,4-diphenyltetrazine was obtained in 27% yield by the reaction of N-phenylsydnone with phosphorus pentasulfide in dry toluene in a sealed tube at 120 °C for 6 h (Scheme 8.116) [17, 151].

The thiopyran ring was synthesized by the reaction of diketone, containing an α,β-unsaturated unit, with phosphorus pentasulfide in pyridine (dry) for 3–5 h at rt (Scheme 8.117) [152]. On the other hand, trithiapentalene was provided by the reaction of same ketone with phosphorus pentasulfide in refluxing xylene [17, 153].

The reaction of diethyl oxomalonate with phosphorus pentasulfide provided thioxomalonate by a selective thionation of oxo group (Scheme 8.118). The adducts were provided by trapping the thioxomalonate in situ with cyclopentadiene, 2,3-dimethylbuta-1,3-diene, and anthracene, respectively. The intramolecular hetero-Diels–Alder reactions of α,β-unsaturated thioketones were performed [154, 155].

An intramolecular Diels–Alder reaction of α,β-unsaturated ketone with phosphorus pentasulfide in carbon disulfide provided cycloadduct, possibly via thione intermediate (Scheme 8.119) [17, 154].

The thiopyran was obtained in 56–63% yield by the reaction of two lactam groups with phosphorus pentasulfide in refluxing pyridine for 10 h (Scheme 8.120) [17, 156].

The reaction of thiobarbituric acid derivative with a phosphorus pentasulfide–pyridine complex synthesized dimers containing dithiino and thiophene moieties, respectively (Scheme 8.121) [157]. A mixture of products was obtained in 65%

Scheme 8.116 Synthesis of 1,4-diphenyltetrazine

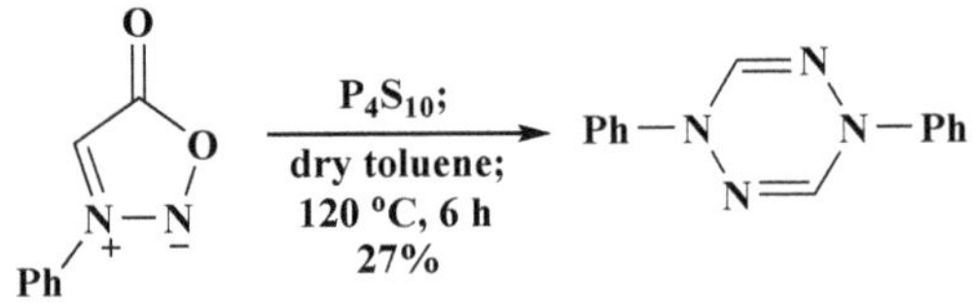

Scheme 8.117 Synthesis of thiopyrans and dithiolo-1,2-dithioles

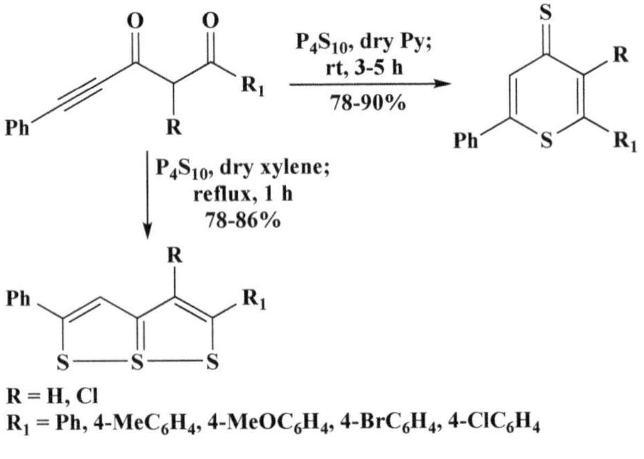

R = H, Cl
R_1 = Ph, 4-MeC_6H_4, 4-$MeOC_6H_4$, 4-BrC_6H_4, 4-ClC_6H_4

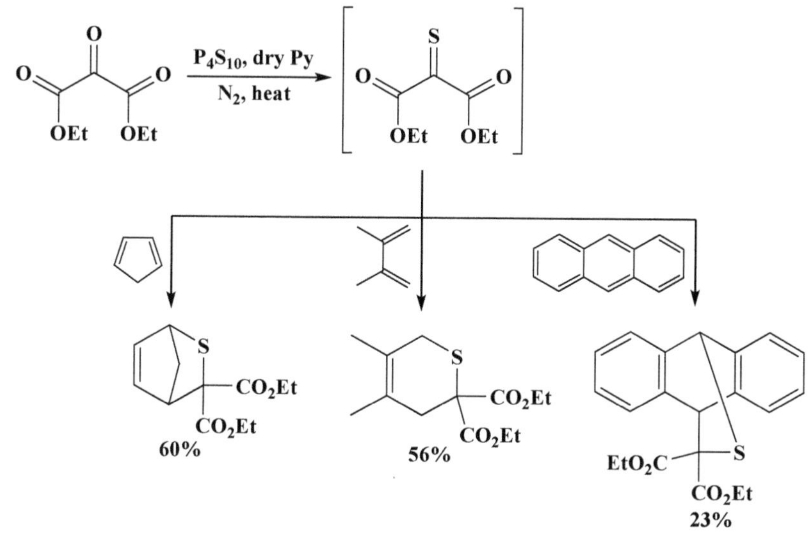

8.3 P$_4$S$_{10}$ in Heterocycle Synthesis

Scheme 8.118 Synthesis of thiopyrans

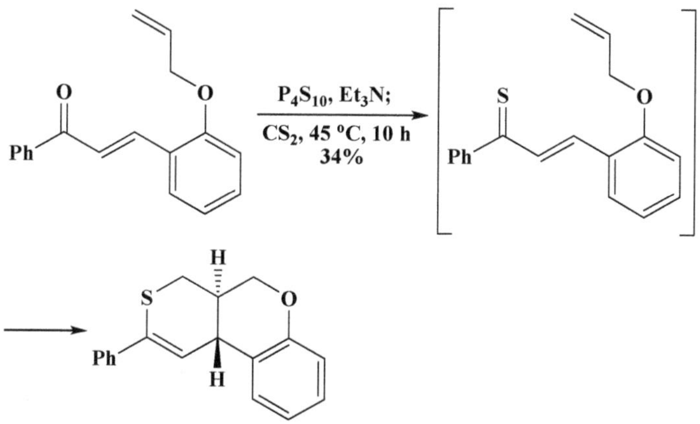

Scheme 8.119 Synthesis of thiopyranochromene

yield when the reaction was conducted in chlorobenzene for 3 d at 130 °C [17].

The 1,2,4-trithiolane in less than 1% yield along with 1,1′-bis(thiobenzoyl)ferrocene in 40% yield was obtained by thionation of 1,1′-dibenzoferrocene with P$_4$S$_{10}$ in a refluxing mixture of dichlorobenzene/ethanol (1:1) for 1 h (Scheme 8.122) [158]. The first step involved the exchange of carbonyl oxygen with sulfur to provide the dithione and then addition of hydrogen sulfide to two thioketones synthesized dithiol, which was oxidized to provide the 1,2,4-trithiolane [17].

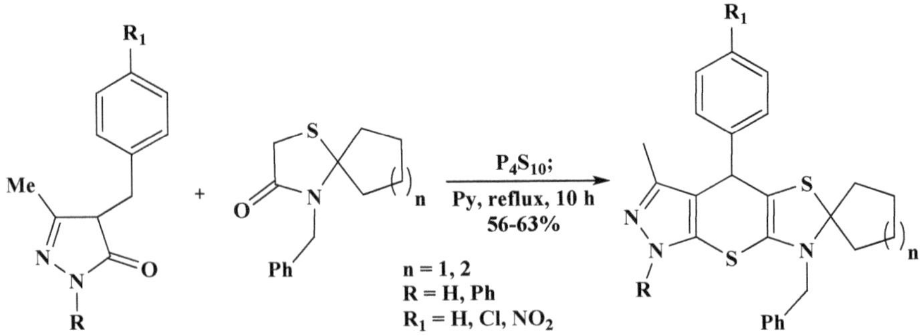

Scheme 8.120 Synthesis of thiopyrans

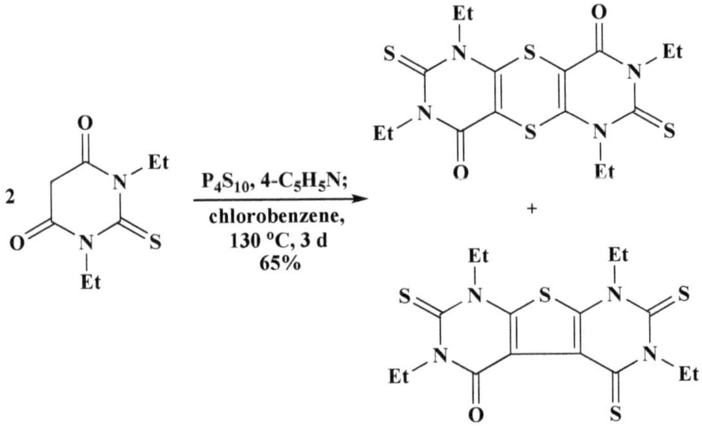

Scheme 8.121 Synthesis of dithiin and thiophene

Unexpected products including the addition of the part of phosphorus pentasulfide or dimerization were obtained in the case where reactive functional units were close enough to offer the reaction with CO groups. An addition product was provided by the reaction of tetrabutyl ammonium salt of camphor with phosphorus pentasulfide in refluxing toluene (Scheme 8.123) [159]. The addition product was afforded by the reaction of 1,3-diketone with phosphorus pentasulfide and lithium carbonate in o-dichlorobenzene at 100 °C (Scheme 8.124) [160]. The dimers and 1,2,3,4-tetrathiins were synthesized by the reaction of oxo sulfenyl chlorides with phosphorus pentasulfide in refluxing toluene for 10 h (Scheme 8.125) [17, 161].

The benzoxanzinones were converted into benzothiazinthiones in good yields on reaction with phosphorus pentasulfide (Scheme 8.126) [162]. The mechanism involved the initial thionation of CO group to afford the benzoxazinones, which were rearranged to benzothiazinones and then second thionation synthesized benzothiazinthiones [17].

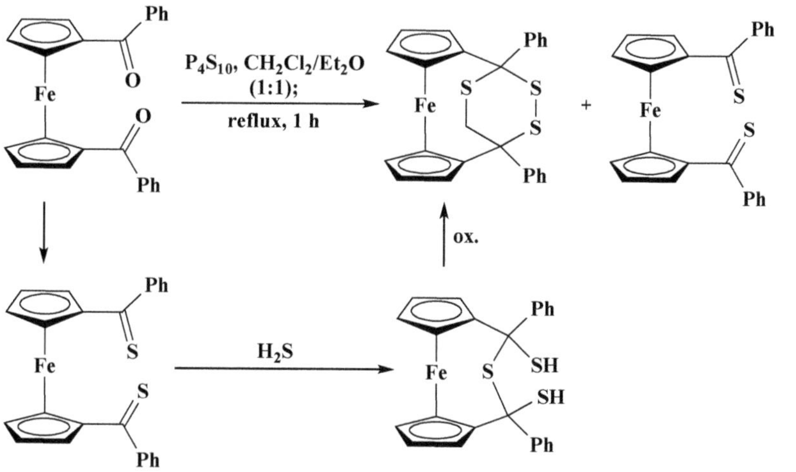

8.3 P$_4$S$_{10}$ in Heterocycle Synthesis

Scheme 8.122 Synthesis of 1,2,4-trithiolane and 1,1'-bis(thiobenzoyl)ferrocene

Scheme 8.123 Synthesis of (6R,8aR)-5,5-dimethyltetrahydro-4H-2,4a-epithio-6,8a-methanobenzo[d][1,3,2]dithiaphosphinine-2-sulfide

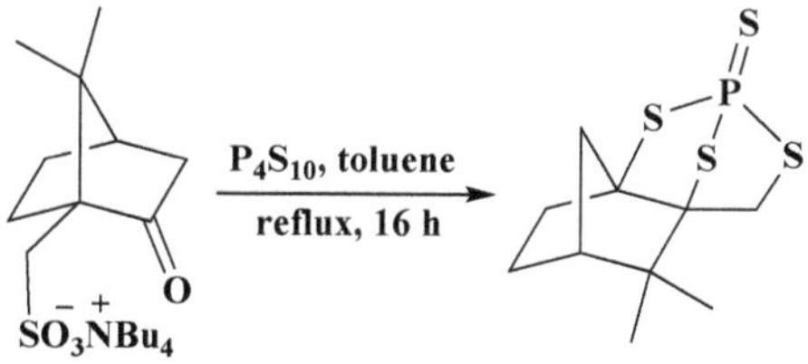

Scheme 8.124 Synthesis of trithiadiphosphatricyclooctane disulfide

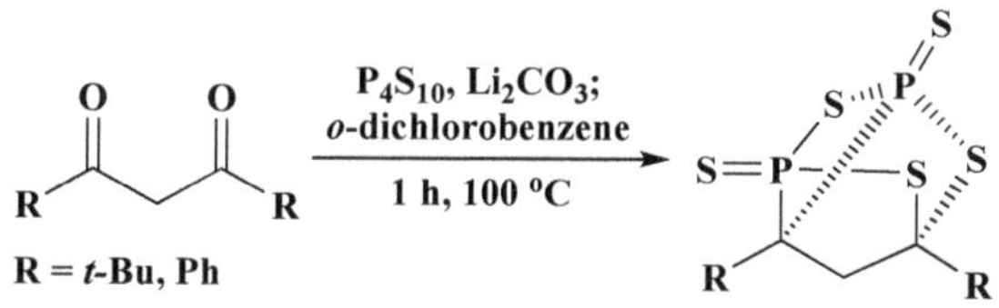

Different results were reported when the ester was reacted with phosphorus pentasulfide in different solvents like xylene (Scheme 8.127) and pyridine (Scheme 8.128) [163]. While the dithiolactone was obtained in 25–30% yield by the reaction of ester with phosphorus pentasulfide in refluxing xylene for 1.5 h, the thionolactone was obtained in 50–70% yield by performing the same reaction in refluxing pyridine [17].

A ring-closure reaction occurred by the reaction of thienothiophene, bearing esters and amide groups *ortho* to each other, with phosphorus pentasulfide in pyridine [164]. Two compounds: the ring-closure and nonring-closure products were afforded by the reaction of carboxylate starting compound with phosphorus pentasulfide in refluxing pyridine for 20 h (Scheme 8.129). On the other hand, only ring-closure products

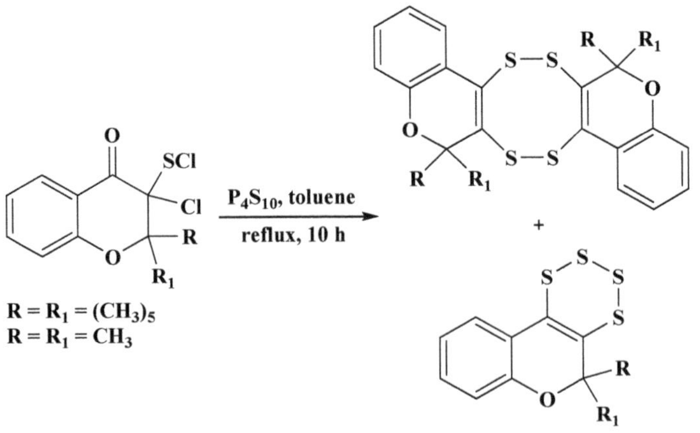

Scheme 8.125 Synthesis of tetrathiocinodichromene and tetrathiinochromene

Scheme 8.126 Synthesis of benzothiazinthiones

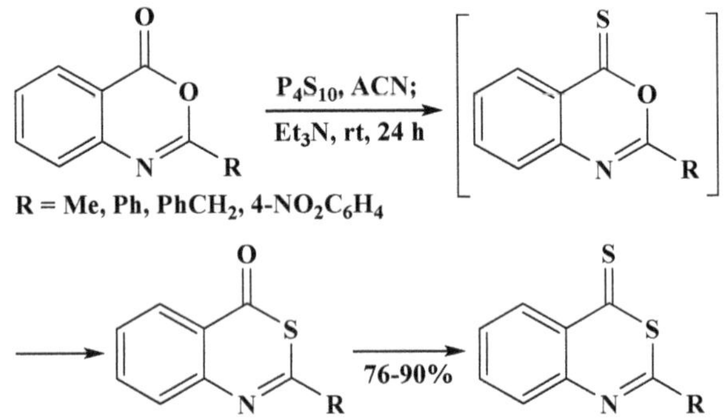

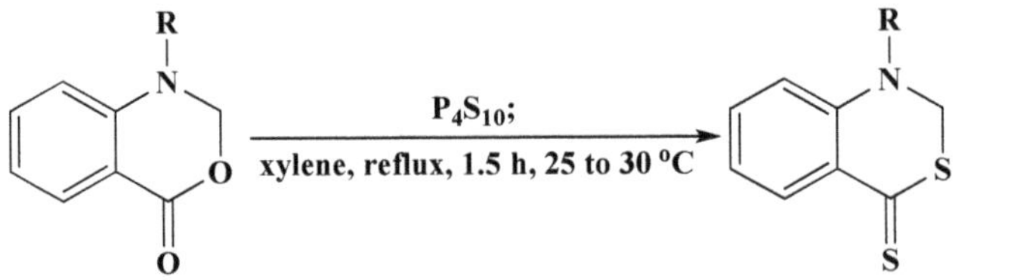

R = Me, Ph, PhCH$_2$, 4-NO$_2$C$_6$H$_4$

Scheme 8.127 Synthesis of benzo-1,3-thiazines

R = Me, Et, Ph, C$_6$H$_5$CH$_2$, 2-MeC$_6$H$_4$, 2-MeC$_6$H$_4$, 3-MeC$_6$H$_4$, 4-MeC$_6$H$_4$, 2-MeOC$_6$H$_4$, 4-MeOC$_6$H$_4$, 2-ClC$_6$H$_4$, 3-ClC$_6$H$_4$, 4-ClC$_6$H$_4$

were afforded by the same reactions of amide starting compound with phosphorus pentasulfide (Scheme 8.130) [17].

The reaction of benzyl arylimines with phosphorus pentasulfide in boiling toluene or xylene provided fused benzo-1,4-thiazines. The reaction mechanism involved the

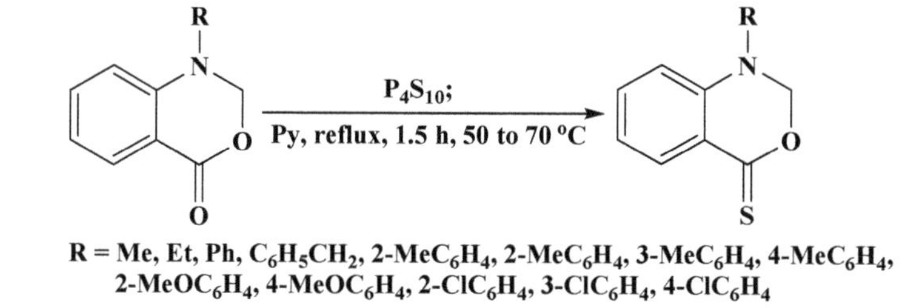

8.3 P$_4$S$_{10}$ in Heterocycle Synthesis

Scheme 8.128 Synthesis of benzo-1,3-oxazines

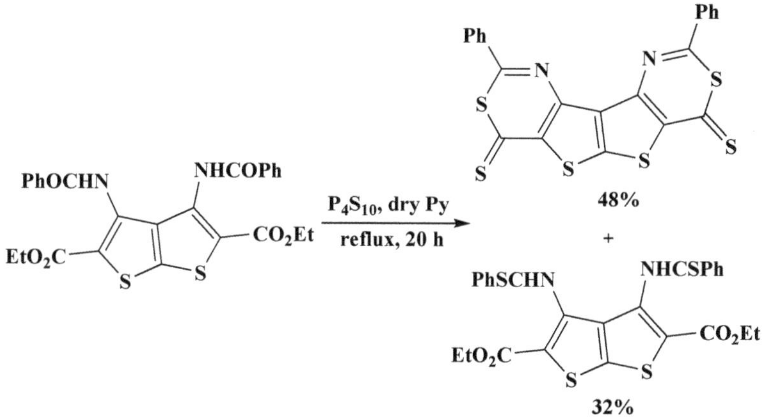

R = Me, Et, Ph, C$_6$H$_5$CH$_2$, 2-MeC$_6$H$_4$, 2-MeC$_6$H$_4$, 3-MeC$_6$H$_4$, 4-MeC$_6$H$_4$, 2-MeOC$_6$H$_4$, 4-MeOC$_6$H$_4$, 2-ClC$_6$H$_4$, 3-ClC$_6$H$_4$, 4-ClC$_6$H$_4$

Scheme 8.129 Synthesis of thienothiophenes

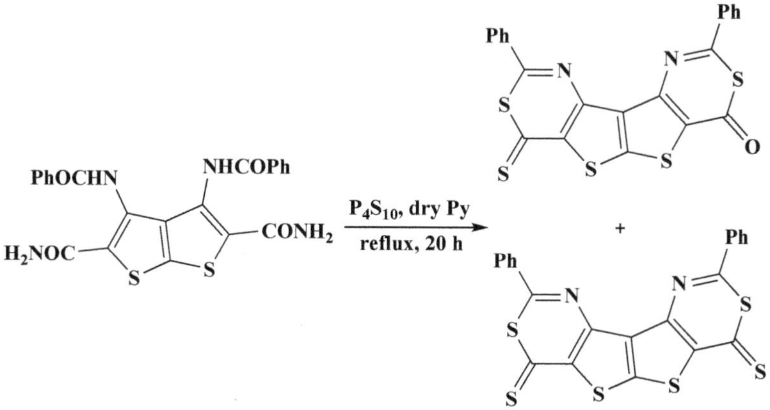

Scheme 8.130 Synthesis of thienothiophenes

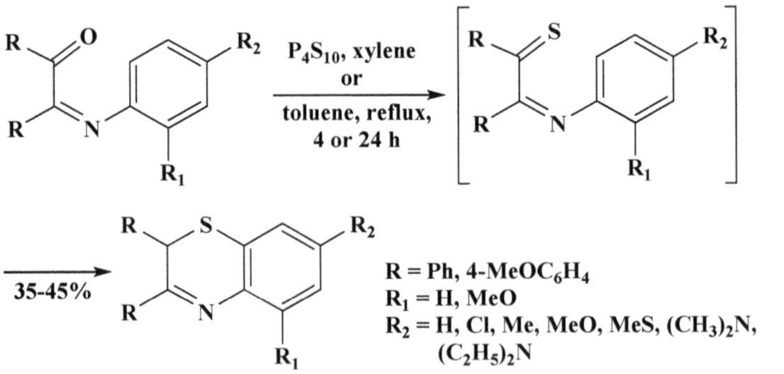

Scheme 8.131 Synthesis of benzo-1,4-thiazines

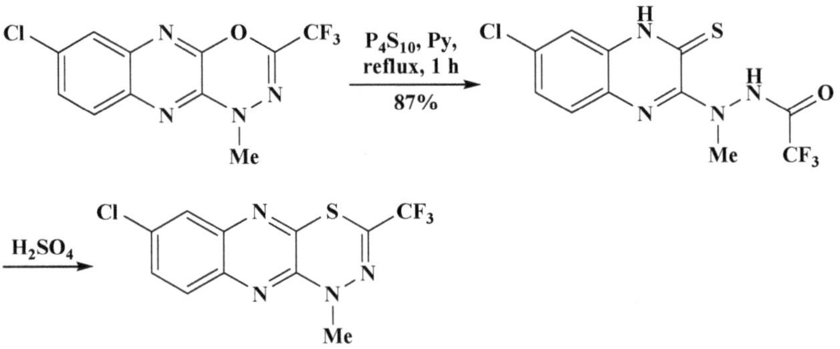

R = Ph, 4-MeOC$_6$H$_4$
R$_1$ = H, MeO
R$_2$ = H, Cl, Me, MeO, MeS, (CH$_3$)$_2$N, (C$_2$H$_5$)$_2$N

Scheme 8.132 Synthesis of quinoxalinothiadiazine

thionation of CO group followed by a cyclization procedure (Scheme 8.131) [17, 40].

A ring-opening of oxadiazino moiety occurred when reacted with phosphorus pentasulfide in refluxing pyridine for 1 h (Scheme 8.132) [165]. One oxo group and one thione group were formed in the product, which was reacted with sulfuric acid to afford the thiadiazino ring [17].

8.3.3 Synthesis of Seven-Membered Heterocycles

The reaction of isatin with phosphorus pentasulfide in pyridine afforded pentathiepino[6,7-*b*]indole (Scheme 8.133) [128, 166].

The pentathiole was formed in 7% yield by reacting isatin with phosphorus pentasulfide in refluxing pyridine, rather than its corresponding thiolactam (Scheme 8.134) [167, 168]. On the other hand, the coupling product was isolated along with indirubin,

Scheme 8.133 Synthesis of pentathiepino[6,7-b]indole

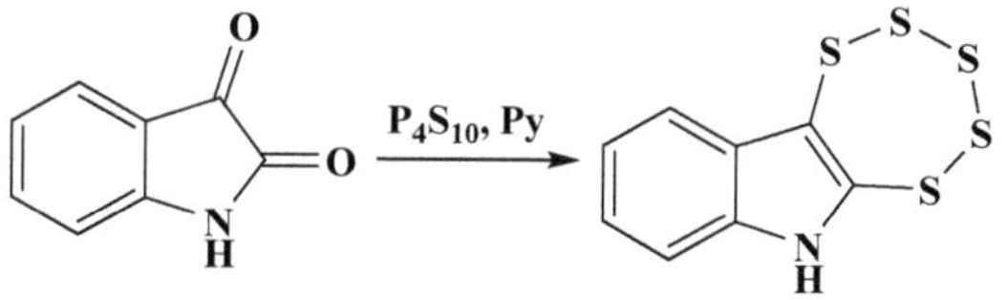

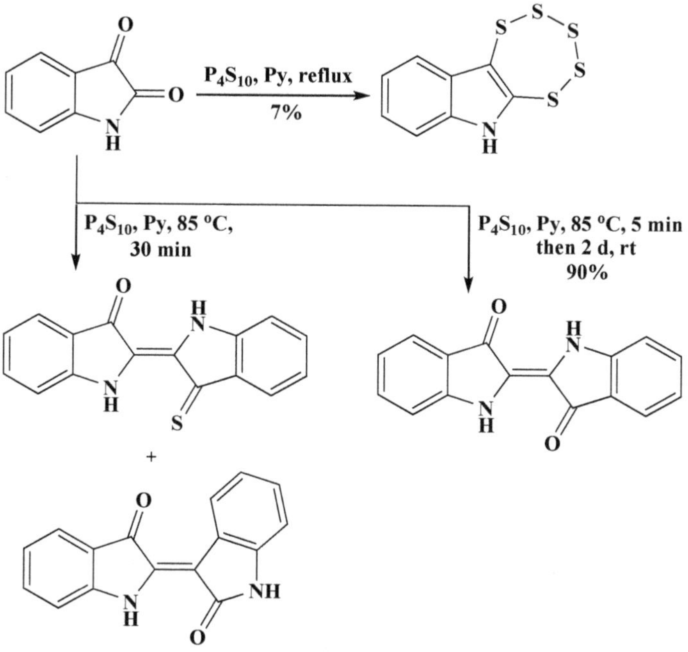

Scheme 8.134 Synthesis of biindolinylidenes

when the reaction was carried out at 85 °C for 30 min, and heating isatin at 85 °C for 5 min and leaving at rt for 2 d provided another coupling product in 90% yield [17].

Generally, the synthesis of thioamides involved the thionation of amides with two reagents: P_4S_{10} [135] and LR [169]. The reaction started with the thionation of easily accessible lactam (Scheme 8.135). Two compounds were obtained in very good yield upon exposure of lactam to 0.25 eq. phosphorus pentasulfide and 1.7 eq. HMDO (Curphey reagent). The minor product was characterized as the desired thio-lactam. The hydrogen bonding to nitrogen activated the bridged lactams, which played a prominent role in these reactions. The medium-bridged bicyclic lactams, having an internal double bond, exhibit enhanced reactivity toward C-NC(O) bond hydrogenolysis [170]. The cleavage of C-NC(S) bond in bridged thioamide having a [4.3.1] ring system was suggested by Aube and Szostak [171, 172].

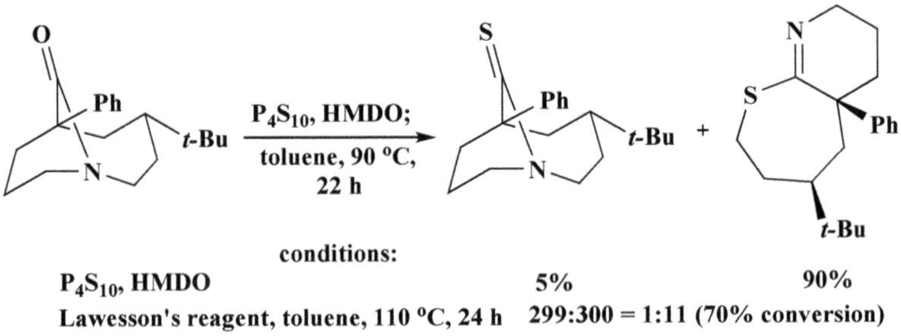

conditions:

P$_4$S$_{10}$, HMDO 5% 90%
Lawesson's reagent, toluene, 110 °C, 24 h 299:300 = 1:11 (70% conversion)

Scheme 8.135 Synthesis of (1S,4S)-4-(t-butyl)-6-phenyl-1-azabicyclo[4.3.1]decane-10-thione and thiepinopyridine

References

1. (a) N. Kaur. 2019. Ionic liquids: A versatile medium for the synthesis of six-membered two nitrogen containing heterocycles. Curr. Org. Chem. 23: 76–96. (b) N. Kaur. 2018. Photochemical reactions for the synthesis of six-membered O-heterocycles. Curr. Org. Synth. 15: 298–320. (c) N. Kaur. 2018. Perspectives of ionic liquids applications for the synthesis of five and six-membered O,N-heterocycles. Synth. Commun. 48: 473–495. (d) N. Kaur. 2019. Metal and non-metal catalysts in the synthesis of five-membered S-heterocycles. Curr. Org. Synth. 16: 258–275. (e) N. Kaur. 2015. Six-membered N-heterocycles: Microwave-assisted synthesis. Synth. Commun. 45: 1–34. (f) N. Kaur and D. Kishore. 2014. Synthesis of oxadiazolo, pyrimido, imidazolo and benzimidazolo containing derivatives of 1,4-benzodiazepin-5-(4'-methylpiperazinyl)-carboxamide through phenylamino spacer. Synth. Commun. 44: 2789–2796. (g) R. Chauhan, N. Kaur, Rajendra, S. Sharma and J. Dwivedi. 2015. Application of chalcone in synthesis of 1-(1,5-benzodiazepino) substituted analogues of indole. Rasayan J. Chem. 8: 115–122.
2. (a) N. Kaur, P. Grewal, P. Bhardwaj, M. Devi, N. Ahlawat and Y. Verma. 2020. Synthesis of five-membered N-heterocycles using silver metal. Synth. Commun. 49: 3058–3100. (b) N. Kaur, M. Devi, Y. Verma, P. Grewal, N.K. Jangid and J. Dwivedi. 2019. Seven and higher-membered oxygen heterocycles: Metal and non-metal. Synth. Commun. 49: 1508–1542. (c) N. Kaur. 2019. Applications of palladium dibenzylideneacetone as catalyst in the synthesis of five-membered N-heterocycles. Synth. Commun. 49: 1205–1230. (d) N. Kaur, N. Ahlawat, P. Bhardwaj, Y. Verma, P. Grewal and N.K. Jangid. 2020. Ag-Mediated synthesis of six-membered N-heterocycles. Synth. Commun. 50: 753–795. (e) N. Kaur. 2018. Ultrasound assisted synthesis of six-membered N-heterocycles. Mini Rev. Org. Chem. 15: 520–536. (f) N. Kaur and D. Kishore. 2014. Synthetic strategies applicable in the synthesis of privileged scaffold: 1,4-Benzodiazepine. Synth. Commun. 44: 1375–1413. (g) N. Kaur and D. Kishore. 2014. Peroxy acids: Role in organic synthesis. Synth. Commun. 44: 721–747.
3. (a) N. Kaur, N. Ahlawat, Y. Verma, P. Grewal, P. Bhardwaj and N.K. Jangid. 2020. Cu-Assisted C-N bond formations in six-membered N-heterocycle synthesis. Synth. Commun. 50: 1075–1132. (b) N. Kaur, Y. Verma, P. Grewal, N. Ahlawat, P. Bhardwaj and N.K. Jangid. 2020. Palladium acetate assisted synthesis of five-membered N-polyheterocycles. Synth. Commun. 50: 1567–1621. (c) N. Kaur. 2015. Review on the synthesis of six-membered N,N-heterocycles by microwave irradiation. Synth. Commun. 45: 1145–1182. (d) N. Kaur, N. Ahlawat, Y. Verma, P. Grewal, P. Bhardwaj and N.K. Jangid. 2020. Metal and organo-complex promoted synthesis of fused five-membered O-heterocycles. Synth. Commun. 50: 457–505. (e) N. Kaur, N. Ahlawat, P. Bhardwaj, Y. Verma, P. Grewal and N.K. Jangid. 2020. Synthesis of five-membered N-heterocycles using Rh based metal catalysts. Synth. Commun. 50: 137–160. (f) N. Kaur and D. Kishore. 2014. Synthesis of 2-(oxadiazolo, pyrimido, imidazolo, and

benzimidazolo) substituted analogues of 1,4-benzodiazepin-5-carboxamides linked through a phenoxyl bridge. J. Chem. Sci. 126: 1861–1867.
4. M.C. Demarcq. 1990. Kinetic and mechanistic aspects of the redox dissociation of tetraphosphorus decasulphide in solution. J. Chem. Soc. Dalton Trans. 1: 35–39.
5. M.C. Demarcq. 1991. Reactivity of technical phosphorus pentasulfide. J. Ind. Eng. Chem. Res. 30: 1906–1911.
6. L. Andrews, G.G. Reynolds, Z. Mielke and M. McCluskey. 1990. Infrared spectra of tetraphosphorous decasulfide and its decomposition products in solid argon. Inorg. Chem. 29: 5222–5225.
7. R. Blachnik, J. Matthiesen, A. Muller, H. Nowottnick, H. Reuter and Z. Krystallogr. 1998. Refinement of the crystal structure of 1,3,5,7-thioxo-2,4,6,8,9,10-hexathia-1,3,5,7-tetraphosphatricyclo[3.3.1.1^{3,7}]decane, tetraphosphorus decasulfide, P_4S_{10}. New Cryst. Struct. 213: 233–234.
8. T. Bjorholm and H.J. Jakobsen. 1991. Multiphase characterization of phosphorus sulfides by multidimensional and magic angle spinning - ^{31}P NMR spectroscopy. Molecular transformations and exchange pathways at high temperatures. J. Am. Chem. Soc. 113: 27–32.
9. L. Andrews, C. Thompson and M.C. Demarcq. 1992. Infrared spectra of phosphorus sulfides, P_4S_{10}, P_4S_9, and P_4S_7 in solid argon. Inorg. Chem. 31: 3173–3175.
10. Y. Tahri and H. Chermette. 1991. Electronic structures of phosphorus oxide P_4O_{10} and phosphorus sulfides P_4S_{10} and P_4S_7. J. Electron Spectrosc. 56: 51–69.
11. R. Gigli, V. Piacente and P. Scardala. 1990. A study on the sublimation behaviour of solid P_4S_{10}. J. Mater. Sci. Lett. 9: 1148–1149.
12. M. Davis. 1982. Sulfur transfer reagents in heterocyclic synthesis. Adv. Heterocycl. Chem. 30: 47–78.
13. K. Hartke and H.-D. Gerber. 1996. Tetraphosphorus decasulfide, revival of an old thionating agent. J. Prakt. Chem. 338: 763–765.
14. W.M. McGregor and D.C. Sherrington. 1993. Some recent synthetic routes to thioketones and thioaldehydes. Chem. Soc. Rev. 22: 199–204.
15. P. Metzner. 1992. The use of thiocarbonyl compounds in carbon-carbon bond forming reactions. Synthesis 12: 1185–1199.
16. V. Polshettiwar. 2004. Phosphorus pentasulfide (P_4S_{10}). Synlett 12: 2245–2246.
17. T. Ozturk, E. Ertas and O. Mert. 2010. A Berzelius reagent, phosphorus decasulfide (P_4S_{10}), in organic syntheses. Chem. Rev. 110: 3419–3478.
18. Z. Brylewicz and R. Rudnicki. 1994. Investigation of the solubility of commercial tetraphosphorus decasulfide. Phosphorus, Sulfur, Silicon Relat. Elem. 88: 163–167.
19. M.A. Shalaby, C.W. Grote and H. Rapoport. 1996. Thiopeptide synthesis. α-Amino thionoacid derivatives of nitrobenzotriazole as thioacylating agents. J. Org. Chem. 61: 9045–9048.
20. M.A. Shalaby and H. Rapoport. 1996. A general and efficient route to thionoesters via thionoacyl nitrobenzotriazoles. J. Org. Chem. 61: 1065–1070.
21. A.R. Katritzky, R.M. Witek, V. Rodriguez-Garcia, P.P. Mohapatra, J.W. Rogers, J. Cusido, A.A.A. Abdel-Fattah and P.J. Steel. 2005. Benzotriazole-assisted thioacylation. J. Org. Chem. 70: 7866–7881.
22. P. Bilek and J. Slouka. 2002. Cyclocondensation reactions of heterocyclic carbonyl compounds VII. Synthesis of some substituted benzo[1,2,4]triazino[2,3-*a*]benzimidazoles. Heterocycl. Commun. 8: 123–128.
23. V.P. Kruglenko, V.P. Gnidets and M.V. Povstyanoi. 2000. Synthesis of 2-methyl-1,2,4-triazolo[4,3-*d*]-1,2,4-triazino[2,3-*a*]-benzimidazole and 2-methyl-9-phenylimidazo[1,2-*b*]-1,2,4-triazolo[4,3-*d*]-1,2,4-triazine. Chem. Heterocycl. Compd. 36: 103–104.
24. K.M. Dawood and B.F. Abdel-Wahab. 2010. Synthetic routes to benzimidazole-based fused polyheterocycles. ARKIVOC (i): 333–389.
25. P.L. Ling, Z. Xiu, M. Jie, Z.Z. Zhen, Z. Zhe, Z. Yan and W.J. Bo. 2009. Synthesis of furan from allenic sulfide derivatives. Sci. Chin. Ser. B: Chem. 52: 1622–1630.
26. E. Campaigne and W.O. Foye. 1952. The synthesis of 2,5-diarylthiophenes. J. Org. Chem. 17: 1405–1412.

27. R. Mishra, K.K. Jha, S. Kumar and I. Tomer. 2011. Synthesis, properties and biological activity of thiophene: A review. Der Pharma Chem. 3: 38–54.
28. C. Paal. 1885. Synthesis of thiophene and pyrrole derivatives. Chem. Ber. 18: 367–371.
29. C. Tanaka, K. Nasu, N. Yamamoto and M. Shibata. 1982. Pyrolysis of benzyl 2-oxazolecarbamates and benzyl 4-alkylallophanates. Chem. Pharm. Bull. 30: 4195–4198.
30. P. Bradley, P. Sampson and A.J. Seed. 2005. Preliminary communication: The synthesis of new mesogenic 1,3,4-thiadiazole-2-carboxylate esters via a novel ring-closure. Liq. Cryst. Today 14: 15–18.
31. B. Sybo, P. Bradley, A. Grubb, S. Miller, K.J.W. Proctor, L. Clowes, M.R. Lawrie, P. Sampson and A.J. Seed. 2007. 1,3,4-Thiadiazole-2-carboxylate esters: New synthetic methodology for the preparation of an elusive family of self-organizing materials. J. Mater. Chem. 17: 3406–3411.
32. H. Erlenmeyer. 1945. Zur kenntnis der thiazol-2-carbonsäure. Helv. Chim. Acta 28: 924–925.
33. A. Bertram, N. Maulucci, O.M. New, S.M.M. Nor and G. Pattenden. 2007. Synthesis of libraries of thiazole, oxazole and imidazole-based cyclic peptides from azole-based amino acids. A new synthetic approach to bistratamides and didmolamides. Org. Biomol. Chem. 5: 1541–1553.
34. E.H. White, F. McCapra, G.F. Field and W.D. McElroy. 1961. The structure and synthesis of firefly luciferin. J. Am. Chem. Soc. 83: 2402–2403.
35. E.H. White, F. McCapra and G.F. Field. 1963. The structure and synthesis of firefly luciferin. J. Am. Chem. Soc. 85: 337–343.
36. N.P. Prajapati, R.H. Vekariya and H.D. Patel. 2015. Microwave induced facile one-pot access to diverse 2-cyanobenzothiazole - a key intermediate for the synthesis of firefly luciferin. Int. Lett. Chem. Phys. Astron. 44: 81–89.
37. I.T. Bazyl, S.P. Kisil, Y.V. Burgart, V.I. Saloutin and O.N. Chupakhin. 1999. The selective *ortho*-methoxylation of pentafluorobenzoic acid - a new way to tetrafluorosalicylic acid and its derivatives. J. Fluorine Chem. 94: 11–13.
38. Z. Li, A. Dellali, J. Malik, M. Motevalli, R.M. Nix, T. Olukoya, Y. Peng, H. Ye, W.P. Gillin, I. Hernández and P.B. Wyatt. 2013. Luminescent zinc(II) complexes of fluorinated benzothiazol-2-yl substituted phenoxide and enolate ligands. Inorg. Chem. 52: 1379–1387.
39. H.M. Kuo, S.Y. Li, H.S. Sheu and C.K. Lai. 2012. Symmetrical mesogenic 2,5-bis(6-naphthalen-2-yl)-1,3,4-thiadiazoles. Tetrahedron 68: 7331–7337.
40. J.-D. Charrier, C. Landreau, D. Deniaud, F. Reliquet, A. Reliquet and J.C. Meslin. 2001. From benzil arylimines to 2*H*-benzo-1,4-thiazines, benzothiazoles or indoles. Tetrahedron 57: 4195–4202.
41. M. Machaj, M. Pach, A. Wolek, A. Zabrzenska, K. Ostrowska, J. Kalinowska-Tluscik and B. Oleksyn. 2007. Succinonitrile activated by thiating agents as precursor of bis-cyclic amidines, tectons for molecular engineering. Monatsh. Chem. 138: 1273–1277.
42. M. Moghadam, I. Mohammadpoor-Baltork, V. Mirkhani, S. Tanqestaninejad, M.A. Alibeik, B.H. Yousefi and H. Kargar. 2007. Rapid and efficient synthesis of imidazolines and bisimidazolines under microwave and ultrasonic irradiation. Monatsh. Chem. 138: 579–583.
43. E.E. Korshin, L.I. Sabirova, A.G. Akhmadullin and Y.A. Levin. 1994. Aminoamidines. Russ. Chem. B 43: 431–438.
44. C.A. Batty, M.K. Manthey, J. Kirk, M. Manthey and R.I. Christopherson. 1997. Synthesis and exchange reactions of 5-alkyl-2-oxo-6-thioxo-1,2,3,6-hexahydropyrimidine-4-carboxylic acids. J. Heterocycl. Chem. 34: 1355–1367.
45. D.B. Nilov, A.V. Kadushkin, N.P. Soloveva and V.G. Granik. 1995. An unexpected synthesis of 7,8-polymethyleneimidazo-1,3,2-diazaphosphorines - heteroanalogues of mercaptopurine derivatives. Mendeleev Commun. 5: 67–67.
46. J.B. Hester. 1970. Process for preparing scopine ester of di-(2-thienyl)glycolic acid, intermediate in synthesis of tiotropium bromide and novel form thereof. Ger. Patent 2.012.190.
47. J.B. Hester. 1972. 1-Carbolower alkoxy-6-phenyl-4*H*-s-triazolo(1,4)benzodiazepine compounds. US Patent 3701782.
48. J.B. Hester. 1976. 6-Phenyl-4*H*-s-triazolo[4,3-*a*][1,4]benzodiazepines. US Patent 3.987.052.

References

49. G.A. Archer and L.H. Sternbach. 1969. 5-Phenyl-3H-1,4-benzodiazepine-2(1H)-thione and derivatives thereof. US Patent 3.422.091.
50. G.A. Archer and L.H. Sternbach. 1964. Quinazolines and 1,4-benzodiazepines. XVI.1 Synthesis and transformations of 5-phenyl-1,4-benzodiazepine-2-thiones. J. Org. Chem. 29: 231–233.
51. J.B. Hester, A.D. Rudzik and B.V. Kamdar. 1971. 6-Phenyl-4H-s-triazolo[4,3-a][1,4]benzodiazepines which have central nervous system depressant activity. J. Med. Chem. 14: 1078–1081.
52. S. Gronowitz. 1985. Preparation of thiophenes by ring-closure reactions and from other ring systems. Chemistry of heterocyclic compounds. S. Gronowitz (Ed.). Thiophene and its derivatives. John Wiley & Sons, Inc. ISBN 0-471-38120-9 (v. 1), Volume 44.
53. J. O'Brochta and A. Lowy. 1939. Thio compounds derived from aroyl-o-benzoic acids. J. Am. Chem. Soc. 61: 2765–2768.
54. B.J. Morrison and O.C. Musgrave. 2002. Thiones as reactive intermediates in condensations of diketones with aromatics mediated by tetraphosphorus decasulfide. Phosphorus, Sulfur, Silicon Relat. Elem. 177: 2725–2744.
55. F. Freeman, M.Y. Lee, H. Lue, X. Wang and E. Rodriguez. 1994. 1-Thia-Cope rearrangements during the thionation of 2-endo-3-endo-bis(aroyl)bicyclo[2.2.1]hept-5-enes. J. Org. Chem. 50: 3695–3698.
56. J.J. Li and E.J. Corey. 2005. Name reactions in heterocyclic chemistry. John Wiley & Sons.
57. G.H. Degenhart. 2008. Synthesis of conjugated oligomers. PhD Thesis, Universiteit Leiden.
58. T. Eicher and S. Hauptmann. 1995. The chemistry of heterocycles. Georg Thieme Verlag, 76.
59. K. Burger, B. Helmreich and O. Jendrewski. 1994. 3-Trifluoromethylthiophens from hexafluoroacetone. J. Fluorine Chem. 66: 13–14.
60. L.-L. Lai, D.H. Reid, S.-L. Wang and F.-L. Liao. 1994. An unexpected synthesis of thiophene derivatives by thionation of N-phenylacetylthiobenzamides. Heteroat. Chem. 5: 479–486.
61. E.I. Mamedov, A.G. Ismailov, V.G. Ibragimov and R.D. Goyushov. 1983. Synthesis of 1,2-dipyrrylethanes. Chem. Heterocycl. Compd. 19: 1243–1243.
62. E. Koltai and K. Lempert. 1973. The reaction of 2,5,5-triphenyl-2-oxazolin-4-one with phosphorus pentasulfide. Tetrahedron 29: 2795–2796.
63. R. Vanasselt, I. Hoogmartens, D. Vanderzande, J. Gelan, P.E. Froehling, M. Aussems, O. Aagaard and R. Schellekens. 1995. New synthetic routes to poly(isothianaphthene) I. Reaction of phthalic anhydride and phthalide with phosphorus pentasulfide. Synth. Met. 74: 65–70.
64. R. Vanasselt, D. Vanderzande, J. Gelan, P.E. Froehling and O. Aagaard. 1996. New synthetic routes to poly(isothianaphthene). II. Mechanistic aspects of the reactions of phthalic anhydride and phthalide with phosphorus pentasulfide. J. Polym. Sci. 34: 1553–1560.
65. R. Vanasselt, D. Vanderzande, J. Gelan, P.E. Froehling and O. Aagaard. 2000. New synthetic routes to poly(isothianaphthene). Synth. Met. 110: 25–30.
66. M. Huskic, D. Vanderzande and J. Gelan. 1998. Optimization of the reaction of phthalic anhydride with P_4S_{10}. J. Acta Chim. Slov. 45: 389–395.
67. M. Huskic, D. Vanderzande and J. Gelan. 1999. Synthesis of aza-analogues of poly(isothianaphthene). Synth. Met. 99: 143–147.
68. N.M. Cullinane, C.G. Davis and G.I. Davies. 1936. Substitution derivatives of diphenylene sulphide and diphenylenesulphone. J. Chem. Soc. 0: 1435–1437.
69. A. Capan, H. Veisi, A.C. Goren and T. Ozturk. 2012. Concise syntheses, polymers, and properties of 3-arylthieno[3,2-b]thiophenes. Macromolecules 45: 8228–8236.
70. A. Capan and T. Ozturk. 2014. Electrochromic properties of 3-arylthieno[3,2-b]thiophenes. Synth. Met. 188: 100–103.
71. M.E. Cinar and T. Ozturk. 2015. Thienothiophenes, dithienothiophenes, and thienoacenes: Syntheses, oligomers, polymers, and properties. Chem. Rev. 115: 3036–3140.
72. M.P. Cava and G.E.M. Husbands. 1969. Tetraphenylthieno[3,4-c]thiophene. A stable nonclassical thiophene. J. Am. Chem. Soc. 91: 3952–3953.
73. M.P. Cava, M. Behforouz, G.E.M. Husbands and M. Srinivasan. 1973. Nonclassical condensed thiophenes. II. Tetraphenylthieno[3,4-c]thiophene. J. Am. Chem. Soc. 95: 2561–2564.

74. E. Ertas and T. Ozturk. 2004. A new reaction of P$_4$S$_{10}$ and Lawesson's reagent; a new method for the synthesis of dithieno[3,2-*b*;2',3'-*d*]thiophenes. Tetrahedron Lett. 45: 3405–3407.
75. I. Osken, H. Bildirir and T. Ozturk. 2011. Electrochromic behavior of poly(3,5-bis(4-bromophenyl)dithieno[3,2-*b*;2',3'-*d*]thiophene). Thin Solid Films 519: 7707–7711.
76. I. Osken, O. Sahin, T. Ozturk, A.S. Gundogan, H. Bildirir, A. Capan, E. Ertas, M.S. Eroglu, J.D. Wallis and K. Topal. 2012. Selective syntheses of vinylenedithiathiophenes (VDTTs) and dithieno[2,3-*b*;2',3'-*d*]thiophenes (DTTs); building blocks for π-conjugated systems. Tetrahedron 68: 1216–1222.
77. O. Mert, E. Sahin, E. Ertas, T. Ozturk, E.A. Aydin and L. Toppare. 2006. Electrochromic properties of poly(diphenyldithieno[3,2-*b*;2',3'-*d*]thiophene). J. Electroanal. Chem. 591: 53–58.
78. Y. Yagci, O. Sahin, T. Ozturk, S. Marchi, S. Grassini and M. Sangermano. 2011. Synthesis of silver/epoxy nanocomposites by visible light sensitization using highly conjugated thiophene derivatives. React. Funct. Polym. 71: 857–862.
79. M. Ates, I. Osken and T. Ozturk. 2012. Poly(3,5-dithiophene-2-yldithieno[3,2-*b*;2',3'-*d*]thiophene-co-ethylenedioxythiophene)/glassy carbon electrode formation and electrochemical impedance spectroscopic study. J. Electrochem. Soc. 159: E115–E121.
80. I. Osken, E. Sezer, E. Ertas and T. Ozturk. 2014. Electrochemical impedance spectroscopy study of poly[3,5-dithiophene-2-yldithieno[3,2-*b*;2',3'-*d*]thiophene] P(Th$_2$DTT). J. Electroanal. Chem. 726: 36–43.
81. C. Ozen, M. Yurtsever and T. Ozturk. 2011. A theoretical approach to the formation mechanism of diphenyldithieno[3,2-*b*:2',3'-*d*]thiophene from 1,8-diketone, 4,5-bis(benzoylmethylthio)thiophene: A DFT study. Tetrahedron Lett. 67: 6275–6280.
82. S.-C. Lin, F.-D. Yang, J.-S. Shiue, S.-M. Yang and J.-M. Fang. 1998. Indolecarbonyl coupling reactions promoted by samarium diiodide. Application to the synthesis of indole-fused compounds. J. Org. Chem. 63: 2909–2917.
83. E.M. Beccalli, A. Marchesini and T. Pilati. 1993. Synthesis of [*a*] annulated carbazoles from indol-2,3-dione. Tetrahedron 49: 4741–4758.
84. K. Albertshofer, B. Tan and C.F. Barbas. 2012. III. Assembly of spirooxindole derivatives containing four consecutive stereocenters via organocatalytic Michael-Henry cascade reactions. Org. Lett. 14: 1834–1837.
85. S.-H. Cao, X.-C. Zhang, Y. Wei and M. Shi. 2011. Chemoselective reduction of isatin-derived electron-deficient alkenes using alkylphosphanes as reduction reagents. Eur. J. Org. Chem. 14: 2668–2672.
86. J. Levy, D. Royer, J. Guilhem, M. Cesario and C. Pascard. 1987. Benzothieno[2,3-*b*]indole and pyridothiéno[2,3-*b*]indole. Bull. Soc. Chim. Fr. 1: 193–198.
87. S. Sugasawa, S. Satoda and J. Yanagisawa. 1938. Studies on the synthesis of nitrogen-ring-compounds (X). Yakugaku Zasshi 58: 139–141.
88. R.A. Irgashev, A.A. Karmatsky, G.L. Rusinov and V.N. Charushin. 2015. A new and convenient synthetic way to 2-substituted thieno[2,3-*b*]indoles. Beilstein J. Org. Chem. 11: 1000–1007.
89. T. Hino, K. Yamada and S. Akaboshi. 1970. Radiation-protective agents. V. Synthesis and hydrolysis of 2-(2-aminoethylthio) indole derivatives. Chem. Pharm. Bull. 18: 389–391.
90. T. Hino, K. Tsuneoka, M. Nakagawa and S. Akaboshi. 1969. Thiation of oxindoles. Chem. Pharm. Bull. 17: 550–558.
91. M.C. Willis, D. Taylor and A.T. Gillmore. 2006. Palladium-catalysed intramolecular enolate *O*-arylation and thio-enolate *S*-arylation: Synthesis of benzo[*b*]furans and benzo[*b*]thiophenes. Tetrahedron 62: 11513–11520.
92. K.T. Potts and D. McKeough. 1974. Nonclassical heterocycles. II. Thieno[3,4-*c*]pyrazole system. J. Am. Chem. Soc. 96: 4276–4279.
93. A.Y. Ponomareva, D.G. Beresnev, N.A. Itsikson, O.N. Chupakhin and G.L. Rusinov. 2006. Synthesis of [2,3-*b*]thieno- and furoquinoxalines by the reactions of 2-substituted quinoxalines with acetophenones. Mendeleev Commun. 16: 16–18.
94. A. Sakurai and M. Goto. 1968. Die synthese des urothions. Tetrahedron Lett. 9: 2941–2944.

95. M. Goto, A. Sakurai, K. Ohta and H. Yamakami. 1967. Die struktur des urothions. Tetrahedron Lett. 8: 4507–4511.
96. Y.A. Ibrahim, S.A.L. Abdel-Hady, M.A. Badawy and M.A.H. Ghazala. 1978. Synthesis of thieno[2,3-*e*]-1,2,4-triazines. J. Heterocycl. Chem. 19: 913–915.
97. T. Janosik, J. Bergman, B. Stensland and C. Stalhandske. 2002. Thionation of bisindole derivatives with P_4S_{10} or elemental sulfur. J. Chem. Soc. Perkin Trans. 1 3: 330–334.
98. Y. Kitahara and M. Funamizu. 1964. Some derivatives of 2,3-diphenylcyclopropenone. Bull. Chem. Soc. Jpn. 37: 1897–1898.
99. A.W. Krebs. 1965. Cyclopropenylium-verbindungen und cyclopropenone. Angew. Chem. 77: 10–22.
100. G. Laban, J. Fabian and R. Mayer. 1968. Representation and UVS spectrum of diphenylcyclopropenthione. Z. Chem. 8: 414–415.
101. E.N. Cain and R.N. Warrener. 1970. Preparation of 1,3-thiazines: Sulphur analogues of nucleic acid pyrimidine bases. Aust. J. Chem. 23: 51–72.
102. L. Capuano, G. Bolz, R. Burger, V. Burkhardt and V. Huch. 1990. Neue synthesen von 1,3-oxazinen, thiazolen und 1,2-dithiolen. Liebigs Ann. Chem. 3: 239–243.
103. S.O. Lawesson, R. Shabana, J.B. Rasmussen and S.O. Olesen. 1980. Imine chemistry - II. Tetrahedron 36: 3047–3051.
104. S. Moriyamma. T. Karakasa and S. Motoki. 1990. Syntheses and reactions of conjugated dienic thioketones. Bull. Chem. Soc. Jpn. 63: 2540–2548.
105. H. Bartsch and T. Erker. 1992. The Lawesson reagent as selective reducing agent for sulfoxides. Tetrahedron Lett. 33: 199–200.
106. H.J. Prochaska, Y. Yeh, P. Baron and B. Polsky. 1993. Oltipraz, an inhibitor of human immunodeficiency virus type 1 replication. Proc. Natl. Acad. Sci. USA 90: 3953–3957.
107. Ger. Offen Patent N^0 2,430,802 [CA 1975, 82, 1562581].
108. Ger. Qffen Patent N^0 2,460,783 [CA. 1976, 85, 1238991].
109. L. Amoretti, F. Mossini and V. Plazzi. 1968. Preparation and antifungal properties of 1,2-benzisothiazolin-3-thiones and benzo-chloro-substituted 1,2-benzodithiol compounds. Farmaco 23: 583–590.
110. N.F. Haley and M.W. Fichtner. 1980. Efficient and general synthesis of 1,3-dithiole-2-thiones. J. Org. Chem. 45: 175–177.
111. D. Leaver, W.A.H. Robertson and D.M. McKinnon. 1962. The dithiole series. Part I. Synthesis of 1,2- and 1,3-dithiolium salts. J. Chem. Soc. 0: 5104–5109.
112. K.M. Pazdro. 1969. Synthesis of 2,2-bis(4',5-diphenylo-1,3-dithiolylidene). Rocz. Chem. 43: 1089–1089.
113. M. Narita and C.U. Pittman. 1976. Preparation of tetrathiafulvalenes (TTF) and their selenium analogs - tetraselenafulvalenes (TSeF). Synthesis 8: 489–514.
114. E.M. Engler, V.V. Patel, J.R. Andersen, R.R. Schumaker and A.A. Fukushima. 1978. Organic metals. Systematic molecular modifications of hexamethylenetetraheterofulvalene donors. J. Am. Chem. Soc. 100: 3769–3776.
115. N.M. Yousif, A.F.M. Fahmy, M.S. Amine, F.A. Gad and H.H. Syed. 1998. Reactions with α,β-spiroepoxy-alkanones. Part II. Uses of spirooxiranes for the synthesis of condensed heterocycles with potential biological activity. Phosphorus, Sulfur, Silicon Relat. Elem. 133: 13–20.
116. H.H. Sayed and M.A. Ali. 2008. Synthesis of 3-[(4-chloro-phenyl)oxiranyl]thiophen-2-ylpropanone and their reactions with some nucleophiles for antiviral evaluations. Phosphorus, Sulfur, Silicon Relat. Elem. 183: 156–167.
117. A. Haas and H.-W. Praas. 1993. Ergebnisse aus versuchen zur darstellung von bis(trifluormethylsulphanyl)thioketen $(CF_3S)_2C-C-S$. J. Fluorine Chem. 60: 153–164.
118. K. Okuma, S. Shibata, K. Shioji and Y. Yokomori. 2000. A new simple synthesis of *cis*- and *trans*-3,5-di-*tert*-butyl-3,5-diaryl-1,2,4-trithiolanes from ketones and tetraphosphorus decasulfide. Chem. Commun. 16: 1535–1536.
119. P.W. Sheldrake, M. Matteucci and E. McDonald. 2006. Facile generation of a library of 5-aryl-2-arylsulfonyl-1,3-thiazoles. Synlett 3: 460–462.

120. M. Saito and J. Nakayama. 2008. Sulfuration with elemental sulfur or phosphorus pentasulfide under microwave irradiation. Sci. Synth. 39: 621–621.
121. X. Lu, Q. Qi, Y. Xiao, N. Li and B. Fu. 2009. A convenient one-pot synthesis of arene-centered tris(thiazoline) compounds. Heterocycles 78: 1031–1039.
122. Y. Xiang, Q. Teng and C.K. Chu. 1995. Novel C-nucleoside analogs of 1,3-dioxolane: Synthesis of enantiomeric (2'R,4'S)- and (2'S,4'R)-2-[4-(hydroxymethyl)-1,3-dioxolan-2-yl]-1,3-thiazol-4-carboxamide. Tetrahedron Lett. 36: 3781–3784.
123. W.A. Ayer, P.A. Craw, Y.T. Ma and S. Miao. 1992. Synthesis of camalexin and related phytoalexins. Tetrahedron 48: 2919–2924.
124. R.A. Aitken, D.P. Armstrong, R.H.B. Galt and S.T.E. Mesher. 1997. Synthesis and oxidation of chiral 2-thiazolines (4,5-dihydro-1,3-thiazoles). J. Chem. Soc. Perkin Trans. 1 6: 935–944.
125. Q. Qiao, R. Dominique and R. Goodnow. 2008. 2,4-Disubstituted-5-acetoxythiazoles: Useful intermediates for the synthesis of thiazolones and 2,4,5-trisubstituted thiazoles. Tetrahedron Lett. 49: 3682–3686.
126. F.A. Vingiello, M.P. Rorer and M.A. Ogliaruso. 1971. 2-(3- and 4-Benz[α]anthracen-7-ylphenyl)-Δ^2-thiazolines and 2-(3- and 4-anthracen-9-ylphenyl)-Δ^2-thiazolines. A new preparation of thiazolines. Chem. 0: 329–329.
127. L.D. Pinkin, V.G. Dzyubenko, P.I. Abramenko and I.S. Shpileva. 1987. Synthesis of substituted thieno[2,3-d]thiazoles and indolo[3,2-d]thiazoles. Chem. Heterocycl. Compd. 23: 345–352.
128. J.F.M. DaSilva, S.J. Garden and A.C. Pinto. 2001. The chemistry of isatins: A review from 1975 to 1999. J. Braz. Chem. Soc. 12: 273–324.
129. Y.L. Aly, A.A. El-Barbary and A.A. El-Shehawy. 2004. Alkylation of thiohydantoins including synthesis, conformational and configurational studies of some acetylated s-pyranosides. Phosphorus, Sulfur, Silicon Relat. Elem. 179: 185–202.
130. O. Uchikawa, K. Fukatsu, M. Suno, T. Aono and T. Doi. 1996. In vivo biological activity of antioxidative aminothiazole derivatives. Chem. Pharm. Bull. 44: 2070–2077.
131. Z.I. Moskalenko and G.P. Shumelyak. 1974. Action of phosphorus pentasulfide on 5-acetamidothiohydantoins and 4-bromo-5-acetamidopyrazoles. Chem. Heterocycl. Compd. 10: 813–815.
132. D.T. Hurst, S. Atcha and K.L. Marshall. 1991. The synthesis of some thiazolo- and oxazolo[5,4-d]pyrimidines and pyrimidinylureas. II. Aust. J. Chem. 44: 129–134.
133. S. Knapp and D.S. Myers. 2001. α-GlcNAc thioconjugates. J. Org. Chem. 66: 3636–3638.
134. S. Knapp and D.S. Myers. 2002. Synthesis of α-GalNAc thioconjugates from an α-GalNAc mercaptan. J. Org. Chem. 67: 2995–2999.
135. T.J. Curphey. 2002. Thionation with the reagent combination of phosphorus pentasulfide and hexamethyldisiloxane. J. Org. Chem. 67: 6461–6473.
136. S. Knapp, R.A. Huhn and B. Amorelli. 2007. Thionation: GlcNAc-thiazoline triacetate {(3aR,5R,6S,7R,7aR)-5-acetoxymethyl-6,7-diacetoxy-2-methyl-5,6,7,7a-tetrahydro-3aH-pyrano[3,2-d]thiazole}. Org. Synth. 84: 68–76.
137. E.F. Dankova, V.A. Bakulev, A.N. Grishakov and V.S. Mokrushin. 1988. Rearrangement of 5-amino-1,2,3-thiadiazole-4-carbothioamides. Bull. Acad. Sci. USSR 37: 987–989.
138. E.F. Dankova, V.A. Bakulev, M.Y. Kolobov, V.I. Shishkina, Y.B. Yasman and A.T. Lebedev. 1988. Synthesis and properties of 5-amino-1,2,3-thiadiazole-4-carbothioamides. Khim. Geterotsikl. Soedin. 9: 1269–1273.
139. A.A. Fadda, S. Bondock, R. Rabie and H.A. Etman. 2008. Cyanoacetamide derivatives as synthons in heterocyclic synthesis. Turk. J. Chem. 32: 259–286.
140. H. Agirbas and K. Kahraman. 1998. A convenient synthesis of 3,4-disubstituted-1,2,4-thiadiazole-5(4H)-thiones. Phosphorus, Sulfur, Silicon Relat. Elem. 134/135: 381–389.
141. F.E. Dayan, G. Meazza, F. Bettarini, E. Signorini, P. Piccardi, J.G. Romagni and S.O. Duke. 2001. Synthesis, herbicidal activity, and mode of action of IR 5790. J. Agric. Food Chem. 49: 2302–2307.
142. V. Polshettiwar and R.S. Varma. 2008. Greener and rapid access to bio-active heterocycles: One-pot solvent-free synthesis of 1,3,4-oxadiazoles and 1,3,4-thiadiazoles. Tetrahedron Lett. 49: 879–883.

143. K.H. Pilgram. 1988. Erratum. Phosphorus, Sulfur, Silicon Relat. Elem. 36: 139–139.
144. K.M. Dawood, N.M. Elwan and B.F. Abdel-Wahab. 2011. Recent advances on the synthesis of azoles, azines and azepines fused to benzimidazole. ARKIVOC (i): 111–195.
145. Z. Zur and E. Dykman. 1975. ChemInform Abstract: Chlorination of phosphorus pentasulphide. Chem. Ind. London 6: 436–436.
146. I.M. Rafiqul, K. Shimada, S. Aoyagi, Y. Fujisawa and Y. Takikawa. 2004. Novel conversion of 6H-1,3,5-oxathiazine S-oxides into 3H-1,2,4-dithiazoles by treating with Lawesson's reagent. Heteroat. Chem. 15: 208–215.
147. B. Renfroe, C. Harrington and G.R. Proctor. 1984. Azepines Part 1. A. Rosowsky (Ed.). The chemistry of heterocyclic compounds. John Wiley & Sons, Inc. ISBN 0-471-01878-3 (v. 1), Volume 43.
148. J.P. Michael, A.S. Parsons and R. Hunter. 1989. Synthesis of two pyrrolidine alkaloids, peripentadenine and dinorperipentadenine. Tetrahedron Lett. 30: 4879–4880.
149. G.L. Morgans. 2008. New routes to arylated azabicyclic systems from enaminones, and applications to alkaloid synthesis. PhD Thesis, University of the Witwatersrand.
150. P.S. Cheesman. 1996. The synthesis of 3,5-disubstituted indolizidines. MSc Dissertation, University of the Witwatersrand.
151. W. Baker, W.D. Ollis and V.D. Poole. 1950. 1,4-Diaryl-1,4-dihydro-1,2,4,5-tetrazines and derived substances. J. Chem. Soc. 0: 3389–3394.
152. M.G. Marei. 1993. A new synthesis of 4H-thiopyran-4-thiones from acetylenic β-diketones. Phosphorus, Sulfur, Silicon Relat. Elem. 81: 101–109.
153. M.G. Marei and M.M. Mishrikey. 1992. A new synthesis of 6-thiathiophthenes from acetylenic β-diketones. Phosphorus, Sulfur, Silicon Relat. Elem. 73: 229–234.
154. T. Saito, M. Nagashima, T. Karakasa and S. Motoki. 1990. Intramolecular hetero Diels-Alder reaction of 1-thiabutadienes, 1-aryl-3-[2-(alkenyloxy)phenyl]propene-1-thiones and 2-[2-(alkenyloxy)benzylidene]-3,4-dihydronaphthalene-1(2H)-thiones. J. Chem. Soc. Chem. Commun. 23: 1665–1667.
155. G.W. Kirby and W.M. McGregor. 1990. The transient dienophile diethyl thioxomalonate and its S-oxide (sulphine) formed by retro-Diels-Alder cleavage reactions. J. Chem. Soc. Perkin Trans. 1 11: 3175–3181.
156. A.A. Alahmadi. 1997. One-flask synthesis of some new spirothiazolopyranopyrazole, spirothiazolodihydropyridinopyrazole and spirothiazolothiopyranopyrazole derivatives as antimicrobial agents. Phosphorus, Sulfur, Silicon Relat. Elem. 122: 121–132.
157. S.M. Mobin, M. Tauqeer, A. Mohammad, V. Mishra and P. Kumari. 2016. Thiophene-containing thiolato dimers, oxygen inserted Cu(II) complex, crystal structures, molecular docking and theoretical studies. J. Coord. Chem. 69: 2015–2023.
158. G. Ferguson. 1990. Thionation of 1,1'-dibenzoylferrocene: Crystal and molecular structure of 1,4-diphenyl-1,4-epithio-2,3-dithia[4](1,1')ferrocenophane. J. Chem. Soc. Dalton Trans. 12: 3697–3700.
159. R. Echarri, M.I. Matheu, C. Claver, S. Castillion, A. Alvarez-Larena and J.F. Piniella. 1997. A new reaction of tetrabutylammonium camphorsulfonate with P_4S_{10}. Synthesis and crystal structure of the first chiral tetrathiophosphate derivative. Tetrahedron Lett. 38: 6457–6460.
160. S. Flanagan, H.A. Luten and W.S. Rees. 1998. Reaction of β-diketones with P_4S_{10} to produce novel alkyl-phosphorus-sulfur clusters. Inorg. Chem. 37: 6093–6095.
161. M.I. Hegab. 2007. New reactions of β-oxo sulfenyl chlorides with 2,4-bis(4-methoxyphenyl)-1,3,2,4-dithiadiphosphetane-2,4-disulfide and phosphorus pentasulfide. Acta Chim. Solv. 54: 545–550.
162. B. Dash, E.K. Dora and C.S. Panda. 1982. Triethylamine solubilised phosphorous pentasulphide as thiation reagent: A novel route to totally thiated heterocycles. Heterocycles 19: 2093–2098.
163. L. Legrand and N. Lozach. 1987. ChemInform Abstract: Heterocyclic sulfur compounds. Part 105. 1,4-Dihydro-2H-3,1,2λ^5-benzothiazaphosphorine-2,4-dithiones. A new synthetic route to 2-arylaminothiobenzamides and to 2-(2-arylaminophenyl)-4,5-dihydrothiazoles (or oxazoles). Prep. Org. Chem. Fr. 18: 15.

164. H. Abdel-Ghany and A. Khodairy. 2000. Synthesis of polyfused thieno(2,3-*b*)thiophenes. Part 2: Synthesis of thienopyrimidine, thienothiazine, thienopyrrolopiperazine derivatives. Phosphorus, Sulfur, Silicon Relat. Elem. 166: 45–56.
165. H.S. Kim, E.A. Kim, G. Jeong, Y.T. Park, Y.S. Hong, Y. Okamoto and Y. Kurasawa. 1998. Synthesis of 4*H*-1,3,4-oxadiazino[5,6-*b*]quinoxalines from 2-substituted quinoxaline 4-oxides. J. Heterocycl. Chem. 35: 445–450.
166. J.J. Brunet, R. Chauvin, F. Kindela and D. Neibecker. 1994. Potassium tetracarbonylhydridoferrate: A reagent for the selective reduction of carbonyl groups. Tetrahedron Lett. 35: 8801–8804.
167. J. Bergman and C. Stalhandske. 1994. Transformation of isatin with P_4S_{10} to pentathiepino[6,7-*b*]indole in one step. Tetrahedron Lett. 35: 5279–5282.
168. K. Jouve and J. Bergman. 2003. Oxidative cyclization of *N*-methyl- and *N*-benzoylpyridylthioureas. Preparation of new thiazolo[4,5-*b*] and [5,4-*b*]pyridine derivatives. J. Heterocycl. Chem. 40: 261–268.
169. T. Ozturk, E. Ertas and O. Mert. 2007. Use of Lawesson's reagent in organic syntheses. Chem. Rev. 107: 5210–5278.
170. M. Szostak and J. Aube. 2009. Direct synthesis of medium-bridged twisted amides via a transannular cyclization strategy. Org. Lett. 11: 3878–3881.
171. M. Szostak and J. Aube. 2009. Synthesis and rearrangement of a bridged thioamide. Chem. Commun. 46: 7122–7124.
172. M. Szostak and J. Aube. 2013. The chemistry of bridged lactams and related heterocycles. Chem. Rev. 113: 5701–5765.

Conclusion

The heterocyclic compounds are center of focus in the field of medicinal research due to their valuable medicinal properties. The rich structural diversity encountered in these compounds, along with their biological and pharmaceutical importance, have encouraged more than 100 years of research aiming at developing efficient, economical, and selective synthetic approaches for such compounds. Modern developments in discovery and process chemistry emphasize novel sustainable synthetic pathways, needing fast and ecologically acceptable substitutes to the classical approaches. The development of sustainable synthetic processes to substitute the efficient but slightly outdated classical approaches started few decades ago and such approaches are in high demand till date.

Numerous new thionating reagents have been prepared and utilized for the formation of organosulfur compounds in the past years. The LR has become now an indispensable reagent to transform the oxo groups to thio groups. Additional reagent which has been utilized effectively to achieve the thiations is P_4S_{10}.

Lightning Source UK Ltd.
Milton Keynes UK
UKHW020624150922
408905UK00002B/18